D1611644

CAMBRIDGE STUDIES IN
ADVANCED MATHEMATICS 13

EDITORIAL BOARD
D. J. H. GARLING D. GORENSTEIN T. TOM DIECK P. WALTERS

Eigenvalues and s-Numbers

Eigenvalues and s-Numbers

ALBRECHT PIETSCH

Friedrich Schiller University, Jena, GDR

The right of the University of Cambridge to print and sell all manner of books was granted by Henry VIII in 1534. The University has printed and published continuously since 1584.

CAMBRIDGE UNIVERSITY PRESS

CAMBRIDGE
LONDON NEW YORK NEW ROCHELLE
MELBOURNE SYDNEY

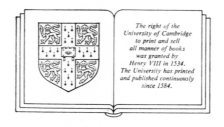

Published in the socialist countries
Akademische Verlagsgesellschaft Geest & Portig K.-G., Leipzig

Published in the non-socialist countries
Press Syndicate of the University of Cambridge
The Pitt Building, Trumpington Street, Cambridge CB2 1RP
32 East 57th Street, New York, NY 10022, USA
10, Stamford Road, Oakleigh, Melbourne 3166, Australia

© Akademische Verlagsgesellschaft Geest & Portig K.-G., Leipzig 1987
Licensed edition for Cambridge University Press, 1987
Printed in the German Democratic Republic

Library of Congress cataloguing in publication data

Pietsch, A. (Albrecht)
Eigenvalues and s-numbers.

(Cambridge studies in advanced mathematics ; 13)
Bibliography: p.
Includes index.
1. Operator theory. 2. Banach spaces. 3. s-numbers.
4. Eigenvalues. I. Title. II. Series.
QA329.P5 1987 515.7′24 86-829

British Library cataloguing in publication data

Pietsch, Albrecht
Eigenvalues and s-numbers. – (Cambridge studies in advanced mathematics)
1. Matrices 2. Eigenvalues
I. Title
512.9′434 QA193

ISBN 0-521-32532-3

CB-SHK
SCIMON

Preface

The idea to write a monograph about *Eigenvalue distributions* arose almost immediately after I had finished the manuscript of *Operator ideals*. It was my intention to show how the abstract theory of operators on Banach spaces can successfully be applied in classical analysis. Fortunately enough, several important open problems were solved by W. B. Johnson, H. König, B. Maurey and J. R. Retherford in 1977 and presented at a conference held in Leipzig. Since that time the underlying techniques have been drastically improved so that now most of the proofs are streamlined (quoted from a letter of H. König).

Many unnamed colleagues, friends and pupils had helped in the production of this work. In particular, I would like to thank H. Jarchow (Zürich) and D. J. H. Garling (Cambridge) for their critical remarks. The latter has also tried to revise my "Queen's English". I am especially grateful to the co-workers of my seminar for stimulating discussions of the manuscript as well as for reading parts of the galley-proofs.

During the writing of this book I had the opportunity to visit several mathematical departments and to use their libraries. Among others, I am indebted to the Steklov Mathematical Institute in Moscow, the Banach Center in Warsaw, the Sonderforschungsbereich 72 in Bonn and to the Forschungsinstitut für Mathematik in Zürich. Last, but not least, I obtained a great deal of support from my home university in Jena.

I am very grateful to friends and institutions for providing an excellent Xerox service. Furthermore, I wish to express my special thanks to Mrs. G. Girlich for very careful work in producing the typescript of this book.

Many classical treatises on integral equations, written by D. Hilbert, M. Bôcher, G. Kowalewski, E. Hellinger, O. Toeplitz and F. Smithies, were published either in Cambridge or in Leipzig. Therefore I regard it as a good omen that this monograph will be jointly issued by publishing houses located in these cities. Moreover, in my opinion such undertakings are valuable contributions by scientists and editors to the realization of a peaceful coexistence of mankind.

Obviously, it is a hard job to be married to a mathematician. Such human beings quite often sit in a chair and do nothing, while the housework remains undone. In gratitude for her sympathetic understanding this book is dedicated

TO MY WIFE.

Jena (GDR), July 1985

Albrecht Pietsch

Contents

Introduction

At the turn of this century, I. Fredholm created the determinant theory of integral operators. Subsequently, D. Hilbert developed the theory of bilinear forms in infinitely many unknowns. In 1918, F. Riesz published his famous paper on compact operators (vollstetige Transformationen) which was based on these ideas. In particular, he proved that such operators have at most a countable set of eigenvalues which, arranged in a sequence, tend to zero. Nothing was said about the rate of this convergence.

On the other hand, I. Schur had already observed in 1909 that the eigenvalue sequence of an integral operator induced by a continuous kernel is square summable. This fact indicates that something got lost within the framework of Riesz's theory. The following problem therefore arises:

Find conditions on an operator T which guarantee that the eigenvalue sequence $(\lambda_n(T))$ belongs to a certain subset of c_0, such as l_r with $0 < r < \infty$.

In the context of integral operators such conditions were given in terms of integrability and differentiability properties of the underlying kernel. Roughly speaking, the following rule of thumb holds:

The smoother the kernel, the quicker the convergence.

Corresponding results for abstract operators were found relatively late. The decisive theorem is due to H. Weyl (1949). Using s-numbers, he proved that, for every compact operator T on a complex Hilbert space,

$$(s_n(T)) \in l_r \quad \text{implies} \quad (\lambda_n(T)) \in l_r.$$

The first result in the Banach space setting goes back to A. Grothendieck (1955) who showed that the eigenvalue sequence of every nuclear operator is square summable. The extension of Weyl's theorem from Hilbert to Banach spaces was an open problem for many years. The real breakthrough came at the end of the seventies when H. König (1978) proved that

$$(a_n(T)) \in l_r \quad \text{implies} \quad (\lambda_n(T)) \in l_r.$$

Here

$$a_n(T) := \inf \{\|T - L\| : \operatorname{rank}(L) < n\}$$

is the n-th approximation number. Afterwards, based on the theory of operator ideals, many striking results on eigenvalue distributions of abstract operators on Banach spaces were established and successfully applied to various types of integral operators. This demonstrates in a convincing way how tools from modern operator theory yield new results in classical analysis which could not be proved by old-fashioned techniques.

The basic concept in the classical approach to the theory of eigenvalue distribution was that of a Fredholm denominator. This is an entire function $\delta(\zeta, T)$ whose zeros are related to the eigenvalues of the operator T by the formula $\zeta\lambda = -1$. Then

$$|\delta(\zeta, T)| \leq M \exp(\mu\, |\zeta|^r) \quad \text{implies} \quad |\lambda_n(T)| \leq c n^{-1/r}.$$

Another important problem, first raised by J. von Neumann (1932), is the following:

Characterize those operators which admit a trace.

The classical connection between traces and eigenvalues is given by the formula

$$\operatorname{trace}(T) = \sum_{n=1}^{\infty} \lambda_n(T),$$

which only makes sense in the case when the sequence $(\lambda_n(T))$ is absolutely summable. We further note that the Taylor coefficients of $\delta(\zeta, T)$ can be expressed in terms of traces (Plemelj's formulas).

It is the intention of this monograph to give an extensive presentation of results connected with the problems mentioned above. The following table provides a rough impression of the interplay of the most important concepts:

Abstract theory

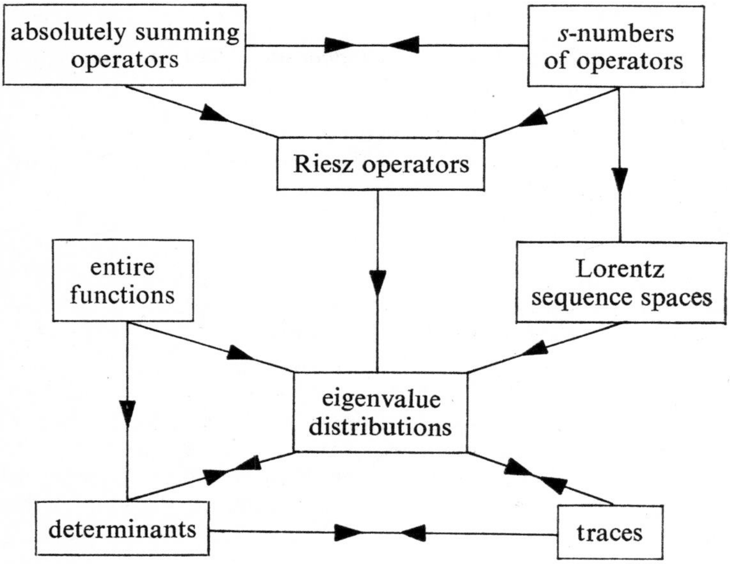

Applications

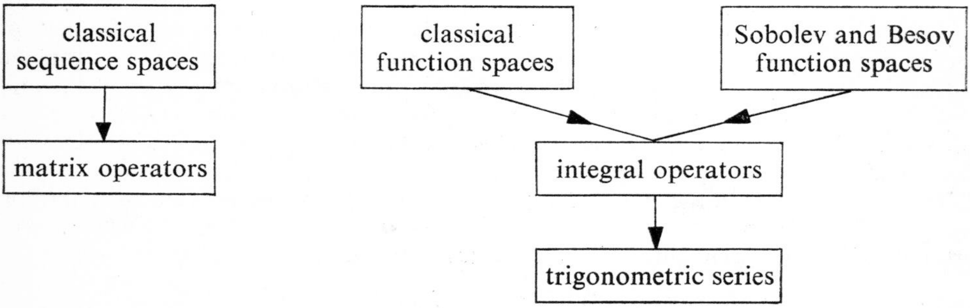

The highlights of this monograph are the eigenvalue theorems for Weyl operators and for absolutely r-summing operators as well as their applications to integral operators induced by kernels of Hille-Tamarkin or Besov type. A nearly geodesic path to these results may be found by studying only those paragraphs indicated by (*). This means that the book contains a sub-book (including less than one third of the material treated in Chapters 1–6) which can serve as a starting point for first reading or as the basis of a seminar.

Another feature of this monograph is an extensive historical survey, which is given in the final chapter. In addition, most of the results are provided with references in the text.

In a sense this treatise is a supplement to my monograph *Operator ideals*. Those readers interested to learn more about the abstract background should consult that book. On the other hand, I have carefully tried to keep this presentation self-contained.

The material is organized as follows:
The smallest units are the paragraphs. These are denoted by symbols consisting of three natural numbers. For instance, 3.6.2 refers to the eigenvalue theorem stated in Chapter 3, Section 6, Paragraph 2. Every paragraph contains precisely one definition, theorem, proposition, lemma or example. Sometimes a remark is added.

Preliminaries

In this introductory chapter we summarize some well-known definitions and explain certain terminology used throughout this treatise. Furthermore, we list various results which are employed without proofs.

It is assumed that the reader already has a basic knowledge in the theory of Banach spaces, linear and multilinear algebra, general topology and measure theory. An elementary background from complex analysis is also needed. This means that the greater part of the book can be understood by any average graduate student.

We further assume some knowledge in interpolation theory (real method). In order to read the last but one chapter it is desirable, but not necessary, to be well-grounded in the theory of Sobolev and Besov spaces.

A. Operators on finite dimensional linear spaces

In this section we summarize some basic results from linear algebra which are the background and model for all of the following considerations. Proofs may be found, for example, in the textbook of P. R. Halmos (HAL).

All linear spaces are considered over the complex field.

A.1. Finite dimensional linear spaces

A.1.1. We denote by $l(n)$ the n-dimensional linear space of all complex-valued vectors $x = (\xi_1, \ldots, \xi_n)$.

The algebraic operations are defined coordinatewise.

A.1.2. Let $(u_1, \ldots, u_n)$ be any basis of a finite dimensional linear space E. Then every element $x \in E$ admits a unique representation

$$x = \sum_{i=1}^{n} \xi_i u_i \quad \text{with} \quad \xi_1, \ldots, \xi_n \in \mathbb{C},$$

and

$$U : (\xi_i) \to \sum_{i=1}^{n} \xi_i u_i$$

defines an isomorphism between $l(n)$ and E.

A.2. Operators and matrices

A.2.1. By an **operator** T on a finite dimensional linear space E we always mean a linear map from E into itself.

A.2.2. Let $M = (\mu_{ij})$ be any (n, n)-matrix. Then

$$M_{\text{op}} : (\eta_j) \to \left(\sum_{j=1}^{n} \mu_{ij}\eta_j \right)$$

defines an operator on $l(n)$. In this way we get a one-to-one correspondence between (n, n)-matrices and operators on $l(n)$.

A.2.3. Let T be an operator on a finite dimensional linear space E. Then, for every fixed basis $(u_1, \ldots, u_n)$, there exists a so-called **representing matrix** $M = (\mu_{ij})$

determined by

$$Tu_j = \sum_{i=1}^{n} \mu_{ij} u_i \quad \text{for} \quad j = 1, \ldots, n.$$

With the operator M_{op} induced by this matrix, we have the diagram

$$\begin{array}{ccc} E & \xrightarrow{\;T\;} & E \\ U^{-1} \downarrow & & \uparrow U \\ l(n) & \xrightarrow[M_{\text{op}}]{} & l(n) \end{array},$$

where U is the isomorphism defined in A.1.2.

A.2.4. The representing matrix M of a given operator T depends on the underlying basis $(u_1, \ldots, u_n)$. If N is the representing matrix with respect to another basis $(v_1, \ldots, v_n)$, then there exists an invertible matrix X such that $N = X^{-1}MX$. The matrix $X = (\xi_{ij})$ is determined by

$$v_j = \sum_{i=1}^{n} \xi_{ij} u_i.$$

Note that the **identity operator**, denoted by I_n or I, is always represented by the **unit matrix** $I(n) = (\delta_{ij})$.

A.2.5. We now emphazise the extremely important fact that, by a clever choice of the underlying basis, it can be arranged that the representing matrix of a given operator takes a nice and handy form.

For every operator T on a finite dimensional linear space E there exists a basis $(u_1, \ldots, u_n)$ such that the representing matrix has upper **triangular form**:

$$M = \begin{pmatrix} \mu_{11} & \cdots & \mu_{1n} \\ & \ddots & \vdots \\ 0 & & \mu_{nn} \end{pmatrix}.$$

This means that $\mu_{ij} = 0$ whenever $i > j$.

In the particular case when E is equipped with a scalar product we may even find an orthogonal basis (**Schur basis**) with this property.

A.2.6. The preceding result can be improved further.

For every operator T on a finite dimensional linear space E there exists a basis $(u_1, \ldots, u_n)$ such that the representing matrix has the **Jordan canonical form**:

$$M = \begin{pmatrix} M_1 & & 0 \\ & \ddots & \\ 0 & & M_k \end{pmatrix}.$$

The basic **Jordan blocks** M_h are matrices, where a complex number μ_h is repeated on the principal diagonal, with ones just above this diagonal and zeros everywhere else:

$$M_h = \begin{pmatrix} \mu_h, & 1 & & & 0 \\ & \mu_h, & 1 & & \\ & & \ddots & \ddots & \\ & & & \mu_h, & 1 \\ 0 & & & & \mu_h \end{pmatrix}.$$

A.3. Traces

A.3.1. The **trace** of an (n, n)-matrix $M = (\mu_{ij})$ is defined by

$$\operatorname{trace}(M) := \sum_{i=1}^{n} \mu_{ii}.$$

In the sequel we use the standard properties of this quantity without further explanation. The commutation formula

$$\operatorname{trace}(XM) = \operatorname{trace}(MX),$$

which holds for arbitrary (n, n)-matrices M and X, is of particular interest.

A.3.2. Let T be an operator on a finite dimensional space. If M and N are representing matrices with respect to different bases, then there exists an invertible matrix X such that $N = X^{-1}MX$. Hence

$$\operatorname{trace}(N) = \operatorname{trace}(X^{-1}MX) = \operatorname{trace}(XX^{-1}M) = \operatorname{trace}(M).$$

Thus all representing matrices have the same trace, and this common value is defined to be the **trace** of the operator T.

A.4. Determinants

A.4.1. The **determinant** of an (n, n)-matrix $M = (\mu_{ij})$ is defined by

$$\det(M) := \sum_{\pi} \operatorname{sign}(\pi)\, \mu_{1\pi(1)} \cdots \mu_{n\pi(n)},$$

where the sum is taken over all permutations π of $\{1, \ldots, n\}$, and $\operatorname{sign}(\pi)$ denotes the signature of π.

In the sequel we use the standard properties of this quantity without further explanation. The multiplication formula

$$\det(MN) = \det(M)\det(N),$$

which holds for arbitrary (n, n)-matrices M and N, is of particular interest.

A.4.2. If the matrix $M = (\mu_{ij})$ has triangular form, then

$$\det(M) = \mu_{11} \cdots \mu_{nn}.$$

A.4.3. A matrix M is invertible if and only if $\det(M) \neq 0$.

A.4.4. Let T be an operator on a finite dimensional linear space. If M and N are representing matrices with respect to different bases, then there exists an invertible matrix X such that $N = X^{-1}MX$. Hence

$$\det(N) = \det(X^{-1}MX) = \det(X^{-1})\det(M)\det(X) = \det(M).$$

Thus all representing matrices have the same determinant, and this common value is defined to be the **determinant** of the operator T.

A.4.5. We now establish **Hadamard's inequality** which is one of the main tools in proving the existence of determinants of infinite matrices and kernels.

Lemma (J. Hadamard 1893: a). Let $M = (\mu_{ij})$. Then

$$|\det(M)| \leq \prod_{i=1}^{n} \left(\sum_{j=1}^{n} |\mu_{ij}|^2 \right)^{1/2}.$$

Proof. If $\det(M) = 0$, then we have nothing to prove. Therefore it may be assumed that $\det(M) \neq 0$. Then the rows

$$m_i := (\mu_{i1}, \ldots, \mu_{in}) \quad \text{for } i = 1, \ldots, n$$

are linearly independent. Applying Schmidt's orthonormalization procedure with respect to the canonical scalar product defined on $l(n)$, we obtain the vectors

$$y_h := (\eta_{h1}, \ldots, \eta_{hn}) \quad \text{for } h = 1, \ldots, n$$

which can be viewed as the rows of a unitary matrix Y. Obviously,

$$m_i = \sum_{h=1}^{n} \xi_{ih} y_h \quad \text{for } i = 1, \ldots, n,$$

where $X = (\xi_{ih})$ is a triangular matrix. It follows from $(y_h, y_k) = \delta_{hk}$ that $\xi_{ih} = (m_i, y_h)$. Hence

$$|\xi_{ii}| = \left| \sum_{j=1}^{n} \mu_{ij} \eta_{ij}^* \right| \leqq \left(\sum_{j=1}^{n} |\mu_{ij}|^2 \right)^{1/2} \left(\sum_{j=1}^{n} |\eta_{ij}|^2 \right)^{1/2},$$

and we obtain

$$|\det(M)| = |\det(XY)| = |\det(X) \det(Y)| \leqq \prod_{i=1}^{n} |\xi_{ii}| \leqq \prod_{i=1}^{n} \left(\sum_{j=1}^{n} |\mu_{ij}|^2 \right)^{1/2}.$$

A.5. Eigenvalues

A.5.1. A complex number λ_0 is an **eigenvalue** of the operator T if there exists $x \in E$ such that

$$Tx = \lambda_0 x \quad \text{and} \quad x \neq o.$$

Remark. The eigenvalues of a matrix M are defined to be the eigenvalues of the operator M_{op} induced by M on $l(n)$.

A.5.2. Note that λ_0 is an eigenvalue of T if and only if $\lambda_0 I - T$ fails to be invertible. Hence the eigenvalues coincide with the zeros of the **characteristic polynomial**

$$\pi(\lambda, T) := \det(\lambda I - T) = \lambda^n + \alpha_1 \lambda^{n-1} + \ldots + \alpha_{n-1}\lambda + \alpha_n.$$

The coefficients are given by the formula

$$\alpha_k := \frac{(-1)^k}{k!} \sum_{i_1=1}^{n} \ldots \sum_{i_k=1}^{n} \det \begin{pmatrix} \mu_{i_1 i_1} & \ldots & \mu_{i_1 i_k} \\ \vdots & & \vdots \\ \mu_{i_k i_1} & \ldots & \mu_{i_k i_k} \end{pmatrix},$$

where the right-hand sum is taken over all principal (k, k)-minors of any representing matrix $M = (\mu_{ij})$.

A.5.3. The algebraic **multiplicity** of an eigenvalue is defined to be its order as a zero of $\pi(\lambda, T)$. If every eigenvalue is counted according to this multiplicity, then we get a set $(\lambda_1, \ldots, \lambda_n)$ arranged in any order we please.

A.5.4. Writing the characteristic polynomial as a product of linear factors yields the **determinant formula**:

$$\det(\lambda I - T) = \prod_{i=1}^{n} (\lambda - \lambda_i).$$

Equating the coefficients of λ^{n-1}, we obtain the **trace formula**:

$$\operatorname{trace}(T) = \sum_{i=1}^{n} \lambda_i .$$

A.5.5. Let $M = (\mu_{ij})$ be any triangular matrix representing an operator T. Then it follows from

$$\det(\lambda I - T) = \prod_{i=1}^{n} (\lambda - \mu_{ii})$$

that the principal diagonal $(\mu_{11}, \ldots, \mu_{nn})$ coincides with the set of eigenvalues $(\lambda_1, \ldots, \lambda_n)$.

B. Spaces and operators

Some basic knowledge on Banach space theory is an indispensable assumption for reading this treatise. We use (DUN) and (TAY) as standard references.

Throughout the monograph H denotes a complex Hilbert space.

B.1. Operators on quasi-Banach spaces

B.1.1. By a **quasi-norm** $\|\cdot\|$ defined on a linear space E we mean a real-valued function with the following properties:

(1) Let $x \in E$. Then $\|x\| = 0$ if and only if $x = o$.

(2) $\|x + y\| \leqq c_E[\|x\| + \|y\|]$ for $x, y \in E$. Here $c_E \geqq 1$ is a constant.

(3) $\|\lambda x\| = |\lambda| \, \|x\|$ for $x \in E$ and $\lambda \in \mathbb{C}$.

In order to indicate the underlying linear space E, we sometimes replace $\|x\|$ by the more specific symbol $\|x \mid E\|$. Formula (2) is called the **quasi-triangle inequality.**

If $c_E = 1$, then $\|\cdot\|$ is said to be a **norm.** In this case, (2) passes into the well-known **triangle inequality.**

B.1.2. Every quasi-norm $\|\cdot\|$ given on a linear space E induces a metrizable topology such that the algebraic operations are continuous. A fundamental system of neighbourhoods of the zero element is formed by the subsets εU with $\varepsilon > 0$, where

$$U := \{x \in E : \|x\| \leqq 1\}$$

is the **closed unit ball.**

B.1.3. A **quasi-Banach space** is a linear space E equipped with a quasi-norm $\|\cdot\|$ which becomes complete with respect to the associated metrizable topology. This means that all Cauchy sequences are convergent.

In the most important case when $\|\cdot\|$ is a norm we call E a **Banach space.**

B.1.4. Quasi-norms $\|\cdot\|_1$ and $\|\cdot\|_2$ defined on a linear space E are said to be **equivalent** if they induce the same topology. This happens if and only if there exists a constant $c > 0$ such that

$$\|x\|_1 \leqq c\|x\|_2 \quad \text{and} \quad \|x\|_2 \leqq c\|x\|_1 \quad \text{for all } x \in E.$$

Let E be a linear space with a quasi-norm $\|\cdot \mid E\|$. If $\varrho > 0$, then ϱE denotes the same linear space endowed, however, with the equivalent quasi-norm

$$\|x \mid \varrho E\| := \varrho\|x \mid E\| \quad \text{for all } x \in E.$$

B.1.5. By a ***p*-norm** ($0 < p \leqq 1$) given on a linear space E we mean a quasi-norm satisfying the ***p*-triangle inequality:**

$$\|x + y\|^p \leqq \|x\|^p + \|y\|^p \quad \text{for all } x, y \in E.$$

If so, then the quasi-triangle inequality holds with $c_E = 2^{1/p-1}$. Every p-norm is also a q-norm for $0 < q < p \leqq 1$.

Putting

$$d(x, y) := \|x - y\|^p \quad \text{for all } x, y \in E,$$

we see that every p-norm defines a metric on E.

B.1.6. A linear space E is complete with respect to the metrizable topology induced by a p-norm if and only if the condition

$$\sum_{i=1}^{\infty} \|x_i\|^p < \infty$$

implies that the series formed by the elements $x_1, x_2, \ldots \in E$ converges in E.

B.1.7. A quasi-Banach space is called a ***p*-Banach space** if its quasi-norm is even a p-norm.

B.1.8. Let $\|\cdot\|$ be any quasi-norm on a linear space E such that

$$\|x + y\| \leqq c_E[\|x\| + \|y\|] \quad \text{for all } x, y \in E.$$

If p is determined by $2^{1/p-1} = c_E$, then there exists on E an equivalent p-norm. This means that every quasi-Banach space can be viewed as a p-Banach space for all sufficiently small exponents p.

B.1.9. Let E and F be quasi-Banach spaces. By an **operator** T from E into F we mean a bounded linear map. Then the quantity

$$\|T\| := \sup\{\|Tx\| : x \in U\}$$

is finite.

In order to indicate the underlying linear spaces E and F, we sometimes replace $\|T\|$ by the more specific symbol $\|T: E \to F\|$.

B.1.10. Suppose that E_0 and E_1 are quasi-Banach spaces such that E_0 is a linear subset of E_1 in the purely algebraic sense and that

$$\|x \mid E_1\| \leqq c\|x \mid E_0\| \quad \text{for all } x \in E_0.$$

Then the identity map from E_0 into E_1 is called an **embedding operator.**

We say that a quasi-Banach space E lies between E_0 and E_1 if

$$E_0 \xrightarrow{I_0} E \xrightarrow{I_1} E_1,$$

where I_0 and I_1 are embedding operators.

B.2. Operators on Banach spaces

From now on, unless the contrary is explicitly stated, E, F and G denote complex Banach spaces with the closed unit balls U, V and W, respectively.

B.2.1. Let $\mathfrak{L}(E, F)$ denote the set of all operators from E into F. Defining the algebraic operations in the canonical way and using the ordinary **operator norm**

$$\|T\| := \sup\{\|Tx\| : x \in U\},$$

we see that $\mathfrak{L}(E, F)$ becomes a Banach space.

To simplify notation, we write $\mathfrak{L}(E)$ instead of $\mathfrak{L}(E, E)$.

Furthermore,

$$\mathfrak{L} := \bigcup_{E,F} \mathfrak{L}(E, F)$$

stands for the class of all operators acting between arbitrary Banach spaces

B.2.2. For every operator $T \in \mathfrak{L}(E, F)$ we define the **null space**

$$N(T) := \{x \in E : Tx = o\}$$

and the **range**

$$M(T) := \{Tx : x \in E\}.$$

Both of these subsets are linear, and $N(T)$ is always closed.

B.2.3. The **identity operator** of a Banach spaces E is denoted by I_E. If there is no risk of confusion, then we simply use the symbol I.

B.2.4. We call

$$ST : x \to Tx \to STx$$

the **product** of the operators $T \in \mathfrak{L}(E, F)$ and $S \in \mathfrak{L}(F, G)$. Observe that

$$ST \in \mathfrak{L}(E, G) \quad \text{and} \quad \|ST\| \leqq \|S\| \, \|T\|.$$

The m-th **power** T^m of $T \in \mathfrak{L}(E)$ is inductively defined by $T^{m+1} := T^m T$ and $T^0 := I$.

B.2.5. An operator $T \in \mathfrak{L}(E, F)$ is **invertible** if there exists $X \in \mathfrak{L}(F, E)$ such that $XT = I_E$ and $TX = I_F$. If so, then the **inverse operator** X, usually denoted by T^{-1}, is uniquely determined.

According to the bounded inverse theorem (DUN, II.2.2) and (TAY, IV.5.8), an operator $T \in \mathfrak{L}(E, F)$ is invertible if and only if it is one-to-one and onto: $N(T) = \{o\}$ and $M(T) = F$.

An invertible operator $T \in \mathfrak{L}(E, F)$ is also referred to as an **isomorphism** between E and F. If $\|T\| = \|T^{-1}\| = 1$, then T is said to be a **metric isomorphism**.

B.2.6. An **injection** $J \in \mathfrak{L}(E, F)$ is a one-to-one operator with closed range. In this case,

$$\|x\|_J := \|Jx\| \quad \text{for all } x \in E$$

defines an equivalent norm on E. If, in addition, we have $\|\cdot\| = \|\cdot\|_J$, then J is said to be a **metric injection.**

B.2.7. By a **subspace** M of a Banach space E we mean a closed linear subset. Obviously, M becomes a Banach space with respect to the norm obtained by restriction.

The **canonical injection** from M into E is denoted by J_M^E.

B.2.8. A **surjection** $Q \in \mathfrak{L}(E, F)$ is an operator which maps E onto F. In this case,

$$\|y\|_Q := \inf\{\|x\| : x \in E, \quad Qx = y\} \quad \text{for all } y \in F$$

defines an equivalent norm on F. If, in addition, we have $\|\cdot\| = \|\cdot\|_Q$, then Q is said to be a **metric surjection.**

B.2.9. Let N be any subspace of a Banach space E. Then the **quotient space** E/N consists of all equivalence classes $x + N$. It turns out that E/N becomes a Banach space with respect to the norm

$$\|x + N\| := \inf\{\|x + y\| : y \in N\}.$$

The **canonical surjection** from E onto E/N is denoted by Q_N^E.

B.2.10. A **projection** $P \in \mathfrak{L}(E)$ is an operator such that $P^2 = P$.

B.2.11. A Banach space E is the **direct sum** of the subspaces M and N if $E = M + N$ and $M \cap N = \{o\}$. In this case, we write $E = M \oplus N$. The above conditions mean that every element $x \in E$ admits a unique decomposition $x = u + v$ with $u \in M$ and $v \in N$. Then $P \in \mathfrak{L}(E)$ defined by $Px := u$ is called the **projection of** E **onto** M **along** N. Conversely, every projection $P \in \mathfrak{L}(E)$ determines a direct sum decomposition $E = M(P) \oplus N(P)$; see (DUN, VI.3.1) and (TAY, IV.12.2).

B.2.12. Let M be any subspace of a Hilbert space H. Then

$$M^\perp := \{x \in H : (x, y) = 0 \quad \text{for all } y \in M\}$$

is called the **orthogonal complement** of M. It turns out that $H = M \oplus M^\perp$. The projection P from H onto M along $M^\perp$ is said to be the **orthogonal projection** from H onto M. Note that $\|P\| = 1$ whenever $M \neq \{o\}$.

B.3. Duality

B.3.1. By a **functional** a defined on a Banach space E we mean a bounded linear form. The value of the functional a at the element x is denoted by $\langle x, a\rangle$.

B.3.2. Let E' denote the set of all functionals on a Banach space E. Defining the algebraic operations in the canonical way and using the norm

$$\|a\| := \sup\{|\langle x, a\rangle| : x \in U\},$$

we see that E' becomes a Banach space. We refer to E' as the **dual Banach space.**

The closed unit ball of E' is denoted by U^0.

B.3.3. The **bidual Banach space** E'' is defined to be the dual of E'. Assigning to every element $x \in E$ the functional

$$f_x : a \to \langle x, a\rangle,$$

we obtain the **evaluation operator** K_E from E into E'' which is a metric injection, because

$$\|x\| = \sup\{|\langle x, a\rangle| : a \in U^0\}.$$

B.3.4. The **dual operator** $T' \in \mathfrak{L}(F', E')$ of $T \in \mathfrak{L}(E, F)$ is determined by

$$\langle x, T'b\rangle = \langle Tx, b\rangle \quad \text{for all } x \in E \text{ and } b \in F'.$$

Observe that $\|T'\| = \|T\|$.

B.3.5. The **bidual operator** $T'' \in \mathfrak{L}(E'', F'')$ is defined to be the dual of $T' \in \mathfrak{L}(F', E')$. The following diagram holds:

$$\begin{array}{ccc} E & \xrightarrow{\quad T \quad} & F \\ K_E \downarrow & & \downarrow K_F \\ E'' & \xrightarrow[\quad T'' \quad]{} & F'' \end{array} \quad .$$

B.3.6. The concepts of an injection and a surjection are dual to each other. This means that

J is an injection if and only if J' is a surjection,
Q is a surjection if and only if Q' is an injection.

B.3.7. The dual of a Hilbert space can be described as follows.

Every element $y \in H$ determines a functional

$$y^* : x \to (x, y)$$

on H such that $\|y^*\| = \|y\|$. Moreover, all functionals $a \in H'$ can be obtained in this way. Hence

$$C_H : y \to y^*$$

yields a conjugate-linear one-to-one correspondence between H and H'.

B.3.8. Let H and K be Hilbert spaces. The **adjoint operator** $T^* \in \mathfrak{L}(K, H)$ of $T \in \mathfrak{L}(H, K)$ is determined by

$$(x, T^*y) = (Tx, y) \quad \text{for all } x \in H \text{ and } y \in K.$$

The relationship between adjoint and dual operators is described by the diagram

$$\begin{array}{ccc} K & \xrightarrow{\quad T^* \quad} & H \\ C_K \downarrow & & \downarrow C_H \\ K' & \xrightarrow[\quad T' \quad]{} & H' \end{array} \quad .$$

B.4. Finite operators on Banach spaces

B.4.1. The **dimension** of a finite dimensional linear space M is denoted by $\dim(M)$.

B.4.2. Let N be a linear subset of a linear space E such that the quotient space E/N is finite dimensional. Then the **codimension** of N in E is defined by the formula $\operatorname{codim}(N) := \dim(E/N)$.

B.4.3. Let $a \in E'$ and $y \in F$. Then

$$a \otimes y : x \to \langle x, a \rangle\, y$$

yields an operator from E into F. Note that $\|a \otimes y\| = \|a\|\,\|y\|$.

B.4.4. An operator $T \in \mathfrak{L}(E, F)$ is said to be **finite** if there exists a so-called **finite representation**

$$T = \sum_{i=1}^{n} a_i \otimes y_i,$$

where $a_1, \ldots, a_n \in E'$ and $y_1, \ldots, y_n \in F$. This means that

$$Tx = \sum_{i=1}^{n} \langle x, a_i \rangle y_i \quad \text{for all } x \in E.$$

We define **rank** (T) to be the smallest number of summands which are required in the above representation. Note that

$$\operatorname{rank}(T) = \operatorname{codim}(N(T)) = \dim(M(T)).$$

B.4.5. It easily follows from A.2.5 that every finite operator $T \in \mathfrak{L}(E)$ admits a representation

$$T = \sum_{i=1}^{n} a_i \otimes x_i$$

such that the matrix $(\langle x_i, a_j \rangle)$ has upper triangular form:

$$\langle x_i, a_j \rangle = 0 \quad \text{if} \quad i > j.$$

B.4.6. We prove here an auxiliary result about finite operators factoring through a Hilbert space H.

Lemma. Suppose that the finite operator $T \in \mathfrak{L}(E, F)$ can be written in the form $T = YA$, where $A \in \mathfrak{L}(E, H)$ and $Y \in \mathfrak{L}(H, F)$. Then there exists an orthogonal projection $P \in \mathfrak{L}(H)$ such that

$$T = YPA \quad \text{and} \quad \operatorname{rank}(P) = \operatorname{rank}(T).$$

Proof. Let P denote the orthogonal projection from H onto

$$M := Y^{-1}[M(T)] \cap N(Y)^{\perp}.$$

Given $x \in E$, we consider the decomposition

$$Ax = f_0 + f \quad \text{with} \quad f_0 \in N(Y) \quad \text{and} \quad f \in N(Y)^{\perp}.$$

Since $Tx = YAx = Yf$, we have $f \in Y^{-1}[M(T)]$. Therefore $f \in M$. This proves that Y maps M onto $M(T)$. Moreover, the restriction of Y to M is one-to-one. Hence

$$\operatorname{rank}(P) = \dim(M) = \dim(M(T)) = \operatorname{rank}(T).$$

Furthermore, it follows from $M(P) \subseteq N(Y)^{\perp}$ that

$$Ax - f = f_0 \in N(Y) \subseteq M(P)^{\perp} = N(P).$$

Hence $PAx = Pf = f$, which in turn implies that

$$Tx = Yf = YPAx \quad \text{for all} \quad x \in E.$$

B.4.7. A Banach space E has the **approximation property** if, given any precompact subset K of E and $\varepsilon > 0$, there exists a finite operator $L \in \mathfrak{L}(E)$ such that

$$\|x - Lx\| \leqq \varepsilon \quad \text{for all } x \in K.$$

Roughly speaking, the overwhelming majority of all popular Banach spaces enjoy this property. However, there are also prominent counterexamples: $\mathfrak{L}(l_2)$; see A. Szankowski (1981) and 4.7.6 (Remark).

C. Sequence and function spaces

We only require some elementary facts on classical sequence and function spaces. Concerning the measure-theoretic background the reader is referred to P. R. Halmos's textbook (HAM).

An introduction to the theory of Lorentz sequence spaces is given in Section 2.1. Some basic properties of Sobolev and Besov spaces are summarized in Section 6.4.

C.1. Classical sequence spaces

C.1.1. In the following we consider **complex-valued families** $x = (\xi_i)$ indexed by the elements of an arbitrary set I. In the case when this set is finite or countably infinite we speak of **vectors** and **sequences**, respectively.

All algebraic operations concerning families are defined coordinatewise.

C.1.2. Given any index set I, we denote by $\mathfrak{F}(I)$ the collection of its finite subsets.

C.1.3. The set of all bounded complex-valued families $x = (\xi_i)$ on a given index set I is denoted by $l_\infty(I)$. It turns out that $l_\infty(I)$ is a Banach space under the norm

$$\|x \mid l_\infty(I)\| := \sup \{|\xi_i| : i \in I\}.$$

C.1.4. We say that a family $x = (\xi_i)$ vanishes at infinity if, given $\varepsilon > 0$, there exists $\mathrm{i} \in \mathfrak{F}(I)$ such that

$$|\xi_i| \leqq \varepsilon \quad \text{whenever} \quad i \notin \mathrm{i}.$$

The subspace of $l_\infty(I)$ formed by these families is denoted by $c_0(I)$.

C.1.5. A complex-valued family (ξ_i) with $i \in I$ is said to be **summable** if there exists a complex number σ with the following property:

Given $\varepsilon > 0$, we can find $\mathrm{i}_0 \in \mathfrak{F}(I)$ such that

$$\Big|\sigma - \sum_{\mathrm{i}} \xi_i\Big| \leqq \varepsilon \quad \text{whenever } \mathrm{i} \in \mathfrak{F}(I) \quad \text{and} \quad \mathrm{i} \supseteqq \mathrm{i}_0.$$

When this is so, then σ is called the **sum** of (ξ_i), and we write

$$\sigma = \sum_I \xi_i.$$

Note that the families (ξ_i) and $(|\xi_i|)$ are simultaneously summable or not.

C.1.6. Let $0 < r < \infty$. We denote by $l_r(I)$ the collection of all complex-valued families $x = (\xi_i)$ such that

$$\|x \mid l_r(I)\| := \Big(\sum_I |\xi_i|^r\Big)^{1/r}$$

is finite. Such families are called ***r*-summable**. Note that $l_r(I)$ is a Banach space if $1 \leqq r < \infty$ and an r-Banach space if $0 < r < 1$. The quasi-triangle inequality (**Minkowski's inequality**) holds with the constant $c_r := \max(2^{1/r-1}, 1)$.

C.1.7. Let $\mathbb{Z}_n := \{1, \ldots, n\}$ and $\mathbb{N} := \{1, 2, \ldots\}$. To simplify notation, we write $l_r(n)$ and l_r instead of $l_r(\mathbb{Z}_n)$ and $l_r(\mathbb{N})$, respectively. The closed unit balls of these spaces are denoted by $U_r(n)$ and U_r. Furthermore, the symbol $c_0(\mathbb{N})$ is replaced by c_0.

C.1.8. Let $0 < p,q < \infty$ and $1/r = 1/p + 1/q$. In the case of complex-valued families **Hölder's inequality** states that

$$\left(\sum_I |\xi_i\eta_i|^r\right)^{1/r} \leqq \left(\sum_I |\xi_i|^p\right)^{1/p} \left(\sum_I |\eta_i|^q\right)^{1/q}$$

for $x = (\xi_i) \in l_p(I)$ and $y = (\eta_i) \in l_q(I)$.

C.1.9. Given r with $1 < r < \infty$, the **dual exponent** r' is defined by $1/r + 1/r' = 1$. Furthermore, the exponents 1 and ∞ are considered to be dual to each other.

C.1.10. Let $1 \leqq r < \infty$. Then, for every family $y = (\eta_i) \in l_{r'}(I)$, the rule

$$x = (\xi_i) \to \langle x, y \rangle := \sum_I \xi_i \eta_i$$

defines a functional on $l_r(I)$. In this way the dual Banach space $l_r(I)'$ can be identified with $l_{r'}(I)$. Analogously, we have $c_0(I)' = l_1(I)$.

C.2. Direct sums of Banach spaces

C.2.1. Let $1 \leqq r \leqq \infty$. Given any family of Banach spaces E_i with $i \in I$, the **direct sum** $[l_r(I), E_i]$ consists of all families (x_i) such that $x_i \in E_i$ and $(\|x_i\|) \in l_r(I)$. Note that $[l_r(I), E_i]$ is a Banach space with respect to the norm

$$\|(x_i) \mid [l_r(I), E_i]\| := \left(\sum_I \|x_i\|^r\right)^{1/r}$$

for $1 \leqq r < \infty$. If $r = \infty$, then we let

$$\|(x_i) \mid [l_\infty(I), E_i]\| := \sup\{\|x_i\| : i \in I\}.$$

In the case of the countable index set $\mathbb{N}$ we use the shortened symbol $[l_r, E_i]$. The direct sum of finitely many Banach spaces $E_1, \ldots, E_n$ is denoted by $E_1 \oplus \ldots \oplus E_n$. Here it is unnecessary to indicate the exponent r, since all linear spaces $[l_r(\mathbb{Z}_n), E_i]$ coincide, while the corresponding norms are equivalent.

C.2.2. For any fixed index $k \in I$ we define the following operators:

The **canonical injection** J_k transforms every element $x \in E_k$ into the family (x_i) with $x_k := x$ and $x_i := o$ for $i \neq k$.

The **canonical surjection** Q_k assigns to every family (x_i) its k-th coordinate x_k.

Then

$$E_k \xrightarrow{J_k} [l_r(I), E_i] \xrightarrow{Q_k} E_k$$

is the identity operator of E_k and $P_k := J_k Q_k$ defines a projection on $[l_r(I), E_i]$.

C.3. Classical function spaces

C.3.1. In the following we consider **complex-valued functions** f defined on a set X, which is assumed to be either a compact Hausdorff space or a σ-finite measure space.

All algebraic operations concerning functions are defined pointwise.

C.3.2. The set of all continuous complex-valued functions f defined on a compact Hausdorff space X is denoted by $C(X)$. As is well-known, $C(X)$ becomes a Banach space under the norm

$$\|f \mid C\| := \sup\{|f(\xi)| : \xi \in X\}.$$

In particular, we write $C(0, 1)$ if X is the unit interval $[0, 1]$.

C.3.3. By a **measure space** (X, μ) we mean a set X together with a measure μ which is defined on a σ-algebra formed by subsets of X. Once and for all, we assume (X, μ) to be σ-finite.

The most important example is the unit interval $[0, 1]$ equipped with the Lebesgue measure. The discrete situation treated in the previous section appears if we deal with the counting measure on a countable index set.

C.3.4. Unless the contrary is explicitly stated, we identify measurable complex-valued functions f and g which are equal almost everywhere:

$$\mu\{\xi \in X : f(\xi) \neq g(\xi)\} = 0.$$

C.3.5. Let $L_\infty(X, \mu)$ denote the set of all (equivalence classes of) measurable complex-valued functions f which are essentially bounded on X. This means that there exists a constant $c \geqq 0$ with $\mu\{\xi \in X : |f(\xi)| \geqq c\} = 0$. Setting

$$\|f \mid L_\infty\| := \text{ess-sup}\,\{|f(\xi)| : \xi \in X\} := \inf c,$$

we get a norm which turns $L_\infty(X, \mu)$ into a Banach space.

C.3.6. Let $0 < r < \infty$. We denote by $L_r(X, \mu)$ the collection of all (equivalence classes of) measurable complex-valued functions f such that

$$\|f \mid L_r\| := \left(\int_X |f(\xi)|^r \, d\mu(\xi)\right)^{1/r}$$

is finite. Such functions are called ***r*-integrable**. In the case $r = 1$ we simply speak of **integrable** functions. Note that $L_r(X, \mu)$ is a Banach space if $1 \leqq r < \infty$ and an r-Banach space if $0 < r < 1$. The quasi-triangle inequality (**Minkowski's inequality**) holds with the constant $c_r := \max(2^{1/r-1}, 1)$.

C.3.7. To simplify notation, we often write L_r instead of $L_r(X, \mu)$. If the underlying measure space is the unit interval equipped with the Lebesgue measure, then we use the symbol $L_r(0, 1)$.

C.3.8. Let $0 < p,q < \infty$ and $1/r = 1/p + 1/q$. In the case of complex-valued functions **Hölder's inequality** (HAY, p. 140) states that

$$\left(\int_X |f(\xi)\, g(\xi)|^r \, d\mu(\xi)\right)^{1/r} \leqq \left(\int_X |f(\xi)|^p \, d\mu(\xi)\right)^{1/p} \left(\int_X |g(\xi)|^q \, d\mu(\xi)\right)^{1/q}$$

for $f \in L_p(X, \mu)$ and $g \in L_q(X, \mu)$.

C.3.9. Let $1 \leqq r < \infty$. Then, for every function $g \in L_{r'}(X, \mu)$, the rule

$$f \to \langle f, g \rangle := \int_X f(\xi)\, g(\xi) \, d\mu(\xi)$$

defines a functional on $L_r(X, \mu)$. In this way the dual Banach space $L_r(X, \mu)'$ can be identified with $L_{r'}(X, \mu)$.

C.3.10. Finally, we state **Jessen's inequality**.

Let K be any complex-valued function defined and measurable on the product $(X \times Y, \mu \times \nu)$ of the measure spaces (X, μ) and (Y, ν). If $0 < p \leqq q < \infty$, then

$$\left(\int_Y \left[\int_X |K(\xi, \eta)|^p \, d\mu(\xi)\right]^{q/p} d\nu(\eta)\right)^{1/q} \leqq \left(\int_X \left[\int_Y |K(\xi, \eta)|^q \, d\nu(\eta)\right]^{p/q} d\mu(\xi)\right)^{1/p}.$$

Remark. The assertion follows immediately by applying Minkowski's inequality for integrals (HAY, p. 148) with the exponent $r := q/p$ to $|K(\xi, \eta)|^p$; see (HAY, pp. 31 and 150) and B. Jessen (1931).

C.4. The metric extension property

C.4.1. A Banach space F has the **metric extension property** if, given any metric injection $J \in \mathfrak{L}(E, E_0)$, every operator $T \in \mathfrak{L}(E, F)$ admits a norm-preserving extension $T_0 \in \mathfrak{L}(E_0, F)$. This means that

$$T = T_0 J \quad \text{and} \quad \|T_0\| = \|T\|.$$

C.4.2. A Banach space possesses the metric extension property if and only if it is metrically isomorphic to a Banach space $C(X)$, the compact Hausdorff space X being totally disconnected; see M. Hasumi (1958). The most elementary examples are the Banach spaces $l_\infty(I)$ on an arbitrary index set I; see also 1.5.4 (Remark).

C.4.3. Let E be any Banach space. Given $x \in E$, the equation

$$f_x(a) := \langle x, a \rangle \quad \text{for all } a \in U^0$$

defines a family $f_x \in l_\infty(U^0)$. Setting $J_E x := f_x$, we obtain a metric injection from E into $l_\infty(U^0)$. In this way every Banach space can be canonically embedded into a Banach space with the metric extension property.

D. Operator ideals

The theory of operator ideals is extensively presented in (PIE). We use this monograph as a standard reference. A concise introduction to this subject is given in (JAR). Operator ideals on Hilbert spaces are treated in (GOH), (RIN) and (SCE). We also refer to (DUN, XI.6 and XI.9), (REE) and (SIM).

D.1. Quasi-Banach operator ideals

D.1.1. Suppose that, for every pair of Banach spaces E and F, we are given a subset $\mathfrak{A}(E, F)$ of $\mathfrak{L}(E, F)$. The class

$$\mathfrak{A} := \bigcup_{E,F} \mathfrak{A}(E, F)$$

is said to be an **operator ideal** if the following conditions are satisfied:

(O_1) $\quad a \otimes y \in \mathfrak{A}(E, F)$ for $a \in E'$ and $y \in F$.

(O_2) $\quad S + T \in \mathfrak{A}(E, F)$ for $S, T \in \mathfrak{A}(E, F)$.

(O_3) $\quad YTX \in \mathfrak{A}(E_0, F_0)$ for $X \in \mathfrak{L}(E_0, E)$, $T \in \mathfrak{A}(E, F)$, $Y \in \mathfrak{L}(F, F_0)$.

Since (O_3) implies that

$$\lambda T \in \mathfrak{A}(E, F) \quad \text{for} \quad T \in \mathfrak{A}(E, F) \quad \text{and} \quad \lambda \in \mathbb{C},$$

every component $\mathfrak{A}(E, F)$ is a linear subset of $\mathfrak{L}(E, F)$.

To simplify notation, we write $\mathfrak{A}(E)$ instead of $\mathfrak{A}(E, E)$.

D.1.2. A function $\|\cdot \mid \mathfrak{A}\|$ which assigns to every operator $T \in \mathfrak{A}$ a non-negative number $\|T \mid \mathfrak{A}\|$ is called a **quasi-norm** on the operator ideal $\mathfrak{A}$ if it has the following properties:

(Q_1) $\quad \|a \otimes y \mid \mathfrak{A}\| = \|a\| \, \|y\|$ for $a \in E'$ and $y \in F$.

(Q_2) $\quad \|S + T \mid \mathfrak{A}\| \leqq c_{\mathfrak{A}}[\|S \mid \mathfrak{A}\| + \|T \mid \mathfrak{A}\|]$ for $S, T \in \mathfrak{A}(E, F)$.
Here $c_{\mathfrak{A}} \geqq 1$ is a constant.

(Q_3) $\quad \|YTX \mid \mathfrak{A}\| \leqq \|Y\| \, \|T \mid \mathfrak{A}\| \, \|X\|$ for $X \in \mathfrak{L}(E_0, E)$, $T \in \mathfrak{A}(E, F)$, $Y \in \mathfrak{L}(F, F_0)$.

In order to indicate the underlying Banach spaces E and F, we sometimes replace $\|T \mid \mathfrak{A}\|$ by the more specific symbol $\|T: E \to F \mid \mathfrak{A}\|$.

Since (Q_3) implies that

$$\|\lambda T \mid \mathfrak{A}\| = |\lambda| \, \|T \mid \mathfrak{A}\| \quad \text{for} \quad T \in \mathfrak{A}(E, F) \quad \text{and} \quad \lambda \in \mathbb{C},$$

the function $\|\cdot \mid \mathfrak{A}\|$ indeed yields a quasi-norm in the sense of B.1.1.

In the case $c_{\mathfrak{A}} = 1$ we simply speak of a **norm.**

Remark. By using (Q_2) and (Q_3) only, it follows that

$$\|a \otimes y \mid \mathfrak{A}\| = c\|a\| \, \|y\| \quad \text{for} \quad a \in E' \quad \text{and} \quad y \in F,$$

where $c := \|I_{\mathbb{C}} \mid \mathfrak{A}\|$. Hence property (Q_1) affects the normalization $c = 1$.

D.1.3. Let $\|\cdot \mid \mathfrak{A}\|$ be a quasi-norm on an operator ideal $\mathfrak{A}$. Then, as shown in (PIE, 6.1.4), we have

$$\|T\| \leqq \|T \mid \mathfrak{A}\| \quad \text{for all} \quad T \in \mathfrak{A}(E, F).$$

D.1.4. On a given operator ideal $\mathfrak{A}$ there exist various quasi-norms, good ones and bad ones. However, a quasi-norm $\|\cdot \mid \mathfrak{A}\|$ really fits if all components $\mathfrak{A}(E, F)$ are complete with respect to the induced metrizable topology. If so, then $\mathfrak{A}$ is said to be a **quasi-Banach operator ideal.** In the particular case when we can find even a norm having this property, then $\mathfrak{A}$ is called a **Banach operator ideal.**

It turns out that the quasi-norm of a quasi-Banach operator ideal is unique up to equivalence; see (PIE, 6.1.8). Therefore we may refer to a quasi-Banach operator deal without specifying its quasi-norm explicitly.

D.1.5. An operator ideal $\mathfrak{A}$ is **closed** if all components $\mathfrak{A}(E, F)$ are closed linear subsets of $\mathfrak{L}(E, F)$. This means that $\mathfrak{A}$ becomes a Banach operator ideal by using the ordinary operator norm.

D.1.6. A ***p*-norm** $(0 < p \leqq 1)$ defined on an operator ideal $\mathfrak{A}$ is a quasi-norm which, in addition to (Q_1) and (Q_3), satisfies the p-triangle inequality:

$$\|S + T \mid \mathfrak{A}\|^p \leqq \|S \mid \mathfrak{A}\|^p + \|T \mid \mathfrak{A}\|^p \quad \text{for} \quad S, T \in \mathfrak{A}(E, F).$$

If so, then (Q_2) holds with $c_{\mathfrak{A}} := 2^{1/p-1}$.

D.1.7. By a ***p*-Banach operator ideal** we mean an operator ideal $\mathfrak{A}$ equipped with a p-norm such that all components $\mathfrak{A}(E, F)$ are complete.

D.1.8. Let $\|\cdot \mid \mathfrak{A}\|$ be any quasi-norm on an operator ideal $\mathfrak{A}$ such that

$$\|S + T \mid \mathfrak{A}\| \leqq c_{\mathfrak{A}}[\|S \mid \mathfrak{A}\| + \|T \mid \mathfrak{A}\|] \quad \text{for} \quad S, T \in \mathfrak{A}(E, F).$$

If p is determined by $2^{1/p-1} = c_{\mathfrak{A}}$, then there exists on $\mathfrak{A}$ an equivalent p-norm; see (PIE, 6.2.5). This means that every quasi-Banach operator ideal can be viewed as a p-Banach operator ideal for all sufficiently small exponents p.

D.1.9. Let $\mathfrak{A}$ and $\mathfrak{B}$ be quasi-Banach operator ideals with $\mathfrak{A} \subseteq \mathfrak{B}$. Then, by the closed graph theorem, there exists a constant $c \geqq 1$ such that

$$\|T \mid \mathfrak{B}\| \leqq c\|T \mid \mathfrak{A}\| \quad \text{for all} \quad T \in \mathfrak{A}(E, F).$$

For a direct proof we refer to (PIE, 6.1.6).

D.1.10. An operator $T \in \mathfrak{L}(E, G)$ belongs to the **product** $\mathfrak{B} \circ \mathfrak{A}$ of the operator ideals $\mathfrak{A}$ and $\mathfrak{B}$ if it can be written in the form $T = BA$, where $A \in \mathfrak{A}(E, F)$ and $B \in \mathfrak{B}(F, G)$. Here F is any suitable Banach space. Obviously, $\mathfrak{B} \circ \mathfrak{A}$ is an operator ideal.

Given quasi-norms $\|\cdot \mid \mathfrak{A}\|$ and $\|\cdot \mid \mathfrak{B}\|$, we put

$$\|T \mid \mathfrak{B} \circ \mathfrak{A}\| := \inf \{\|B \mid \mathfrak{B}\| \, \|A \mid \mathfrak{A}\|\},$$

the infimum being taken over all possible factorizations. In this way $\mathfrak{B} \circ \mathfrak{A}$ becomes a quasi-Banach operator ideal whenever $\mathfrak{A}$ and $\mathfrak{B}$ are complete; see (PIE, 7.1.2).

The m-th **power** of an operator ideal $\mathfrak{A}$ is inductively defined by $\mathfrak{A}^{m+1} := \mathfrak{A}^m \circ \mathfrak{A}$ and $\mathfrak{A}^\circ := \mathfrak{L}$.

D.1.11. An operator $T \in \mathfrak{L}(E, F)$ belongs to the **sum** $\mathfrak{A} + \mathfrak{B}$ of the operator ideals $\mathfrak{A}$ and $\mathfrak{B}$ if it can be written in the form $T = A + B$, where $A \in \mathfrak{A}(E, F)$ and $B \in \mathfrak{B}(E, F)$. Obviously, $\mathfrak{A} + \mathfrak{B}$ is an operator ideal.

Given quasi-norms $\|\cdot \mid \mathfrak{A}\|$ and $\|\cdot \mid \mathfrak{B}\|$, we put

$$\|T \mid \mathfrak{A} + \mathfrak{B}\| := \inf \{\|A \mid \mathfrak{A}\| + \|B \mid \mathfrak{B}\|\},$$

the infimum being taken over all possible decompositions. In this way $\mathfrak{A} + \mathfrak{B}$ becomes a quasi-Banach operator ideal whenever $\mathfrak{A}$ and $\mathfrak{B}$ are complete.

D.1.12. For every operator ideal $\mathfrak{A}$ the **dual operator ideal** $\mathfrak{A}'$ is defined by

$$\mathfrak{A}'(E, F) := \{T \in \mathfrak{L}(E, F) : T' \in \mathfrak{A}(F', E')\}.$$

Given any quasi-norm $\|\cdot \mid \mathfrak{A}\|$, we put

$$\|T \mid \mathfrak{A}'\| := \|T' \mid \mathfrak{A}\|.$$

In this way $\mathfrak{A}'$ becomes a quasi-Banach operator ideal whenever $\mathfrak{A}$ is complete.

D.1.13. A quasi-Banach operator ideal $\mathfrak{A}$ is **approximative** if $\mathfrak{F}(E, F)$, the set of finite operators, is dense in every component $\mathfrak{A}(E, F)$.

Let $\mathfrak{A}$ be any quasi-Banach operator ideal. An operator $T \in \mathfrak{A}(E, F)$ is said to be **$\mathfrak{A}$-approximable** if there exists a sequence of operators $T_n \in \mathfrak{F}(E, F)$ such that

$$\lim_n \|T - T_n \mid \mathfrak{A}\| = 0.$$

The class of these operators is the so-called **approximative kernel** $\mathfrak{A}^{(a)}$ of $\mathfrak{A}$. Obviously, $\mathfrak{A}^{(a)}(E, F)$ can be viewed as the closed hull of $\mathfrak{F}(E, F)$ in $\mathfrak{A}(E, F)$. Therefore, restricting the quasi-norm of $\mathfrak{A}$ to $\mathfrak{A}^{(a)}$, we get an approximative quasi-Banach operator ideal.

D.1.14. A quasi-Banach operator ideal $\mathfrak{A}$ is said to be **injective** if, given any metric injection $J \in \mathfrak{L}(F, F_0)$,

$$T \in \mathfrak{L}(E, F) \text{ and } JT \in \mathfrak{A}(E, F_0) \text{ imply } T \in \mathfrak{A}(E, F) \text{ and } \|T \mid \mathfrak{A}\| = \|JT \mid \mathfrak{A}\|.$$

This property means that the answer to the question whether or not an operator $T \in \mathfrak{L}(E, F)$ belongs to the ideal $\mathfrak{A}$ is independent of the size of the target space F.

D.1.15. A quasi-Banach operator ideal $\mathfrak{A}$ is said to be **surjective** if, given any metric surjection $Q \in \mathfrak{L}(E_0, E)$,

$$T \in \mathfrak{L}(E, F) \text{ and } TQ \in \mathfrak{A}(E_0, F) \text{ imply } T \in \mathfrak{A}(E, F) \text{ and } \|T \mid \mathfrak{A}\| = \|TQ \mid \mathfrak{A}\|.$$

Note that injectivity and surjectivity are dual concepts.

D.2. Some examples of operator ideals

D.2.1. Recall that, according to B.4.4, an operator $T \in \mathfrak{L}(E, F)$ is **finite** if it can be written in the form

$$T = \sum_{i=1}^{n} a_i \otimes y_i,$$

where $a_1, \ldots, a_n \in E'$ and $y_1, \ldots, y_n \in F$. The set of these operators is denoted by $\mathfrak{F}(E, F)$. It easily turns out that $\mathfrak{F}$ is the smallest operator ideal.

D.2.2. An operator $T \in \mathfrak{L}(E, F)$ is said to be **approximable** if there exists a sequence of operators $T_n \in \mathfrak{F}(E, F)$ such that

$$\lim_n \|T - T_n\| = 0.$$

The set of these operators is denoted by $\mathfrak{G}(E, F)$ or $\mathfrak{L}^{(a)}(E, F)$. Obviously, $\mathfrak{G}$ is the smallest closed operator ideal.

Note that $\mathfrak{G} = \mathfrak{G}'$. However, $\mathfrak{G}$ is neither injective nor surjective.

D.2.3. An operator $T \in \mathfrak{L}(E, F)$ is **compact** if it maps the closed unit ball of E into a precompact subset of F. The set of these operators is denoted by $\mathfrak{K}(E, F)$. As is well-known, $\mathfrak{K}$ is a closed operator ideal; see (DUN, VI.5.4) and (TAY, V.7.1 and V.7.2).

Note that $\mathfrak{K} = \mathfrak{K}'$. Moreover, $\mathfrak{K}$ is injective and surjective, but it fails to be approximative.

D.2.4. We call $T \in \mathfrak{L}(E, F)$ a **Hilbert operator** if it factors through a Hilbert space. This means that $T = BA$, where $A \in \mathfrak{L}(E, H)$ and $B \in \mathfrak{L}(H, F)$. The set of these operators is denoted by $\mathfrak{H}(E, F)$.

For $T \in \mathfrak{H}(E, F)$ we define

$$\|T \mid \mathfrak{H}\| := \inf \{\|B\| \; \|A\|\},$$

the infimum being taken over all possible factorizations. It can easily be shown that $\mathfrak{H}$ is a Banach operator ideal; see (PIE, 6.6.2).

Note that $\mathfrak{H} = \mathfrak{H}'$. Moreover, $\mathfrak{H}$ is injective and surjective.

D.3. Extensions of operator ideals

D.3.1. The concept of an operator ideal

$$\mathfrak{A} := \bigcup_{E,F} \mathfrak{A}(E, F)$$

also makes sense if E and F range only over a certain subclass of Banach spaces. Of particular interest is the case

$$\mathfrak{A} := \bigcup_{H,K} \mathfrak{A}(H, K),$$

where H and K are arbitrary Hilbert spaces.

D.3.2. Let $\mathfrak{A}$ be an operator ideal on the class of Hilbert spaces. An operator ideal $\mathfrak{A}_0$ defined on the class of all Banach spaces is said to be an **extension** of $\mathfrak{A}$ if

$$\mathfrak{A}_0(H, K) = \mathfrak{A}(H, K) \quad \text{for all } H \text{ and } K.$$

In the case when $\mathfrak{A}_0$ and $\mathfrak{A}$ are quasi-Banach operator ideals such that the corresponding quasi-norms coincide on all components $\mathfrak{A}_0(H, K) = \mathfrak{A}(H, K)$ we speak of a **metric extension.**

D.3.3. Every operator ideal $\mathfrak{A}$ defined on the class of Hilbert spaces admits a smallest and a largest extension which, in general, differ from each other considerably; see (PIE, 15.6).

E. Tensor products

From the theory of tensor products we only need the basic definitions and some elementary facts which are to be found in (DIL), (KÖT) and (JAR).

E.1. Algebraic tensor products

E.1.1. Let $\varepsilon(E, F)$ denote the Banach space of all bounded bilinear forms z defined on the cartesian product $E' \times F'$. The norm is given by

$$\|z \mid \varepsilon\| := \sup\{|z(a, b)| : a \in U^0, b \in V^0\},$$

where U^0 and V^0 are the closed unit balls of E' and F', respectively.

E.1.2. The **algebraic tensor product** of the Banach spaces E and F is the linear subset $E \otimes F$ of $\varepsilon(E, F)$ spanned by all bounded bilinear functionals

$$x \otimes y : (a, b) \to \langle x, a\rangle \langle y, b\rangle$$

with $x \in E$ and $y \in F$. This means that every element $z \in E \otimes F$ admits a representation

$$(*) \qquad z = \sum_{i=1}^{n} x_i \otimes y_i .$$

Alternatively, $E \otimes F$ may be viewed as the collection of all formal sums (*), with the understanding that

$$\lambda(x \otimes y) = (\lambda x) \otimes y = x \otimes (\lambda y),$$

$$(x_1 + x_2) \otimes y = x_1 \otimes y + x_2 \otimes y,$$
$$x \otimes (y_1 + y_2) = x \otimes y_1 + x \otimes y_2 .$$

E.1.3. Let $S \in \mathfrak{L}(E, E_0)$ and $T \in \mathfrak{L}(F, F_0)$. Given $z \in \varepsilon(E, F)$, then

$$z_0 : (a_0, b_0) \to z(S'a_0, T'b_0)$$

defines a bounded bilinear functional $z_0 \in \varepsilon(E_0, F_0)$. Setting $\varepsilon(S, T)\, z := z_0$, we obtain an operator $\varepsilon(S, T) \in \mathfrak{L}(\varepsilon(E, F), \varepsilon(E_0, F_0))$ with $\|\varepsilon(S, T)\| = \|S\|\,\|T\|$. Observe that $\varepsilon(S, T)\,(x \otimes y) = Sx \otimes Ty$.

E.1.4. The **algebraic tensor product** of the operators S and T is the linear map $S \otimes T$ from $E \otimes F$ into $E_0 \otimes F_0$ induced by $\varepsilon(S, T)$.

E.2. Banach tensor products

E.2.1. A non-negative function $\alpha: z \to \|z \mid \alpha\|$ defined on all algebraic tensor products $E \otimes F$ is said to be a **tensor norm** if it has the following properties:

(TN$_1$) $\|x \otimes y \mid \alpha\| = \|x\| \, \|y\|$ for $x \in E$ and $y \in F$.

(TN$_2$) $\|u + v \mid \alpha\| \leqq \|u \mid \alpha\| + \|v \mid \alpha\|$ for $u, v \in E \otimes F$.

(TN$_3$) $\|(S \otimes T)\, z \mid \alpha\| \leqq \|S\| \, \|T\| \, \|z \mid \alpha\|$ for $S \in \mathfrak{L}(E, E_0)$, $T \in \mathfrak{L}(F, F_0)$ and $z \in E \otimes F$.

E.2.2. Next we give the most important examples.
There exists a schmallest tensor norm ε which is defined by

$$\|z \mid \varepsilon\| := \sup\{|z(a, b)| : a \in U^0, b \in V^0\}.$$

There exists a largest tensor norm ν which is defined by

$$\|z \mid \nu\| := \inf\left\{\sum_{i=1}^{n} \|x_i\| \, \|y_i\|\right\},$$

the infimum being taken over all representations

$$z = \sum_{i=1}^{n} x_i \otimes y_i.$$

E.2.3. The Banach space $E \tilde{\otimes}_\alpha F$ obtained by completing $E \otimes F$ with respect to a given tensor norm α is called the **α-tensor product** of the Banach spaces E and F.

E.2.4. Let $S \in \mathfrak{L}(E, E_0)$ and $T \in \mathfrak{L}(F, F_0)$. Given any tensor norm α, then the continuous extension $S \tilde{\otimes}_\alpha T \in \mathfrak{L}(E \tilde{\otimes}_\alpha F, E_0 \tilde{\otimes}_\alpha F_0)$ of $S \otimes T$ is said to be the **α-tensor product** of the operators S and T.

E.2.5. If H and K are Hilbert spaces, then there exists a scalar product on the algebraic tensor product $H \otimes K$ which is uniquely determined by the following property:

$$(x_1 \otimes y_1, x_2 \otimes y_2) = (x_1, x_2)(y_1, y_2) \text{ for } x_1, x_2 \in H \text{ and } y_1, y_2 \in K.$$

The corresponding tensor norm (defined on the class of Hilbert spaces only) is denoted by σ.

E.2.6. To illustrate the preceding definition, we note that the continuous and linear extension of the map

$$(\xi_i) \otimes (\eta_j) \to (\xi_i \eta_j)$$

s a metric isomorphism between $l_2(I) \tilde{\otimes}_\sigma l_2(J)$ and $l_2(I \times J)$.

E.3. Tensor stability of operator ideals

E.3.1. An operator ideal $\mathfrak{A}$ is said to be **stable with respect to a tensor norm** α if

$$S \in \mathfrak{A}(E, E_0) \quad \text{and} \quad T \in \mathfrak{A}(F, F_0) \quad \text{imply} \quad S \tilde{\otimes}_\alpha T \in \mathfrak{A}(E \tilde{\otimes}_\alpha F, E_0 \tilde{\otimes}_\alpha F_0).$$

E.3.2. From the theoretical viewpoint the following consequence of the closed graph theorem is very important. In all concrete cases, however, this statement is a by-product obtained free of charge within the proof of tensor stability.

Let $\mathfrak{A}$ be a quasi-Banach operator ideal which is stable with respect to a tensor norm α. Then there exists a constant $c \geqq 1$ such that

$$\|S \tilde{\otimes}_\alpha T \mid \mathfrak{A}\| \leqq c\|S \mid \mathfrak{A}\| \, \|T \mid \mathfrak{A}\| \quad \text{for } S \in \mathfrak{A}(E, E_0) \quad \text{and} \quad T \in \mathfrak{A}(F, F_0).$$

F. Interpolation theory

In order to prove some fundamental theorems in their optimal version, we need a few results from interpolation theory (real method) which are to be found in (BER) and (TRI); see also (BUB).

F.1. Intermediate spaces

F.1.1. A **quasi-Banach interpolation couple** (E_0, E_1) consists of two quasi-Banach spaces E_0 and E_1 which are continuously embedded into a Hausdorff topological linear space X.

F.1.2. With every quasi-Banach interpolation couple (E_0, E_1) we may associate the quasi-Banach spaces $E_0 \cap E_1$ and $E_0 + E_1$ endowed with the quasi-norms

$$\|x \mid E_0 \cap E_1\| := \max\{\|x \mid E_0\|, \|x \mid E_1\|\}$$

and

$$\|x \mid E_0 + E_1\| := \inf\{\|x_0 \mid E_0\| + \|x_1 \mid E_1\| : x_0 + x_1 = x\},$$

respectively. Then we have the following diagram, where the arrows indicate embedding operators:

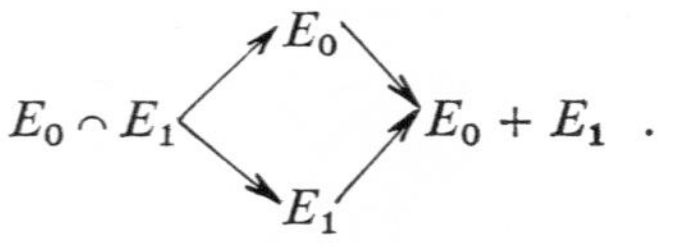

Remark. From now on the quasi-Banach space $E_0 + E_1$ can play the role of the Hausdorff topological linear space X which is only required to fix the mutual position of E_0 and E_1.

F.1.3. A quasi-Banach space E is said to be **intermediate** with respect to the quasi-Banach interpolation couple (E_0, E_1) if it lies between $E_0 \cap E_1$ and $E_0 + E_1$:

$$E_0 \cap E_1 \longrightarrow E \longrightarrow E_0 + E_1 .$$

F.2. Interpolation methods

F.2.1. By an **interpolation method** we mean a **functor** Φ which assigns to every quasi-Banach interpolation couple (E_0, E_1) an intermediate quasi-Banach space $(E_0, E_1)_\Phi$ having the so-called **interpolation property**:

Let (E_0, E_1) and (F_0, F_1) be arbitrary quasi-Banach interpolation couples. If an operator $T \in \mathfrak{L}(E_0 + E_1, F_0 + F_1)$ transforms E_0 into F_0 and E_1 into F_1, then it induces an operator from $(E_0, E_1)_\Phi$ into $(F_0, F_1)_\Phi$. In this case, there exists a constant $c \geqq 1$ such that

$$\|T : (E_0, E_1)_\Phi \to (F_0, F_1)_\Phi\| \leqq c \max\{\|T : E_0 \to F_0\|, \|T : E_1 \to F_1\|\}.$$

F.2.2. Let $0 < \theta < 1$. An interpolation method Φ is said to be of **type** θ if, in addition to the above properties, we have

$$\|T : (E_0, E_1)_\Phi \to (F_0, F_1)_\Phi\| \leqq c\|T : E_0 \to F_0\|^{1-\theta} \, \|T : E_1 \to F_1\|^\theta .$$

F.3. Real interpolation

F.3.1. Given any quasi-Banach interpolation couple (E_0, E_1), for $x \in E_0 + E_1$ and $\tau > 0$, the **Peetre K-functional** is defined by

$$K(\tau, x, E_0, E_1) := \inf \{\|x_0 \mid E_0\| + \tau\|x_1 \mid E_1\| : x_0 + x_1 = x\} .$$

If there is no risk of confusion, then we write $K(\tau, x)$ instead of $K(\tau, x, E_0, E_1)$.

Obviously, $K(\tau, x)$ yields a one-parameter family of equivalent quasi-norms on $E_0 + E_1$ which, for fixed $x \in E_0 + E_1$, depend continuously on τ.

F.3.2. We now define the so-called **real interpolation method**, where the adjective "real" refers to the nature of the construction and not to the underlying quasi-Banach spaces which (as always throughout this book) are supposed to be complex.

Let $0 < \theta < 1$ and $0 < w < \infty$. For every quasi-Banach interpolation couple (E_0, E_1) we denote by $(E_0, E_1)_{\theta,w}$ the collection of all elements $x \in E_0 + E_1$ such that the expression

$$\|x \mid (E_0, E_1)_{\theta,w}\| := \left(\int_0^\infty |\tau^{-\theta} K(\tau, x)|^w \frac{d\tau}{\tau} \right)^{1/w}$$

is finite. In the limiting case $w = \infty$ the same is assumed for

$$\|x \mid (E_0, E_1)_{\theta,\infty}\| := \sup_\tau \tau^{-\theta} K(\tau, x).$$

It can easily be seen that $(E_0, E_1)_{\theta,w}$ becomes a quasi-Banach space with respect to the quasi-norm just defined. Furthermore, as shown in (BER, 3.1.2 and 3.11.2) and (TRI, 1.3.3), the functor

$$\Phi_{\theta,w} : (E_0, E_1) \to (E_0, E_1)_{\theta,w}$$

defines an interpolation method of type θ.

F.3.3. The real interpolation method possesses the **reiteration property**:

Let $0 < \Sigma_0 < \Sigma_1 < 1$, $0 < \theta < 1$ and $0 < w_0, w_1, w \leqq \infty$. If we have $\Sigma = (1 - \theta)\, \Sigma_0 + \theta\, \Sigma_1$, then

$$((E_0, E_1)_{\Sigma_0,w_0}, (E_0, E_1)_{\Sigma_1,w_1})_{\theta,w} = (E_0, E_1)_{\Sigma,w} .$$

Proofs of this formula are given in (BER, 3.5.3 and 3.11.5) and (TRI, 1.10.2).

F.3.4. We now describe the most important example which was the starting point of interpolation theory. For this purpose, let (X, μ) be a measure space. Then the function spaces $L_r(X, \mu)$ with $0 < r \leqq \infty$ are continuously embedded into the complete metric linear space $L_0(X, \mu)$ formed by all (equivalence classes of) measurable complex-valued functions. In this way $(L_{r_0}(X, \mu), L_{r_1}(X, \mu))$ can be viewed as a quasi-Banach interpolation couple.

Let $0 < r_0 < r_1 \leqq \infty$ and $0 < \theta < 1$. If $1/r = (1 - \theta)/r_0 + \theta/r_1$, then

$$(L_{r_0}(X, \mu), L_{r_1}(X, \mu))_{\theta,r} = L_r(X, \mu).$$

The quasi-norms on both sides are equivalent.

F.3.5. Finally, we state **Lyapunov's inequality** which is closely related to the preceding result; see (HAY, pp. 27 and 146).

Let $0 < r_0 < r_1 \leqq \infty$ and $0 < \theta < 1$. If $1/r = (1 - \theta)/r_0 + \theta/r_1$, then

$$\|f \,|\, L_r\| \leqq \|f \,|\, L_{r_0}\|^{1-\theta} \, \|f \,|\, L_{r_1}\|^{\theta} \quad \text{for all } f \in L_{r_0}(X, \mu) \cap L_{r_1}(X, \mu).$$

F.4. Interpolation of quasi-Banach operator ideals

F.4.1. Let $\mathfrak{A}_0$ and $\mathfrak{A}_1$ be quasi-Banach operator ideals. Then for arbitrary Banach spaces E and F, the components $\mathfrak{A}_0(E, F)$ and $\mathfrak{A}_1(E, F)$ are continuously embedded in $\mathfrak{L}(E, F)$. This means that $(\mathfrak{A}_0(E, F), \mathfrak{A}_1(E, F))$ is a quasi-Banach interpolation couple. Hence, given any interpolation functor Φ, we may define

$$(\mathfrak{A}_0, \mathfrak{A}_1)_\Phi := \bigcup_{E,F} (\mathfrak{A}_0(E, F), \mathfrak{A}_1(E, F))_\Phi.$$

It easily follows that $(\mathfrak{A}_0, \mathfrak{A}_1)_\Phi$ is a quasi-Banach operator ideal.

G. Inequalities

We use the classical treatise of G. H. Hardy / J. E. Littlewood / G. Pólya (HAY) as a standard reference on inequalities. For the convenience of the reader a proof of Khintchine's inequality is given in Section G.2.

Further important inequalities are already stated above:

Hadamard's inequality (A.4.5),
Minkowski's inequality (C.1.6), (C.3.6),
Hölder's inequality (C.1.8), (C.3.8),
Lyapunov's inequality (F.3.5),
Jessen's inequality (C.3.10).

G.1. The inequality of means

G.1.1. The **inequality of means** (HAY, pp. 16–21) states that

$$\left(\prod_{k=1}^{n} \alpha_k\right)^{1/n} \leqq \left(\frac{1}{n}\sum_{k=1}^{n} \alpha_k\right) \quad \text{for } \alpha_1, \ldots, \alpha_n \geqq 0.$$

G.1.2. For completeness, we mention the following supplement of the preceding result:

$$\left(\prod_{k=1}^{n} \alpha_k\right)^{1/n} = \lim_{p \to 0} \left(\frac{1}{n}\sum_{k=1}^{n} \alpha_k^p\right)^{1/p} \quad \text{for } \alpha_1, \ldots, \alpha_n \geqq 0.$$

Note that

$$\left(\frac{1}{n}\sum_{k=1}^{n} \alpha_k^p\right)^{1/p} \leqq \left(\frac{1}{n}\sum_{k=1}^{n} \alpha_k^q\right)^{1/q} \quad \text{if } 0 < p < q < \infty.$$

G.2. The Khintchine inequality

G.2.1. The **Rademacher functions** ϱ_k with $k = 1, 2, \ldots$ are defined by

$$\varrho_k(\tau) := \operatorname{sign}(\sin 2^k \pi\tau) \quad \text{if } 0 \leqq \tau \leqq 1 .$$

This means that

$$\varrho_k(\tau) = (-1)^{h+1} \quad \text{if } \frac{h-1}{2^k} < \tau < \frac{h}{2^k} \quad \text{and} \quad h = 1, \ldots, 2^k .$$

G.2.2. Obviously, (ϱ_k) is an orthonormal sequence in $L_2(0, 1)$. We have, however, a much stronger property.

Lemma. Let $p_1, \ldots, p_n \in \mathbb{N}$ and $k_1, \ldots, k_n \in \mathbb{N}$, where $k_1 < \ldots < k_n$. Then

$$\int_0^1 \varrho_{k_1}^{p_1}(\tau) \ldots \varrho_{k_n}^{p_n}(\tau)\, d\tau = \begin{cases} 1 & \text{if all } p_i \text{ are even,} \\ 0 & \text{otherwise.} \end{cases}$$

Proof. Every factor of the integrand for which p_i is even can be cancelled, since it equals 1 almost everywhere. Thus we may assume that p_n is odd. Divide (0,1) into intervals of length 2^{-k_n}. Then the factors $\varrho_{k_i}^{p_i}(\tau)$ with $i < n$ are constant on the union of the $(2h-1)$-th and the $2h$-th interval. On the other hand, $\varrho_{k_n}^{p_n}(\tau)$ takes the values $+1$ and -1, respectively. This implies that the integral over each of these pairs of intervals vanishes and so does the integral over (0,1).

G.2.3. We are now in a position to establish **Khintchine's inequality** which is one of the basic tools in the theory of absolutely summing operators.

Lemma (A. Ya. Khintchine 1923). If $0 < r < \infty$, then there exist constants a_r and b_r such that

$$a_r \left(\sum_{k=1}^n |\xi_k|^2 \right)^{1/2} \leqq \left(\int_0^1 \left| \sum_{k=1}^n \xi_k \varrho_k(\tau) \right|^r d\tau \right)^{1/r} \leqq b_r \left(\sum_{k=1}^n |\xi_k|^2 \right)^{1/2}$$

for all finite families of complex numbers $\xi_1, \ldots, \xi_n$.

Proof. Since the Rademacher functions are orthonormal, Parseval's equation states that

$$\left(\int_0^1 \left| \sum_{k=1}^n \xi_k \varrho_k(\tau) \right|^2 d\tau \right)^{1/2} = \left(\sum_{k=1}^n |\xi_k|^2 \right)^{1/2} .$$

Hence the above inequality holds by taking

$$a_r = 1 \quad \text{if} \quad 2 \leqq r < \infty \quad \text{and} \quad b_r = 1 \quad \text{if} \quad 0 < r \leqq 2 .$$

Next we treat the special case in which $r = 2s$ with $s = 1, 2, \ldots$ is an even number. Let $\xi_1, \ldots, \xi_n$ be real. Applying the n-nomial formula, by G.2.2, we obtain

$$\int_0^1 \left| \sum_{k=1}^n \xi_k \varrho_k(\tau) \right|^{2s} d\tau = \sum \frac{(2s)!}{(2s_1)! \ldots (2s_n)!} |\xi_1|^{2s_1} \ldots |\xi_n|^{2s_n}$$

and

$$\Big(\sum_{k=1}^{n} |\xi_k|^2\Big)^s = \sum \frac{s!}{s_1! \dots s_n!} |\xi_1|^{2s_1} \dots |\xi_n|^{2s_n},$$

where the right-hand sums are taken over all combinations $(s_1, \dots, s_n)$ such that $s_1 + \dots + s_n = s$. Comparing the n-nomial coefficients, we see that

$$\Big(\int_0^1 \Big|\sum_{k=1}^{n} \xi_k \varrho_k(\tau)\Big|^{2s} d\tau\Big)^{1/2s} \leqq b_{2s} \Big(\sum_{k=1}^{n} |\xi_k|^2\Big)^{1/2},$$

where

$$b_{2s} := \max \Big[\frac{(2s)!}{s!} \frac{s_1!}{(2s_1)!} \dots \frac{s_n!}{(2s_n)!}\Big]^{1/2s}.$$

Next it follows from

$$2^m \leqq \frac{(2m)!}{m!} \leqq (2m)^m$$

that

$$b_{2s} \leqq [(2s)^s \, 2^{-s_1 - \dots - s_n}]^{1/2s} \leqq s^{1/2}.$$

This proves the right-hand Khintchine inequality for even exponents and real coefficients.

The complex case can be treated by splitting into real and imaginary parts.

Next, given an arbitrary exponent r with $2 < r < \infty$, we choose $s \in \mathbb{N}$ such that $r \leqq 2s$. Then the required inequality holds with $b_r := b_{2s}$.

It remains to verify the left-hand Khintchine inequality for $0 < r < 2$. To this end, the parameter θ is defined by

$$1/2 = (1 - \theta)/r + \theta/8.$$

Applying Lyapunov's inequality and the preceding results, we get

$$\Big(\sum_{k=1}^{n} |\xi_k|^2\Big)^{1/2} = \Big(\int_0^1 \Big|\sum_{k=1}^{n} \xi_k \varrho_k(\tau)\Big|^2 d\tau\Big)^{1/2}$$

$$\leqq \Big(\int_0^1 \Big|\sum_{k=1}^{n} \xi_k \varrho_k(\tau)\Big|^r d\tau\Big)^{(1-\theta)/r} \Big(\int_0^1 \Big|\sum_{k=1}^{n} \xi_k \varrho_k(\tau)\Big|^8 d\tau\Big)^{\theta/8}$$

$$\leqq \Big(\int_0^1 \Big|\sum_{k=1}^{n} \xi_k \varrho_k(\tau)\Big|^r d\tau\Big)^{(1-\theta)/r} b_8 \Big(\sum_{k=1}^{n} |\xi_k|^2\Big)^{\theta/2}.$$

Hence

$$\Big(\sum_{k=1}^{n} |\xi_k|^2\Big)^{1/2} \leqq 2^{1/(1-\theta)} \Big(\int_0^1 \Big|\sum_{k=1}^{n} \xi_k \varrho_k(\tau)\Big|^r d\tau\Big)^{1/r}.$$

Remark. From now on we assume that the constants a_r and b_r are chosen best possible. Although these optimal values are of sporting interest only, they have been computed by the efforts of several mathematicians. The final result is due to U. Haagerup (1982) who also described the historical development.

As observed by B. Tomaszewski (see S. J. Szarek 1976) the values of a_r and b_r are the same for the real and complex case.

G.3. Asymptotic estimates

G.3.1. Let (α_n) and (β_n) be non-negative real-valued sequences. Then

$$\alpha_n \prec \beta_n \text{ means that } \alpha_n \leqq c\beta_n \text{ for } n = 1, 2, \ldots,$$

where the constant $c > 0$ may depend on various parameters but not on the index n.
We write

$$\alpha_n \asymp \beta_n \text{ if } \alpha_n \prec \beta_n \text{ and } \beta_n \prec \alpha_n.$$

G.3.2. The following examples illustrate the above definition:
Let $-\infty < \beta < +\infty$. Then

$$\sum_{h=0}^{k} \frac{1}{2^{h\alpha}(h+1)^\beta} \asymp \frac{1}{2^{k\alpha}(k+1)^\beta} \qquad \text{if } \alpha < 0$$

and

$$\sum_{h=k+1}^{\infty} \frac{1}{2^{h\alpha}(h+1)^\beta} \asymp \frac{1}{2^{k\alpha}(k+1)^\beta} \qquad \text{if } \alpha > 0.$$

This implies that

$$\sum_{m=1}^{n} \frac{1}{m^{\alpha+1}(1+\log m)^\beta} \asymp \frac{1}{n^{\alpha+1}(1+\log n)^\beta} \qquad \text{if } \alpha < 0$$

and

$$\sum_{m=n+1}^{\infty} \frac{1}{m^{\alpha+1}(1+\log m)^\beta} \asymp \frac{1}{n^{\alpha+1}(1+\log n)^\beta} \qquad \text{if } \alpha > 0.$$

G.3.3. Next we state a consequence of **Stirling's formula** (TIT, 1.87):

$$n! \asymp n^{n+1/2} \exp(-n).$$

In this connection we refer to the elementary estimate

$$\frac{\xi^n}{n!} \leqq \exp(\xi) \quad \text{for all } \xi > 0$$

which, by taking $\xi := n$, implies that

$$n! \geqq n^n \exp(-n).$$

CHAPTER 1

Absolutely summing operators

This chapter contains an introduction to the theory of absolutely summing operators. We mainly concentrate on such results as are relevant for later applications within the theory of s-numbers and eigenvalues.

We begin by summarizing some elementary facts about absolutely and weakly r-summable families in Banach spaces. Next the basic properties of absolutely (r, s)-summing operators are established, $1 \leqq s \leqq r < \infty$. These operators constitute a Banach ideal, denoted by $\mathfrak{P}_{r,s}$.

Of special importance are the operator ideals $\mathfrak{P}_r := \mathfrak{P}_{r,r}$ and $\mathfrak{P}_{r,2}$. In particular, we stress the outstanding role of $\mathfrak{P}_2$, which has become an indispensable tool not only for the purpose of this monograph but also for Banach space theory as a whole. Among other circumstances this is due to the fact that the absolutely 2-summing operators share many properties with the famous Hilbert-Schmidt operators which have proved to be quite useful in the Hilbert space setting. We also deal with the powers of the operator ideal $\mathfrak{P}_2$.

Diagonal operators from l_p into l_q are investigated. On the one hand, these operators yield instructive examples for exploring the relationship between different operator ideals. On the other hand, they play an important part within the theory of eigenvalue distributions of operators induced by infinite matrices.

In the last section we treat ideals of p-nuclear and $(r, 2)$-nuclear operators for $0 < p \leqq 1$ and $0 < r \leqq 2$, respectively.

1.1. Summable sequences

Throughout this section we assume that $1 \leqq r < \infty$.

1.1.1.* Let I be any index set, and denote by $\mathfrak{F}(I)$ the collection of its finite subsets.

1.1.2.* An E-valued family (x_i) with $i \in I$ is said to be **summable** if there exists an element $s \in E$ with the following property:

Given $\varepsilon > 0$, we can find $\mathrm{i}_0 \in \mathfrak{F}(I)$ such that

$$\Big\| s - \sum_{\mathrm{i}} x_i \Big\| \leqq \varepsilon \quad \text{whenever } \mathrm{i} \in \mathfrak{F}(I) \quad \text{and} \quad \mathrm{i} \supseteq \mathrm{i}_0 .$$

When this is so, then s is called the **sum** of (x_i), and we write

$$s = \sum_I x_i .$$

1.1.3.* An E-valued family (x_i) with $i \in I$ is said to be **absolutely r-summable** if $(\|x_i\|) \in l_r(I)$. The set of these families is denoted by $[l_r(I), E]$.

For $(x_i) \in [l_r(I), E]$ we define

$$\|(x_i) \mid [l_r(I), E]\| := \Big(\sum_I \|x_i\|^r\Big)^{1/r}.$$

If there is no risk of confusion, then we use the shortened symbol $\|(x_i) \mid l_r\|$. Moreover, we write $[l_r, E]$ instead of $[l_r(\mathbb{N}), E]$.

1.1.4.* We state the following fact without proof.

Proposition. $[l_r(I), E]$ is a Banach space.

1.1.5.* The next result is also straightforward.

Proposition. Let $T \in \mathfrak{L}(E, F)$. Then

$$T(I) : (x_i) \to (Tx_i)$$

defines an operator from $[l_r(I), E]$ into $[l_r(I), F]$. Moreover, we have

$$\|(Tx_i) \mid l_r\| \leqq \|T\| \, \|(x_i) \mid l_r\| \quad \text{for} \quad (x_i) \in [l_r(I), E].$$

1.1.6.* We now state an important formula which is proved in (BER, 5.6.1) and (TRI, 1.18.1). In order to include the limiting case, we let $[l_\infty(I), E]$ denote the Banach space of all bounded E-valued families (x_i) with $i \in I$; see C.1.3.

Interpolation proposition (J. L. Lions / J. Peetre 1964). Let $1 \leqq r_0 < r_1 \leqq \infty$ and $0 < \theta < 1$. If $1/r = (1 - \theta)/r_0 + \theta/r_1$, then

$$([l_{r_0}(I), E], [l_{r_1}(I), E])_{\theta, r} = [l_r(I), E].$$

The norms on both sides are equivalent.

1.1.7.* An E-valued family (x_i) with $i \in I$ is said to be **weakly r-summable** if $(\langle x_i, a\rangle) \in l_r(I)$ for all $a \in E'$. The set of these families is denoted by $[w_r(I), E]$.

For $(x_i) \in [w_r(I), E]$ we define

$$\|(x_i) \mid [w_r(I), E]\| := \sup\left\{\left(\sum_I |\langle x_i, a\rangle|^r\right)^{1/r} : a \in U^0\right\},$$

where U^0 is the closed unit ball of E'. If there is no risk of confusion, then we use the shortened symbol $\|(x_i) \mid w_r\|$. Moreover, we always write $[w_r, E]$ instead of $[w_r(\mathbb{N}), E]$.

Remark. It follows from the principle of uniform boundedness, (DUN, II.3.21) and (TAY, IV.1.2), that $\|(x_i) \mid w_r\|$ is indeed finite.

1.1.8.* We state the following fact without proof.

Proposition. $[w_r(I), E]$ is a Banach space.

1.1.9.* The next result is an analogue of 1.1.5.

Proposition. Let $T \in \mathfrak{L}(E, F)$. Then

$$T(I) : (x_i) \to (Tx_i)$$

defines an operator from $[w_r(I), E]$ into $[w_r(I), F]$. Moreover, we have

$$\|(Tx_i) \mid w_r\| \leqq \|T\| \, \|(x_i) \mid w_r\| \quad \text{for} \quad (x_i) \in [w_r(I), E].$$

Proof. Let U^0 and V^0 denote the closed unit balls of E' and F', respectively. Then

$$\begin{aligned}
\|(Tx_i) \mid w_r\| &= \sup\left\{\left(\sum_I |\langle Tx_i, b\rangle|^r\right)^{1/r} : b \in V^0\right\} \\
&= \sup\left\{\left(\sum_I |\langle x_i, T'b\rangle|^r\right)^{1/r} : b \in V^0\right\} \\
&\leqq \sup\left\{\left(\sum_I |\langle x_i, a\rangle|^r\right)^{1/r} : a \in \|T\| \, U^0\right\} \\
&= \|T\| \, \|(x_i) \mid w_r\|.
\end{aligned}$$

1.1.10.* Obviously, every absolutely r-summable family is necessarily weakly r-summable.

Proposition. $[l_r(I), E] \subseteqq [w_r(I), E]$.

Remark. The famous Dvoretzky-Rogers theorem states that, for an infinite index set I, equality holds only in finite dimensional Banach spaces. Thus there is indeed a significant difference between the two concepts of r-summability. This fact has initiated the theory of absolutely r-summing operators which is the main subject of this chapter.

1.1.11.* Next we mention a classical result.

Example. Every orthonormal family (u_i) in a Hilbert space is weakly 2-summable, and we have $\|(u_i) \mid w_2\| = 1$.

Proof. The conclusion can be deduced from Bessel's inequality:

$$\sum_I |(u_i, x)|^2 \leqq \|x\|^2 \quad \text{for all } x \in H.$$

1.1.12.* It follows immediately from the definition of a weakly r-summable family (x_i) that

$$(x_i)_{\text{op}} : a \to (\langle x_i, a \rangle)$$

defines an operator from E' into $l_r(I)$ with

$$\|(x_i)_{\text{op}} : E' \to l_r(I)\| = \|(x_i) \mid w_r\|.$$

We now show that, for $1 < r < \infty$, there exists $X \in \mathfrak{L}(l_{r'}(I), E)$ such that $(x_i)_{\text{op}} = X'$.

Lemma (R. S. Phillips 1940). Let $1 < r < \infty$. Given $(x_i) \in [w_r(I), E]$,

$$X : (\alpha_i) \to \sum_I \alpha_i x_i$$

defines an operator from $l_{r'}(I)$ into E such that $x_i = Xe_i$ for $i \in I$. This correspondence yields a metric isomorphism between $[w_r(I), E]$ and $\mathfrak{L}(l_{r'}(I), E)$.

Proof. Obviously, the expression

$$X(\alpha_i) := \sum_I \alpha_i x_i$$

makes sense for those families (α_i) which have a finite number of non-vanishing coordinates. Then it follows from

$$|\langle X(\alpha_i), a \rangle| = \Big|\sum_I \alpha_i \langle x_i, a \rangle\Big| \leqq \Big(\sum_I |\alpha_i|^{r'}\Big)^{1/r'} \Big(\sum_I |\langle x_i, a \rangle|^r\Big)^{1/r}$$

for $a \in E'$ that

$$\|X(\alpha_i)\| \leqq \|(x_i) \mid w_r\| \, \|(\alpha_i) \mid l_{r'}\|.$$

Thus we have already obtained a continuous linear map on a dense linear subset of $l_{r'}(I)$. Take its continuous extension, and observe that $X(\alpha_i)$ is given by the same formula for arbitrary families $(\alpha_i) \in l_{r'}(I)$. Clearly, $x_i = Xe_i$ and $\|X\| \leqq \|(x_i) \mid w_r\|$.

Conversely, we show that every operator $X \in \mathfrak{L}(l_{r'}(I), E)$ can be obtained in this way. Note that (e_i) is weakly r-summable in $l_{r'}(I)$ and $\|(e_i) \mid w_r\| = 1$. Now it follows from 1.1.9 that $(x_i) := (Xe_i)$ is a weakly r-summable family with $\|(x_i) \mid w_r\| \leqq \|X\|$.

1.1.13.* As a corollary of the Hahn-Banach theorem we have

$$\|x\| = \sup\{|\langle x, a \rangle| : a \in U^0\} \quad \text{for all } x \in E.$$

For various purposes it is desirable that $\|x\|$ can be obtained by taking the right-hand supremum only over a certain subset A of U^0, the smaller the better. This observation leads to the following definition.

A subset A of E' is said to be **norming** if

$$\|x\| = \sup\{|\langle x, a\rangle| : a \in A\} \quad \text{for all } x \in E.$$

For example, given any compact Hausdorff space X, the collection of all Dirac evaluation functionals $\delta_\xi : f \to f(\xi)$ with $\xi \in X$ is a norming subset of $C(X)'$. Identifying δ_ξ and ξ, we denote this subset by X. Moreover, for every Banach space E, the image of U with respect to the evaluation operator K_E defined in B.3.3 is a norming subset of E''.

1.1.14.* The following lemma shows how the concept just introduced enters the theory of weakly r-summable families.

Lemma. Suppose that A is any norming subset of E'. Let $x_1, \ldots, x_n \in E$. Then

$$\sup\left\{\left(\sum_{i=1}^{n} |\langle x_i, a\rangle|^r\right)^{1/r} : a \in A\right\} = \sup\left\{\left\|\sum_{i=1}^{n} \alpha_i x_i\right\| : \sum_{i=1}^{n} |\alpha_i|^{r'} \leqq 1\right\}$$

if $1 < r < \infty$ and

$$\sup\left\{\sum_{i=1}^{n} |\langle x_i, a\rangle| : a \in A\right\} = \sup\left\{\left\|\sum_{i=1}^{n} \alpha_i x_i\right\| : |\alpha_i| \leqq 1\right\}.$$

1.1.15.* Taking into account that the right-hand expressions in the preceding lemma do not depend on A, we are led to the following equation.

Lemma. Suppose that A is any norming subset of E'. Let $x_1, \ldots, x_n \in E$. Then

$$\|(x_i) \mid w_r\| = \sup\left\{\left(\sum_{i=1}^{n} |\langle x_i, a\rangle|^r\right)^{1/r} : a \in A\right\}.$$

1.1.16. Finally, we illustrate the previous results in some concrete Banach spaces.

Example. Let $f_1, \ldots, f_n \in C(X)$, where X is any compact Hausdorff space. Then

$$\|(f_i) \mid w_r\| = \sup\left\{\left(\sum_{i=1}^{n} |f_i(\xi)|^r\right)^{1/r} : \xi \in X\right\}.$$

Proof. The assertion follows from the fact mentioned in 1.1.13 that X can be viewed as a norming subset of $C(X)'$.

1.1.17. Example. Let $f_1, \ldots, f_n \in L_\infty(X, \mu)$, where (X, μ) is any measure space. Then

$$\|(f_i) \mid w_r\| = \text{ess-sup}\left\{\left(\sum_{i=1}^{n} |f_i(\xi)|^r\right)^{1/r} : \xi \in X\right\}.$$

Proof. Given $a = (\alpha_1, \ldots, \alpha_n) \in U_{r'}(n)$, by 1.1.14, there exists a null set $N(a)$ such that

$$\left|\sum_{i=1}^{n} \alpha_i f_i(\xi)\right| \leqq \|(f_i) \mid w_r\| \quad \text{for all } \xi \in X \setminus N(a).$$

Choose any countable dense subset A of $U_{r'}(n)$, and denote by N the union of all $N(a)$ with $a \in A$. Applying a continuity argument, we obtain

$$\left|\sum_{i=1}^{n} \alpha_i f_i(\xi)\right| \leqq \|(f_i) \mid w_r\| \quad \text{for all } \xi \in X \setminus N,$$

where the null set N no longer depends on $a \in U_{r'}(n)$. Hence

$$\left(\sum_{i=1}^{n} |f_i(\xi)|^r\right)^{1/r} \leqq \|(f_i) \mid w_r\| \quad \text{for almost all } \xi \in X.$$

This proves that

$$\text{ess-sup}\left\{\left(\sum_{i=1}^{n} |f_i(\xi)|^r\right)^{1/r} : \xi \in X\right\} \leqq \|(f_i) \mid w_r\|.$$

The reverse inequality is obvious.

1.2. Absolutely (r, s)-summing operators

Throughout this section we assume that $1 \leqq s \leqq r < \infty$.

1.2.1.* An operator $T \in \mathfrak{L}(E, F)$ is called **absolutely** (r, s)**-summing** if there exists a constant $c \geqq 0$ such that

$$\left(\sum_{i=1}^{n} \|Tx_i\|^r\right)^{1/r} \leqq c \sup\left\{\left(\sum_{i=1}^{n} |\langle x_i, a\rangle|^s\right)^{1/s} : a \in U^0\right\}$$

for every finite family of elements $x_1, \ldots, x_n \in E$. The set of these operators is denoted by $\mathfrak{P}_{r,s}(E, F)$.

For $T \in \mathfrak{P}_{r,s}(E, F)$ we define

$$\|T \mid \mathfrak{P}_{r,s}\| := \inf c,$$

the infimum being taken over all constants $c \geqq 0$ for which the above inequality holds.

1.2.2.* The term "absolutely (r, s)-summing" is derived from the fact that such operators send weakly s-summable families into absolutely r-summable ones. More precisely, we have the following criterion.

Proposition (A. Pietsch 1967). Let I be any infinite index set. An operator $T \in \mathfrak{L}(E, F)$ is absolutely (r, s)-summing if and only if

$$T(I) : (x_i) \to (Tx_i)$$

defines an operator from $[w_s(I), E]$ into $[l_r(I), F]$. When this is so, then

$$\|T \mid \mathfrak{P}_{r,s}\| = \|T(I) : [w_s(I), E] \to [l_r(I), F]\|.$$

Proof. Suppose that $T \in \mathfrak{P}_{r,s}(E, F)$ and $(x_i) \in [w_s(I), E]$. Then we have

$$\left(\sum_{\mathfrak{i}} \|Tx_i\|^r\right)^{1/r} \leqq \|T \mid \mathfrak{P}_{r,s}\| \sup\left\{\left(\sum_{\mathfrak{i}} |\langle x_i, a\rangle|^s\right)^{1/s} : a \in U^0\right\}$$

for all $\mathfrak{i} \in \mathfrak{F}(I)$. Passing to the limit as $\mathfrak{i} \to I$ yields

$$\|(Tx_i) \mid [l_r(I), F]\| \leqq \|T \mid \mathfrak{P}_{r,s}\| \, \|(x_i) \mid [w_s(I), E]\|.$$

This proves that

$$\|T(I) : [w_s(I), E] \to [l_r(I), F]\| \leqq \|T \mid \mathfrak{P}_{r,s}\|.$$

The reverse inequality is obvious.

We now suppose that $T \in \mathfrak{L}(E, F)$ fails to be absolutely (r, s)-summing. Then, for $h = 1, 2, \ldots$, we can find $x_{h1}, \ldots, x_{hn_h} \in E$ such that

$$\sum_{k=1}^{n_h} \|Tx_{hk}\|^r \geqq 1 \quad \text{and} \quad \sup\left\{\sum_{k=1}^{n_h} |\langle x_{hk}, a\rangle|^s : a \in U^0\right\} \leqq 2^{-h}.$$

Put $x_{hk} := o$ if $k > n_h$. Then the double sequence (x_{hk}) is weakly s-summable, because

$$\sum_{h=1}^{\infty} \sum_{k=1}^{\infty} |\langle x_{hk}, a\rangle|^s \leqq \sum_{h=1}^{\infty} 2^{-h} = 1 \quad \text{for} \quad a \in U^0.$$

On the other hand, we conclude from

$$\sum_{h=1}^{\infty} \sum_{k=1}^{\infty} \|Tx_{hk}\|^r \geqq 1 + 1 + \ldots$$

that (Tx_{hk}) cannot be absolutely r-summable. Therefore $T(\mathbb{N} \times \mathbb{N})$ fails to map $[w_s(\mathbb{N} \times \mathbb{N}), E]$ into $[l_r(\mathbb{N} \times \mathbb{N}), F]$.

To finish the proof, we observe that every infinite index set I contains a copy of $\mathbb{N} \times \mathbb{N}$ as a subset. Hence the family (x_{hk}) can be extended to a family (x_i) with $i \in I$ by putting $x_i := o$ whenever $i \notin \mathbb{N} \times \mathbb{N}$. Thus there exists $(x_i) \in [w_s(I), E]$ such, that $(Tx_i) \notin [l_r(I), F]$.

1.2.3.* We now establish the basic result of this section.

Theorem (S. Kwapień 1968). $\mathfrak{P}_{r,s}$ is an injective Banach operator ideal.

Proof. Let $T \in \mathfrak{P}_{r,s}(E, F)$. By definition, we have

$$\left(\sum_{i=1}^{n} \|Tx_i\|^r\right)^{1/r} \leqq \|T \mid \mathfrak{P}_{r,s}\| \sup\left\{\left(\sum_{i=1}^{n} |\langle x_i, a\rangle|^s\right)^{1/s} : a \in U^0\right\}$$

for $x_1, \ldots, x_n \in E$. Taking $n = 1$, we obtain

$$\|Tx\| \leqq \|T \mid \mathfrak{P}_{r,s}\| \sup\{|\langle x, a\rangle| : a \in U^0\} \quad \text{for} \quad x \in E.$$

This proves that

$$\|T\| \leqq \|T \mid \mathfrak{P}_{r,s}\|.$$

Next we consider an operator of the form $a_0 \otimes y_0$ with $a_0 \in E'$ and $y_0 \in F$. Then it follows from

$$\left(\sum_{i=1}^{n} \|(a_0 \otimes y_0)\, x_i\|^r\right)^{1/r} \leqq \left(\sum_{i=1}^{n} |\langle x_i, a_0\rangle|^s\right)^{1/s} \|y_0\|$$

$$\leqq \|a_0\|\, \|y_0\| \sup\left\{\left(\sum_{i=1}^{n} |\langle x_i, a\rangle|^s\right)^{1/s} : a \in U^0\right\}$$

that

$$\|a_0 \otimes y_0 \mid \mathfrak{P}_{r,s}\| \leqq \|a_0\|\, \|y_0\|.$$

Hence

$$a_0 \otimes y_0 \in \mathfrak{P}_{r,s}(E, F) \quad \text{and} \quad \|a_0 \otimes y_0 \mid \mathfrak{P}_{r,s}\| = \|a_0\|\, \|y_0\|.$$

We now assume that $S, T \in \mathfrak{P}_{r,s}(E, F)$. Then

$$\left(\sum_{i=1}^{n} \|(S + T)\, x_i\|^r\right)^{1/r} \leqq \left(\sum_{i=1}^{n} \|Sx_i\|^r\right)^{1/r} + \left(\sum_{i=1}^{n} \|Tx_i\|^r\right)^{1/r}$$

$$\leqq [\|S \mid \mathfrak{P}_{r,s}\| + \|T \mid \mathfrak{P}_{r,s}\|] \sup\left\{\left(\sum_{i=1}^{n} |\langle x_i, a\rangle|^s\right)^{1/s} : a \in U^0\right\}$$

for $x_1, \ldots, x_n \in E$. Therefore

$$S + T \in \mathfrak{P}_{r,s}(E, F) \quad \text{and} \quad \|S + T \mid \mathfrak{P}_{r,s}\| \leqq \|S \mid \mathfrak{P}_{r,s}\| + \|T \mid \mathfrak{P}_{r,s}\|.$$

If $X \in \mathfrak{L}(E_0, E)$, $T \in \mathfrak{P}_{r,s}(E, F)$ and $Y \in \mathfrak{L}(F, F_0)$, then we deduce from 1.1.5 and 1.1.9 that

$$YTX \in \mathfrak{P}_{r,s}(E_0, F_0) \quad \text{and} \quad \|YTX \mid \mathfrak{P}_{r,s}\| \leqq \|Y\| \, \|T \mid \mathfrak{P}_{r,s}\| \, \|X\|.$$

Finally, let (T_k) be a Cauchy sequence in $\mathfrak{P}_{r,s}(E, F)$. Then it follows from

$$\|T_h - T_k\| \leqq \|T_h - T_k \mid \mathfrak{P}_{r,s}\|$$

that (T_k) has a limit $T \in \mathfrak{L}(E, F)$ with respect to the operator norm. Given $\varepsilon > 0$ we choose k_0 such that

$$\|T_h - T_k \mid \mathfrak{P}_{r,s}\| \leqq \varepsilon \quad \text{for } h > k \geqq k_0.$$

This means that

$$\left(\sum_{i=1}^{n} \|(T_h - T_k)\, x_i\|^r\right)^{1/r} \leqq \varepsilon \sup\left\{\left(\sum_{i=1}^{n} |\langle x_i, a\rangle|^s\right)^{1/s} : a \in U^0\right\}$$

for $x_1, \ldots, x_n \in E$. Letting $h \to \infty$, we obtain

$$\left(\sum_{i=1}^{n} \|(T - T_k)\, x_i\|^r\right)^{1/r} \leqq \varepsilon \sup\left\{\left(\sum_{i=1}^{n} |\langle x_i, a\rangle|^s\right)^{1/s} : a \in U^0\right\}.$$

Hence

$$T - T_k \in \mathfrak{P}_{r,s}(E, F) \quad \text{and} \quad \|T - T_k \mid \mathfrak{P}_{r,s}\| \leqq \varepsilon \quad \text{for } k \geqq k_0.$$

Thus T is also the limit of (T_k) with respect to the norm of $\mathfrak{P}_{r,s}$.

The injectivity of the Banach operator ideal $\mathfrak{P}_{r,s}$ is obvious.

Remark. The assertion could be deduced by identifying $\mathfrak{P}_{r,s}(E, F)$ with a subspace of $\mathfrak{L}([w_s, E], [l_r, F])$ as described in 1.2.2.

1.2.4.* We now investigate how the operator ideals $\mathfrak{P}_{r,s}$ depend on the parameters r and s. First a trivial inclusion is stated.

Proposition (S. Kwapień 1968). Let $1 \leqq s_1 \leqq s_0 \leqq r_0 \leqq r_1 < \infty$. Then $\mathfrak{P}_{r_0,s_0} \subseteqq \mathfrak{P}_{r_1,s_1}$. Moreover, we have

$$\|T \mid \mathfrak{P}_{r_1,s_1}\| \leqq \|T \mid \mathfrak{P}_{r_0,s_0}\| \quad \text{for all } T \in \mathfrak{P}_{r_0,s_0}(E, F).$$

1.2.5.* The next result turns out to be slightly more complicated.

Proposition (S. Kwapień 1968). Let $1 \leqq r_0 \leqq r_1 < \infty$ and $1 \leqq s_0 \leqq s_1 < \infty$. If $1/r_0 - 1/r_1 = 1/s_0 - 1/s_1$, then $\mathfrak{P}_{r_0,s_0} \subseteqq \mathfrak{P}_{r_1,s_1}$. Moreover, we have

$$\|T \mid \mathfrak{P}_{r_1,s_1}\| \leqq \|T \mid \mathfrak{P}_{r_0,s_0}\| \quad \text{for all } T \in \mathfrak{P}_{r_0,s_0}(E, F).$$

Proof. Define $1/r := 1/r_0 - 1/r_1$ and $1/s := 1/s_0 - 1/s_1$. By assumption, we obtain $r = s$. Let $T \in \mathfrak{P}_{r_0,s_0}(E, F)$, $x_1, \ldots, x_n \in E$ and $\alpha_1, \ldots, \alpha_n \in \mathbb{C}$. Then Hölder's inequality yields

$$\left(\sum_{i=1}^{n} \|\alpha_i T x_i\|^{r_0}\right)^{1/r_0} \leqq \|T \mid \mathfrak{P}_{r_0,s_0}\| \sup\left\{\left(\sum_{i=1}^{n} |\langle \alpha_i x_i, a\rangle|^{s_0}\right)^{1/s_0} : a \in U^0\right\}$$

$$\leqq \|T \mid \mathfrak{P}_{r_0,s_0}\| \sup\left\{\left(\sum_{i=1}^{n} |\langle x_i, a\rangle|^{s_1}\right)^{1/s_1} : a \in U^0\right\} \left(\sum_{i=1}^{n} |\alpha_i|^s\right)^{1/s}.$$

Taking $\alpha_i := \|Tx_i\|^{r_1/r}$, we conclude that

$$\sum_{i=1}^{n} \|\alpha_i T x_i\|^{r_0} = \sum_{i=1}^{n} |\alpha_i|^r = \sum_{i=1}^{n} \|Tx_i\|^{r_1}.$$

Thus

$$\left(\sum_{i=1}^{n} \|Tx_i\|^{r_1}\right)^{1/r_1} \leqq \|T \mid \mathfrak{P}_{r_0,s_0}\| \sup\left\{\left(\sum_{i=1}^{n} |\langle x_i, a\rangle|^{s_1}\right)^{1/s_1} : a \in U^0\right\}.$$

This proves that

$$T \in \mathfrak{P}_{r_1,s_1}(E, F) \quad \text{and} \quad \|T \mid \mathfrak{P}_{r_1,s_1}\| \leqq \|T \mid \mathfrak{P}_{r_0,s_0}\|.$$

1.2.6. The following formula is very important. In order to include the limiting case, we let $\mathfrak{P}_{\infty,s} := \mathfrak{L}$.

Interpolation theorem (H. König 1978). Let $1 \leqq s \leqq r_0 < r_1 \leqq \infty$ and $0 < \theta < 1$. If $1/r = (1-\theta)/r_0 + \theta/r_1$, then

$$(\mathfrak{P}_{r_0,s}, \mathfrak{P}_{r_1,s})_{\theta,r} \subseteqq \mathfrak{P}_{r,s}.$$

Proof. Given $(x_i) \in [w_s, E]$, we define by $X(T) := (Tx_i)$ an operator

$$X : \mathfrak{L}(E, F) \to [w_s, F].$$

Note that

$$X : \mathfrak{P}_{r_0,s}(E, F) \to [l_{r_0}, F]$$

and

$$X : \mathfrak{P}_{r_1,s}(E, F) \to [l_{r_1}, F].$$

The interpolation property now implies that

$$X : (\mathfrak{P}_{r_0,s}(E, F), \mathfrak{P}_{r_1,s}(E, F))_{\theta,r} \to ([l_{r_0}, F], [l_{r_1}, F])_{\theta,r}.$$

Hence the assertion follows from 1.1.6.

1.2.7.* In order to prove the following result, we need a criterion which is also of interest in its own right.

Lemma. An operator $T \in \mathfrak{L}(E, F)$ is absolutely $(r, 2)$-summing if and only if $TX \in \mathfrak{P}_{r,2}(l_2, F)$ for all $X \in \mathfrak{L}(l_2, E)$. When this is so, then

$$\|T \mid \mathfrak{P}_{r,2}\| = \sup\{\|TX \mid \mathfrak{P}_{r,2}\| : X \in \mathfrak{L}(l_2, E),\ \|X\| \leqq 1\}.$$

Proof. It follows from 1.1.12 that, given $(x_i) \in [w_2, E]$,

$$X : (\alpha_i) \to \sum_{i=1}^{\infty} \alpha_i x_i$$

defines an operator from l_2 into E with $\|X\| = \|(x_i) \mid w_2\|$. If (e_i) denotes the standard basis of l_2, then $\|(e_i) \mid w_2\| = 1$, by 1.1.11. Moreover, $x_i = Xe_i$. We now assume that $T \in \mathfrak{L}(E, F)$ has the property stated above. Writing

$$\|T \mid \mathfrak{P}_{r,2}\|_0 := \sup\{\|TX \mid \mathfrak{P}_{r,2}\| : X \in \mathfrak{L}(l_2, E),\ \|X\| \leqq 1\},$$

we obtain

$$\|(Tx_i) \mid l_r\| = \|(TXe_i) \mid l_r\| \leqq \|TX \mid \mathfrak{P}_{r,2}\| \, \|(e_i) \mid w_2\| \leqq \|T \mid \mathfrak{P}_{r,2}\|_0 \, \|(x_i) \mid w_2\|.$$

Hence

$$T \in \mathfrak{P}_{r,2}(E, F) \quad \text{and} \quad \|T \mid \mathfrak{P}_{r,2}\| \leqq \|T \mid \mathfrak{P}_{r,2}\|_0 .$$

The converse is obvious.

1.2.8.* We now establish an important inequality. The following proof is due to S. Kwapień (unpublished).

Lemma. Let $T \in \mathfrak{L}(E, F)$. Then

$$\|T \mid \mathfrak{P}_{r,2}\| \leqq n^{1/r}\|T\| \quad \text{whenever rank } (T) \leqq n.$$

Proof. Let $x_1, \ldots, x_m \in l_2(n)$, and denote the closed unit ball of $l_2(n)$ by $U_2(n)$. Obviously,

$$\left(\sum_{i=1}^{m} \|x_i \mid l_2(n)\|^2\right)^{1/2} = \left(\sum_{j=1}^{n}\sum_{i=1}^{m} |(x_i, e_j)|^2\right)^{1/2}$$
$$\leqq n^{1/2} \sup\left\{\left(\sum_{i=1}^{m} |(x_i, a)|^2\right)^{1/2} : a \in U_2(n)\right\}.$$

Therefore

$$\|I : l_2(n) \to l_2(n) \mid \mathfrak{P}_2\| \leqq n^{1/2}.$$

In the case $2 < r < \infty$, we define θ by $1/r = (1 - \theta)/2 + \theta/\infty$. Then

$$\left(\sum_{i=1}^{m} \|x_i \mid l_2(n)\|^r\right)^{1/r} \leqq \left(\sum_{i=1}^{m} \|x_i \mid l_2(n)\|^2\right)^{(1-\theta)/2} \left(\max_i \|x_i \mid l_2(n)\|\right)^{\theta}.$$

Note that

$$\max_i \|x_i \mid l_2(n)\| \leqq \sup\left\{\left(\sum_{i=1}^{m} |(x_i, a)|^2\right)^{1/2} : a \in U_2(n)\right\}.$$

Combining the preceding inequalities, we obtain

$$\left(\sum_{i=1}^{m} \|x_i \mid l_2(n)\|^r\right)^{1/r} \leqq n^{(1-\theta)/2} \sup\left\{\left(\sum_{i=1}^{m} |(x_i, a)|^2\right)^{1/2} : a \in U_2(n)\right\}.$$

This shows that

$$\|I : l_2(n) \to l_2(n) \mid \mathfrak{P}_{r,2}\| \leqq n^{1/r}.$$

We now treat an arbitrary operator $T \in \mathfrak{L}(E, F)$ with rank $(T) \leqq n$. Given $X \in \mathfrak{L}(l_2, E)$, we put $N := N(TX)$. Consider the factorization $TX = SQ$, where $S \in \mathfrak{L}(l_2/N, F)$ is the operator induced by TX and Q denotes the canonical surjection from l_2 onto l_2/N. Note that $\|S\| = \|TX\|$. Since l_2/N is a Hilbert space with

$$\dim(l_2/N) = \operatorname{codim}(N) = \operatorname{rank}(TX) \leqq \operatorname{rank}(T) \leqq n,$$

it follows from the preceding result that

$$\|TX \mid \mathfrak{P}_{r,2}\| \leqq \|S\| \, \|I : l_2/N \to l_2/N \mid \mathfrak{P}_{r,2}\| \, \|Q\| \leqq n^{1/r}\|T\| \, \|X\|.$$

This yields the assertion, by 1.2.7.

1.3. Absolutely r-summing operators

Throughout this section we assume that $1 \leqq r < \infty$.

1.3.1.* An operator $T \in \mathfrak{L}(E, F)$ is called **absolutely r-summing** if there exists a constant $c \geqq 0$ such that

$$\left(\sum_{i=1}^{n} \|Tx_i\|^r\right)^{1/r} \leqq c \sup\left\{\left(\sum_{i=1}^{n} |\langle x_i, a\rangle|^r\right)^{1/r} : a \in U^0\right\}$$

for every finite family of elements $x_1, \ldots, x_n \in E$. The set of these operators is denoted by $\mathfrak{P}_r(E, F)$. In the special case $r = 1$ we simply speak of an **absolutely summing** operator, and write $\mathfrak{P}(E, F)$ instead of $\mathfrak{P}_1(E, F)$.

For $T \in \mathfrak{P}_r(E, F)$ we define

$$\|T \mid \mathfrak{P}_r\| := \inf c,$$

the infimum being taken over all constants $c \geqq 0$ for which the above inequality holds.

1.3.2.* Taking into account that $\mathfrak{P}_r = \mathfrak{P}_{r,r}$, we may specialize some results proved in the preceding section. For example, 1.2.3 reads as follows.

Theorem (A. Pietsch 1967). $\mathfrak{P}_r$ is an injective Banach operator ideal.

1.3.3.* The next statement can be derived from 1.2.5.

Proposition (A. Pietsch 1967). Let $1 \leqq r_0 \leqq r_1 < \infty$. Then $\mathfrak{P}_{r_0} \subseteq \mathfrak{P}_{r_1}$. Moreover, we have

$$\|T \mid \mathfrak{P}_{r_1}\| \leqq \|T \mid \mathfrak{P}_{r_0}\| \quad \text{for all } T \in \mathfrak{P}_{r_0}(E, F).$$

Remark. From Example 1.6.3, it can easily be seen that the above inclusion is strict whenever $r_0 < r_1$.

1.3.4.* We now take a first step towards the main theorem of this section.

Proposition (A. Persson / A. Pietsch 1969). Let $T \in \mathfrak{L}(E, F)$. If there exists an absolutely r-summable sequence (a_k) in E' such that

$$\|Tx\| \leqq \left(\sum_{k=1}^{\infty} |\langle x, a_k\rangle|^r\right)^{1/r} \quad \text{for all} \quad x \in E,$$

then

$$T \in \mathfrak{P}_r(E, F) \quad \text{and} \quad \|T \mid \mathfrak{P}_r\| \leqq \left(\sum_{k=1}^{\infty} \|a_k\|^r\right)^{1/r}.$$

Proof. For $x_1, \ldots, x_n \in E$ we have

$$\left(\sum_{i=1}^{n} \|Tx_i\|^r\right)^{1/r} \leqq \left(\sum_{i=1}^{n} \sum_{k=1}^{\infty} |\langle x_i, a_k\rangle|^r\right)^{1/r} = \left(\sum_{k=1}^{\infty} \sum_{i=1}^{n} |\langle x_i, a_k\rangle|^r\right)^{1/r}$$
$$\leqq \left(\sum_{k=1}^{\infty} \|a_k\|^r\right)^{1/r} \sup\left\{\left(\sum_{i=1}^{n} |\langle x_i, a\rangle|^r\right)^{1/r} : a \in U^0\right\}.$$

Remark. By a diagonal procedure, it can easily be seen that every operator with the above property is compact. On the other hand, in view of Examples 1.3.8 and 1.6.4, there are absolutely r-summing operators which fail to be compact. Hence the inequality

$$\|Tx\| \leqq \left(\sum_{k=1}^{\infty} |\langle x, a_k\rangle|^r\right)^{1/r} \quad \text{for all } x \in E$$

must be modified in order to obtain a necessary and sufficient condition. This can be achieved by using an integral instead of the infinite sum:

$$\|Tx\| \leqq \left(\int_A |\langle x, a\rangle|^r \, d\mu(a)\right)^{1/r} \quad \text{for all } x \in E.$$

1.3.5.* We now establish the announced criterion which dominates the whole theory of absolutely r-summing operators. To this end, some additional notation is required.

First of all let $l_\infty^{\text{real}}(I)$ denote the linear space of all bounded real-valued families on a given index set I. Furthermore, if A is any bounded subset of a dual Banach space E', then we assign to every element $x \in E$ the bounded complex-valued family $f_x := (\langle x, a\rangle)$ with $a \in A$.

Domination theorem (A. Pietsch 1967). Let A be any norming subset of E'. An operator $T \in \mathfrak{L}(E, F)$ is absolutely r-summing if and only if there exist a positive normalized linear form m on $l_\infty^{\text{real}}(A)$ and a constant $c \geqq 0$ such that

(*) $\|Tx\| \leqq cm(|f_x|^r)^{1/r}$ for all $x \in E$.

When this is so, then

$$\|T \mid \mathfrak{P}_r\| = \inf c,$$

where the infimum is taken over all constants $c \geqq 0$ for which a linear form m with the required properties can be found.

Proof. Assume that $T \in \mathfrak{P}_r(E, F)$, and let $c := \|T \mid \mathfrak{P}_r\|$. Define

$$s(f) := \inf \left\{\sup \left[f(a) + c^r \sum_{i=1}^{n} |\langle x_i, a\rangle|^r : a \in A\right] - \sum_{i=1}^{n} \|Tx_i\|^r\right\},$$

the infimum being taken over all finite families of elements $x_1, \ldots, x_n \in E$. Easy computations show that s is a positive homogeneous and subadditive functional on the real linear space $l_\infty^{\text{real}}(A)$:

(1) $s(\lambda f) = \lambda s(f)$ for $\lambda > 0$ and $f \in l_\infty^{\text{real}}(A)$,

(2) $s(f + g) \leqq s(f) + s(g)$ for $f, g \in l_\infty^{\text{real}}(A)$.

Moreover, we have

(3) $\inf [f(a) : a \in A] \leqq s(f) \leqq \sup [f(a) : a \in A]$.

The Hahn-Banach theorem, (DUN, II.3.10) and (TAY, I.10.4), yields a linear form m such that

$$m(f) \leqq s(f) \quad \text{for all } f \in l_\infty^{\text{real}}(A).$$

If $f \geqq 0$, then it follows from (3) that

$$m(-f) \leqq s(-f) \leqq \sup [-f(a) : a \in A] \leqq 0.$$

Hence the linear form m is positive. Moreover,

$$m(-1) \leqq s(-1) = -1 \quad \text{and} \quad m(1) \leqq s(1) = 1$$

imply $m(1) = 1$. Thus the linear form m is normalized. Finally, we observe that

$$m(-c^r \,|f_x|^r) \leqq s(-c^r|f_x|^r)$$
$$\leqq \sup [-c^r|f_x(a)|^r + c^r|\langle x, a\rangle|^r : a \in A] - \|Tx\|^r.$$

Therefore, because $c = \|T \mid \mathfrak{P}_r\|$ and $f_x(a) = \langle x, a\rangle$, we have indeed

$$\|Tx\| \leqq \|T \mid \mathfrak{P}_r\| \, m(|f_x|^r)^{1/r} \quad \text{for all } x \in E.$$

Conversely, for $x_1, \ldots, x_n \in E$, it follows from (*) that

$$\Big(\sum_{i=1}^{n} \|Tx_i\|^r\Big)^{1/r} \leqq c\,\Big(\sum_{i=1}^{n} m(|f_{x_i}|^r)\Big)^{1/r} = c\Big(m\Big(\sum_{i=1}^{n} |f_{x_i}|^r\Big)\Big)^{1/r}$$
$$\leqq c \sup\Big\{\Big(\sum_{i=1}^{n} |\langle x_i, a\rangle|^r\Big)^{1/r} : a \in A\Big\}.$$

Hence

$$T \in \mathfrak{P}_r(E, F) \quad \text{and} \quad \|T \mid \mathfrak{P}_r\| \leqq c.$$

Remark. We stress the fact that, for $T \in \mathfrak{P}_r(E, F)$, the inequality (*) can be satisfied with a constant $c \geqq 0$ equal to $\|T \mid \mathfrak{P}_r\|$.

1.3.6.* Next we rephrase the domination theorem in a measure theoretic setting. For this purpose, some preliminaries are needed.

Let $C^{\text{real}}(X)$ denote the linear space of all continuous real-valued functions defined on a given compact Hausdorff space X. Then the Riesz representation theorem, (DUN, IV.6.3) and (HAM, p. 247), asserts that every positive linear form m on $C^{\text{real}}(X)$ can be written as an integral:

$$m(f) = \int_X f(\xi)\, d\mu(\xi) \quad \text{for all } f \in C^{\text{real}}(X),$$

where μ is a (regular) Borel measure on X. The underlying σ-algebra is generated by the open subsets of X.

We know from the Alaoglu-Bourbaki theorem, (DUN, V.4.2) and (TAY, III.10.2), that the closed unit ball U^0 of any dual Banach space E' is compact in the weak topology generated by E. The defining semi-norms are given by

$$p_M(a) := \sup\{|\langle x, a\rangle| : x \in M\} \quad \text{for } a \in E',$$

where M ranges over all finite subsets of E.

Domination theorem (measure theoretic version). Let A be any norming subset of E' which is compact in the weak topology generated by E. An operator $T \in \mathfrak{L}(E, F)$ is absolutely r-summing if and only if there exist a normalized Borel measure μ on A and a constant $c \geqq 0$ such that

$$\|Tx\| \leqq c\Big(\int_A |\langle x, a\rangle|^r\, d\mu(a)\Big)^{1/r} \quad \text{for all } x \in E.$$

When this is so, then

$$\|T \mid \mathfrak{P}_r\| = \inf c,$$

where the infimum is taken over all constants $c \geqq 0$ for which a measure μ with the required properties can be found.

Proof. If $T \in \mathfrak{L}(E, F)$ is absolutely r-summing, then we choose a positive normalized linear functional m on $l_\infty^{\text{real}}(A)$ as described in the preceding paragraph. Next we restrict this functional to the subspace $C^{\text{real}}(A)$. Therefore, according to the introductory remarks, there exists a normalized Borel measure μ such that

$$m(f) = \int_A f(a)\, d\mu(a) \quad \text{for all } f \in C^{\text{real}}(A).$$

Obviously, μ has the required property.

The sufficiency of the above condition follows in the same way as in the previous proof.

1.3.7. Finally, we state the domination theorem in a special but rather illustrative situation.

Proposition. Let X be any compact Hausdorff space. An operator $T \in \mathfrak{L}(C(X), F)$ is absolutely r-summing if and only if there exists a finite Borel measure μ on X such that

$$\|Tf\| \leqq \left(\int_X |f(\xi)|^r \, d\mu(\xi)\right)^{1/r} \quad \text{for all } f \in C(X).$$

When this is so, then

$$\|T \mid \mathfrak{P}_r\| = \inf \mu(X)^{1/r},$$

where the infimum is taken over all possible measures μ.

1.3.8. We now consider some important examples which are the prototypes of absolutely r-summing operators.

Example (A. Pietsch 1967). Let μ be a normalized Borel measure on any compact Hausdorff space X. Then the canonical operator I from $C(X)$ into $L_r(X, \mu)$ is absolutely r-summing, and

$$\|I : C \to L_r \mid \mathfrak{P}_r\| = 1.$$

Proof. From 1.3.7 we deduce that $I \in \mathfrak{P}_r(C, L_r)$. Moreover,

$$1 \leqq \|I : C \to L_r\| \leqq \|I : C \to L_r \mid \mathfrak{P}_r\| \leqq 1.$$

1.3.9. **Example** (A. Persson/A. Pietsch 1969). Let (X, μ) be any normalized measure space. Then the embedding operator I from $L_\infty(X, \mu)$ into $L_r(X, \mu)$ is absolutely r-summing, and

$$\|I : L_\infty \to L_r \mid \mathfrak{P}_r\| = 1.$$

Proof. From 1.1.17 we easily see that $I \in \mathfrak{P}_r(L_\infty, L_r)$. Moreover,

$$1 \leqq \|I : L_\infty \to L_r\| \leqq \|I : L_\infty \to L_r \mid \mathfrak{P}_r\| \leqq 1.$$

Remark. If (X, μ) is an arbitrary measure space, then it can be shown by the same method that the multiplication operator

$$M_t : f(\xi) \to t(\xi) f(\xi)$$

associated with any function $t \in L_r(X, \mu)$ is absolutely r-summing from $L_\infty(X, \mu)$ into $L_r(X, \mu)$, and

$$\|M_t : L_\infty \to L_r \mid \mathfrak{P}_r\| = \|f | L_r\|;$$

see (PIE, 17.3.8). For the special case of diagonal operators we refer to 1.6.3.

1.3.10. Next we establish a very important result.

Multiplication theorem (A. Pietsch 1967). Let $1 \leqq p, q, r < \infty$. If we have $1/p + 1/q = 1/r$, then $\mathfrak{P}_p \circ \mathfrak{P}_q \subseteq \mathfrak{P}_r$. More precisely,

$$\|ST \mid \mathfrak{P}_r\| \leqq \|S \mid \mathfrak{P}_p\| \, \|T \mid \mathfrak{P}_q\| \quad \text{for all } T \in \mathfrak{P}_q(E, F) \text{ and } S \in \mathfrak{P}_p(F, G).$$

Proof. We are going to show that

$$\left(\sum_{i=1}^n \|STx_i\|^r\right)^{1/r} \leqq \|S \mid \mathfrak{P}_p\| \, \|T \mid \mathfrak{P}_q\| \sup\left\{\left(\sum_{i=1}^n |\langle x_i, a\rangle|^r\right)^{1/r} : a \in U^0\right\}$$

for every finite family of elements $x_1, \ldots, x_n \in E$.

By 1.3.6, there exists a normalized Borel measure μ on U^0 such that

$$\|Tx\| \leqq \|T \mid \mathfrak{P}_q\| \left(\int_{U^0} |\langle x, a\rangle|^q \, d\mu(a)\right)^{1/q} \quad \text{for all } x \in E.$$

Put

$$\xi := \sup\left\{\left(\sum_{i=1}^n |\langle x_i, a\rangle|^r\right)^{1/r} : a \in U^0\right\}$$

and

$$\varrho_i := \left(\int_{U^0} |\langle x_i, a\rangle|^r \, d\mu(a)\right)^{1/q}.$$

Without loss of generality, it may be assumed that $Tx_i \neq o$. Then it follows that $\varrho_i > 0$, and we can define $x_i^0 := \varrho_i^{-1} x_i$. Choose any finite family of complex numbers $\alpha_1, \ldots, \alpha_n$ such that

$$\sum_{i=1}^n |\alpha_i|^{p'} \leqq 1.$$

Applying the threefold Hölder inequality with the exponents r', q and p, for $a \in U^0$, we get

$$\begin{aligned}
\left|\left\langle \sum_{i=1}^n \alpha_i x_i^0, a\right\rangle\right| &\leqq \sum_{i=1}^n |\alpha_i|^{p'/r'} \, |\alpha_i|^{p'/q} \, \varrho_i^{-1} |\langle x_i, a\rangle|^{r/q} \, |\langle x_i, a\rangle|^{r/p} \\
&\leqq \left(\sum_{i=1}^n |\alpha_i|^{p'}\right)^{1/r'} \left(\sum_{i=1}^n |\alpha_i|^{p'} \varrho_i^{-q} |\langle x_i, a\rangle|^r\right)^{1/q} \left(\sum_{i=1}^n |\langle x_i, a\rangle|^r\right)^{1/p} \\
&\leqq \left(\sum_{i=1}^n |\alpha_i|^{p'} \varrho_i^{-q} |\langle x_i, a\rangle|^r\right)^{1/q} \xi^{r/p}.
\end{aligned}$$

Hence

$$\begin{aligned}
\left\|\sum_{i=1}^n \alpha_i T x_i^0\right\| &= \left\|T\left(\sum_{i=1}^n \alpha_i x_i^0\right)\right\| \\
&\leqq \|T \mid \mathfrak{P}_q\| \left(\int_{U^0} \left|\left\langle \sum_{i=1}^n \alpha_i x_i^0, a\right\rangle\right|^q d\mu(a)\right)^{1/q} \\
&\leqq \|T \mid \mathfrak{P}_q\| \left(\sum_{i=1}^n |\alpha_i|^{p'} \varrho_i^{-q} \int_{U^0} |\langle x_i, a\rangle|^r \, d\mu(a)\right)^{1/q} \xi^{r/p} \\
&\leqq \|T \mid \mathfrak{P}_q\| \, \xi^{r/p}.
\end{aligned}$$

By Lemma 1.1.14, this means that

$$\sup\left\{\left(\sum_{i=1}^n |\langle Tx_i^0, b\rangle|^p\right)^{1/p} : b \in V^0\right\} \leqq \|T \mid \mathfrak{P}_q\| \, \xi^{r/p}.$$

Therefore

$$\left(\sum_{i=1}^n \|STx_i^0\|^p\right)^{1/p} \leqq \|S \mid \mathfrak{P}_p\| \, \|T \mid \mathfrak{P}_q\| \, \xi^{r/p}.$$

Finally, another application of Hölder's inequality yields

$$\left(\sum_{i=1}^{n} \|STx_i\|^r\right)^{1/r} = \left(\sum_{i=1}^{n} \|\varrho_i STx_i^0\|^r\right)^{1/r}$$

$$\leqq \left(\sum_{i=1}^{n} \|STx_i^0\|^p\right)^{1/p} \left(\sum_{i=1}^{n} \varrho_i^q\right)^{1/q} \leqq \|S \mid \mathfrak{P}_p\| \, \|T \mid \mathfrak{P}_q\| \, \xi^{r/p+r/q}$$

$$\leqq \|S \mid \mathfrak{P}_p\| \, \|T \mid \mathfrak{P}_q\| \sup\left\{\left(\sum_{i=1}^{n} |\langle x_i, a\rangle|^r\right)^{1/r} : a \in U^0\right\}.$$

This proves that

$$ST \in \mathfrak{P}_r(E, G) \quad \text{and} \quad \|ST \mid \mathfrak{P}_r\| \leqq \|S \mid \mathfrak{P}_p\| \, \|T \mid \mathfrak{P}_q\|.$$

1.3.11.* The following stability property is quite useful for the theory of eigenvalue distributions.

Theorem (J. R. Holub 1970/74). The Banach operator ideal $\mathfrak{P}_r$ is stable with respect to the tensor norm ε. More precisely,

$$\|S \tilde{\otimes}_\varepsilon T \mid \mathfrak{P}_r\| \leqq \|S \mid \mathfrak{P}_r\| \, \|T \mid \mathfrak{P}_r\| \quad \text{for } S \in \mathfrak{P}_r(E, E_0) \text{ and } T \in \mathfrak{P}_r(F, F_0).$$

Proof. The domination theorem 1.3.6 yields normalized Borel measures μ and ν on U^0 and V^0 such that

$$\|Sx\| \leqq \|S \mid \mathfrak{P}_r\| \left(\int_{U^0} |\langle x, a\rangle|^r \, d\mu(a)\right)^{1/r} \quad \text{for all } x \in E$$

and

$$\|Ty\| \leqq \|T \mid \mathfrak{P}_r\| \left(\int_{V^0} |\langle y, b\rangle|^r \, d\nu(b)\right)^{1/r} \quad \text{for all } y \in F,$$

respectively. Let $x_1, \ldots, x_n \in E$ and $y_1, \ldots, y_n \in F$. If $\|a_0\| \leqq 1$ and $\|b_0\| \leqq 1$, then

$$\left|\sum_{i=1}^{n} \langle Sx_i, a_0\rangle \langle Ty_i, b_0\rangle\right| \leqq \left\|S\left(\sum_{i=1}^{n} x_i \langle Ty_i, b_0\rangle\right)\right\|$$

$$\leqq \|S \mid \mathfrak{P}_r\| \left(\int_{U^0} \left|\left\langle \sum_{i=1}^{n} x_i \langle Ty_i, b_0\rangle, a\right\rangle\right|^r d\mu(a)\right)^{1/r}$$

$$\leqq \|S \mid \mathfrak{P}_r\| \left(\int_{U^0} \left\|T\left(\sum_{i=1}^{n} \langle x_i, a\rangle \, y_i\right)\right\|^r d\mu(a)\right)^{1/r}$$

$$\leqq \|S \mid \mathfrak{P}_r\| \, \|T \mid \mathfrak{P}_r\| \left(\int_{U^0} \left[\int_{V^0} \left|\sum_{i=1}^{n} \langle x_i, a\rangle \langle y_i, b\rangle\right|^r d\nu(b)\right] d\mu(a)\right)^{1/r}.$$

This means that

$$\|(S \otimes T)z \mid \varepsilon\| \leqq \|S \mid \mathfrak{P}_r\| \, \|T \mid \mathfrak{P}_r\| \left(\int_{U^0} \int_{V^0} |z(a, b)|^r d\mu(a) d\nu(b)\right)^{1/r}$$

for all $z \in E \otimes F$. Since $E \tilde{\otimes}_\varepsilon F$ is the closed hull of $E \otimes F$ in $\varepsilon(E, F)$, by continuous extension, we obtain

$$\|(S \tilde{\otimes}_\varepsilon T)\, z \mid \varepsilon\| \leqq \|S \mid \mathfrak{P}_r\| \, \|T \mid \mathfrak{P}_r\| \left(\int_{U^0} \int_{V^0} |z(a, b)|^r \, d\mu(a) \, d\nu(b)\right)^{1/r}$$

for all $z \in E \tilde{\otimes}_\varepsilon F$. From the definition of the tensor norm ε we infer that $U^0 \times V^0$ is a norming subset of $\varepsilon(E \times F)'$. Hence, by the elementary part of the domination theorem,

$$S\tilde{\otimes}_\varepsilon T \in \mathfrak{P}_r(E\tilde{\otimes}_\varepsilon F, E_0\tilde{\otimes}_\varepsilon F_0) \quad \text{and} \quad \|S\tilde{\otimes}_\varepsilon T \mid \mathfrak{P}_r\| \leqq \|S \mid \mathfrak{P}_r\| \, \|T \mid \mathfrak{P}_r\|.$$

1.3.12. In the following we need some information about the local structure of the classical Banach function spaces.

Lemma. Given $f_1, \ldots, f_n \in L_r(X, \mu)$ and $\varepsilon > 0$, there exist $Q \in \mathfrak{L}(L_r, l_r(m))$ and $J \in \mathfrak{L}(l_r(m), L_r)$ such that $\|Q\| \leqq 1$, $\|J\| \leqq 1$ and $\|f_i - JQf_i \mid L_r\| \leqq \varepsilon$ for $i = 1, \ldots, n$.

Proof. Choose simple functions $s_1, \ldots, s_n \in L_r(X, \mu)$ such that

$$\|f_i - s_i \mid L_r\| \leqq \varepsilon/2 \quad \text{for} \quad i = 1, \ldots, n.$$

Write

$$s_i = \sum_{j=1}^{m} \alpha_{ij} h_j,$$

where $h_1, \ldots, h_m$ are characteristic functions of pairwise disjoint measurable subsets $X_1, \ldots, X_m$ of X. We now define the required operators by

$$Q := \sum_{j=1}^{m} \mu(X_j)^{-1/r'} h_j \otimes e_j \quad \text{and} \quad J := \sum_{j=1}^{m} \mu(X_j)^{-1/r} e_j \otimes h_j.$$

Obviously, $\|Q\| \leqq 1$ and $\|J\| \leqq 1$. Moreover, it follows from $JQh_j = h_j$ that $JQs_i = s_i$. Hence

$$\|f_i - JQf_i \mid L_r\| \leqq \|f_i - s_i \mid L_r\| + \|JQs_i - JQf_i \mid L_r\| \leqq 2\|f_i - s_i \mid L_r\| \leqq \varepsilon.$$

This completes the proof.

1.3.13. Let I denote the embedding operator from l_2 into c_0. Then I' and I'' can be identified with the embedding operators from l_1 into l_2 and from l_2 into l_∞, respectively. As proved in 1.6.4, we have $I' \in \mathfrak{P}_r(l_1, l_2)$. On the other hand, $I \notin \mathfrak{P}_r(l_2, c_0)$ and $I'' \notin \mathfrak{P}_r(l_2, l_\infty)$. This observation shows that the property of being absolutely r-summing neither goes over from an operator to its dual nor conversely. The situation becomes more favourable for operators acting between special Banach spaces.

Proposition (A. Persson 1969, P. Saphar 1970, J. S. Cohen 1973). Let $T \in \mathfrak{L}(E, L_r)$. Then $T' \in \mathfrak{P}_r(L_r', E')$ implies $T \in \mathfrak{P}_r(E, L_r)$ and $\|T \mid \mathfrak{P}_r\| \leqq \|T' \mid \mathfrak{P}_r\|$.

Proof. We first consider an operator $T \in \mathfrak{L}(E, l_r(m))$. Then, for $x \in E$,

$$\|Tx \mid l_r(m)\| = \left(\sum_{k=1}^{m} |\langle Tx, e_k\rangle|^r\right)^{1/r} = \left(\sum_{k=1}^{m} |\langle x, T'e_k\rangle|^r\right)^{1/r}.$$

Hence it follows from 1.3.4 that

$$\|T \mid \mathfrak{P}_r\| \leqq \left(\sum_{k=1}^{m} \|T'e_k\|^r\right)^{1/r} \leqq \|T' \mid \mathfrak{P}_r\| \, \|(e_r) \mid w_r\| = \|T' \mid \mathfrak{P}_r\|.$$

We now treat the general case when $T' \in \mathfrak{P}_r(L_r', E')$. Given $x_1, \ldots, x_n \in E$ and $\varepsilon > 0$, by the preceding lemma, there exist $Q \in \mathfrak{L}(L_r, l_r(m))$ and $J \in \mathfrak{L}(l_r(m), L_r)$ such that $\|Q\| \leqq 1$, $\|J\| \leqq 1$ and

$$\|Tx_i - JQTx_i \mid L_r\| \leqq \frac{\varepsilon}{1+\varepsilon} \|Tx_i \mid L_r\| \quad \text{for} \quad i = 1, \ldots, n.$$

Since $QT \in \mathfrak{L}(E, l_r(m))$, from the first part of the proof we know that

$$\|QT \mid \mathfrak{P}_r\| \leqq \|(QT)' \mid \mathfrak{P}_r\| \leqq \|T' \mid \mathfrak{P}_r\|.$$

Moreover, since

$$\|Tx_i \mid L_r\| \leqq \|JQTx_i \mid L_r\| + \|Tx_i - JQTx_i \mid L_r\|$$
$$\leqq \|QTx_i \mid l_r(m)\| + \frac{\varepsilon}{1+\varepsilon} \|Tx_i \mid L_r\|,$$

we have

$$\|(Tx_i) \mid l_r\| \leqq (1+\varepsilon) \|(QTx_i) \mid l_r\| \leqq (1+\varepsilon) \|QT \mid \mathfrak{P}_r\| \, \|(x_i) \mid w_r\|.$$

This proves that

$$T \in \mathfrak{P}_r(E, L_r) \quad \text{and} \quad \|T \mid \mathfrak{P}_r\| \leqq \|T' \mid \mathfrak{P}_r\|.$$

1.3.14. We now turn to the very special case in which the operators act from an arbitrary Banach space into a Hilbert space. Here we use the Rademacher functions $\varrho_1, \ldots, \varrho_m$ defined in G.2.1. Moreover, a_r and b_r are the constants appearing in Khintchine's inequality G.2.3.

Lemma (A. Pietsch 1972: c). Let $T \in \mathfrak{L}(E, l_2(m))$. Then

$$a_r \|T \mid \mathfrak{P}_r\| \leqq \left(\int_0^1 \left\| \sum_{k=1}^m T' e_k \varrho_k(\tau) \right\|^r \mathrm{d}\tau \right)^{1/r} \leqq b_r \|T' \mid \mathfrak{P}_r\|.$$

Proof. For $x_1, \ldots, x_n \in E$, by Khintchine's inequality, we have

$$\left(\sum_{i=1}^n \|Tx_i \mid l_2(m)\|^r \right)^{1/r} = \left(\sum_{i=1}^n \left[\sum_{k=1}^m |\langle Tx_i, e_k \rangle|^2 \right]^{r/2} \right)^{1/r}$$
$$\leqq a_r^{-1} \left(\sum_{i=1}^n \int_0^1 \left| \sum_{k=1}^m \langle x_i, T' e_k \, \varrho_k(\tau) \rangle \right|^r \mathrm{d}\tau \right)^{1/r}$$
$$= a_r^{-1} \left(\int_0^1 \sum_{i=1}^n \left| \langle x_i, \sum_{k=1}^m T' e_k \varrho_k(\tau) \rangle \right|^r \mathrm{d}\tau \right)^{1/r}$$
$$\leqq a_r^{-1} \left(\int_0^1 \left\| \sum_{k=1}^m T' e_k \varrho_k(\tau) \right\|^r \mathrm{d}\tau \right)^{1/r} \sup \left\{ \left(\sum_{i=1}^n |\langle x_i, a \rangle|^r \right)^{1/r} : a \in U^0 \right\}.$$

This proves the left-hand inequality.

The domination theorem yields a normalized Borel measure μ on $U_2(m)$, the closed unit ball of $l_2(m)$, such that

$$\|T'b\| \leqq \|T' \mid \mathfrak{P}_r\| \left(\int_{U_2(m)} |\langle y, b \rangle|^r \, \mathrm{d}\mu(y) \right)^{1/r} \text{ for all } b \in l_2(m).$$

We now obtain the right-hand inequality as follows:

$$\left(\int_0^1 \left\| \sum_{k=1}^m T' e_k \varrho_k(\tau) \right\|^r \mathrm{d}\tau \right)^{1/r} \leqq$$
$$\leqq \|T' \mid \mathfrak{P}_r\| \left(\int_0^1 \int_{U_2(m)} \left| \langle y, \sum_{k=1}^m e_k \varrho_k(\tau) \rangle \right|^r \mathrm{d}\mu(y) \, \mathrm{d}\tau \right)^{1/r}$$
$$= \|T' \mid \mathfrak{P}_r\| \left(\int_{U_2(m)} \int_0^1 \left| \sum_{k=1}^m \langle y, e_k \rangle \, \varrho_k(\tau) \right|^r \mathrm{d}\tau \, \mathrm{d}\mu(y) \right)^{1/r}$$
$$\leqq b_r \|T' \mid \mathfrak{P}_r\| \left(\int_{U_2(m)} \left(\sum_{k=1}^m |\langle y, e_k \rangle|^2 \right)^{r/2} \mathrm{d}\mu(y) \right)^{1/r} \leqq b_r \|T' \mid \mathfrak{P}_r\|.$$

This completes the proof.

Remark. In place of Rademacher functions one may use independent Gaussian variables.

1.3.15. Proposition (S. Kwapień 1970). Let $T \in \mathfrak{L}(E, H)$, where H is a Hilbert space. Then $T' \in \mathfrak{P}_r(H', E')$ implies $T \in \mathfrak{P}_s(E, H)$ for $1 \leqq r, s < \infty$.

Proof. In view of 1.3.3, it is enough to treat the case $1 \leqq s \leqq r < \infty$. Given $x_1, \ldots, x_n \in E$, we denote the linear span of $Tx_1, \ldots, Tx_n$ by M. Let $Q \in \mathfrak{L}(H, M)$ be the orthogonal projection from H onto M. Identifying the finite dimensional Hilbert space M with $l_2(m)$, we may apply the preceding lemma to the operator QT. Hence

$$a_s \|QT \mid \mathfrak{P}_s\| \leqq \left(\int_0^1 \left\| \sum_{k=1}^m (QT)' \, e_k \varrho_k(\tau) \right\|^s \mathrm{d}\tau \right)^{1/s}$$

$$\leqq \left(\int_0^1 \left\| \sum_{k=1}^m (QT)' e_k \varrho_k(\tau) \right\|^r \mathrm{d}\tau \right)^{1/r} \leqq b_r \|(QT)' \mid \mathfrak{P}_r\| .$$

Since $\|Q\| \leqq 1$, it follows that

$$\|QT \mid \mathfrak{P}_s\| \leqq a_s^{-1} b_r \|T' \mid \mathfrak{P}_r\| .$$

We now obtain

$$\|(Tx_i) \mid l_s\| = \|(QTx_i) \mid l_s\| \leqq a_s^{-1} b_r \|T' \mid \mathfrak{P}_r\| \, \|(x_i) \mid w_s\| .$$

This proves that

$$T \in \mathfrak{P}_s(E, H) \quad \text{and} \quad \|T \mid \mathfrak{P}_s\| \leqq a_s^{-1} b_r \|T' \mid \mathfrak{P}_r\| .$$

1.3.16. Applying the preceding proposition twice, we arrive at a result which is, at first sight, very surprising.

Theorem (A. Pełczyński 1967, A. Pietsch 1967). Let H and K be Hilbert spaces. Then all components $\mathfrak{P}_r(H, K)$ with $1 \leqq r < \infty$ coincide, the norms $\|\cdot \mid \mathfrak{P}_r\|$ being equivalent.

Remark. The best possible constants in the estimates between $\|\cdot \mid \mathfrak{P}_2\|$ and $\|\cdot \mid \mathfrak{P}_r\|$ were determined by D. J. H. Garling (1970).

1.4. Hilbert-Schmidt operators

Throughout this section H and K are complex Hilbert spaces.

1.4.1.* First we state an immediate consequence of Parseval's equation.

Lemma (R. Schatten/J. von Neumann 1946). Let (u_i) with $i \in I$ and (v_j) with $j \in J$ be orthonormal bases of H and K, respectively. Then, for every operator $T \in \mathfrak{L}(E, F)$, we have

$$\sum_I \|Tu_i\|^2 = \sum_I \sum_J |(Tu_i, v_j)|^2 = \sum_J \sum_I |(u_i, T^* v_j)|^2 = \sum_J \|T^* v_j\|^2 ,$$

where the sums may or may not be finite.

1.4.2.* We call $T \in \mathfrak{L}(E, F)$ a **Hilbert-Schmidt operator** if $(\|Tu_i\|^2)$ is summable for some orthonormal basis (u_i) of H. The set of these operators is denoted by $\mathfrak{S}(H, K)$.

The preceding lemma implies that, for $T \in \mathfrak{S}(H, K)$, the expression

$$\|T \mid \mathfrak{S}\| := \Big(\sum_I \|Tu_i\|^2\Big)^{1/2}$$

does not depend on the special choice of (u_i).

1.4.3.* The following theorem is well-known. We leave its straightforward proof as an exercise to the reader.

Theorem (R. Schatten/J. von Neumann 1946). $\mathfrak{S}$ is a Banach operator ideal on the class of Hilbert spaces. Furthermore, given any orthonormal basis (u_i) of H, the formula

$$(S, T) := \sum_I (Su_i, Tu_i)$$

defines an inner product on $\mathfrak{S}(H, K)$.

Remark. Thus $\mathfrak{S}$ is indeed a Hilbert operator ideal which means, by definition, that its norm can be derived from an inner product.

1.4.4.* The next result states that all Hilbert-Schmidt operators are approximable.

Proposition (E. Schmidt 1907: a). The Banach operator ideal $\mathfrak{S}$ is approximative.

Proof. Let (u_i) with $i \in I$ be any orthonormal basis of H. Given $T \in \mathfrak{S}(H, K)$ and $\varepsilon > 0$, we choose $\mathfrak{i} \in \mathfrak{F}(I)$ such that

$$\Big(\sum_{I \setminus \mathfrak{i}} \|Tu_i\|^2\Big)^{1/2} \leqq \varepsilon .$$

Define the orthogonal projection

$$P_{\mathfrak{i}} := \sum_{\mathfrak{i}} u_i^* \otimes u_i$$

which has finite rank. Then

$$\|T - TP_{\mathfrak{i}} \mid \mathfrak{S}\| = \Big(\sum_{I \setminus \mathfrak{i}} \|Tu_i\|^2\Big)^{1/2} \leqq \varepsilon .$$

1.4.5.* The concept of a Hilbert-Schmidt operator has proved to be very useful in treating various problems in Hilbert spaces. The following theorem explains why absolutely 2-summing operators can be considered as a fairly good substitute in the Banach space setting.

Theorem (A. Pietsch 1967). An operator $T \in \mathfrak{L}(H, K)$ is absolutely 2-summing if and only if it is Hilbert-Schmidt. If so, then we have $\|T \mid \mathfrak{P}_2\| = \|T \mid \mathfrak{S}\|$.

Proof. Let $T \in \mathfrak{P}_2(H, K)$. If (u_i) with $i \in I$ is any orthonormal basis of H, then 1.1.11 and 1.2.2 imply that

$$T \in \mathfrak{S}(H, K) \quad \text{and} \quad \Big(\sum_I \|Tu_i\|^2\Big)^{1/2} \leqq \|T \mid \mathfrak{P}_2\| .$$

In order to verify the converse, we consider in K any orthonormal basis (v_j) with $j \in J$. Let $x_1, \ldots, x_n \in H$. Using Parseval's equation, we now obtain

$$\Big(\sum_{i=1}^n \|Tx_i\|^2\Big)^{1/2} = \Big(\sum_{i=1}^n \sum_J |(Tx_i, v_j)|^2\Big)^{1/2} = \Big(\sum_J \sum_{i=1}^n |(x_i, T^*v_j)|^2\Big)^{1/2}$$
$$\leqq \Big(\sum_J \|T^*v_j\|^2\Big)^{1/2} \sup\Big\{\Big(\sum_{i=1}^n |(x_i, a)|^2\Big)^{1/2} : a \in U\Big\}.$$

Consequently,

$$T \in \mathfrak{P}_2(H, K) \quad \text{and} \quad \|T \mid \mathfrak{P}_2\| \leqq \Big(\sum_J \|T^* v_j\|^2\Big)^{1/2}.$$

1.4.6.* A matrix $M = (\mu_{ij})$ is said to be of **Hilbert-Schmidt type** if it belongs to $l_2(I \times J)$. Then

$$M_{\mathrm{op}} : (\eta_j) \to \Big(\sum_J \mu_{ij} \eta_j\Big)$$

defines an operator from $l_2(J)$ into $l_2(I)$.

1.4.7.* Using the preceding definition, we may rephrase Theorem 1.4.5 in terms of matrices.

Proposition. An operator from $l_2(J)$ into $l_2(I)$ is absolutely 2-summing if and only if it is induced by a Hilbert-Schmidt matrix $M = (\mu_{ij})$. If so, then we have

$$\|M_{\mathrm{op}} \mid \mathfrak{P}_2\| = \Big(\sum_I \sum_J |\mu_{ij}|^2\Big)^{1/2}.$$

1.4.8.* We now consider integral operators on Hilbert function spaces. For this purpose, let (X, μ) and (Y, ν) be σ-finite measure spaces.

A kernel K is said to be of **Hilbert-Schmidt type** if it belongs to $L_2(X \times Y, \mu \times \nu)$. Then

$$K_{\mathrm{op}} : g(\eta) \to \int_Y K(\xi, \eta)\, g(\eta)\, d\nu(\eta)$$

defines an operator from $L_2(Y, \nu)$ into $L_2(X, \mu)$.

1.4.9.* We now establish a counterpart of 1.4.7.

Proposition. An operator from $L_2(Y, \nu)$ into $L_2(X, \mu)$ is absolutely 2-summing if and only if it is induced by a Hilbert-Schmidt kernel K. If so, then we have

$$\|K_{\mathrm{op}} \mid \mathfrak{P}_2\| = \Big(\int_X \int_Y |K(\xi, \eta)|^2\, d\mu(\xi)\, d\nu(\eta)\Big)^{1/2}.$$

Proof (sketch). Let (u_i) with $i \in I$ and (v_j) with $j \in J$ be orthonormal bases of $L_2(X, \mu)$ and $L_2(Y, \nu)$, respectively. Observe that the functions $u_i(\xi)\, v_j(\eta)$ form an orthonormal basis of $L_2(X \times Y, \mu \times \nu)$.

If $T \in \mathfrak{P}_2(L_2(Y, \nu), L_2(X, \mu))$, then

$$\Big(\sum_I \sum_J |(Tv_j, u_i)|^2\Big)^{1/2} = \|T \mid \mathfrak{P}_2\|$$

is finite. Hence we may define the kernel

$$K(\xi, \eta) := \sum_I \sum_J (Tv_j, u_i)\, u_i(\xi)\, v_j(\eta),$$

where the right-hand sum converges in mean square. Note that K is of Hilbert-Schmidt type and

$$\Big(\int_X \int_Y |K(\xi, \eta)|^2\, d\mu(\xi)\, d\nu(\eta)\Big)^{1/2} = \|T \mid \mathfrak{P}_2\|.$$

Moreover, K induces the given operator T.

The converse is obvious.

1.4.10.* We conclude this section with an elementary observation; see B.3.8.

Proposition (R. Schatten/J. von Neumann 1946). The following are equivalent for $T \in \mathfrak{L}(H, K)$:

$$T \in \mathfrak{S}(H, K), \quad T^* \in \mathfrak{S}(K, H), \quad T' \in \mathfrak{S}(K', H').$$

The corresponding norms coincide.

1.5. Absolutely 2-summing operators

1.5.1.* In the case $r = 2$ the domination theorem leads to another result which has far-reaching consequences.

Factorization theorem (A. Pietsch 1967). Every operator $T \in \mathfrak{P}_2(E, F)$ admits a factorization

$$\begin{array}{ccc} E & \xrightarrow{\quad T \quad} & F \\ {\scriptstyle A}\downarrow & & \uparrow{\scriptstyle Y} \\ l_\infty(I) & \xrightarrow[\quad S \quad]{} & H \end{array}$$

such that the following conditions are satisfied:

(1) I is an index set.

(2) H is a Hilbert space.

(3) $A \in \mathfrak{L}(E, l_\infty(I))$ and $\|A\| = 1$, $Y \in \mathfrak{L}(H, F)$ and $\|Y\| = \|T \mid \mathfrak{P}_2\|$.

(4) $S \in \mathfrak{P}_2(l_\infty(I), H)$ and $\|S \mid \mathfrak{P}_2\| = 1$.

In particular, if T is finite, then it can be arranged that $\dim(H) = \operatorname{rank}(T)$.

Proof. Take U^0 as the underlying index set. Applying the domination theorem 1.3.5, we find a positive normalized linear form m on $l_\infty^{\text{real}}(U^0)$ such that

$$\|Tx\| \leqq \|T \mid \mathfrak{P}_2\| \, m(|f_x|^2)^{1/2} \quad \text{for all } x \in E.$$

Obviously, m can be extended canonically to the complex Banach space $l_\infty(U^0)$. Form the quotient space $l_\infty(U^0)/N$, where

$$N := \{f \in l_\infty(U^0) : m(|f|^2) = 0\},$$

and denote by $\hat{f} := f + N$ the equivalence class determined by f. Then

$$(\hat{f}, \hat{g}) := m(fg^*) \quad \text{for} \quad f, g \in l_\infty(U^0)$$

yields a well-defined inner product on $l_\infty(U^0)/N$, and the desired Hilbert space can be constructed by completion.

Next we define the operators A, S and Y. Put $Ax := f_x$, where $f_x(a) := \langle x, a \rangle$ for $a \in U^0$. Let S be the canonical operator from $l_\infty(U^0)$ into H given by $Sf := \hat{f}$. By analogy with Examples 1.3.8 and 1.3.9 we see that

$$S \in \mathfrak{P}_2(l_\infty(U^0), H) \quad \text{and} \quad \|S \mid \mathfrak{P}_2\| = 1.$$

The definition of Y is slightly more involved. Observe first that $\hat{f}_{x_1} = \hat{f}_{x_2}$ implies $Tx_1 = Tx_2$, because

$$\|Tx_1 - Tx_2\| \leqq \|T \mid \mathfrak{P}_2\| \, m(|f_{x_1} - f_{x_2}|^2)^{1/2} \quad \text{for } x_1, x_2 \in E.$$

Hence the map Y_0 which sends $\hat{f}_x$ into Tx is well-defined on $M(SA)$. It follows from

$$\|Y_0\hat{f}_x\| = \|Tx\| \leqq \|T \mid \mathfrak{P}_2\| \, m(|f_x|^2)^{1/2} = \|T \mid \mathfrak{P}_2\| \, \|\hat{f}_x \mid H\|$$

that Y_0 is continuous: $\|Y_0\| \leqq \|T \mid \mathfrak{P}_2\|$. Thus there exists a continuous extension $\overline{Y}_0$ on $H_0 := \overline{M(SA)}$. Next we consider the orthogonal projection $Q \in \mathfrak{L}(H, H_0)$ from H onto H_0, and set $Yf := \overline{Y}_0 Qf$ for $f \in H$. Then

$$\|Y\| \leqq \|\overline{Y}_0\| \, \|Q\| \leqq \|T \mid \mathfrak{P}_2\| \quad \text{and} \quad YSAx = YSf_x = \overline{Y}_0 Q\hat{f}_x = Y_0\hat{f}_x = Tx.$$

We already know that $\|A\| = 1$ and $\|S \mid \mathfrak{P}_2\| = 1$. Furthermore,

$$\|T \mid \mathfrak{P}_2\| = \|YSA \mid \mathfrak{P}_2\| \leqq \|Y\| \, \|S \mid \mathfrak{P}_2\| \, \|A\| \leqq \|Y\| \leqq \|T \mid \mathfrak{P}_2\|$$

yields $\|Y\| = \|T \mid \mathfrak{P}_2\|$. Hence $T = YSA$ is the desired factorization.

Finally, we observe that $M(T) \subseteq M(Y) \subseteq \overline{M(T)}$. Thus, in the case when T has finite rank, it follows that $M(Y) = M(T)$. Passing from H to $H/N(Y)$ we find a factorization through a rank (T)-dimensional Hilbert space; see also B.4.6.

1.5.2.* If $l_\infty(U^0)$ is replaced by $C(U^0)$, then the preceding proof becomes more transparent, since the abstract construction of the Hilbert space H turns into the classical procedure which relates $L_2(U^0, \mu)$ to $C(U^0)$; see 1.3.6.

Factorization theorem (measure theoretic version). Every operator $T \in \mathfrak{P}_2(E, F)$ admits a factorization

$$\begin{array}{ccc} E & \xrightarrow{\quad T \quad} & F \\ {\scriptstyle A}\downarrow & & \uparrow{\scriptstyle Y} \\ C(X) & \xrightarrow[\quad I \quad]{} & L_2(X, \mu) \end{array}$$

such that the following conditions are satisfied:

(1) X is a compact Hausdorff space.

(2) μ is a normalized Borel measure on X.

(3) $A \in \mathfrak{L}(E, C(X))$ and $\|A\| = 1$, $Y \in \mathfrak{L}(L_2(X, \mu), F)$ and $\|Y\| = \|T \mid \mathfrak{P}_2\|$

(4) I is the canonical operator from $C(X)$ into $L_2(X, \mu)$.

1.5.3.* The preceding factorization theorems imply a useful multiplication formula.

Proposition. $\mathfrak{P}_2 = \mathfrak{H} \circ \mathfrak{P}_2$. The norms on both sides are equal.

1.5.4.* Absolutely 2-summing operators have another remarkable property.

Extension theorem. Let $J \in \mathfrak{L}(E, E_0)$ be any metric injection. Then every operator $T \in \mathfrak{P}_2(E, F)$ admits an extension $T_0 \in \mathfrak{P}_2(E_0, F)$ which preserves the $\mathfrak{P}_2$-norm. This means that $T = T_0 J$ and $\|T_0 \mid \mathfrak{P}_2\| = \|T \mid \mathfrak{P}_2\|$.

Proof. Consider a factorization $T = YSA$ as described in 1.5.1. Since $l_\infty(I)$ possesses the metric extension property, there exists $A_0 \in \mathfrak{L}(E_0, l_\infty(I))$ such that $A = A_0 J$ and $\|A_0\| = 1$. Then $T_0 := YSA_0$ is the extension we are looking for.

Remark. The idea of the preceding proof is due to G. Jameson. Previously, one had used a factorization of the form

$$\begin{array}{ccc} E & \xrightarrow{\quad T \quad} & F \\ {\scriptstyle A}\downarrow & & \uparrow{\scriptstyle Y} \\ L_\infty(X, \mu) & \xrightarrow[\quad I \quad]{} & L_2(X, \mu) \end{array}$$

which follows immediately from 1.5.2. However, to verify the metric extension property of the complex Banach space $L_\infty(X, \mu)$ is by no means as easy as in the case of $l_\infty(I)$; see (DIL, pp. 228 and 230) and (KÖN, 1.c.2).

1.5.5. We now establish an extremely important lemma due to M. I. Kadets/ M. G. Snobar (1971). The original proof was based on a result of F. John (1948) concerning ellipsoids of minimal volume containing the closed unit ball of a given finite dimensional Banach space. The following approach goes back to S. Kwapień.

Lemma. Let M be any n-dimensional subspace of E. Then there exists a projection $P \in \mathfrak{L}(E)$ such that $M(P) = M$ and $\|P\| \leqq n^{1/2}$.

Proof. It follows from 1.2.8 that $\|I_M \mid \mathfrak{P}_2\| \leqq n^{1/2}$. Thus, by 1.5.4, there exists $T \in \mathfrak{L}(E, M)$ such that

$$TJ_M^E = I_M \quad \text{and} \quad \|T\| \leqq \|T \mid \mathfrak{P}_2\| \leqq n^{1/2}.$$

Then $P := J_M^E T$ is the projection we are looking for.

1.5.6. In the remainder of this section we investigate the operator ideals $(\mathfrak{P}_2)^m$ with $m = 1, 2, \ldots$ For this purpose, the following auxiliary result is needed.

Lemma. Suppose that the quasi-Banach operator ideal $\mathfrak{A}$ is p-normed. Then $\mathfrak{A} \circ \mathfrak{P}_2$ is q-normed, where $1/q = 1/p + 1/2$.

Proof. Given $S, T \in \mathfrak{A} \circ \mathfrak{P}_2(E, F)$ and $\varepsilon > 0$, we choose factorizations $S = XA$ and $T = YB$ such that

$$A \in \mathfrak{P}_2(E, M), \; X \in \mathfrak{A}(M, F), \quad B \in \mathfrak{P}_2(E, N), \; Y \in \mathfrak{A}(N, F),$$

$$\|A \mid \mathfrak{P}_2\| \leqq [(1 + \varepsilon)\, \|S \mid \mathfrak{A} \circ \mathfrak{P}_2\|]^{q/2}, \quad \|B \mid \mathfrak{P}_2\| \leqq [(1 + \varepsilon)\, \|T \mid \mathfrak{A} \circ \mathfrak{P}_2\|]^{q/2},$$

$$\|X \mid \mathfrak{A}\| \leqq [(1 + \varepsilon)\, \|S \mid \mathfrak{A} \circ \mathfrak{P}_2\|]^{q/p}, \quad \|Y \mid \mathfrak{A}\| \leqq [(1 + \varepsilon)\, \|T \mid \mathfrak{A} \circ \mathfrak{P}_2\|]^{q/p}.$$

Consider the direct sum $M \oplus N$ endowed with the euclidean norm. The corresponding injections and surjections are denoted by J_M, J_N and Q_M, Q_N, respectively. Then, for $x_1, \ldots, x_n \in E$, we have

$$\left(\sum_{i=1}^{n} \|(J_M A + J_N B)\, x_i\|^2\right)^{1/2} = \left(\sum_{i=1}^{n} (\|Ax_i\|^2 + \|Bx_i\|^2)\right)^{1/2}$$

$$\leqq (\|A \mid \mathfrak{P}_2\|^2 + \|B \mid \mathfrak{P}_2\|^2)^{1/2} \sup\left\{\left(\sum_{i=1}^{n} |\langle x_i, a\rangle|^2\right)^{1/2} : a \in U^0\right\}.$$

Therefore

$$\|J_M A + J_N B \mid \mathfrak{P}_2\| \leqq (\|A \mid \mathfrak{P}_2\|^2 + \|B \mid \mathfrak{P}_2\|^2)^{1/2}$$
$$\leqq (1 + \varepsilon)^{q/2} (\|S \mid \mathfrak{A} \circ \mathfrak{P}_2\|^q + \|T \mid \mathfrak{A} \circ \mathfrak{P}_2\|^q)^{1/2}.$$

Moreover,

$$\|XQ_M + YQ_N \mid \mathfrak{A}\| \leqq (\|X \mid \mathfrak{A}\|^p + \|Y \mid \mathfrak{A}\|^p)^{1/p}$$
$$\leqq (1 + \varepsilon)^{q/p} (\|S \mid \mathfrak{A} \circ \mathfrak{P}_2\|^q + \|T \mid \mathfrak{A} \circ \mathfrak{P}_2\|^q)^{1/p}.$$

Combining these inequalities, from

$$S + T = (XQ_M + YQ_N)(J_M A + J_N B)$$

we deduce that

$$\begin{aligned}\|S + T \mid \mathfrak{A} \circ \mathfrak{P}_2\| &\leqq \|XQ_M + YQ_N \mid \mathfrak{A}\| \, \|J_M A + J_N B \mid \mathfrak{P}_2\| \\ &\leqq (1 + \varepsilon)(\|S \mid \mathfrak{A} \circ \mathfrak{P}_2\|^q + \|T \mid \mathfrak{A} \circ \mathfrak{P}_2\|^q)^{1/q}.\end{aligned}$$

Letting $\varepsilon \to 0$ yields the q-triangle inequality.

1.5.7. Using the preceding lemma, we get the following result by induction.

Proposition. The quasi-Banach operator ideal $(\mathfrak{P}_2)^m$ is $2/(m+1)$-normed.

1.5.8. **Proposition.** The quasi-Banach operator ideal $(\mathfrak{P}_2)^m$ is injective.

Proof. The assertion follows from 1.3.2 and the trivial fact that the product of injective quasi-Banach operator ideals is injective, as well.

1.5.9. **Proposition.** The quasi-Banach operator ideal $(\mathfrak{P}_2)^{m+1}$ is approximative.

Proof. Obviously, it is enough to treat the case $m = 1$. From 1.5.1 we know that every operator $T \in (\mathfrak{P}_2)^2(E, F)$ admits a factorization $T = YSA$ such that $A \in \mathfrak{P}_2(E, H)$, $S \in \mathfrak{P}_2(H, K)$ and $Y \in \mathfrak{L}(K, F)$, where H and K are Hilbert spaces. Recall that $\mathfrak{P}_2(H, K) = \mathfrak{S}(H, K)$, by 1.4.5. Since $\mathfrak{S}(H, K)$ is approximative, there exists a sequence of operators $S_k \in \mathfrak{F}(H, K)$ converging to S with respect to the Hilbert-Schmidt norm. Finally, we see from

$$\|YSA - YS_k A \mid (\mathfrak{P}_2)^2\| \leqq \|Y\| \, \|S - S_k \mid \mathfrak{P}_2\| \, \|A \mid \mathfrak{P}_2\|$$

that the finite operators $T_k := YS_k A$ converge to $T = YSA$ in $(\mathfrak{P}_2)^2(E, F)$.

Remark. As shown in 1.6.4, we have $I \in \mathfrak{P}_2(l_1, l_2)$. Since $I \notin \mathfrak{K}(l_1, l_2)$, the Banach operator ideal $\mathfrak{P}_2$ fails to be approximative.

1.5.10. **Proposition.** $(\mathfrak{P}_2')^{m+1} \subseteqq (\mathfrak{P}_2)^m$.

Proof. By definition, every operator $T \in (\mathfrak{P}_2')^{m+1}(E, F)$ can be written as a product $T = T_{m+1} \ldots T_1$ of operators $T_k \in \mathfrak{P}_2'(E_{k-1}, E_k)$, where $E_0 = E$ and $E_{m+1} = F$. Next we choose factorizations $T_k' = A_k X_k$ such that $X_k \in \mathfrak{P}_2(E_k', H_k)$ and $A_k \in \mathfrak{L}(H_k, E_{k-1}')$, where $H_1, \ldots, H_{m+1}$ are Hilbert spaces. In view of 1.3.15, it now follows from $X_k A_{k+1} \in \mathfrak{P}_2(H_{k+1}, H_k)$ that $A_{k+1}' X_k' \in \mathfrak{P}_2(H_k', H_{k+1}')$. Observe that this conclusion could also be inferred from 1.4.5 and 1.4.10. Hence

$$K_F T = T'' K_E = X_{m+1}'(A_{m+1}' X_m') \ldots (A_2' X_1') \, A_1' K_E \in (\mathfrak{P}_2)^m (E, F'').$$

Finally, the injectivity of $(\mathfrak{P}_2)^m$ yields $T \in (\mathfrak{P}_2)^m (E, F)$.

1.6. Diagonal operators

Throughout this section we assume that $1 \leqq p, q \leqq \infty$. Moreover, e_k denotes either the k-th unit sequence or the k-th coordinate functional.

1.6.1.* In the following we are concerned with **diagonal operators**

$$D_t : (\xi_k) \to (\tau_k \xi_k)$$

induced by complex-valued sequences $t = (\tau_k)$ and acting from l_p into l_q. Without loss of generality, it may be assumed that $\tau_k \geqq 0$. If $1 \leqq p \leqq q \leqq \infty$, then the embedding operator from l_p into l_q is denoted by I.

1.6.2.* First we state an immediate consequence of Hölder's inequality. Recall that $\xi_+ := \max(\xi, 0)$ for $\xi \in \mathbb{R}$.

Lemma. Let $1/r = (1/q - 1/p)_+$. Then

$D_t \in \mathfrak{L}(l_p, l_q)$ if and only if $t \in l_r$.

1.6.3.* Next a basic example, related to 1.3.9, is treated.

Example (A. Persson / A. Pietsch 1969). Let $1 \leqq q < \infty$. Then

$D_t \in \mathfrak{P}_q(l_\infty, l_q)$ if and only if $t \in l_q$.

Proof. Assume that $t \in l_q$. Since

$$\|D_t x \mid l_q\| = \left(\sum_{k=1}^{\infty} |\langle x, \tau_k e_k\rangle|^q\right)^{1/q} \quad \text{for all} \quad x \in l_\infty,$$

it follows from 1.3.4 that

$$D_t \in \mathfrak{P}_q(l_\infty, l_q) \quad \text{and} \quad \|D_t : l_\infty \to l_q \mid \mathfrak{P}_q\| \leqq \|t \mid l_q\|.$$

The converse is obvious, by 1.6.2.

1.6.4.* We now give an application of the domination theorem in which the underlying Borel measure can be described explicitly.

Example (A. Grothendieck 1956: b). $I \in \mathfrak{P}(l_1, l_2)$.

Proof. Let $r(\tau) := (\varrho_i(\tau))$, where ϱ_i denotes the i-th Rademacher function. The function r defined in this way maps the interval $[0, 1]$ into the closed unit ball U_∞ of l_∞. If μ is the image of the Lebesgue measure, then Khintchine's inequality G.2.3 yields

$$\begin{aligned}\|x \mid l_2\| = \left(\sum_{i=1}^{\infty} |\langle x, e_i\rangle|^2\right)^{1/2} &\leqq a_1^{-1} \int_0^1 \left|\sum_{i=1}^{\infty} \langle x, e_i\rangle\, \varrho_i(\tau)\right| d\tau \\ &= a_1^{-1} \int_0^1 |\langle x, r(\tau)\rangle|\, d\tau = a_1^{-1} \int_{U_\infty} |\langle x, a\rangle|\, d\mu(a)\end{aligned}$$

for all $x \in l_1$. This completes the proof, by 1.3.6.

Remark. The example under consideration can be deduced from the famous **Grothendieck inequality** which implies that all operators from l_1 into l_2 are absolutely summing. More precisely, there exists a constant $c_G \geqq 1$ such that

$$\|T \mid \mathfrak{P}\| \leqq c_G \|T\| \quad \text{for all } T \in \mathfrak{L}(l_1, l_2).$$

For proofs of this fundamental theorem the reader is referred to (DIE, p. 181), (LIN, Vol. 2.b.6) and (PIE, 22.4.4). The best possible value of the so-called Grothendieck constant c_G is still unknown.

1.6.5.* The following result is classical.

Example (W. Orlicz 1933). $I \in \mathfrak{P}_{2,1}(l_p, l_p)$ if $1 \leqq p \leqq 2$.

Proof. For $x_1, \ldots, x_n \in l_p$ it follows from Jessen's and Khintchine's inequalities that

$$\begin{aligned}\left(\sum_{i=1}^{n} \|x_i \mid l_p\|^2\right)^{1/2} &= \left(\sum_{i=1}^{n} \left(\sum_{j=1}^{\infty} |\langle x_i, e_j\rangle|^p\right)^{2/p}\right)^{1/2} \\ &\leqq \left(\sum_{j=1}^{\infty} \left(\sum_{i=1}^{n} |\langle x_i, e_j\rangle|^2\right)^{p/2}\right)^{1/p} \leqq a_p^{-1}\left(\sum_{j=1}^{\infty} \int_0^1 \left|\sum_{i=1}^{n} \langle x_i, e_j\rangle \varrho_i(\tau)\right|^p d\tau\right)^{1/p} \\ &= a_p^{-1}\left(\int_0^1 \left\|\sum_{i=1}^{n} x_i \varrho_i(\tau) \mid l_p\right\|^p d\tau\right)^{1/p} \leqq a_p^{-1} \sup\left\{\left\|\sum_{i=1}^{n} \alpha_i x_i \mid l_p\right\| : |\alpha_i| \leqq 1\right\}.\end{aligned}$$

Thus we have $I \in \mathfrak{P}_{2,1}(l_p, l_p)$, by 1.1.14.

1.6.6.* We now give a more general example which includes both 1.6.4 and 1.6.5 as special cases.

Example (G. Bennett 1973, B. Carl 1974). Let $1 \leqq p \leqq q \leqq 2$. Then $I \in \mathfrak{P}_{s,1}(l_p, l_q)$, where $1/s = 1/p - 1/q + 1/2$.

Proof. We first assume that $q = 2$. If $p = 1$ and $p = 2$, then the assertion is proved in 1.6.4 and 1.6.5, respectively. Thus it suffices to deal with the case $1 < p < 2$. Define

$$1/u := p - 1 \quad \text{and} \quad 1/v := 2(p-1)/p.$$

Then

$$1/u' = 2 - p \quad \text{and} \quad 1/v' = (2-p)/p.$$

Let $x_1, \ldots, x_n \in l_p$. Note that

$$\sum_{i=1}^{n} \|x_i \mid l_2\|^p = \sum_{i=1}^{n} \sum_{j=1}^{\infty} |\langle x_i, e_j \rangle|^2 \, \|x_i \mid l_2\|^{p-2}$$
$$= \sum_{j=1}^{\infty} \sum_{i=1}^{n} |\langle x_i, e_j \rangle|^{2/u} \, |\langle x_i, e_j \rangle|^{2/u'} \, \|x_i \mid l_2\|^{-1/u'}.$$

Applying Hölder's inequality twice, we obtain

$$\sum_{i=1}^{n} \|x_i \mid l_2\|^p \leqq \sum_{j=1}^{\infty} \left(\sum_{i=1}^{n} |\langle x_i, e_j \rangle|^2 \right)^{1/u} \left(\sum_{i=1}^{n} |\langle x_i, e_j \rangle|^2 \, \|x_i \mid l_2\|^{-1} \right)^{1/u'}$$
$$\leqq \left(\sum_{j=1}^{\infty} \left(\sum_{i=1}^{n} |\langle x_i, e_j \rangle|^2 \right)^{p/2} \right)^{1/v} \left(\sum_{j=1}^{\infty} \left(\sum_{i=1}^{n} |\langle x_i, e_j \rangle|^2 \, \|x_i \mid l_2\|^{-1} \right)^{p} \right)^{1/v'}.$$

The two factors in the last expression are now treated separately. First, by Khintchine's inequality G.2.3, it follows that

$$\left(\sum_{j=1}^{\infty} \left(\sum_{i=1}^{n} |\langle x_i, e_j \rangle|^2 \right)^{p/2} \right)^{1/p} \leqq a_p^{-1} \left(\sum_{j=1}^{\infty} \int_0^1 \left| \sum_{i=1}^{n} \langle x_i, e_j \rangle \, \varrho_i(\tau) \right|^p d\tau \right)^{1/p}$$
$$= a_p^{-1} \left(\int_0^1 \left\| \sum_{i=1}^{n} x_i \varrho_i(\tau) \mid l_p \right\|^p d\tau \right)^{1/p}$$
$$\leqq a_p^{-1} \sup \left\{ \left\| \sum_{i=1}^{n} \alpha_i x_i \mid l_p \right\| : |\alpha_i| \leqq 1 \right\}$$
$$= a_p^{-1} \sup \left\{ \sum_{i=1}^{n} |\langle x_i, a \rangle| : a \in U_{p'} \right\}.$$

Next, the second term is estimated. Using Khintchine's inequality once again, for all $a = (\alpha_j) \in l_{p'}$ we have

$$\sum_{j=1}^{\infty} \alpha_j \sum_{i=1}^{n} |\langle x_i, e_j \rangle|^2 \, \|x_i \mid l_2\|^{-1}$$
$$= \sum_{i=1}^{n} \sum_{j=1}^{\infty} \alpha_j |\langle x_i, e_j \rangle| \, |\langle x_i, e_j \rangle| \, \|x_i \mid l_2\|^{-1}$$
$$\leqq \sum_{i=1}^{n} \left(\sum_{j=1}^{\infty} |\alpha_j \langle x_i, e_j \rangle|^2 \right)^{1/2} \left(\sum_{j=1}^{\infty} |\langle x_i, e_j \rangle|^2 \, \|x_i \mid l_2\|^{-2} \right)^{1/2}$$
$$\leqq a_1^{-1} \sum_{i=1}^{n} \int_0^1 \left| \sum_{j=1}^{\infty} \alpha_j \langle x_i, e_j \rangle \, \varrho_j(\tau) \right| d\tau \leqq a_1^{-1} \int_0^1 \sum_{i=1}^{n} |\langle x_i, \mathrm{ar}(\tau) \rangle| \, d\tau,$$

where $\operatorname{ar}(\tau) = (\alpha_j \varrho_j(\tau))$. This implies that

$$\left(\sum_{j=1}^{\infty}\left(\sum_{i=1}^{n} |\langle x_i, e_j\rangle|^2 \, \|x_i \mid l_2\|^{-1}\right)^p\right)^{1/p} \leqq a_1^{-1} \sup\left\{\sum_{i=1}^{n} |\langle x_i, a\rangle| : a \in U_{p'}\right\}.$$

Combining the preceding inequalities, we obtain

$$\left(\sum_{i=1}^{n} \|x_i \mid l_2\|^p\right)^{1/p}$$

$$\leqq \left(\sum_{j=1}^{\infty}\left(\sum_{i=1}^{n} |\langle x_i, e_j\rangle|^2\right)^{p/2}\right)^{1/pv} \left(\sum_{j=1}^{\infty}\left(\sum_{i=1}^{n} |\langle x_i, e_j\rangle|^2 \, \|x_i \mid l_2\|^{-1}\right)^p\right)^{1/pv'}$$

$$\leqq a_p^{-1/v} a_1^{-1/v'} \sup\left\{\sum_{i=1}^{n} |\langle x_i, a\rangle| : a \in U_{p'}\right\}.$$

Hence $I \in \mathfrak{P}_{p,1}(l_p, l_2)$.

Finally, we treat the case $1 \leqq p < q < 2$ by an easy interpolation argument. Define θ by

$$1/q = (1 - \theta)/2 + \theta/p.$$

Then it follows from $1/s = 1/p - 1/q + 1/2$ that

$$1/s = (1 - \theta)/p + \theta/2.$$

As already shown,

$$I \in \mathfrak{P}_{p,1}(l_p, l_2) \quad \text{and} \quad I \in \mathfrak{P}_{2,1}(l_p, l_p).$$

Since

$$\left(\sum_{i=1}^{n} \|x_i \mid l_q\|^s\right)^{1/s} \leqq \left(\sum_{i=1}^{n} [\|x_i \mid l_2\|^{1-\theta} \, \|x_i \mid l_p\|^\theta]^s\right)^{1/s}$$

$$\leqq \left(\sum_{i=1}^{n} \|x_i \mid l_2\|^p\right)^{(1-\theta)/p} \left(\sum_{i=1}^{n} \|x_i \mid l_p\|^2\right)^{\theta/2}$$

for $x_1, \ldots, x_n \in l_p$, we have

$$\|I : l_p \to l_q \mid \mathfrak{P}_{s,1}\| \leqq \|I : l_p \to l_2 \mid \mathfrak{P}_{p,1}\|^{1-\theta} \, \|I : l_p \to l_p \mid \mathfrak{P}_{2,1}\|^\theta.$$

Hence $I \in \mathfrak{P}_{s,1}(l_p, l_q)$.

1.6.7.* Next we state an immediate consequence of the preceding result which is crucial for later applications

Example (G. Bennett 1973, B. Carl 1974). Let $1 \leqq p \leqq q \leqq \infty$. Then $I \in \mathfrak{P}_{r,2}(l_p, l_q)$, where

$$1/r := \begin{cases} 1/p - 1/q & \text{if } 1 \leqq p \leqq q \leqq 2, \\ 1/p - 1/2 & \text{if } 1 \leqq p \leqq 2 \leqq q \leqq \infty, \\ 0 & \text{if } 2 \leqq p \leqq q \leqq \infty. \end{cases}$$

Proof. If $1 \leqq p \leqq q \leqq 2$, then the assertion follows from 1.2.5 and 1.6.6. If $1 \leqq p \leqq 2 \leqq q \leqq \infty$, then we conclude from $I \in \mathfrak{P}_{p,1}(l_p, l_2)$ that $I \in \mathfrak{P}_{p,1}(l_p, l_q)$. Hence $I \in \mathfrak{P}_{r,2}(l_p, l_q)$ with $1/r = 1/p - 1/2$. The case $2 \leqq p \leqq q \leqq \infty$ is trivial.

1.6.8. Example. $D_t \in (\mathfrak{P}_2)^m\,(l_2, l_1)$ if and only if $t \in l_{2/(m+1)}$.

Proof. If $t = (\tau_k) \in l_{2/(m+1)}$, then $s := (\tau_k^{1/(m+1)}) \in l_2$. Now it follows from $D_s \in \mathfrak{P}_2(l_2, l_2)$ and $D_s \in \mathfrak{L}(l_2, l_1)$ that $D_t = D_s D_s^m \in (\mathfrak{P}_2)^m(l_2, l_1)$.

By 1.6.4, the embedding operator from l_1 into l_2 is absolutely 2-summing. Hence $D_t \in (\mathfrak{P}_2)^m(l_2, l_1)$ implies that $D_t \in (\mathfrak{P}_2)^{m+1}(l_2, l_2)$. Since D_t has the eigenvalue sequence t, we may infer from 3.7.3 that $t \in l_{2/(m+1)}$.

1.6.9. Example. $D_t \in (\mathfrak{P}_2)^m\,(l_\infty, l_2)$ if and only if $t \in l_{2/m}$.

Proof. If $t = (\tau_k) \in l_{2/m}$, then $s := (\tau_k^{1/m}) \in l_2$. Now it follows from $D_s \in \mathfrak{P}_2(l_\infty, l_2)$ and $D_s \in \mathfrak{P}_2(l_2, l_2)$ that $D_t = D_s^{m-1} D_s \in (\mathfrak{P}_2)^m(l_\infty, l_2)$.

Conversely, $D_t \in (\mathfrak{P}_2)^m(l_\infty, l_2)$ implies $D_t \in (\mathfrak{P}_2)^m(l_2, l_2)$. Since D_t has the eigenvalue sequence t, we may infer from 3.7.3 that $t \in l_{2/m}$.

1.6.10. Example. $D_t \in (\mathfrak{P}_2)^m\,(l_1, l_2)$ if and only if $t \in l_{2/(m-1)}$.

Proof. In view of 1.6.4, every bounded sequence induces an absolutely 2-summing operator from l_1 into l_2. Therefore the case $m = 1$ is trivial.

If $t = (\tau_k) \in l_{2/(m-1)}$, then $s := (\tau_k^{1/(m-1)}) \in l_2$. Now it follows from $D_s \in \mathfrak{P}_2(l_2, l_2)$ that $D_t = D_s^{m-1} \in (\mathfrak{P}_2)^{m-1}(l_2, l_2)$. Hence $D_t \in (\mathfrak{P}_2)^m(l_1, l_2)$, by 1.6.4.

Let $D_t \in (\mathfrak{P}_2)^m(l_1, l_2)$. Since $D_s \in \mathfrak{L}(l_2, l_1)$ for all $s = (\sigma_k) \in l_2$, we obtain immediately $D_{st} = D_s D_t \in (\mathfrak{P}_2)^m(l_1, l_1)$. Hence, as in the previous proofs, we may conclude that $st \in l_{2/m}$. This implies that $t \in l_{2/(m-1)}$.

1.7. Nuclear operators

Throughout this section we assume that $0 < p \leqq 1$ and $0 < r \leqq 2$.

1.7.1. An operator $T \in \mathfrak{L}(E, F)$ is said to be ***p*-nuclear** if there exists a so-called ***p*-nuclear representation**

$$T = \sum_{i=1}^{\infty} a_i \otimes y_i$$

such that $a_1, a_2, \ldots \in E'$, $y_1, y_2, \ldots \in F$ and $(\|a_i\|\ \|y_i\|) \in l_p$. The set of these operators is denoted by $\mathfrak{N}_p(E, F)$. In the special case $p = 1$ we simply speak of a **nuclear** operator, and write $\mathfrak{N}(E, F)$ instead of $\mathfrak{N}_1(E, F)$.

For $T \in \mathfrak{N}_p(E, F)$ we define

$$\|T \mid \mathfrak{N}_p\| := \inf \left(\sum_{i=1}^{\infty} \|a_i\|^p\ \|y_i\|^p \right)^{1/p},$$

the infimum being taken over all representations described above.

1.7.2. The following result can easily be verified (PIE, 6.3.2 and 6.7.2).

Theorem (R. Oloff 1969). $\mathfrak{N}_p$ is the smallest p-Banach operator ideal.

1.7.3. We now give a very important characterization which relates p-nuclear operators to diagonal operators (PIE, 18.1.3).

Factorization theorem (GRO, Chap. II, p. 12). An operator $T \in \mathfrak{L}(E, F)$ is p-nuclear if and only if it admits a factorization

$$\begin{array}{ccc} E & \xrightarrow{\ T\ } & F \\ {\scriptstyle A}\big\downarrow & & \big\uparrow{\scriptstyle Y} \\ l_\infty & \xrightarrow[\ D_t\ ]{} & l_1 \end{array}$$

such that the following conditions are satisfied:

(1) D_t is a diagonal operator with $t \in l_p$.

(2) $A \in \mathfrak{L}(E, l_\infty)$ and $Y \in \mathfrak{L}(l_1, F)$.

When this is so, then

$$\|T \mid \mathfrak{N}_p\| = \inf \{\|Y\| \, \|t \mid l_p\| \, \|A\|\},$$

where the infimum is taken over all possible factorizations.

1.7.4. We now provide an auxiliary result.

Lemma. Assume that

$$\tau_i \geqq 0 \quad \text{and} \quad \sum_{i=1}^{\infty} \tau_i < \infty .$$

Given $\varepsilon > 0$, there exists (α_i) such that

$$0 < \alpha_i \leqq 1, \ \lim_i \alpha_i = 0 \quad \text{and} \quad \sum_{i=1}^{\infty} \alpha_i^{-1} \tau_i \leqq (1 + \varepsilon) \sum_{i=1}^{\infty} \tau_i .$$

Proof. Let $0 = i_0 < i_1 < \dots$ such that

$$\sum_{i_n+1}^{\infty} \tau_i \leqq \frac{\varepsilon}{4^n} \sum_{i=1}^{\infty} \tau_i \quad \text{for} \quad n = 1, 2, \dots$$

Define

$$\alpha_i = 2^{-n} \text{ if } i_n < i \leqq i_{n+1} \quad \text{and} \quad n = 0, 1, \dots$$

Then

$$\sum_{i=1}^{\infty} \alpha_i^{-1} \tau_i = \sum_{1}^{i_1} \tau_i + \sum_{n=1}^{\infty} 2^n \sum_{i_n+1}^{i_{n+1}} \tau_i \leqq \left(1 + \varepsilon \sum_{n=1}^{\infty} 2^{-n}\right) \sum_{i=1}^{\infty} \tau_i .$$

This completes the proof.

1.7.5. **Proposition.** The quasi-Banach operator ideal $\mathfrak{N}_p$ is approximative. More precisely, we have $\mathfrak{N}_p = \mathfrak{G} \circ \mathfrak{N}_p \circ \mathfrak{G}$.

Proof. Given $T \in \mathfrak{N}_p(E, F)$, we consider a factorization $T = Y D_t A$ as described in 1.7.3. In view of 1.7.4, the sequence $t \in l_p$ can be written in the form $t = ysa$ such that $s \in l_p$ and $a, y \in c_0$. Then it follows that $T = (Y D_y) D_s (D_a A)$, where $D_a A \in \mathfrak{G}(E, l_\infty)$ $D_s \in \mathfrak{N}_p(l_\infty, l_1)$ and $Y D_y \in \mathfrak{G}(l_1, F)$. Hence $T \in \mathfrak{G} \circ \mathfrak{N}_p \circ \mathfrak{G}(E, F)$.

1.7.6. Next we mention a fundamental result which is well-known as Auerbach's lemma; see (BAN, p. 238) and (PIE, B.4.8). First proofs were published by A. E. Taylor (1947) and A. F. Ruston (1962).

Lemma. Suppose that the Banach space E is n-dimensional. Then there exist $x_1, \dots, x_n \in E$ and $a_1, \dots, a_n \in E'$ such that

$$\|x_i\| = 1, \quad \|a_i\| = 1 \quad \text{and} \quad \langle x_i, a_j \rangle = \delta_{ij} .$$

1.7.7. We now prove an analogue of 1.2.8.

Lemma (A. F. Ruston 1962). Let $T \in \mathfrak{L}(E, F)$. Then

$$\|T \mid \mathfrak{N}_p\| \leqq n^{1/p} \|T\| \quad \text{whenever} \quad \operatorname{rank}(T) \leqq n .$$

Proof. We first treat the identity operator of an n-dimensional Banach space E. Choosing $x_1, \ldots, x_n \in E$ and $a_1, \ldots, a_n \in E'$ as in Auerbach's lemma yields the representation

$$I_E = \sum_{i=1}^{n} a_i \otimes x_i .$$

Hence $\|I_E \mid \mathfrak{N}_p\| \leqq n^{1/p}$.

Let $T \in \mathfrak{L}(E, F)$ be any finite operator. Put $M = M(T)$, and consider the canonical factorization $T = J_M^F T_0$, where $T_0 \in \mathfrak{L}(E, M)$ is the operator induced by T. Since $\|T_0\| = \|T\|$ and $\dim(M) = \operatorname{rank}(T)$, we get

$$\|T \mid \mathfrak{N}_p\| \leqq \|J_M^F\| \, \|I_M \mid \mathfrak{N}_p\| \, \|T_0\| \leqq n^{1/p} \|T\| .$$

Remark. If $\dim(E) = n$, then it follows from $\operatorname{trace}(I_E) = n$ that $\|I_E \mid \mathfrak{N}\| = n$. However, for $0 < p < 1$, the value of $\|I_E \mid \mathfrak{N}_p\|$ depends on the geometric structure of E.

1.7.8. An operator $T \in \mathfrak{L}(E, F)$ is said to be $(r, 2)$-**nuclear** if there exists a so-called $(r, 2)$-**nuclear representation**

$$T = \sum_{i=1}^{\infty} a_i \otimes y_i$$

such that $(a_i) \in [l_r, E']$ and $(y_i) \in [w_2, F]$. The set of these operators is denoted by $\mathfrak{N}_{r,2}(E, F)$. In the special case $r = 2$ we simply speak of a 2-**nuclear** operator, and write $\mathfrak{N}_2(E, F)$ instead of $\mathfrak{N}_{2,2}(E, F)$.

For $T \in \mathfrak{N}_{r,2}(E, F)$ we define

$$\|T \mid \mathfrak{N}_{r,2}\| := \inf \{\|(a_i) \mid l_r\| \, \|(y_i) \mid w_2\|\},$$

the infimum being taken over all representations described above.

Remark. In (PIE) the $(r, 2)$-nuclear operators are called $(r, 2, 1)$-nuclear.

1.7.9. For a proof of the following result we refer to (PIE, 18.1.2).

Theorem. $\mathfrak{N}_{r,2}$ is a p-Banach operator ideal, where $1/p = 1/r + 1/2$.

1.7.10. Several good properties of $(r, 2)$-nuclear operators are due to the fact that they factor nicely through the Hilbert space l_2.

Factorization theorem. An operator $T \in \mathfrak{L}(E, F)$ is $(r, 2)$-nuclear if and only if it admits a factorization

$$\begin{array}{ccc} E & \xrightarrow{\;T\;} & F \\ A \downarrow & & \uparrow Y \\ l_\infty & \xrightarrow[\;D_t\;]{} & l_2 \end{array}$$

such that the following conditions are satisfied:

(1) D_t is a diagonal operator with $t \in l_r$.

(2) $A \in \mathfrak{L}(E, l_\infty)$ and $Y \in \mathfrak{L}(l_2, F)$.

When this is so, then

$$\|T \mid \mathfrak{N}_{r,2}\| = \inf \{\|Y\| \, \|t \mid l_r\| \, \|A\|\},$$

where the infimum is taken over all possible factorizations.

1.7.11. The next result is analogous to 1.7.5.

Proposition. The quasi-Banach operator ideal $\mathfrak{N}_{r,2}$ is approximative. More precisely, we have $\mathfrak{N}_{r,2} = \mathfrak{G} \circ \mathfrak{N}_{r,2} \circ \mathfrak{G}$.

1.7.12. We now establish a stability property which turns out to be quite useful for the theory of eigenvalue distributions.

Theorem (J. R. Holub 1970/74). The quasi-Banach operator ideal $\mathfrak{N}_{r,2}$ is stable with respect to the tensor norm ε. More precisely,

$$\|S \tilde{\otimes}_\varepsilon T \mid \mathfrak{N}_{r,2}\| \leqq \|S \mid \mathfrak{N}_{r,2}\| \, \|T \mid \mathfrak{N}_{r,2}\|$$

for $S \in \mathfrak{N}_{r,2}(E, E_0)$ and $T \in \mathfrak{N}_{r,2}(F, F_0)$.

Proof. Given $\delta > 0$, we consider representations

$$S = \sum_{i=1}^{\infty} a_i \otimes x_i \quad \text{and} \quad T = \sum_{j=1}^{\infty} b_j \otimes y_j$$

such that

$$\|(a_i) \mid [l_r, E']\| \leqq \|S \mid \mathfrak{N}_{r,2}\| \quad \text{and} \quad \|(b_j) \mid [l_r, F']\| \leqq \|T \mid \mathfrak{N}_{r,2}\|$$

$$\|(x_i) \mid [w_2, E_0]\| \leqq 1 + \delta \quad \text{and} \quad \|(y_j) \mid [w_2, F_0]\| \leqq 1 + \delta .$$

Obviously,

$$\begin{aligned} \|(a_i \otimes b_j) \mid [l_r(\mathbb{N} \times \mathbb{N}), (E \tilde{\otimes}_\varepsilon F)']\| &\leqq \|(a_i) \mid [l_r, E']\| \, \|(b_j) \mid [l_r, F']\| \\ &\leqq \|S \mid \mathfrak{N}_{r,2}\| \, \|T \mid \mathfrak{N}_{r,2}\| . \end{aligned}$$

Furthermore, for all $(\gamma_{ij}) \in l_2(\mathbb{N} \times \mathbb{N})$, we have

$$\begin{aligned} &\left\| \sum_{i=1}^{\infty} \sum_{j=1}^{\infty} \gamma_{ij} x_i \otimes y_j \right\| \\ &= \sup \left\{ \left| \sum_{i=1}^{\infty} \sum_{j=1}^{\infty} \gamma_{ij} \langle x_i, a \rangle \langle y_i, b \rangle \right| : \|a\| \leqq 1, \|b\| \leqq 1 \right\} \\ &\leqq (1 + \delta)^2 \left(\sum_{i=1}^{\infty} \sum_{j=1}^{\infty} |\gamma_{ij}|^2 \right)^{1/2} . \end{aligned}$$

Thus, in view of 1.1.14, the double sequence $(x_i \tilde{\otimes} y_j)$ is weakly 2-summable in $E_0 \tilde{\otimes}_\varepsilon F_0$ and

$$\|(x_i \otimes y_j) \mid [w_2(\mathbb{N} \times \mathbb{N}), E_0 \tilde{\otimes}_\varepsilon F_0]\| \leqq (1 + \delta)^2 .$$

Therefore

$$S \tilde{\otimes}_\varepsilon T = \sum_{i=1}^{\infty} \sum_{j=1}^{\infty} (a_i \otimes b_j) \otimes (x_i \otimes y_j)$$

is an $(r, 2)$-nuclear operator. Moreover, letting $\delta \to 0$ yields

$$\|S \tilde{\otimes}_\varepsilon T \mid \mathfrak{N}_{r,2}\| \leqq \|S \mid \mathfrak{N}_{r,2}\| \, \|T \mid \mathfrak{N}_{r,2}\| .$$

1.7.13. The following result is obvious.

Proposition. If $1/p = 1/r + 1/2$, then $\mathfrak{N}_p \subseteq \mathfrak{N}_{r,2}$.

Remark. Note that $\mathfrak{N}_p(l_2) \neq \mathfrak{N}_{r,2}(l_2)$, by 2.11.26 and 2.11.27. Therefore the above inclusion is strict.

1.7.14. Next we investigate the relationship between the Banach operator ideals $\mathfrak{P}_2$ and $\mathfrak{N}_2$.

Theorem (A. Pietsch 1968). $\mathfrak{P}_2^{(a)} = \mathfrak{N}_2$. The norms on both sides are equal.

Proof. Let $T \in \mathfrak{N}_2(E, F)$. Given $\varepsilon > 0$, there exists a representation

$$T = \sum_{i=1}^{\infty} a_i \otimes y_i$$

such that

$$\|(a_i) \mid l_2\| \leqq (1 + \varepsilon) \|T \mid \mathfrak{N}_2\| \quad \text{and} \quad \|(y_i) \mid w_2\| \leqq 1.$$

Then, for $x \in E$ and $b \in F'$, we have

$$|\langle Tx, b\rangle| = \left|\sum_{i=1}^{\infty} \langle x, a_i\rangle \langle y_i, b\rangle\right| \leqq \left(\sum_{i=1}^{\infty} |\langle x, a_i\rangle|^2\right)^{1/2} \left(\sum_{i=1}^{\infty} |\langle y_i, b\rangle|^2\right)^{1/2}.$$

This implies that

$$\|Tx\| \leqq \left(\sum_{i=1}^{\infty} |\langle x, a_i\rangle|^2\right)^{1/2} \quad \text{for all } x \in E.$$

Hence, by 1.3.4,

$$T \in \mathfrak{P}_2(E, F) \quad \text{and} \quad \|T \mid \mathfrak{P}_2\| \leqq \|(a_i) \mid l_2\| \leqq (1 + \varepsilon) \|T \mid \mathfrak{N}_2\|.$$

Letting $\varepsilon \to 0$ yields

$$\|T \mid \mathfrak{P}_2\| \leqq \|T \mid \mathfrak{N}_2\| \quad \text{for all } T \in \mathfrak{N}_2(E, F).$$

This conclusion can also be inferred from 1.7.10 and 1.6.3.

We now assume that $T \in \mathfrak{F}(E, F)$. By 1.5.2, there exists a factorization

$$\begin{array}{ccc} E & \xrightarrow{\quad T \quad} & F \\ {\scriptstyle A}\downarrow & & \uparrow{\scriptstyle Y} \\ C(X) & \xrightarrow[\quad I \quad]{} & L_2(X, \mu) \end{array}$$

such that $\|A\| = \|T \mid \mathfrak{P}_2\|$ and $\|Y\| = 1$. Moreover, as remarked in 1.5.1, we may arrange that Y has finite rank. Let

$$Y = \sum_{i=1}^{m} f_i \otimes y_i$$

be any representation. Given $\varepsilon > 0$, we choose simple functions $s_1, \ldots, s_m$ such that

$$\sum_{i=1}^{m} \|f_i - s_i \mid L_2\| \, \|y_i\| \leqq \varepsilon.$$

Define

$$Z := \sum_{i=1}^{m} s_i \otimes y_i.$$

Note that

$$\|Y - Z\| \leqq \|Y - Z \mid \mathfrak{N}_2\| \leqq \|Y - Z \mid \mathfrak{N}\| \leqq \varepsilon$$

and

$$\|Z\| \leqq \|Y\| + \|Z - Y\| \leqq 1 + \varepsilon.$$

Write

$$s_i = \sum_{j=1}^{n} \alpha_{ij} h_j,$$

where $h_1, \dots, h_n$ are characteristic functions of pairwise disjoint measurable subsets $X_1, \dots, X_n$ of X. If

$$u_j := \mu(X_j)^{-1/2} h_j \quad \text{and} \quad z_j := \mu(X_j)^{1/2} \sum_{i=1}^{m} \alpha_{ij} y_i,$$

then

$$Z = \sum_{j=1}^{n} u_j \otimes z_j \quad \text{and} \quad Zu_j = z_j.$$

Since (u_j) is an orthonormal family in $L_2(X, \mu)$, by 1.1.9 and 1.1.11, we have

$$\|(z_j) \mid w_2\| \leqq \|Z\| \, \|(u_j) \mid w_2\| \leqq 1 + \varepsilon.$$

Recall that I denotes the canonical operator from $C(X)$ into $L_2(X, \mu)$. Thus $I'u_j$ is the functional defined by u_j on $C(X)$, and we obtain

$$\|(I'u_j) \mid l_2\| = \left(\sum_{j=1}^{n} \|u_j \mid L_1\|^2\right)^{1/2} = \left(\sum_{j=1}^{n} \mu(X_j)\right)^{1/2} \leqq \mu(X)^{1/2} = 1.$$

In view of the preceding results, it follows from

$$ZI = \sum_{j=1}^{n} I'u_j \otimes z_j$$

that

$$\|ZI \mid \mathfrak{N}_2\| \leqq 1 + \varepsilon.$$

Next we get

$$\|YI \mid \mathfrak{N}_2\| \leqq \|ZI \mid \mathfrak{N}_2\| + \|(Y - Z) I \mid \mathfrak{N}_2\| \leqq 1 + 2\varepsilon.$$

Therefore

$$\|T \mid \mathfrak{P}_2\| \leqq \|T \mid \mathfrak{N}_2\| = \|YIA \mid \mathfrak{N}_2\| \leqq (1 + 2\varepsilon) \, \|A\| \leqq (1 + 2\varepsilon) \, \|T \mid \mathfrak{P}_2\|.$$

Letting $\varepsilon \to 0$ yields

$$\|T \mid \mathfrak{P}_2\| = \|T \mid \mathfrak{N}_2\| \quad \text{for all } T \in \mathfrak{F}(E, F).$$

Finally, we consider any operator $T \in \mathfrak{P}_2^{(a)}(E, F)$. Then, by definition, there exists a sequence of operators $T_k \in \mathfrak{F}(E, F)$ converging to T with respect to the norm of $\mathfrak{P}_2$. We see from

$$\|T_h - T_k \mid \mathfrak{N}_2\| = \|T_h - T_k \mid \mathfrak{P}_2\|$$

that (T_k) is a Cauchy sequence in the Banach operator ideal $\mathfrak{N}_2(E, F)$. Since T is the only possible limit, it follows that $T \in \mathfrak{N}_2(E, F)$. Moreover, we have

$$\|T \mid \mathfrak{N}_2\| = \lim_k \|T_k \mid \mathfrak{N}_2\| = \lim_k \|T_k \mid \mathfrak{P}_2\| = \|T \mid \mathfrak{P}_2\|.$$

1.7.15. The following result can be traced back to the the famous "Resumé" of

A. Grothendieck (1956:b) which contains the statement that every operator of the form $C \to L_2 \to C \to L_2$ is nuclear.

Theorem (H. Jarchow / R. Ott 1982). $(\mathfrak{P}_2)^2 = \mathfrak{H} \circ \mathfrak{N}$. The quasi-norms on both sides are equal.

Proof. We first assume that $T \in \mathfrak{N}_2(E, F)$ and $S \in \mathfrak{P}_2(F, G)$. Given $\varepsilon > 0$, there exists a representation

$$T = \sum_{i=1}^{\infty} a_i \otimes y_i$$

such that

$$\|(a_i) \mid l_2\| \leqq (1 + \varepsilon)\, \|T \mid \mathfrak{N}_2\| \quad \text{and} \quad \|(y_i) \mid w_2\| \leqq 1 .$$

Hence

$$ST = \sum_{i=1}^{\infty} a_i \otimes Sy_i$$

is a nuclear operator with

$$\|ST \mid \mathfrak{N}\| \leqq \sum_{i=1}^{\infty} \|a_i\|\, \|Sy_i\| \leqq (1 + \varepsilon)\, \|S \mid \mathfrak{P}_2\|\, \|T \mid \mathfrak{N}_2\| .$$

This proves that $\mathfrak{P}_2 \circ \mathfrak{N}_2 \subseteq \mathfrak{N}$. By 1.7.11, we have $\mathfrak{N}_2 = \mathfrak{N}_2 \circ \mathfrak{G}$. Hence it follows from 1.7.14 that

$$\mathfrak{N}_2 \circ \mathfrak{P}_2 \subseteq \mathfrak{N}_2 \circ \mathfrak{G} \circ \mathfrak{P}_2 \subseteq \mathfrak{N}_2 \circ \mathfrak{P}_2^{(a)} \subseteq \mathfrak{P}_2 \circ \mathfrak{N}_2 \subseteq \mathfrak{N} .$$

Moreover, 1.4.4 and 1.4.5 imply that $\mathfrak{H} \circ \mathfrak{P}_2 \circ \mathfrak{H} \subseteq \mathfrak{P}_2^{(a)} = \mathfrak{N}_2$.

Combining these results, we deduce from 1.5.3 that

$$(\mathfrak{P}_2)^2 = \mathfrak{H} \circ \mathfrak{P}_2 \circ \mathfrak{H} \circ \mathfrak{P}_2 \subseteq \mathfrak{H} \circ \mathfrak{N}_2 \circ \mathfrak{P}_2 \subseteq \mathfrak{H} \circ \mathfrak{N} .$$

Carefully looking at the quasi-norms involved, we obtain

$$\|T \mid \mathfrak{H} \circ \mathfrak{N}\| \leqq \|T \mid (\mathfrak{P}_2)^2\| \quad \text{for all } T \in (\mathfrak{P}_2)^2 (E, F) .$$

Conversely, let $T \in \mathfrak{H} \circ \mathfrak{N} (E, F)$. Given $\varepsilon > 0$, there exists a factorization

$$\begin{array}{ccccc} E & & \xrightarrow{\quad T \quad} & & F \\ {\scriptstyle A}\downarrow & & & & \uparrow{\scriptstyle Y} \\ l_\infty & \xrightarrow[D_t]{} & l_1 & \xrightarrow[B]{} & H \end{array}$$

such that

$$\|Y\|\, \|B\|\, \|t \mid l_1\|\, \|A\| \leqq (1 + \varepsilon)\, \|T \mid \mathfrak{H} \circ \mathfrak{N}\| .$$

Choose $s \in l_2$ with $s^2 = t$. Then, by 1.6.3, we have

$$D_s \in \mathfrak{P}_2(l_\infty, l_2) \quad \text{and} \quad D_s \in \mathfrak{P}_2'(l_2, l_1) .$$

Hence, in view of 1.3.13, it follows that

$$T = Y(BD_s)\, D_s A \in (\mathfrak{P}_2)^2 (E, F)$$

and

$$\|T \mid (\mathfrak{P}_2)^2\| \leqq \|Y\| \, \|BD_s \mid \mathfrak{P}_2\| \, \|D_s \mid \mathfrak{P}_2\| \, \|A\|$$
$$\leqq \|Y\| \, \|B\| \, \|s \mid l_2\|^2 \, \|A\| \leqq (1 + \varepsilon) \, \|T \mid \mathfrak{H} \circ \mathfrak{N}\| .$$

This completes the proof.

1.7.16. The following (strict) inclusion is obvious.

Proposition. $\mathfrak{N}_{2/m,2} \subseteqq (\mathfrak{P}_2)^m$.

1.7.17. We conclude this section with a fundamental result; see 1.5.5.

Lemma (D. J. H. Garling / Y. Gordon 1971). Let N be any n-codimensional subspace of E. Then, given $\varepsilon > 0$, there exists a projection $P \in \mathfrak{L}(E)$ such that $N(P) = N$ and $\|P\| \leqq (1 + \varepsilon) \, n^{1/2}$.

Proof. It follows from 1.2.8 and 1.7.14 that

$$\|I'_{E/N} \mid \mathfrak{N}_2\| = \|I'_{E/N} \mid \mathfrak{P}_2\| \leqq n^{1/2} .$$

Thus, identifying E/N and $(E/N)''$, we can find a representation

$$I_{E/N} = \sum_{i=1}^{\infty} \hat{a}_i \otimes \hat{x}_i$$

such that

$$\|(\hat{a}_i) \mid [w_2, (E/N)']\| \leqq 1 \quad \text{and} \quad \|(\hat{x}_i) \mid [l_2, E/N]\| \leqq (1 + \varepsilon)^{1/2} n^{1/2} .$$

Choose $x_i \in E$ such that

$$Q_N^E x_i = \hat{x}_i \quad \text{and} \quad \|x_i\| \leqq (1 + \varepsilon)^{1/2} \, \|\hat{x}_i\| .$$

Define

$$S := \sum_{i=1}^{\infty} \hat{a}_i \otimes x_i .$$

Then

$$\|(\hat{a}_i) \mid [w_2, (E/N)']\| \leqq 1 \quad \text{and} \quad \|(x_i) \mid [l_2, E]\| \leqq (1 + \varepsilon) \, n^{1/2}$$

imply that

$$\|S\| = \|S'\| \leqq \|S' \mid \mathfrak{N}_2\| \leqq (1 + \varepsilon) \, n^{1/2} .$$

We now show that $P := SQ_N^E$ is the desired projection. To this end, observe that $Q_N^E S = I_{E/N}$. Therefore

$$P^2 = S(Q_N^E S) \, Q_N^E = SQ_N^E = P .$$

Next we deduce from

$$SQ_N^E = P \quad \text{and} \quad Q_N^E P = (Q_N^E S) \, Q_N^E = Q_N^E$$

that $N(P) = N(Q_N^E) = N$. This completes the proof.

1.7.18. Finally, we give an example of a non-trivial nuclear representation.

Lemma (A. Grothendieck 1956: c).
$\|I : l_1(m) \to l_\infty(m) \mid \mathfrak{N}\| = 1$.

Proof. Note that

$$I = 2^{-m} \sum_{e} e \otimes e,$$

where the sum is taken over all vectors $e = (\varepsilon_1, \ldots, \varepsilon_m)$ with $\varepsilon_k = \pm 1$. Therefore

$$\|I : l_1(m) \to l_\infty(m) \mid \mathfrak{N}\| \leqq 2^{-m} \sum_{e} \|e \mid l_1(m)'\| \, \|e \mid l_\infty(m)\| = 1 .$$

The reverse inequality is obvious.

1.7.19. Example. $D_t \in \mathfrak{N}(l_1, l_\infty)$ if and only if $t \in c_0$.

Proof. Assume that $t = (\tau_k) \in c_0$, and let D_n denote the diagonal operator induced by the finite sequence $t_n := (\tau_1, \ldots, \tau_n, 0, \ldots)$. By the preceding lemma, we have

$$\|D_m - D_n \mid \mathfrak{N}\| \leqq \|t_m - t_n \mid l_\infty\| \quad \text{whenever } m > n .$$

Thus (D_n) is a Cauchy sequence in $\mathfrak{N}(l_1, l_\infty)$. Since D_t is the only candidate for the limit, it follows that $D_t \in \mathfrak{N}(l_1, l_\infty)$.

Conversely, $D_t \in \mathfrak{N}(l_1, l_\infty)$ implies $D_t \in \mathfrak{K}(l_1, l_\infty)$. Hence $t \in c_0$.

CHAPTER 2

s-Numbers

This chapter is devoted to the theory of s-numbers and the operator ideals determined by them.

First of all, we provide some basic facts about Lorentz sequence spaces.

Next the concept of s-numbers is introduced axiomatically. It turns out that there exist many possibilities of assigning to every operator T a certain sequence of numbers $s_1(T) \geqq s_2(T) \geqq \ldots \geqq 0$ which characterizes its degree of approximability or compactness. In this way the quasi-Banach operator ideals

$$\mathfrak{L}_{r,w}^{(s)} := \{T \in \mathfrak{L} : (n^{1/r-1/w} s_n(T)) \in l_w\}$$

are obtained.

In the following sections we treat several important examples of s-numbers. The simplest ones are the approximation numbers. Next we deal with the Gel'fand and Weyl numbers. Their dual counterparts, called Kolmogorov and Chang numbers, are introduced, as well. As the last example we mention the Hilbert numbers which are less interesting for later applications within the theory of eigenvalue distributions.

Next we investigate the relationship between the operator ideals $\mathfrak{P}_{r,2}$ and $\mathfrak{L}_{r,w}^{(s)}$. Generalized approximation numbers associated with an arbitrary quasi-Banach operator ideal are treated subsequently.

As in the preceding chapter, we also consider diagonal operators from l_p into l_q which yield instructive examples. Furthermore, we establish inclusions between operator ideals $\mathfrak{L}_{p,w}^{(s)}$ and $\mathfrak{L}_{q,w}^{(t)}$ determined by different s-numbers.

The last section is concerned with operators on Hilbert spaces. In this special situation the s-numbers are uniquely determined by their axiomatic properties. Therefore, by restricting $\mathfrak{L}_{r,w}^{(s)}$ to the class of Hilbert spaces, for any choice of s-numbers we obtain one and the same class. These are the famous Schatten von Neumann ideals $\mathfrak{S}_{r,w}$ which have played an important role as the historical starting point of the whole theory. Conversely, given any operator ideal defined on the class of Hilbert spaces, we look for extensions to the class of all Banach spaces. In particular, the largest extension of $\mathfrak{S}_{r,w}$ is investigated.

Some results about operators on Hilbert space are used in the earlier sections. For this reason, it is a good idea, for the first reading at least, to proceed to Section 11 immediately after Section 4 and Proposition 2.9.4.

If not otherwise specified, we assume throughout this chapter that $0 < r < \infty$ and $0 < w \leqq \infty$. All proofs are carried out for $0 < w < \infty$, while the obvious modifications in the case $w = \infty$ are left to the reader. Moreover, m and n denote arbitrary natural numbers.

2.1. Lorentz sequence spaces

2.1.1.* A complex-valued family $x = (\xi_i)$ on a given index set I is said to be **finite** if it possesses only finitely many non-zero coordinates. The cardinality of $\{i \in I : \xi_i \neq 0\}$ is denoted by card (x).

2.1.2.* The n-th **approximation number** of $x \in l_\infty(I)$ is defined by

$$a_n(x) := \inf \{\|x - u \mid l_\infty(I)\| : u \in l_\infty(I),\ \text{card}\,(u) < n\}.$$

In the particular case when $x = (\xi_i)$ is a sequence such that $|\xi_1| \geqq |\xi_2| \geqq \ldots \geqq 0$ we have $a_n(x) = |\xi_n|$. Therefore $(a_n(x))$ is often called the **non-increasing rearrangement** of x.

Remark. Note that

$$a_n(x) = \inf\{c \geqq 0 : \operatorname{card}(i \in I : |\xi_i| \geqq c) < n\}.$$

2.1.3.* Next we list some basic properties of the approximation numbers just defined.

Proposition.

(1) $$\|x \mid l_\infty(I)\| = a_1(x) \geqq a_2(x) \geqq \ldots \geqq 0 \quad \text{for } x \in l_\infty(I).$$

(2) $$a_{m+n-1}(x + y) \leqq a_m(x) + a_n(y) \quad \text{for } x, y \in l_\infty(I).$$

(3) $$a_{m+n-1}(xy) \leqq a_m(x)\, a_n(y) \quad \text{for } x, y \in l_\infty(I).$$

(4) If $\operatorname{card}(x) < n$, then $a_n(x) = 0$.

2.1.4.* The **Lorentz space** $l_{r,w}(I)$ consists of all complex-valued families $x = (\xi_i)$ such that $(n^{1/r-1/w}a_n(x)) \in l_w$.

For $x \in l_{r,w}(I)$ we define

$$\|x \mid l_{r,w}(I)\| := \|(n^{1/r-1/w}a_n(x)) \mid l_w\|.$$

More explicitly,

$$\|x \mid l_{r,w}(I)\| := \left(\sum_{n=1}^{\infty} [n^{1/r-1/w}a_n(x)]^w\right)^{1/w}$$

if $0 < w < \infty$ and

$$\|x \mid l_{r,\infty}(I)\| := \sup\{n^{1/r}a_n(x) : n \in \mathbb{N}\}.$$

To simplify notation, we write $l_{r,w}$ instead of $l_{r,w}(\mathbb{N})$.

Remark. The above definition makes also sense for $r = \infty$.

2.1.5.* We state the following result without proof.

Theorem. $l_{r,w}(I)$ is a quasi-Banach spaces.

2.1.6.* In order to make subsequent considerations easier to understand, we formulate a tautology.

Proposition. Let $x \in l_\infty(I)$. Then

$$x \in l_{r,w}(I) \quad \text{if and only if} \quad (a_n(x)) \in l_{r,w}.$$

Moreover,

$$\|x \mid l_{r,w}(I)\| = \|(a_n(x)) \mid l_{r,w}\|.$$

2.1.7.* The next result is closely related to Hardy's inequality; see (HAY, p. 239).

Proposition. Assume that $0 < p < r < \infty$. Let $x \in l_\infty(I)$. Then

$$x \in l_{r,w}(I) \text{ if and only if } \left(\left(\frac{1}{n}\sum_{k=1}^{n} a_k(x)^p\right)^{1/p}\right) \in l_{r,w}.$$

Moreover,

$$\|x \mid l_{r,w}(I)\|_p^{\text{mean}} := \left\| \left(\left(\frac{1}{n}\sum_{k=1}^{n} a_k(x)^p\right)^{1/p}\right) \middle| l_{r,w} \right\|$$

defines an equivalent quasi-norm on $l_{r,w}(I)$.

Proof. It follows from $a_1(x) \geqq \ldots \geqq a_n(x)$ that

$$a_n(x) \leqq \left(\frac{1}{n}\sum_{k=1}^{n} a_k(x)^p\right)^{1/p}.$$

Consequently,

$$\|x \mid l_{r,w}(I)\| \leqq \|x \mid l_{r,w}(I)\|_p^{\text{mean}}.$$

Hence

$$\left(\left(\frac{1}{n}\sum_{k=1}^{n} a_k(x)^p\right)^{1/p}\right) \in l_{r,w} \quad \text{implies} \quad x \in l_{r,w}(I).$$

To verify the converse, we define $1/u := 1/p + 1/w$. Choose any number ϱ with $1/r < \varrho < 1/p$. We deduce from

$$k a_k(x)^p \leqq \sum_{h=1}^{n} a_h(x)^p \quad \text{for } k = 1, \ldots, n$$

and Hölder's inequality that

$$\left(\sum_{k=1}^{n} a_k(x)^p\right)^{1/u} = \left(\sum_{k=1}^{n} [k^{-\varrho}(k^{\varrho-1/w} a_k(x))\,(k a_k(x)^p)^{1/w}]^u\right)^{1/u}$$
$$\leqq \left(\sum_{k=1}^{n} k^{-\varrho p}\right)^{1/p} \left(\sum_{k=1}^{n} [k^{\varrho-1/w} a_k(x)]^w\right)^{1/w} \left(\sum_{h=1}^{n} a_h(x)^p\right)^{1/w}.$$

Therefore, by G.3.2,

$$\left(\sum_{k=1}^{n} a_k(x)^p\right)^{1/p} \leqq c_0 n^{1/p-\varrho} \left(\sum_{k=1}^{n} [k^{\varrho-1/w} a_k(x)]^w\right)^{1/w}.$$

Applying G.3.2 again, we finally obtain

$$\|x \mid l_{r,w}(I)\|_p^{\text{mean}} = \left(\sum_{n=1}^{\infty} \left[n^{1/r-1/w} \left(\frac{1}{n}\sum_{k=1}^{n} a_k(x)^p\right)^{1/p}\right]^w\right)^{1/w}$$
$$\leqq c_0 \left(\sum_{n=1}^{\infty} n^{w/r-1-w\varrho} \left[\sum_{k=1}^{n} k^{w\varrho-1} a_k(x)^w\right]\right)^{1/w}$$
$$= c_0 \left(\sum_{k=1}^{\infty} k^{w\varrho-1} a_k(x)^w \left[\sum_{n=k}^{\infty} n^{w/r-w\varrho-1}\right]\right)^{1/w}$$
$$\leqq c \left(\sum_{k=1}^{\infty} k^{w/r-1} a_k(x)^w\right)^{1/w} = c\|x \mid l_{r,w}(I)\|.$$

This shows that

$$x \in l_{r,w}(I) \quad \text{implies} \quad \left(\left(\frac{1}{n}\sum_{k=1}^{n} a_k(x)^p\right)^{1/p}\right) \in l_{r,w}.$$

Remark. If $1 \leqq p < r < \infty$ and $1 \leqq w \leqq \infty$, then $\|\cdot \mid l_{r,w}(I)\|_p^{\text{mean}}$ is even a norm.

2.1.8.* The following result can be viewed as the limiting case of the preceding proposition as $p \to 0$; see G.1.2.

Proposition. Let $x \in l_\infty(I)$. Then

$$x \in l_{r,w}(I) \quad \text{if and only if} \quad \left(\left(\prod_{k=1}^{n} a_k(x)\right)^{1/n}\right) \in l_{r,w}.$$

Moreover,

$$\|x \mid l_{r,w}(I)\|_0^{\text{mean}} := \left\| \left(\left(\prod_{k=1}^{n} a_k(x) \right)^{1/n} \right) \Big| \, l_{r,w} \right\|$$

defines an equivalent quasi-norm on $l_{r,w}(I)$.

Proof. It follows from $a_1(x) \geqq \ldots \geqq a_n(x)$ that

$$a_n(x) \leqq \left(\prod_{k=1}^{n} a_k(x) \right)^{1/n}.$$

Consequently,

$$\|x \mid l_{r,w}(I)\| \leqq \|x \mid l_{r,w}(I)\|_0^{\text{mean}}.$$

Hence

$$\left(\left(\prod_{k=1}^{n} a_k(x) \right)^{1/n} \right) \in l_{r,w} \quad \text{implies} \quad x \in l_{r,w}(I).$$

Conversely, we know from the inequality of means that

$$\left(\prod_{k=1}^{n} a_k(x) \right)^{1/n} \leqq \left(\frac{1}{n} \sum_{k=1}^{n} a_k(x)^p \right)^{1/p}.$$

Fix any p with $0 < p < r$. Then, by 2.1.7,

$$\|x \mid l_{r,w}(I)\|_0^{\text{mean}} \leqq \|x \mid l_{r,w}(I)\|_p^{\text{mean}} \leqq c \| \, x \mid l_{r,w}(I)\| .$$

This shows that

$$x \in l_{r,w}(I) \quad \text{implies} \quad \left(\left(\prod_{k=1}^{n} a_k(x) \right)^{1/n} \right) \in l_{r,w}.$$

Remark. The implication just established can also be deduced from Carleman's inequality

$$\sum_{n=1}^{\infty} \left(\prod_{k=1}^{n} \alpha_k \right)^{1/n} \leqq e \sum_{n=1}^{\infty} \alpha_n \quad \text{for} \quad \alpha_1, \alpha_2, \ldots \geqq 0;$$

see (HAY, p. 249).

2.1.9.* In the following it is shown that we only need to know the approximation numbers $a_n(x)$ on a certain subset of indices in order to decide whether or not a given family $x \in l_\infty(I)$ belongs to some Lorentz space. First we mention the case in which n runs through an arithmetic progression. For later applications it is enough to deal with the subset of odd numbers.

Proposition. Let $x \in l_\infty(I)$. Then

$$x \in l_{r,w}(I) \quad \text{if and only if} \quad (a_{2n-1}(x)) \in l_{r,w}.$$

Moreover,

$$\|x \mid l_{r,w}(I)\|^{\text{odd}} := \|(a_{2n-1}(x)) \mid l_{r,w}\|$$

defines an equivalent quasi-norm on $l_{r,w}(I)$.

Proof. Note that

$$(2n-1)^{1/r-1/w} \leqq (2n)^{1/r-1/w} = 2^{1/r-1/w} n^{1/r-1/w} \quad \text{if} \quad r \leqq w$$

and

$$(2n)^{1/r-1/w} \leqq (2n-1)^{1/r-1/w} \leqq n^{1/r-1/w} \quad \text{if} \quad r \geqq w.$$

Therefore we have

$$\begin{aligned} \|x \mid l_{r,w}(I)\| &= \left(\sum_{n=1}^{\infty} [n^{1/r-1/w} a_n(x)]^w\right)^{1/w} \\ &= \left(\sum_{n=1}^{\infty} [(2n-1)^{1/r-1/w} a_{2n-1}(x)]^w + \sum_{n=1}^{\infty} [(2n)^{1/r-1/w} a_{2n}(x)]^w\right)^{1/w} \\ &\leqq \max(2^{1/r}, 2^{1/w}) \left(\sum_{n=1}^{\infty} [n^{1/r-1/w} a_{2n-1}(x)]^w\right)^{1/w} \\ &= \max(2^{1/r}, 2^{1/w}) \, \|x \mid l_{r,w}(I)\|^{\text{odd}}. \end{aligned}$$

Thus

$$(a_{2n-1}(x)) \in l_{r,w} \quad \text{implies} \quad (a_n(x)) \in l_{r,w}.$$

Conversely, it follows from $a_{2n-1}(x) \leqq a_n(x)$ that

$$\begin{aligned} \|x \mid l_{r,w}(I)\|^{\text{odd}} &= \left(\sum_{n=1}^{\infty} [n^{1/r-1/w} a_{2n-1}(x)]^w\right)^{1/w} \\ &\leqq \left(\sum_{n=1}^{\infty} [n^{1/r-1/w} a_n(x)]^w\right)^{1/w} = \|x \mid l_{r,w}(I)\|. \end{aligned}$$

Consequently,

$$(a_n(x)) \in l_{r,w} \quad \text{implies} \quad (a_{2n-1}(x)) \in l_{r,w}.$$

2.1.10.* Next we consider the case in which the index n runs through a geometric progression (q^k) with $k = 0, 1, \ldots$ and any fixed $q = 2, 3, \ldots$
This result is closely related to Cauchy's condensation test.

Proposition. Let $x \in l_\infty(I)$. Then

$$x \in l_{r,w}(I) \quad \text{if and only if} \quad (q^{k/r} a_{q^k}(x)) \in l_w.$$

Moreover,

$$\|x \mid l_{r,w}(I)\|_q^{\text{geom}} := \| (q^{k/r} a_{q^k}(x)) \mid l_w \|$$

defines an equivalent quasi-norm on $l_{r,w}(I)$.

Proof. Let

$$U_k := \{n : q^k \leqq n < q^{k+1}\}.$$

It follows from

$$\sum_{U_k} n^{w/r-1} \asymp q^{kw/r}$$

that

$$\|x \mid l_{r,w}(I)\| = \left(\sum_{k=0}^{\infty} \sum_{U_k} [n^{1/r-1/w} a_n(x)]^w\right)^{1/w}$$

$$\leqq \left(\sum_{k=0}^{\infty} \left[\sum_{U_k} n^{w/r-1}\right] a_{q^k}(x)^w\right)^{1/w}$$

$$\leqq c \left(\sum_{k=0}^{\infty} [q^{k/r} a_{q^k}(x)]^w\right)^{1/w}$$

$$= c\, \|x \mid l_{r,w}(I)\|_q^{\text{geom}}.$$

Consequently,

$$(q^{k/r} a_{q^k}(x)) \in l_w \quad \text{implies} \quad x \in l_{r,w}(I).$$

To verify the converse, let

$$V_k := \{n : q^k < n \leqq q^{k+1}\}.$$

Then

$$\sum_{V_k} n^{w/r-1} \asymp q^{(k+1)w/r}.$$

We therefore obtain

$$\|x \mid l_{r,w}(I)\|_q^{\text{geom}} = \left(\sum_{k=0}^{\infty} [q^{k/r} a_{q^k}(x)]^w\right)^{1/w}$$

$$\leqq \left(a_1(x)^w + \sum_{k=0}^{\infty} [q^{(k+1)/r} a_{q^{k+1}}(x)]^w\right)^{1/w}$$

$$\leqq c \left(a_1(x)^w + \sum_{k=0}^{\infty} \left[\sum_{V_k} n^{w/r-1}\right] a_{q^{k+1}}(x)^w\right)^{1/w}$$

$$\leqq c \left(a_1(x)^w + \sum_{k=0}^{\infty} \sum_{V_k} [n^{1/r-1/w} a_n(x)]^w\right)^{1/w}$$

$$= c\, \|x \mid l_{r,w}(I)\|.$$

Hence

$$x \in l_{r,w}(I) \quad \text{implies} \quad (q^{k/r} a_{q^k}(x)) \in l_w.$$

2.1.11.* As an immediate consequence of the preceding proposition we see that the 2-parameter scale of Lorentz spaces is lexicographically ordered.

Proposition.

$l_{r_0,w_0}(I) \subseteq l_{r_1,w_1}(I)$ for $0 < r_0 < r_1 < \infty$ and arbitrary w_0, w_1,

$l_{r,w_0}(I) \subseteq l_{r,w_1}(I)$ for arbitrary r and $0 < w_0 < w_1 \leqq \infty$.

2.1.12.* The following example shows that the inclusions just stated are strict. If $0 < r < \infty$ and $\alpha \geqq 0$, then we let

$$c_{r,\alpha} := \left(1; \ldots; \frac{1}{2^{k/r}(k+1)^\alpha}, \ldots, \frac{1}{2^{k/r}(k+1)^\alpha}; \ldots\right),$$

where the k-th term is repeated 2^k-times.

Applying 2.1.10 with $q = 2$, we see immediately that

$$c_{r,\alpha} \in l_{r,w} \quad \text{if and only if} \quad \alpha w > 1 \quad \text{and } c_{r,0} \in l_{r,\infty}.$$

The same result holds for the sequences

$$d_{r,\alpha} := \left(\frac{1}{n^{1/r}(1 + \log n)^\alpha}\right).$$

2.1.13.* Let $l_{p,u}(I) \circ l_{q,v}(I)$ denote the collection of all complex-valued families x which can be written in the form

$$x = (\alpha_i \beta_i) \quad \text{with} \quad a = (\alpha_i) \in l_{p,u}(I) \quad \text{and} \quad b = (\beta_i) \in l_{q,v}(I).$$

Using this notation, we formulate a useful multiplication formula.

Proposition. If $1/p + 1/q = 1/r$ and $1/u + 1/v = 1/w$, then

$$l_{p,u}(I) \circ l_{q,v}(I) = l_{r,w}(I).$$

2.1.14.* Finally, we describe how the scale of Lorentz spaces behaves under real interpolation; see (BER, 5.3.1) and (TRI, 1.18.3).

Interpolation proposition (J. Peetre 1963). Let $0 < r_0 < r_1 < \infty$, $0 < w_0, w_1, w \leqq \infty$ and $0 < \theta < 1$. If $1/r = (1 - \theta)/r_0 + \theta/r_1$, then

$$(l_{r_0,w_0}(I), l_{r_1,w_1}(I))_{\theta,w} = l_{r,w}(I).$$

The quasi-norms on both sides are equivalent.

2.2. Axiomatic theory of s-numbers

2.2.1.* A rule

$$s: T \to (s_n(T)),$$

which assigns to every operator a scalar sequence, is said to be an **s-scale** if the following conditions are satisfied:

(S_1) $\|T\| = s_1(T) \geqq s_2(T) \geqq \ldots \geqq 0$ for $T \in \mathfrak{L}(E, F)$.

(S_2) $s_{m+n-1}(S + T) \leqq s_m(S) + s_n(T)$ for $S, T \in \mathfrak{L}(E, F)$.

(S_3) $s_n(YTX) \leqq \|Y\|\, s_n(T)\, \|X\|$ for $X \in \mathfrak{L}(E_0, E)$, $T \in \mathfrak{L}(E, F)$, $Y \in \mathfrak{L}(F, F_0)$.

(S_4) If rank $(T) < n$, then $s_n(T) = 0$.

(S_5) $s_n(I: l_2(n) \to l_2(n)) = 1$.

We call $s_n(T)$ the n-th **s-number** of the operator T.

In order to indicate the underlying Banach spaces E and F, we sometimes replace $s_n(T)$ by the more specific symbol $s_n(T: E \to F)$.

2.2.2.* First of all we show that the converse of (S_4) is true, as well.

Lemma. If $s_n(T) = 0$, then rank $(T) < n$.

Proof. Assume that rank $(T) \geqq n$. Then there exist $x_1, \ldots, x_n \in E$ such that $Tx_1, \ldots, Tx_n$ are linearly independent. Next we choose $b_1, \ldots, b_n \in F'$ with

$\langle Tx_i, b_j\rangle = \delta_{ij}$. Define $X \in \mathfrak{L}(l_2(n), E)$ and $B \in \mathfrak{L}(F, l_2(n))$ by

$$X := \sum_{i=1}^{n} e_i \otimes x_i \quad \text{and} \quad B := \sum_{i=1}^{n} b_i \otimes e_i .$$

Since $BTX = I$ is the identity operator of $l_2(n)$, we have

$$1 = s_n(I) \leqq \|B\|\, s_n(T)\, \|X\| .$$

Hence $s_n(T) > 0$.

2.2.3.* We now observe that $s_n(T)$ depends on T continuously.

Lemma. $|s_n(S) - s_n(T)| \leqq \|S - T\|$ for $S, T \in \mathfrak{L}(E, F)$.

Proof. Taking $m = 1$, it follows from (S_2) that

$$s_n(S) \leqq \|S - T\| + s_n(T).$$

Hence

$$s_n(S) - s_n(T) \leqq \|S - T\| .$$

Interchanging the roles of S and T, we obtain

$$s_n(T) - s_n(S) \leqq \|T - S\| .$$

This proves the desired inequality.

2.2.4.* Let s be any s-scale. An operator $T \in \mathfrak{L}(E, F)$ is said to be of s-**type** $l_{r,w}$ if $(s_n(T)) \in l_{r,w}$. The set of these operators is denoted by $\mathfrak{L}^{(s)}_{r,w}(E, F)$.

For $T \in \mathfrak{L}^{(s)}_{r,w}(E, F)$ we define

$$\|T \mid \mathfrak{L}^{(s)}_{r,w}\| := \|(s_n(T)) \mid l_{r,w}\| .$$

To simplify notation, the index w is omitted whenever $w = r$.

Remark. The above definition makes also sense for $r = \infty$.

2.2.5.* **Theorem** (A. Pietsch 1972: a) $\mathfrak{L}^{(s)}_{r,w}$ is a quasi-Banach operator ideal.

Proof. Let $S, T \in \mathfrak{L}^{(s)}_{r,w}(E, F)$. Then, by 2.1.9,

$$\begin{aligned}
\|(s_n(S + T)) \mid l_{r,w}\| &\leqq c_0 \|(s_n(S + T)) \mid l_{r,w}\|^{\mathrm{odd}} \\
&= c_0 \|(n^{1/r-1/w} s_{2n-1}(S + T) \mid l_w\| \\
&\leqq c_0 \|(n^{1/r-1/w} s_n(S) + n^{1/r-1/w} s_n(T)) \mid l_w\| \\
&\leqq c[\|(n^{1/r-1/w} s_n(S)) \mid l_w\| + \|(n^{1/r-1/w} s_n(T)) \mid l_w\|] \\
&= c[\|(s_n(S)) \mid l_{r,w}\| + \|(s_n(T)) \mid l_{r,w}\|].
\end{aligned}$$

This proves that $S + T \in \mathfrak{L}^{(s)}_{r,w}(E, F)$ and

$$\|S + T \mid \mathfrak{L}^{(s)}_{r,w}\| \leqq c[\|S \mid \mathfrak{L}^{(s)}_{r,w}\| + \|T \mid \mathfrak{L}^{(s)}_{r,w}\|].$$

Since the other ideal properties are evident, we immediately pass to the proof of completeness. Let (T_k) be a Cauchy sequence in $\mathfrak{L}^{(s)}_{r,w}(E, F)$. Then it follows from

$$\|T_h - T_k\| \leqq \|T_h - T_k \mid \mathfrak{L}^{(s)}_{r,w}\|$$

that (T_k) has a limit $T \in \mathfrak{L}(E, F)$ with respect to the operator norm. Given $\varepsilon > 0$, we choose k_0 such that

$$\|T_h - T_k \mid \mathfrak{L}^{(s)}_{r,w}\| \leqq \varepsilon \quad \text{for } h, k \geqq k_0 .$$

This means that

$$\left(\sum_{n=1}^{m} [n^{1/r-1/w} s_n(T_h - T_k)]^w\right)^{1/w} \leqq \varepsilon \quad \text{for } m = 1, 2, \ldots$$

Letting $h \to \infty$, by 2.2.3, we obtain

$$\left(\sum_{n=1}^{m} [n^{1/r-1/w} s_n(T - T_k)]^w\right)^{1/w} \leqq \varepsilon \quad \text{for } m = 1, 2, \ldots$$

Hence

$$T - T_k \in \mathfrak{L}_{r,w}^{(s)}(E, F) \quad \text{and} \quad \|T - T_k \mid \mathfrak{L}_{r,w}^{(s)}\| \leqq \varepsilon \quad \text{for } k \geqq k_0.$$

Thus T is also the limit of (T_k) with respect to the quasi-norm of $\mathfrak{L}_{r,w}^{(s)}$.

2.2.6.* We now establish an immediate consequence of the results stated in 2.1.7, 2.1.8, 2.1.9 and 2.1.10.

Proposition. The expressions

$$\|T \mid \mathfrak{L}_{r,w}^{(s)}\|_p^{\text{mean}} := \left\|\left(\left(\frac{1}{n}\sum_{k=1}^{n} s_k(T)^p\right)^{1/p}\right) \Big| \, l_{r,w}\right\| \quad \text{with } 0 < p < r,$$

$$\|T \mid \mathfrak{L}_{r,w}^{(s)}\|_0^{\text{mean}} := \left\|\left(\left(\prod_{k=1}^{n} s_k(T)\right)^{1/n}\right) \Big| \, l_{r,w}\right\|,$$

$$\|T \mid \mathfrak{L}_{r,w}^{(s)}\|^{\text{odd}} := \|(s_{2n-1}(T)) \mid l_{r,w}\|,$$

$$\|T \mid \mathfrak{L}_{r,w}^{(s)}\|_q^{\text{geom}} := \|(q^{k/r} s_{q^k}(T)) \mid l_w\| \quad \text{with } q = 2, 3, \ldots$$

define equivalent quasi-norms on $\mathfrak{L}_{r,w}^{(s)}$.

Remark. Note that the first of the above quasi-norms fails to be normalized.

2.2.7.* The lexicographical order of Lorentz sequence spaces stated in 2.1.11 can be transferred to the associated operator ideals.

Proposition.

$$\mathfrak{L}_{r_0,w_0}^{(s)} \subseteqq \mathfrak{L}_{r_1,w_1}^{(s)} \quad \text{for } 0 < r_0 < r_1 < \infty \text{ and arbitrary } w_0, w_1,$$

$$\mathfrak{L}_{r,w_0}^{(s)} \subseteqq \mathfrak{L}_{r,w_1}^{(s)} \quad \text{for arbitrary } r \text{ and } 0 < w_0 < w_1 \leqq \infty.$$

2.2.8.* An s-scale s is called **multiplicative** if

$$s_{m+n-1}(ST) \leqq s_m(S)\, s_n(T) \quad \text{for } T \in \mathfrak{L}(E, F) \text{ and } S \in \mathfrak{L}(F, G).$$

2.2.9.* The following result is very important.

Multiplication theorem (A. Pietsch 1972: a). Let s be any multiplicative s-scale. If $1/p + 1/q = 1/r$ and $1/u + 1/v = 1/w$, then

$$\mathfrak{L}_{p,u}^{(s)} \circ \mathfrak{L}_{q,v}^{(s)} \subseteqq \mathfrak{L}_{r,w}^{(s)}.$$

Proof. Let $T \in \mathfrak{L}_{q,v}^{(s)}(E, F)$ and $S \in \mathfrak{L}_{p,u}^{(s)}(F, G)$. Then, by 2.1.9 and Hölder's inequality, we have

$$\begin{aligned}\|(s_n(ST)) \mid l_{r,w}\| &\leqq c\|(s_{2n-1}(ST)) \mid l_{r,w}\| \\ &= c\|(n^{1/r-1/w}s_{2n-1}(ST)) \mid l_w\| \\ &\leqq c\|(n^{1/p-1/u}s_n(S)\, n^{1/q-1/v}s_n(T)) \mid l_w\| \\ &\leqq c\|(n^{1/p-1/u}s_n(S)) \mid l_u\|\ \|(n^{1/q-1/v}s_n(T)) \mid l_v\| \\ &= c\|(s_n(S)) \mid l_{p,u}\|\ \|(s_n(T)) \mid l_{q,v}\| .\end{aligned}$$

This proves that $ST \in \mathfrak{L}^{(s)}_{r,w}(E, G)$.

2.2.10.* Next we establish another useful formula.

Interpolation theorem (H. König 1978). Assume that $0 < r_0 < r_1 < \infty$, $0 < w_0, w_1, w \leqq \infty$ and $0 < \theta < 1$. If $1/r = (1-\theta)/r^0 + \theta/r_1$, then

$$(\mathfrak{L}^{(s)}_{r_0,w_0}, \mathfrak{L}^{(s)}_{r_1,w_1})_{\theta,w} \subseteq \mathfrak{L}^{(s)}_{r,w}.$$

Proof. Given $T \in (\mathfrak{L}^{(s)}_{r_0,\infty}(E, F), \mathfrak{L}^{(s)}_{r_1,\infty}(E, F))_{\theta,w}$, we consider any decomposition

$$T = T_0 + T_1 \quad \text{with} \quad T_0 \in \mathfrak{L}^{(s)}_{r_0,\infty}(E, F) \quad \text{and} \quad T_1 \in \mathfrak{L}^{(s)}_{r_1,\infty}(E, F).$$

Then it follows from

$$\begin{aligned}s_{2n}(T) &\leqq s_{2n-1}(T) \leqq s_n(T_0) + s_n(T_1) \\ &\leqq n^{-1/r_0}[\|T_0 \mid \mathfrak{L}^{(s)}_{r_0,\infty}\| + n^{1/r_0-1/r_1}\|T_1 \mid \mathfrak{L}^{(s)}_{r_1,\infty}\|]\end{aligned}$$

that

$$s_{2n}(T) \leqq n^{-1/r_0}K(n^{1/r_0-1/r_1}, T),$$

where $K(\tau, T)$ denotes the K-functional with respect to the quasi-Banach couple $(\mathfrak{L}^{(s)}_{r_0,\infty}(E, F), \mathfrak{L}^{(s)}_{r_1,\infty}(E, F))$. We now obtain

$$\begin{aligned}\sum_{k=0}^{\infty} [2^{k/r}s_{2^{k+1}}(T)]^w &\leqq \sum_{k=0}^{\infty} [2^{k(1/r-1/r_0)}K(2^{k(1/r_0-1/r_1)}, T)]^w \\ &\leqq c_0 \sum_{k=0}^{\infty} \int_{2^k}^{2^{k+1}} [\sigma^{1/r-1/r_0}K(\sigma^{1/r_0-1/r_1}, T)]^w \frac{d\sigma}{\sigma} \\ &= c \int_1^{\infty} [\tau^{-\theta}K(\tau, T)]^w \frac{d\tau}{\tau} .\end{aligned}$$

Hence $T \in \mathfrak{L}^{(s)}_{r,w}(E, F)$. This proves that

$$(\mathfrak{L}^{(s)}_{r_0,w_0}, \mathfrak{L}^{(s)}_{r_1,w_1})_{\theta,w} \subseteq (\mathfrak{L}^{(s)}_{r_0,\infty}, \mathfrak{L}^{(s)}_{r_1,\infty})_{\theta,w} \subseteq \mathfrak{L}^{(s)}_{r,w}.$$

2.2.11.* Let s be any s-scale. An operator $T \in \mathfrak{L}(E, F)$ is said to be of **s-type c_0** if $(s_n(T)) \in c_0$. The set of these operators is denoted by $\mathfrak{L}^{(s)}(E, F)$.

2.2.12.* **Theorem** (A. Pietsch 1974: a). $\mathfrak{L}^{(s)}$ is a closed operator ideal.

Proof. Since the ideal properties of $\mathfrak{L}^{(s)}$ are obvious, we only verify its closedness. Let (T_k) be any sequence in $\mathfrak{L}^{(s)}(E, F)$ converging to $T \in \mathfrak{L}(E, F)$ with respect to the operator norm. Given $\varepsilon > 0$, we successively fix k and n_0 such that

$$\|T - T_k\| \leqq \varepsilon \quad \text{and} \quad s_n(T_k) \leqq \varepsilon \quad \text{for } n \geqq n_0 .$$

Taking $m = 1$ in (S_2), it follows that

$$s_n(T) \leqq \|T - T_k\| + s_n(T_k) \leqq 2\varepsilon \quad \text{for } n \geqq n_0.$$

Hence $T \in \mathfrak{L}^{(s)}(E, F)$.

2.3. Approximation numbers

2.3.1.* The n-th **approximation number** of $T \in \mathfrak{L}(E, F)$ is defined by

$$a_n(T) := \inf\{\|T - L\| : L \in \mathfrak{F}(E, F), \text{rank}(L) < n\}.$$

2.3.2.* We begin the investigation of these numbers with the following observation.

Lemma. If $\dim(E) \geqq n$, then $a_n(I_E) = 1$.

Proof. Assume that $a_n(I_E) < 1$. Then there exists $L \in \mathfrak{F}(E)$ such that

$$\|I_E - L\| < 1 \quad \text{and} \quad \text{rank}(L) < n.$$

Using Neumann's expansion, we see that $L = I_E - (I_E - L)$ must be invertible. Therefore $\dim(E) = \text{rank}(L) < n$.

2.3.3.* **Theorem** (A. Pietsch 1972: a). The rule

$$a : T \to (a_n(T)),$$

which assigns to every operator the sequence of its approximation numbers, is an s-scale.

Proof. The preceding lemma implies (S_5). The verification of the other properties is an easy exercise.

2.3.4.* The foregoing result can be improved as follows.

Theorem (A. Pietsch 1972: a). The approximation numbers yield the largest s-scale.

Proof. Let $T \in \mathfrak{L}(E, F)$, and fix any natural number n. Given $\varepsilon > 0$, we choose $L \in \mathfrak{F}(E, F)$ such that

$$\|T - L\| \leqq (1 + \varepsilon)\, a_n(T) \quad \text{and} \quad \text{rank}(L) < n.$$

Then, for an arbitrary s-scale s, it follows that

$$s_n(T) \leqq \|T - L\| + s_n(L) \leqq (1 + \varepsilon)\, a_n(T).$$

Letting $\varepsilon \to 0$, we obtain $s_n(T) \leqq a_n(T)$.

2.3.5.* The operators belonging to $\mathfrak{L}^{(a)}_{r,w}$ are said to be of **approximation type** $l_{r,w}$.

2.3.6.* The next result is an immediate consequence of 2.2.5 and 2.3.3.

Theorem (A. Pietsch 1963: b). $\mathfrak{L}^{(a)}_{r,w}$ is a quasi-Banach operator ideal.

2.3.7. We stress the fact that $\mathfrak{L}^{(a)}_r$ with $1 \leqq r < \infty$ fails to be a Banach operator ideal. The following proposition describes this surprising situation more precisely.

Proposition. Let $1 \leqq r < \infty$. Then $p := r/(r + 1)$ is the largest exponent for which there exists an equivalent p-norm on $\mathfrak{L}^{(a)}_r$.

Proof. Carefully looking at the proof of 2.2.5, it can be seen that the original quasi-norm of $\mathfrak{L}^{(a)}_r$ satisfies the quasi-triangle inequality with the constant $c := 2^{1/r}$. Thus, by D.1.8, we may find an equivalent p-norm with $1/p := 1 + 1/r$.

In view of 2.9.10, we have

$$D_t \in \mathfrak{L}_r^{(a)}(l_\infty, l_1) \quad \text{if and only if} \quad t \in l_{p,r}.$$

Moreover, for $0 < q \leq 1$, it turns out that

$$D_t \in \mathfrak{N}_q(l_\infty, l_1) \quad \text{if and only if} \quad t \in l_q.$$

Hence

$$\mathfrak{N}_q(l_\infty, l_1) \nsubseteq \mathfrak{L}_r^{(a)}(l_\infty, l_1) \quad \text{whenever } p < q.$$

By 1.7.2, this implies that there cannot exist any equivalent q-norm on $\mathfrak{L}_r^{(a)}$.
Remark. Unfortunately, we do not know any concrete and non-sophisticated expression which yields a p-norm on $\mathfrak{L}_r^{(a)}$.

2.3.8. We now establish an extremely important characterization.

Representation theorem (A. Pietsch 1980:b, 1981:a). Let $q = 2, 3, \ldots$ An operator $T \in \mathfrak{L}(E, F)$ is of approximation type $l_{r,w}$ if and only if it can be written in the form

$$T = \sum_{k=0}^{\infty} T_k$$

such that $T_k \in \mathfrak{F}(E, F)$, rank $(T_k) \leq q^k$ and $(q^{k/r} \|T_k\|) \in l_w$. Moreover,

$$\|T \mid \mathfrak{L}_{r,w}^{(a)}\|_q^{\mathrm{rep}} := \inf \|(q^{k/r} \|T_k\|) \mid l_w\|,$$

where the infimum is taken over all possible representations, defines an equivalent quasi-norm on $\mathfrak{L}_{r,w}^{(a)}$.

Proof. If $T \in \mathfrak{L}_{r,w}^{(a)}(E, F)$, then we choose $L_k \in \mathfrak{F}(E, F)$ such that

$$\|T - L_k\| \leq 2a_{q^k}(T) \quad \text{and} \quad \text{rank}\,(L_k) < q^k.$$

In the special case when $a_{q^k}(T) = 0$, it follows from 2.2.2 that rank $(T) < q^k$. Hence we may take $L_k := T$. Define

$$T_0 := O, \quad T_1 := O \quad \text{and} \quad T_{k+2} := L_{k+1} - L_k \quad \text{for } k = 0, 1, \ldots$$

Then

$$\text{rank}(T_{k+2}) \leq \text{rank}(L_{k+1}) + \text{rank}(L_k) < q^{k+1} + q^k < q^{k+2}$$

and

$$\|T_{k+2}\| \leq \|L_{k+1} - T\| + \|T - L_k\| \leq 4a_{q^k}(T).$$

By 2.1.10, the latter estimate implies that

$$(q^{k/r} \|T_k\|) \in l_w.$$

Finally, we have

$$T = \lim_k L_k = \sum_{k=0}^{\infty} T_k.$$

Additionally, it turns out that

$$\|T \mid \mathfrak{L}_{r,w}^{(a)}\|_q^{\mathrm{rep}} \leq \|(q^{k/r} \|T_k\|) \mid l_w\|$$
$$\leq c\|(q^{k/r} a_{q^k}(T)) \mid l_w\| = c\|T \mid \mathfrak{L}_{r,w}^{(a)}\|_q^{\mathrm{geom}}.$$

We now verify the sufficiency of the given condition. To this end, assume that $T \in \mathfrak{L}(E, F)$ admits a representation with the properties stated above. Then

$$\operatorname{rank}\left(\sum_{k=0}^{h-1} T_k\right) \leqq \sum_{k=0}^{h-1} q^k < q^h.$$

Fix p and θ such that $0 < p < \min(w, 1)$ and $0 < \theta < 1$. Define s by $1/p = 1/s + 1/w$. Applying Hölder's inequality, we obtain

$$\begin{aligned} a_{q^h}(T) &\leqq \left\| T - \sum_{k=0}^{h-1} T_k \right\| \leqq \sum_{k=h}^{\infty} \|T_k\| \leqq \left(\sum_{k=h}^{\infty} \|T_k\|^p\right)^{1/p} \\ &\leqq \left(\sum_{k=h}^{\infty} q^{-\theta ks/r}\right)^{1/s} \left(\sum_{k=h}^{\infty} q^{\theta kw/r} \|T_k\|^w\right)^{1/w} \\ &\leqq c_0 q^{-\theta h/r} \left(\sum_{k=h}^{\infty} q^{\theta kw/r} \|T_k\|^w\right)^{1/w}. \end{aligned}$$

From this estimate it follows that

$$\begin{aligned} \|T \mid \mathfrak{L}_{r,w}^{(a)}\|_q^{\text{geom}} &= \left(\sum_{h=0}^{\infty} q^{hw/r} a_{q^h}(T)^w\right)^{1/w} \\ &\leqq c_0 \left(\sum_{h=0}^{\infty} q^{(1-\theta)hw/r} \left[\sum_{k=h}^{\infty} q^{\theta kw/r} \|T_k\|^w\right]\right)^{1/w} \\ &= c_0 \left(\sum_{k=0}^{\infty} \left[\sum_{h=0}^{k} q^{(1-\theta)hw/r}\right] q^{\theta kw/r} \|T_k\|^w\right)^{1/w} \\ &\leqq c \left(\sum_{k=0}^{\infty} q^{kw/r} \|T_k\|^w\right)^{1/w}. \end{aligned}$$

By 2.1.10, this proves that $T \in \mathfrak{L}_{r,w}^{(a)}(E, F)$ and

$$\|T \mid \mathfrak{L}_{r,w}^{(a)}\|_q^{\text{geom}} \leqq c \|T \mid \mathfrak{L}_{r,w}^{(a)}\|_q^{\text{rep}}.$$

Remark. For every finite operator $T \in \mathfrak{L}(E, F)$ the representation constructed above has the property that $T_k = O$ whenever k is sufficiently large. We stress the fact the quasi-norm introduced in the representation theorem fails to be normalized.

2.3.9. Proposition (A. Pietsch 1963:b, 1981:a). If $0 < w < \infty$, then the quasi-Banach operator ideal $\mathfrak{L}_{r,w}^{(a)}$ is approximative.

Proof. Consider any representation

$$T = \sum_{k=0}^{\infty} T_k$$

such that $T_k \in \mathfrak{F}(E, F)$, $\operatorname{rank}(T_k) \leqq 2^k$ and $(2^{k/r}\|T_k\|) \in l_w$. Given $\varepsilon > 0$, we choose a natural number n_0 with

$$\left(\sum_{k=n_0}^{\infty} [2^{k/r} \|T_k\|]^w\right)^{1/w} \leqq \varepsilon.$$

Now it follows that

$$\left\| T - \sum_{k=0}^{n} T_k \mid \mathfrak{L}_{r,w}^{(a)} \right\|_2^{\text{rep}} \leqq \varepsilon \quad \text{for } n \geqq n_0.$$

This completes the proof.

Remark. In the case $w = \infty$, the closed hull of $\mathfrak{F}(E, F)$ in $\mathfrak{L}_{r,\infty}^{(a)}$ consists of all operators $T \in \mathfrak{L}(E, F)$ with $(n^{1/r} a_n(T)) \in c_0$.

2.3.10. The representation theorem can be used to establish inclusions between $\mathfrak{L}_{r,w}^{(a)}$ and various quasi-Banach operator ideals.

Embedding theorem (A. Pietsch 1981:a). Let $0 < r < \infty$. Assume that $\mathfrak{A}$ is any w-Banach operator ideal ($0 < w \leqq 1$) such that

$$\|T \mid \mathfrak{A}\| \leqq c n^{1/r} \|T\| \quad \text{whenever} \ \operatorname{rank}(T) \leqq n,$$

where $c > 0$ is a constant. Then $\mathfrak{L}_{r,w}^{(a)} \subseteq \mathfrak{A}$.

Proof. Given $T \in \mathfrak{L}_{r,w}^{(a)}(E, F)$, we choose a representation

$$T = \sum_{k=0}^{\infty} T_k$$

such that $T_k \in \mathfrak{F}(E, F)$, $\operatorname{rank}(T_k) \leqq 2^k$, and $(2^{k/r} \|T_k\|) \in l_w$. Then

$$\|T_k \mid \mathfrak{A}\| \leqq c\, 2^{k/r} \|T_k\|$$

implies that the sequence (T_k) is absolutely w-summable in $\mathfrak{A}$. However, since $\mathfrak{A}$ is complete with respect to the w-norm $\|\cdot \mid \mathfrak{A}\|$, we obtain $T \in \mathfrak{A}(E, F)$.

2.3.11. We now apply the preceding embedding theorem in a special case.

Proposition (A. Pietsch 1963:b). $\mathfrak{L}_p^{(a)} \subseteq \mathfrak{N}_p$ for $0 < p \leqq 1$.

Proof. The assertion follows from 1.7.7 and 2.3.10.

Remark. Considering diagonal operators from l_∞ into l_1, it can easily be shown that the above inclusion is strict; see 2.9.10.

2.3.12. Proposition (A. Pietsch 1963:b). The approximation numbers are multiplicative.

Proof. Let $T \in \mathfrak{L}(E, F)$ and $S \in \mathfrak{L}(F, G)$. Fix any natural numbers m and n. Given $\varepsilon > 0$, we can find $B \in \mathfrak{F}(E, F)$ and $A \in \mathfrak{F}(F, G)$ such that

$$\begin{aligned} \|S - A\| &\leqq (1 + \varepsilon)\, a_m(S) \quad \text{and} \quad \operatorname{rank}(A) < m, \\ \|T - B\| &\leqq (1 + \varepsilon)\, a_n(T) \quad \text{and} \quad \operatorname{rank}(B) < n. \end{aligned}$$

It follows from

$$\operatorname{rank}(AT + (S - A)B) \leqq \operatorname{rank}(A) + \operatorname{rank}(B) < m + n - 1$$

that

$$\begin{aligned} a_{m+n-1}(ST) &\leqq \|ST - (AT + SB - AB)\| = \|(S - A)(T - B)\| \\ &\leqq \|S - A\|\, \|T - B\| \leqq (1 + \varepsilon)^2\, a_m(S)\, a_n(T). \end{aligned}$$

Letting $\varepsilon \to 0$, we arrive at the desired inequality.

2.3.13. For the operator ideals determined by the approximation numbers the inclusion stated in 2.2.9 becomes an equality.

Multiplication theorem (A. Pietsch 1980:b). If $1/p + 1/q = 1/r$ and $1/u + 1/v = 1/w$, then

$$\mathfrak{L}_{p,u}^{(a)} \circ \mathfrak{L}_{q,v}^{(a)} = \mathfrak{L}_{r,w}^{(a)}.$$

The quasi-norms on both sides are equivalent.

Proof. Given $R \in \mathfrak{L}^{(a)}_{r,w}(E, G)$, we consider a representation

$$R = \sum_{k=0}^{\infty} R_k$$

such that $R_k \in \mathfrak{F}(E, G)$, $\operatorname{rank}(R_k) \leqq 2^k$, and $(2^{k/r}\|R_k\|) \in l_w$. Choose factorizations $R_k = S_k T_k$ with $T_k \in \mathfrak{L}(E, F_k)$, $S_k \in \mathfrak{L}(F_k, G)$, $\|R_k\| = \|S_k\|\,\|T_k\|$ and $\dim(F_k) \leqq 2^k$. If $0 < w < \infty$, it can be arranged that

$$2^{k/p}\|S_k\| = (2^{k/r}\|R_k\|)^{w/u} \quad \text{and} \quad 2^{k/q}\|T_k\| = (2^{k/r}\|R_k\|)^{w/v}.$$

Form the Banach space $F := [l_2, F_k]$ as described in C.2.1. Let $J_k \in \mathfrak{L}(F_k, F)$ and $Q_k \in \mathfrak{L}(F, F_k)$ denote the canonical injections and surjections, respectively. Observe that

$$\operatorname{rank}(S_k Q_k) \leqq 2^k \quad \text{and} \quad \operatorname{rank}(J_k T_k) \leqq 2^k,$$
$$(2^{k/p}\|S_k Q_k\|) \in l_u \quad \text{and} \quad (2^{k/q}\|J_k T_k\|) \in l_v.$$

Therefore the operators

$$S := \sum_{k=0}^{\infty} S_k Q_k \quad \text{and} \quad T := \sum_{k=0}^{\infty} J_k T_k$$

are of approximation type $l_{p,u}$, and $l_{q,v}$, respectively. The representation theorem implies that $R = ST \in \mathfrak{L}^{(a)}_{p,u} \circ \mathfrak{L}^{(a)}_{q,v}(E,G)$.

In the case $u = v = w = \infty$ we choose the factorization $R_k = S_k T_k$ such that $\|S_k\| = \|R_k\|^{r/p}$ and $\|T_k\| = \|R_k\|^{r/q}$.

2.3.14. The quasi-Banach operator ideals associated with the approximation numbers behave better with respect to interpolation than those corresponding to arbitrary s-numbers.

Interpolation theorem (J. Peetre / G. Sparr 1972, H. König 1978). Let $0 < r_0 < r_1 < \infty$, $0 < w_0, w_1, w \leqq \infty$ and $0 < \theta < 1$. If $1/r = (1-\theta)/r_0 + \theta/r_1$, then

$$(\mathfrak{L}^{(a)}_{r_0,w_0}, \mathfrak{L}^{(a)}_{r_1,w_1})_{\theta,w} = \mathfrak{L}^{(a)}_{r,w}.$$

The quasi-norms on both sides are equivalent.

Proof. First we establish the formula

$$(\mathfrak{L}^{(a)}_p, \mathfrak{L})_{\theta,w} = \mathfrak{L}^{(a)}_{r,w}$$

for $0 < p < \infty$, $0 < w \leqq \infty$, $0 < \theta < 1$ and $1/r = (1-\theta)/p + \theta/\infty$. Let $T \in \mathfrak{L}(E, F)$, and fix a natural number n. Choose $L \in \mathfrak{F}(E, F)$ such that

$$\|T - L\| \leqq 2a_n(T) \quad \text{and} \quad \operatorname{rank}(L) < n.$$

Then, for $k = 1, \ldots, n$, we have

$$\|T - L\| \leqq 2a_k(T) \quad \text{and} \quad a_k(L) \leqq \|L - T\| + a_k(T) \leqq 3a_k(T).$$

Hence

$$K(n^{1/p}, T) \leqq \|L \mid \mathfrak{L}^{(a)}_p\| + n^{1/p}\|T - L\|$$
$$\leqq \Big(\sum_{k=1}^{n} a_k(L)^p\Big)^{1/p} + 2\Big(\sum_{k=1}^{n} a_k(T)^p\Big)^{1/p} \leqq 5\Big(\sum_{k=1}^{n} a_k(T)^p\Big)^{1/p}.$$

Obviously, $K(\tau, T) \leqq \tau\|T\|$.

From now on we assume that $T \in \mathfrak{L}_{r,w}^{(a)}(E, F)$. Using an elementary discretization argument, it follows from 2.1.7 that

$$\int_0^\infty [\tau^{-\theta} K(\tau, T)]^w \frac{d\tau}{\tau} \leqq \int_0^1 [\tau^{-\theta} K(\tau, T)]^w \frac{d\tau}{\tau} + \int_1^\infty [\tau^{-\theta} K(\tau, T)]^w \frac{d\tau}{\tau}$$

$$\leqq \frac{1}{(1-\theta)\,w} \|T\| + c_1 \sum_{n=1}^\infty [n^{-\theta/p} K(n^{1/p}, T)]^w \frac{(n+1)^{1/p} - n^{1/p}}{n^{1/p}}$$

$$\leqq \frac{1}{(1-\theta)\,w} \|T\| + c_2 \sum_{n=1}^\infty [n^{-\theta/p - 1/w} K(n^{1/p}, T)]^w$$

$$\leqq c_3 \sum_{n=1}^\infty \left[n^{1/r - 1/w} \left(\frac{1}{n} \sum_{k=1}^n a_k(T)^p \right)^{1/p} \right]^w < \infty .$$

Hence $T \in (\mathfrak{L}_p^{(a)}(E, F), \mathfrak{L}(E, F))_{\theta, w}$. This proves that

$$\mathfrak{L}_{r,w}^{(a)} \subseteq (\mathfrak{L}_p^{(a)}, \mathfrak{L})_{\theta, w}.$$

In view of 2.2.10, we even have equality.

Finally, the general case is settled by the reiteration property. Choose p such that $0 < p < r_0 < r_1 < \infty$. Define Σ_0 and Σ_1 by

$$1/r_0 = (1 - \Sigma_0)/p \quad \text{and} \quad 1/r_1 = (1 - \Sigma_1)/p,$$

respectively. Then

$$\mathfrak{L}_{r_0,w_0}^{(a)}(E, F) = (\mathfrak{L}_p^{(a)}(E, F), \mathfrak{L}(E, F))_{\Sigma_0, w_0}$$

and

$$\mathfrak{L}_{r_1,w_1}^{(a)}(E, F) = (\mathfrak{L}_p^{(a)}(E, F), \mathfrak{L}(E, F))_{\Sigma_1, w_1}.$$

Applying F.3.3, we get

$$\begin{aligned}
&(\mathfrak{L}_{r_0,w_0}^{(a)}(E, F), \mathfrak{L}_{r_1,w_1}^{(a)}(E, F))_{\theta, w} = \\
&= ((\mathfrak{L}_p^{(a)}(E, F), \mathfrak{L}(E, F))_{\Sigma_0, w_0}, (\mathfrak{L}_p^{(a)}(E, F), \mathfrak{L}(E, F))_{\Sigma_1, w_1})_{\theta, w} \\
&= (\mathfrak{L}_p^{(a)}(E, F), \mathfrak{L}(E, F))_{\Sigma, w} = \mathfrak{L}_{r,w}^{(a)}(E, F)
\end{aligned}$$

with $1/r = (1 - \Sigma)/p$ and $\Sigma = (1 - \theta)\,\Sigma_0 + \theta \Sigma_1$. Therefore $1/r = (1 - \theta)/r_0 + \theta/r_1$.

2.3.15. As shown by H. König (1984:a) the operator ideals $\mathfrak{L}_{r,w}^{(a)}$ fail to be stable with respect to the tensor norms ε and ν. We conjecture that this negative result even holds for all tensor norms. On the other hand, it turns out that they are almost stable in the following sense.

Proposition (A. Pietsch 1982:b). Let $0 < q < p < \infty$ and $0 < u,\ v \leqq \infty$. If $S \in \mathfrak{L}_{p,u}^{(a)}(E, E_0)$ and $T \in \mathfrak{L}_{q,v}^{(a)}(F, F_0)$, then

$$S \mathbin{\tilde{\otimes}_\alpha} T \in \mathfrak{L}_{p,u}^{(a)}(E \mathbin{\tilde{\otimes}_\alpha} F, F_0 \mathbin{\tilde{\otimes}_\alpha} F_0)$$

for every tensor norm α.

Proof. First we assume that $u = v = 1$. Choose a natural number k such that $\frac{k+1}{p} \leqq \frac{k}{q}$. Then, putting $m := 2^{k+1}$ and $n := 2^k$, we have $m^{1/p} \leqq n^{1/q}$. Next, according to 2.3.8, we consider representations

$$S = \sum_{i=0}^{\infty} S_i \quad \text{and} \quad T = \sum_{j=0}^{\infty} T_j$$

such that

$$\operatorname{rank}(S_i) \leqq m^i \quad \text{and} \quad \operatorname{rank}(T_j) \leqq n^j,$$
$$(m^{i/p}\|S_i\|) \in l_1 \quad \text{and} \quad (n^{j/q}\|T_j\|) \in l_1 .$$

Let $R_0 := O$ and

$$R_{h+1} := \sum_{i+j=h} S_i \mathbin{\tilde{\otimes}_\alpha} T_j \quad \text{for } h = 0, 1, \ldots$$

Then

$$\operatorname{rank}(R_{h+1}) \leqq \sum_{i+j=h} \operatorname{rank}(S_i \mathbin{\tilde{\otimes}_\alpha} T_j) = \sum_{i+j=h} \operatorname{rank}(S_i)\operatorname{rank}(T_j)$$
$$\leqq \sum_{i+j=h} m^i n^j \leqq m^h \sum_{j=0}^{h} \left(\frac{n}{m}\right)^j \leqq m^h \sum_{j=0}^{\infty} \left(\frac{1}{2}\right)^j < m^{h+1}.$$

Moreover, it follows from

$$m^{h/p}\|R_{h+1}\| \leqq \sum_{i+j=h} m^{(i+j)/p}\|S_i \mathbin{\tilde{\otimes}_\alpha} T_j\| \leqq \sum_{i+j=h} m^{i/p}\|S_i\|\, n^{j/q}\, \|T_j\|$$

that

$$\sum_{h=0}^{\infty} m^{(h+1)/p}\|R_{h+1}\| \leqq m^{1/p} \sum_{i=0}^{\infty} m^{i/p}\|S_i\| \sum_{j=0}^{\infty} n^{j/q}\|T_j\| < \infty .$$

Since we have the representation

$$S \mathbin{\tilde{\otimes}_\alpha} T = \sum_{h=0}^{\infty} R_h,$$

Theorem 2.3.8 implies that $S \mathbin{\tilde{\otimes}_\alpha} T$ is of approximation type $l_{p,1}$.

Next the general case is treated. Enlarging q a little bit (if necessary), it can always be arranged that $v = 1$. Choose p_0, p_1 and θ such that $1/p = (1-\theta)/p_0 + \theta/p_1$ and $q < p_0 < p < p_1 < \infty$. The result already verified says that, for $T \in \mathfrak{L}_{q,1}^{(a)}(F, F_0)$ fixed, the map $T^{\otimes}$ which assigns to every operator $S \in \mathfrak{L}(E, E_0)$ the tensor product $S \mathbin{\tilde{\otimes}_\alpha} T \in \mathfrak{L}(E \mathbin{\tilde{\otimes}_\alpha} F, E_0 \mathbin{\tilde{\otimes}_\alpha} F_0)$, acts as follows:

$$T^{\otimes} : \mathfrak{L}_{p_0,1}^{(a)}(E, E_0) \to \mathfrak{L}_{p_0,1}^{(a)}(E \mathbin{\tilde{\otimes}_\alpha} F, E_0 \mathbin{\tilde{\otimes}_\alpha} F_0)$$

and

$$T^{\otimes} : \mathfrak{L}_{p_1,1}^{(a)}(E, E_0) \to \mathfrak{L}_{p_1,1}^{(a)}(E \mathbin{\tilde{\otimes}_\alpha} F, E_0 \mathbin{\tilde{\otimes}_\alpha} F_0).$$

Therefore, by 2.3.14, the interpolation property yields

$$T^{\otimes} : \mathfrak{L}_{p,u}^{(a)}(E, E_0) \to \mathfrak{L}_{p,u}^{(a)}(E \mathbin{\tilde{\otimes}_\alpha} F, E_0 \mathbin{\tilde{\otimes}_\alpha} F_0).$$

Remark. In the above proof, the interpolation argument can be avoided at the cost of some more complicated computations; see A. Pietsch (1982:b).

2.3.16. As shown in (PIE, 11.7.4), we have

$$a_n(T) = a_n(T') \quad \text{for all } T \in \mathfrak{K}(E, F).$$

This formula implies that the operators T and T' possess the same approximation type.

Proposition (C. V. Hutten 1974). $\mathfrak{L}^{(a)}_{r,w} = (\mathfrak{L}^{(a)}_{r,w})'$.

2.4. Gel'fand and Weyl numbers

2.4.1.* An s-scale s is **injective** if, given any metric injection $J \in \mathfrak{L}(F, F_0)$,

$$s_n(T) = s_n(JT) \quad \text{for all } T \in \mathfrak{L}(E, F).$$

2.4.2.* The n-th **Gel'fand number** of $T \in \mathfrak{L}(E, F)$ is defined by

$$c_n(T) := \inf \{\|TJ_M^E\| : \operatorname{codim}(M) < n\}.$$

Recall that J_M^E denotes the canonical injection from the subspace M into the Banach space E. Hence TJ_M^E is the restriction of T to M.

Remark. Observe that the above definition reflects the basic idea of the Fischer-Courant minimax principle; see (DUN, X.4.3).

2.4.3.* The following result can easily be verified.

Theorem (A. Pietsch 1972:a). The rule

$$c : T \to (c_n(T)),$$

which assigns to every operator the sequence of its Gel'fand numbers, is an injective s-scale.

2.4.4.* The operators belonging to $\mathfrak{L}^{(c)}_{r,w}$ are said to be of **Gel'fand type** $l_{r,w}$.

2.4.5.* **Theorem.** $\mathfrak{L}^{(c)}_{r,w}$ is an injective quasi-Banach operator ideal.

2.4.6.* The identity operator from $l_1(m)$ into $l_2(m)$ shows that the Gel'fand numbers may be strictly smaller than the approximation numbers. However, there are important situations in which equality holds.

Proposition. Suppose that H is a Hilbert space. Then

$$c_n(T) = a_n(T) \quad \text{for all } T \in \mathfrak{L}(H, F).$$

Proof. Fix any natural number n, and consider an arbitrary subspace M of H with $\operatorname{codim}(M) < n$. Let $P \in \mathfrak{L}(H)$ denote the orthogonal projection from H onto M, and define $L := T - TP$. Since $M \subseteq N(L)$, it follows that

$$\operatorname{rank}(L) = \operatorname{codim}(N(L)) \leqq \operatorname{codim}(M) < n.$$

Therefore

$$a_n(T) \leqq \|T - L\| = \|TP\| = \|TJ_M^H\|.$$

This implies that $a_n(T) \leqq c_n(T)$. In view of 2.3.4, we even have equality.

2.4.7. The next result is along the same line.

Proposition (A. Pietsch 1974:a). Suppose that F has the metric extension property. Then

$$c_n(T) = a_n(T) \quad \text{for all } T \in \mathfrak{L}(E, F).$$

Proof. Fix any natural number n, and consider an arbitrary subspace M of E with codim $(M) < n$. By hypothesis, there exists a norm preserving extension $T_0 \in \mathfrak{L}(E, F)$ of TJ_M^E. This means that $T_0 J_M^E = TJ_M^E$ and $\|T_0\| = \|TJ_M^E\|$. Define $L := T - T_0$. Since $M \subseteq N(L)$, it follows that

$$\operatorname{rank}(L) = \operatorname{codim}(N(L)) \leqq \operatorname{codim}(M) < n.$$

Therefore

$$a_n(T) \leqq \|T - L\| = \|T_0\| = \|TJ_M^E\|.$$

This implies that $a_n(T) \leqq c_n(T)$. In view of 2.3.4, we even have equality.

2.4.8. **Proposition** (A. Pietsch 1972:a). The Gel'fand numbers are multiplicative.

Proof. Let $T \in \mathfrak{L}(E, F)$ and $S \in \mathfrak{L}(F, G)$. Fix any natural numbers m and n. Given $\varepsilon > 0$, we can find subspaces X and Y of E and F, respectively, such that

$$\|TJ_X^E\| \leqq (1 + \varepsilon)\, c_n(T) \quad \text{and} \quad \operatorname{codim}(X) < n,$$
$$\|SJ_Y^F\| \leqq (1 + \varepsilon)\, c_m(S) \quad \text{and} \quad \operatorname{codim}(Y) < m.$$

Define $M := X \cap T^{-1}(Y)$. Then it follows from

$$\operatorname{codim}(M) \leqq \operatorname{codim}(X) + \operatorname{codim}(Y) < m + n - 1$$

that

$$c_{m+n-1}(ST) \leqq \|STJ_M^E\| \leqq \|SJ_Y^F\|\, \|TJ_X^E\| \leqq (1 + \varepsilon)^2\, c_m(S)\, c_n(T).$$

Letting $\varepsilon \to 0$, we arrive at the desired inequality.

2.4.9. Next we state an immediate corollary of 2.2.9 and 2.4.8.
Multiplication theorem. If $1/p + 1/q = 1/r$ and $1/u + 1/v = 1/w$, then

$$\mathfrak{L}_{p,u}^{(c)} \circ \mathfrak{L}_{q,v}^{(c)} \subseteq \mathfrak{L}_{r,w}^{(c)}.$$

2.4.10.* We now establish a useful criterion.

Proposition (H. E. Lacey 1963). An operator $T \in \mathfrak{L}(E, F)$ is compact if and only if for every $\varepsilon > 0$ there exists a finite codimensional subspace M of E such that $\|TJ_M^E\| \leqq \varepsilon$.

Proof. Assume first that $T \in \mathfrak{K}(E, F)$. Given $\varepsilon > 0$, we choose $y_1, \ldots, y_n \in F$ such that

$$T(U) \subseteq \bigcup_{j=1}^{n} \{y_j + \varepsilon V\}.$$

Next we pick $b_1, \ldots, b_n \in V^0$ with $|\langle y_j, b_j\rangle| = \|y_j\|$. Now it can easily be seen that

$$M := \{x \in E : \langle Tx, b_j\rangle = 0 \quad \text{for } j = 1, \ldots, n\}$$

is the desired finite codimensional subspace. To this end, let $x \in M \cap U$. Then there exists some y_j with $\|Tx - y_j\| \leqq \varepsilon$, and it follows from

$$\|y_j\| = |\langle y_j, b_j\rangle| \leqq |\langle y_j - Tx, b_j\rangle| + |\langle Tx, b_j\rangle| \leqq \varepsilon$$

that

$$\|Tx\| \leqq \|Tx - y_j\| + \|y_j\| \leqq 2\varepsilon .$$

Hence $\|TJ_M^E\| \leqq 2\varepsilon$.

We now assume that the above condition is satisfied. Fix $\varepsilon > 0$, and take any finite codimensional subspace M of E such that $\|TJ_M^E\| \leqq \varepsilon$. Then E can be represented as a direct sum $E = M \oplus N$, where the subspace N is finite dimensional. Let P_M and P_N denote the corresponding projections. Put

$$\delta := \min\left(\frac{\varepsilon}{\|T\|}, 1\right).$$

Since P_N has finite rank, there exists $x_1, \ldots, x_m \in U$ such that

$$P_N(U) \subseteq \bigcup_{i=1}^{m} \{P_N x_i + \delta U\}.$$

Thus, given $x \in U$, we may choose x_i with $\|P_N x - P_N x_i\| \leqq \delta$. Then

$$\|P_M x - P_M x_i\| \leqq \|x - x_i\| + \|P_N x - P_N x_i\| \leqq 2 + \delta \leqq 3 .$$

Hence

$$\|Tx - Tx_i\| \leqq \|TP_M x - TP_M x_i\| + \|TP_N x - TP_N x_i\| \leqq$$
$$\leqq 3\|TJ_M^E\| + \delta\|T\| \leqq 4\varepsilon .$$

This proves that

$$T(U) \subseteq \bigcup_{i=1}^{m} \{Tx_i + 4\varepsilon V\}.$$

Since $\varepsilon > 0$ is arbitrary, we have $T \in \mathfrak{K}(E, F)$.

2.4.11.* In terms of Gel'fand numbers the preceding result reads as follows.

Theorem. $\mathfrak{L}^{(c)} = \mathfrak{K}$.

2.4.12.* Next we show that the Gel'fand numbers of any operator T are determined by the Gel'fand numbers of its restrictions to all finite dimensional subspaces. This means that $c_n(T)$ is a "local quantity".

Proposition (A. Pietsch 1974: b).

$$c_n(T) = \sup\{c_n(TJ_M^E) : \dim(M) < \infty\} \quad \text{for all } T \in \mathfrak{L}(E, F).$$

Proof. First of all, we make the trivial observation that

$$c_n(TJ_M^E) \leqq c_n(T) \quad \text{for all subspaces } M.$$

Therefore the right-hand supremum is less than or equal to $c_n(T)$.

In order to prove equality, it is enough to consider the case when E is infinite dimensional and $n \geqq 2$. Given $\varepsilon > 0$, in every finite dimensional subspace M with $\dim(M) \geqq n$ we choose a subspace $N(M)$ such that

$$\|TJ_{N(M)}^E\| \leqq (1 + \varepsilon)\, c_n(TJ_M^E) \quad \text{and} \quad \dim(M/N(M)) = n - 1 .$$

According to Auerbach's lemma 1.7.6, there exist

$$\hat{x}_1(M), \ldots, \hat{x}_{n-1}(M) \in M/N(M) \quad \text{and} \quad \hat{a}_1(M), \ldots, \hat{a}_{n-1}(M) \in (M/N(M))'$$

such that

$$\|\hat{x}_i(M)\| = 1, \quad \|\hat{a}_i(M)\| = 1 \quad \text{and} \quad \langle \hat{x}_i(M), \hat{a}_j(M) \rangle = \delta_{ij}.$$

Taking appropriate representatives $x_i(M) \in \hat{x}_i(M)$ and extending to E the functionals induced by $\hat{a}_i(M)$ on M, we can find

$$x_1(M), \ldots, x_{n-1}(M) \in M \quad \text{and} \quad a_1(M), \ldots, a_{n-1}(M) \in E'$$

such that

$$\|x_i(M)\| \leqq 2, \quad \|a_i(M)\| = 1 \quad \text{and} \quad \langle x_i(M), a_j(M) \rangle = \delta_{ij}$$

as well as

$$\langle x, a_i(M) \rangle = 0 \quad \text{for all } x \in N(M).$$

Define the projection

$$P(M) := \sum_{i=1}^{n-1} a_i(M) \otimes x_i(M),$$

and note that its range is a complement of $N(M)$ in M.

We know from the Alaoglu-Bourbaki theorem, already mentioned in 1.3.6, that U^0 is compact in the weak topology generated by E. Furthermore, the collection of all finite dimensional subspaces M with $\dim(M) \geqq n$ is directed upwards. Therefore the generalized sequence of $(n-1)$-tuples $(a_1(M), \ldots, a_{n-1}(M))$ has a cluster point

$$(a_1, \ldots, a_{n-1}) \in U^0 \times \ldots \times U^0;$$

see (DUN, I.7). We now define the subspace

$$N := \{x \in E : \langle x, a_i \rangle = 0 \quad \text{for } i = 1, \ldots, n-1\}.$$

Then it follows from

$$P(M)\,x = \sum_{i=1}^{n-1} \langle x, a_i(M) \rangle\, x_i(M)$$

that

$$\lim_M \|P(M)\,x\| = 0 \quad \text{for all } x \in N.$$

Hence, for fixed $x \in N \cap U$, there exists M_0 such that $x \in M_0$ and

$$\|T\|\,\|P(M)\,x\| \leqq \varepsilon \quad \text{for all } M \supseteq M_0.$$

Note that $x - P(M)\,x \in N(M)$ whenever $x \in M$. For $M \supseteq M_0$ we now obtain

$$\begin{aligned}\|TX\| &\leqq \|T(x - P(M)\,x)\| + \|TP(M)\,x\| \\ &\leqq \|TJ^E_{N(M)}\|\,\|x - P(M)\,x\| + \|T\|\,\|P(M)\,x\| \\ &\leqq \|TJ^E_{N(M)}\| + 2\|T\|\,\|P(M)\,x\| \leqq (1+\varepsilon)\,c_n(TJ^E_M) + 2\varepsilon.\end{aligned}$$

This implies that

$$c_n(T) \leqq \|TJ^E_N\| \leqq (1+\varepsilon) \sup\{c_n(TJ^E_M) : \dim(M) < \infty\} + 2\varepsilon.$$

Letting $\varepsilon \to 0$ completes the proof.

2.4.13.* The n-th **Weyl number** of $T \in \mathfrak{L}(E, F)$ is defined by

$$x_n(T) := \sup \{a_n(TX) : X \in \mathfrak{L}(l_2, E), \|X\| \leqq 1\}.$$

Remark. According to 2.4.6 and 2.4.12, the same numbers are obtained by taking the right-hand supremum over all (finite) operators $X \in \mathfrak{L}(H, E)$ with $\|X\| \leqq 1$, where H is an arbitrary infinite dimensional Hilbert space.

2.4.14.* Theorem (A. Pietsch 1980:c). The rule

$$x : T \to (x_n(T)),$$

which assigns to every operators the sequence of its Weyl numbers, is an injective s-scale.

Proof. The injectivity follows from 2.4.3 and 2.4.6. The verification of the other properties is an easy exercise.

2.4.15.* The operators belonging to $\mathfrak{L}_{r,w}^{(x)}$ are said to be of **Weyl type** $l_{r,w}$.

2.4.16.* Theorem. $\mathfrak{L}_{r,w}^{(x)}$ is an injective quasi-Banach operator ideal.

2.4.17.* Proposition (A. Pietsch 1980:c). The Weyl numbers are multiplicative.

Proof. Let $T \in \mathfrak{L}(E, F)$ and $S \in \mathfrak{L}(F, G)$. Fix any natural numbers m and n. Given $X \in \mathfrak{L}(l_2, E)$ and $\varepsilon > 0$, we first choose $B \in \mathfrak{F}(l_2, F)$ such that

$$\|TX - B\| \leqq (1 + \varepsilon)\, a_n(TX) \quad \text{and} \quad \operatorname{rank}(B) < n.$$

Secondly we pick $A \in \mathfrak{F}(l_2, G)$ such that

$$\|S(TX - B) - A\| \leqq (1 + \varepsilon)\, a_m(S(TX - B)) \quad \text{and} \quad \operatorname{rank}(A) < m.$$

It follows from

$$\operatorname{rank}(A + SB) \leqq \operatorname{rank}(A) + \operatorname{rank}(B) < m + n - 1$$

that

$$\begin{aligned} a_{m+n-1}(STX) &\leqq \|STX - SB - A\| \leqq (1 + \varepsilon)\, a_m(S(TX - B)) \\ &\leqq (1 + \varepsilon)\, x_m(S)\, \|TX - B\| \\ &\leqq (1 + \varepsilon)^2\, x_m(S)\, a_n(TX) \\ &\leqq (1 + \varepsilon)^2\, x_m(S)\, x_n(T)\, \|X\|. \end{aligned}$$

Hence

$$x_{m+n-1}(ST) \leqq (1 + \varepsilon)^2\, x_m(S)\, x_n(T).$$

Letting $\varepsilon \to 0$, we obtain the result.

2.4.18.* Next we state an immediate corollary of 2.2.9 and 2.4.17.

Multiplication theorem. If $1/p + 1/q = 1/r$ and $1/u + 1/v = 1/w$, then $\mathfrak{L}_{p,u}^{(x)} \circ \mathfrak{L}_{q,v}^{(x)} \subseteqq \mathfrak{L}_{r,w}^{(x)}$.

2.4.19.* We now compare the Weyl numbers with the Gel'fand numbers.

Proposition (A. Pietsch 1980:c). $x_n(T) \leqq c_n(T)$ for all $T \in \mathfrak{L}(E, F)$.

Proof. By 2.4.6, we have

$$a_n(TX) = c_n(TX) \leqq c_n(T)\, \|X\| \quad \text{for} \quad X \in \mathfrak{L}(l_2, E).$$

This implies the desired inequality.

2.4.20.* Finally, in view of 2.4.13 (Remark), we may formulate the following analogue of 2.4.6.

Proposition. Suppose that H is a Hilbert space. Then

$$x_n(T) = a_n(T) \quad \text{for all } T \in \mathfrak{L}(H, F).$$

2.5. Kolmogorov and Chang numbers

2.5.1. An s-scale s is **surjective** if, given any metric surjection $Q \in \mathfrak{L}(E_0, E)$,

$$s_n(T) = s_n(TQ) \quad \text{for all } T \in \mathfrak{L}(E, F).$$

2.5.2. The n-th **Kolmogorov number** of $T \in \mathfrak{L}(E, F)$ is defined by

$$d_n(T) := \inf\{\|Q_N^F T\| : \dim(N) < n\}.$$

Recall that Q_N denotes the canonical surjection from the Banach space F onto the quotient space F/N.

2.5.3. The following result can easily be verified.

Theorem (A. Pietsch 1972:a). The rule

$$d: T \to (d_n(T)),$$

which assigns to every operator the sequence of its Kolmogorov numbers, is a surjective s-scale.

2.5.4. The operators belonging to $\mathfrak{L}^{(d)}_{r,w}$ are said to be of **Kolmogorov type** $l_{r,w}$.

2.5.5. **Theorem** (I. A. Novosel'skij 1964). $\mathfrak{L}^{(d)}_{r,w}$ is a surjective quasi-Banach operator ideal.

2.5.6. As shown in (PIE, 11.7.7), we have

$$c_n(T) = d_n(T') \quad \text{and} \quad d_n(T) = c_n(T') \quad \text{for all } T \in \mathfrak{K}(E, F).$$

This means that the concepts of Gel'fand and Kolmogorov numbers are dual to each other.

Proposition (A. Pietsch 1974:a). $\mathfrak{L}^{(c)}_{r,w} = (\mathfrak{L}^{(d)}_{r,w})'$ and $\mathfrak{L}^{(d)}_{r,w} = (\mathfrak{L}^{(c)}_{r,w})'$.

2.5.7. The following result is a dual version of 2.4.11.

Theorem. $\mathfrak{L}^{(d)} = \mathfrak{K}$.

2.5.8. The n-th **Chang number** of $T \in \mathfrak{L}(E, F)$ is defined by

$$y_n(T) := \sup\{a_n(YT) : Y \in \mathfrak{L}(F, l_2),\ \|Y\| \leqq 1\}.$$

2.5.9. **Theorem** (A. Pietsch 1980:c). The rule

$$y: T \to (y_n(T)),$$

which assigns to every operator the sequence of its Chang numbers, is a surjective s-scale.

2.5.10. The operators belonging to $\mathfrak{L}^{(y)}_{r,w}$ are said to be of **Chang type** $l_{r,w}$.

2.5.11. **Theorem.** $\mathfrak{L}^{(y)}_{r,w}$ is a surjective quasi-Banach operator ideal.

2.5.12. We state, without proof, that

$$x_n(T) = y_n(T') \quad \text{and} \quad y_n(T) = x_n(T') \quad \text{for all } T \in \mathfrak{L}(E, F).$$

This means that the concepts of Weyl and Chang numbers are dual to each other.

Proposition (A. Pietsch 1980:c). $\mathfrak{L}_{r,w}^{(x)} = (\mathfrak{L}_{r,w}^{(y)})'$ and $\mathfrak{L}_{r,w}^{(y)} = (\mathfrak{L}_{r,w}^{(x)})'$.

2.5.13. The following result is dual to 2.4.19.

Proposition. $y_n(T) \leqq d_n(T)$ for all $T \in \mathfrak{L}(E, F)$.

2.6. Hilbert numbers

2.6.1. The n-th **Hilbert number** of $T \in \mathfrak{L}(E, F)$ is defined by

$$h_n(T) := \sup \left\{ a_n(YTX) : \begin{array}{l} X \in \mathfrak{L}(l_2, E),\ \|X\| \leqq 1 \\ Y \in \mathfrak{L}(F, l_2),\ \|Y\| \leqq 1 \end{array} \right\}.$$

2.6.2. The following result can easily be verified.

Theorem (W. Bauhardt 1977). The rule

$$h : T \to (h_n(T)),$$

which assigns to every operator the sequence of its Hilbert numbers, is an s-scale.

2.6.3. The preceding result can be improved as follows.

Theorem (W. Bauhardt 1977). The Hilbert numbers yield the smallest s-scale.

Proof. Let $T \in \mathfrak{L}(E, F)$. If $X \in \mathfrak{L}(l_2, E)$ and $Y \in \mathfrak{L}(F, l_2)$, then YTX acts on l_2. Hence, by 2.11.9, we have

$$a_n(YTX) = s_n(YTX) \leqq \|Y\|\, s_n(T)\, \|X\|$$

for every s-scale s. The definition of Hilbert numbers now implies that $h_n(T) \leqq s_n(T)$.

2.6.4. The operators belonging to $\mathfrak{L}_{r,w}^{(h)}$ are said to be of **Hilbert type** $l_{r,w}$.

2.6.5. **Theorem.** $\mathfrak{L}_{r,w}^{(h)}$ is a quasi-Banach operator ideal.

2.6.6. As observed in 2.9.19 (Remark), the Hilbert numbers fail to be multiplicative. We have, however, the following property.

Lemma. If $T \in \mathfrak{L}(E, F)$ and $S \in \mathfrak{L}(F, G)$, then

$$h_{m+n-1}(ST) \leqq h_m(S)\, x_n(T) \quad \text{and} \quad h_{m+n-1}(ST) \leqq y_m(S)\, h_n(T).$$

Proof. Fix any natural numbers m and n. Let $X \in \mathfrak{L}(l_2, E)$ and $Z \in \mathfrak{L}(G, l_2)$. Given $\varepsilon > 0$, we first choose $B \in \mathfrak{F}(l_2, F)$ such that

$$\|TX - B\| \leqq (1 + \varepsilon)\, a_n(TX) \quad \text{and} \quad \operatorname{rank}(B) < n.$$

Secondly we pick $A \in \mathfrak{F}(l_2, l_2)$ such that

$$\|ZS(TX - B) - A\| \leqq (1 + \varepsilon)\, a_m(ZS(TX - B)) \quad \text{and} \quad \operatorname{rank}(A) < m.$$

It follows from

$$\operatorname{rank}(A + ZSB) \leqq \operatorname{rank}(A) + \operatorname{rank}(B) < m + n - 1$$

that

$$\begin{aligned} a_{m+n-1}(ZSTX) &\leqq \|ZSTX - ZSB - A\| \leqq (1+\varepsilon)\, a_m(ZS(TX-B)) \\ &\leqq (1+\varepsilon)\, \|Z\|\, h_m(S)\, \|TX - B\| \\ &\leqq (1+\varepsilon)^2\, \|Z\|\, h_m(S)\, a_n(TX) \\ &\leqq (1+\varepsilon)^2\, \|Z\|\, h_m(S)\, x_n(T)\, \|X\|\,. \end{aligned}$$

Hence

$$h_{m+n-1}(ST) \leqq (1+\varepsilon)^2\, h_m(S)\, x_n(T).$$

Letting $\varepsilon \to 0$, we obtain the left-hand inequality.

The other one can be checked by dual considerations.

2.6.7. Reasoning as in the proof of 2.2.9 yields some further multiplication formulas.

Theorem. If $1/p + 1/q = 1/r$ and $1/u + 1/v = 1/w$, then

$$\mathfrak{L}^{(h)}_{p,u} \circ \mathfrak{L}^{(x)}_{q,v} \subseteq \mathfrak{L}^{(h)}_{r,w} \quad \text{and} \quad \mathfrak{L}^{(y)}_{p,u} \circ \mathfrak{L}^{(h)}_{q,v} \subseteq \mathfrak{L}^{(h)}_{r,w}.$$

2.6.8. In view of 2.10.7, we may state the following supplement of the preceding result.

Theorem. Suppose that $0 < p < 2$ or $0 < q < 2$. If $1/p + 1/q - 1/2 = 1/r$ and $1/u + 1/v = 1/w$, then

$$\mathfrak{L}^{(h)}_{p,u} \circ \mathfrak{L}^{(h)}_{q,v} \subseteq \mathfrak{L}^{(h)}_{r,w}.$$

2.6.9. Finally, we mention that

$$h_n(T) = h_n(T') \quad \text{for all } T \in \mathfrak{L}(E, F).$$

This formula implies that the operators T and T' possess the same Hilbert type.

Proposition (W. Bauhardt 1977). $\mathfrak{L}^{(h)}_{r,w} = (\mathfrak{L}^{(h)}_{r,w})'$.

2.7. Absolutely $(r, 2)$-summing operators

Throughout this section H denotes a complex Hilbert space.

2.7.1.* The following lemma due to G. Pisier (see H. König 1980:c) is very elementary but quite useful.

Lemma. Let $T \in \mathfrak{L}(H, F)$. If $a_n(T) > 0$, then for every $\varepsilon > 0$ there exists an orthonormal family $(x_1, \ldots, x_n)$ in H such that

$$a_k(T) \leqq (1+\varepsilon)\, \|Tx_k\| \quad \text{for } k = 1, \ldots, n.$$

Proof. The required family can be constructed by induction. If $x_1, \ldots, x_{n-1}$ have already been found, then we define the subspace

$$M_n := \{x \in H : (x, x_k) = 0 \quad \text{for } k = 1, \ldots, n-1\}.$$

Since $\operatorname{codim}(M_n) < n$, it follows from 2.4.6 that

$$a_n(T) = c_n(T) \leqq \|TJ^H_{M_n}\|\,.$$

Hence there exists $x_n \in M_n$ such that

$$a_n(T) \leqq (1+\varepsilon)\, \|Tx_n\| \quad \text{and} \quad \|x_n\| = 1\,.$$

This completes the proof.

2.7.2.* **Lemma** (H. König 1980:**c**). Let $2 \leqq r < \infty$. Then

$$\left(\sum_{k=1}^{n} a_k(T)^r\right)^{1/r} \leqq \|T \mid \mathfrak{P}_{r,2}\| \quad \text{for all } T \in \mathfrak{P}_{r,2}(H, F).$$

Proof. We may assume that $a_n(T) > 0$. Given $\varepsilon > 0$, there exist $x_1, \ldots, x_n$ as constructed in the preceding lemma. Note that

$$a_k(T) \leqq (1 + \varepsilon) \|Tx_k\| \quad \text{and} \quad \|(x_k) \mid w_2\| = 1.$$

Therefore

$$\left(\sum_{k=1}^{n} a_k(T)^r\right)^{1/r} \leqq (1 + \varepsilon) \left(\sum_{k=1}^{n} \|Tx_k\|^r\right)^{1/r}$$
$$= (1 + \varepsilon) \|(Tx_k) \mid l_r\| \leqq (1 + \varepsilon)\|T \mid \mathfrak{P}_{r,2}\| \, \|(x_k) \mid w_2\|.$$

Letting $\varepsilon \to 0$, we obtain the desired inequality.

2.7.3.* The next result is extremely important.

Lemma (A. Pietsch 1980:c). Let $2 \leqq r < \infty$. Then

$$n^{1/r} x_n(T) \leqq \|T \mid \mathfrak{P}_{r,2}\| \quad \text{for all } T \in \mathfrak{P}_{r,2}(E, F).$$

Proof. If $X \in \mathfrak{L}(l_2, E)$, then 2.7.2 implies that

$$n^{1/r} a_n(TX) \leqq \left(\sum_{k=1}^{n} a_k(TX)^r\right)^{1/r} \leqq \|TX \mid \mathfrak{P}_{r,2}\| \leqq \|T \mid \mathfrak{P}_{r,2}\| \, \|X\|,$$

and the definition of Weyl numbers yields the inequality we are looking for.

2.7.4.* We are now in a position to establish the basic result of this section.

Theorem (A. Pietsch 1980:c). Let $2 \leqq r < \infty$. Then

$$\mathfrak{L}_{r,1}^{(x)} \subseteq \mathfrak{P}_{r,2} \subseteq \mathfrak{L}_{r,\infty}^{(x)}.$$

Proof. From 1.2.8 we know that

$$\|T \mid \mathfrak{P}_{r,2}\| \leqq n^{1/r}\|T\| \quad \text{whenever} \quad \operatorname{rank}(T) \leqq n.$$

Therefore the embedding theorem 2.3.10 implies that $\mathfrak{L}_{r,1}^{(a)} \subseteq \mathfrak{P}_{r,2}$.

We now assume that $T \in \mathfrak{L}_{r,1}^{(x)}(E, F)$. Then

$$TX \in \mathfrak{L}_{r,1}^{(a)}(l_2, F) \subseteq \mathfrak{P}_{r,2}(l_2, F) \quad \text{for all } X \in \mathfrak{L}(l_2, E).$$

Hence $T \in \mathfrak{P}_{r,2}(E, F)$, by 1.2.7. This proves the first inclusion.

The second one is an immediate consequence of 2.7.3.

Remark. The following examples show that, for $r = 2$, the parameters in the first inclusion are the best possible.

From 1.6.8 and 2.9.10 we know that

$$D_t \in \mathfrak{P}_2(l_2, l_1) \quad \text{if and only if} \quad t \in l_1$$

and

$$D_t \in \mathfrak{L}_{2,w}^{(x)}(l_2, l_1) \quad \text{if and only if} \quad t \in l_{1,w}.$$

Therefore

$$\mathfrak{L}_{2,w}^{(x)}(l_2, l_1) \nsubseteq \mathfrak{P}_2(l_2, l_1) \quad \text{whenever } 1 < w \leqq \infty.$$

Let $1/p := 1/r + 1/2$. Then $I \in \mathfrak{P}_{r,2}(l_p, l_2)$, by 1.6.7. Moreover, it can be shown that $x_n(I: l_p \to l_2) \asymp n^{-1/r}$. Hence

$$\mathfrak{P}_{r,2}(l_p, l_2) \nsubseteq \mathfrak{L}^{(x)}_{r,w}(l_p, l_2) \quad \text{whenever } 0 < w < \infty .$$

Thus the second inclusion is sharp for $2 \leqq r < \infty$.

Finally, in view of

$$\mathfrak{L}^{(x)}_{r,1}(l_2) \subset \mathfrak{P}_{r,2}(l_2) \subset \mathfrak{L}^{(x)}_{r,\infty}(l_2),$$

both inclusions are strict.

2.7.5. The preceding result can be improved by interpolation.

Theorem (A. Pietsch 1980:c). Let $2 < r < \infty$. Then

$$\mathfrak{L}^{(x)}_{r} \subseteq \mathfrak{P}_{r,2} \subseteq \mathfrak{L}^{(x)}_{r,\infty} .$$

Proof. Choose r_0, r_1 and θ such that $1/r = (1-\theta)/r_0 + \theta/r_1$, $2 < r_0 < r < r_1 < \infty$ and $0 < \theta < 1$. Then it follows from 2.3.14, 2.7.4 and 1.2.6 that

$$\mathfrak{L}^{(a)}_{r} = (\mathfrak{L}^{(a)}_{r_0,1}, \mathfrak{L}^{(a)}_{r_1,1})_{\theta,r} \subseteq (\mathfrak{P}_{r_0,2}, \mathfrak{P}_{r_1,2})_{\theta,r} \subseteq \mathfrak{P}_{r,2} .$$

We now assume that $T \in \mathfrak{L}^{(x)}_{r}(E, F)$. Then

$$TX \in \mathfrak{L}^{(a)}_{r}(l_2, F) \subseteq \mathfrak{P}_{r,2}(l_2, F) \quad \text{for all } X \in \mathfrak{L}(l_2, E).$$

Hence $T \in \mathfrak{P}_{r,2}(E, F)$, by 1.2.7.

2.7.6. **Proposition** (H. König 1980:c). Let $2 < r < \infty$. Then

$$\mathfrak{L}^{(a)}_{r}(H, F) = \mathfrak{L}^{(x)}_{r}(H, F) = \mathfrak{P}_{r,2}(H, F).$$

Proof. From the preceding theorem we know that

$$\mathfrak{L}^{(a)}_{r}(H, F) \subseteq \mathfrak{L}^{(x)}_{r}(H, F) \subseteq \mathfrak{P}_{r,2}(H, F).$$

Furthermore, Lemma 2.7.2 yields

$$\mathfrak{P}_{r,2}(H, F) \subseteq \mathfrak{L}^{(a)}_{r}(H, F).$$

Thus equality holds.

2.7.7. As an easy application of 2.7.4 we now prove a striking result for absolutely $(r, 2)$-summing operators.

Multiplication theorem (A. Pietsch 1980:c). Let $2 \leqq p, q, s < \infty$. If $1/p + 1/q > 1/s$, then

$$\mathfrak{P}_{p,2} \circ \mathfrak{P}_{q,2} \subseteq \mathfrak{P}_{s,2} .$$

Proof. Writing $1/r := 1/p + 1/q$, we have $r < s$. Hence 2.7.4, 2.4.18 and 2.2.7 imply that

$$\mathfrak{P}_{p,2} \circ \mathfrak{P}_{q,2} \subseteq \mathfrak{L}^{(x)}_{p,\infty} \circ \mathfrak{L}^{(x)}_{q,\infty} \subseteq \mathfrak{L}^{(x)}_{r,\infty} \subseteq \mathfrak{L}^{(x)}_{s,1} \subseteq \mathfrak{P}_{s,2} .$$

Remark. Let $1/p + 1/q = 1/r$ and $2 \leqq r < \infty$. Then, as shown in 2.11.30,

$$\mathfrak{P}_{p,2} \circ \mathfrak{P}_{q,2}(l_2) = \mathfrak{L}^{(a)}_{r,q}(l_2).$$

On the other hand, we know from 2.7.6 that

$$\mathfrak{P}_{r,2}(l_2) = \mathfrak{L}^{(a)}_{r,r}(l_2).$$

Hence

$$\mathfrak{P}_{p,2} \circ \mathfrak{P}_{q,2} \subsetneqq \mathfrak{P}_{r,2}.$$

This proves that, in the preceding theorem, the assumption $1/p + 1/q > 1/s$ is essential.

2.7.8.* Finally, we prove an inequality which looks rather complicated. It has, however, far-reaching applications.

Lemma (A. Pietsch 1980:c). Let $T \in \mathfrak{L}(E, F)$ and $S \in \mathfrak{P}_2(F, G)$. Then

$$\left[\prod_{k=1}^{2n-1} x_k(ST)\right]^{1/(2n-1)} \leqq \exp\left(\frac{1}{2n-1} \|S \mid \mathfrak{P}_2\|^2\right) \left[\prod_{k=1}^{n} x_k(T)\right]^{1/n}.$$

Proof. Since the Weyl numbers are multiplicative, we have

$$x_{2k-1}(ST) \leqq x_k(S)\, x_k(T) \quad \text{and} \quad x_{2k}(ST) \leqq x_k(S)\, x_{k+1}(T).$$

Hence

$$\prod_{k=1}^{2n-1} x_k(ST) = \prod_{k=1}^{n} x_{2k-1}(ST) \prod_{k=1}^{n-1} x_{2k}(ST)$$
$$\leqq \left(\prod_{k=1}^{n} x_k(S) \prod_{k=1}^{n-1} x_k(S)\right) \left(\prod_{k=1}^{n} x_k(T) \prod_{k=1}^{n-1} x_{k+1}(T)\right).$$

Write $\sigma := \|S \mid \mathfrak{P}_2\|^2$. From 2.7.3 we know that $k^{1/2} x_k(S) \leqq \sigma^{1/2}$. Thus

$$\prod_{k=1}^{n} x_k(S) \prod_{k=1}^{n-1} x_k(S) \leqq \left[\frac{\sigma^n}{n!} \frac{\sigma^{n-1}}{(n-1)!}\right]^{1/2} \leqq \exp(\sigma).$$

Furthermore, since

$$\left[\prod_{k=1}^{n} x_k(T)\right]^{1/n} \leqq \|T\|,$$

it follows that

$$\prod_{k=1}^{n} x_k(T) \prod_{k=1}^{n-1} x_{k+1}(T) = \|T\|^{-1} \left[\prod_{k=1}^{n} x_k(T)\right]^2 \leqq \left[\prod_{k=1}^{n} x_k(T)\right]^{(2n-1)/n}.$$

Combining these inequalities, we obtain the desired result.

2.8. Generalized approximation numbers

2.8.1.* Let $\mathfrak{A}$ be any quasi-Banach operator ideal. The n-th **$\mathfrak{A}$-approximation number** of $T \in \mathfrak{A}(E, F)$ is defined by

$$a_n(T \mid \mathfrak{A}) := \inf\{\|T - L \mid \mathfrak{A}\| : L \in \mathfrak{F}(E, F),\ \operatorname{rank}(L) < n\}.$$

2.8.2. The basic properties of these generalized approximation numbers are similar to those of the original ones. Recall that $c_{\mathfrak{A}}$ denotes the constant in the quasi-triangle inequality of the underlying quasi-norm $\|\cdot \mid \mathfrak{A}\|$.

Theorem.

(1) $\|T \mid \mathfrak{A}\| = a_1(T \mid \mathfrak{A}) \geqq a_2(T \mid \mathfrak{A}) \geqq \ldots \geqq 0$ for $T \in \mathfrak{A}(E, F)$.

(2) $a_{m+n-1}(S + T \mid \mathfrak{A}) \leqq c_{\mathfrak{A}}[a_m(S \mid \mathfrak{A}) + a_n(T \mid \mathfrak{A})]$ for $S, T \in \mathfrak{A}(E, F)$.

(3) $a_n(YTX \mid \mathfrak{A}) \leqq \|Y\| \, a_n(T \mid \mathfrak{A}) \, \|X\|$ for $X \in \mathfrak{L}(E_0, E)$, $T \in \mathfrak{A}(E, F)$, $Y \in \mathfrak{L}(F, F_0)$.

(4) If rank $(T) < n$, then $a_n(T \mid \mathfrak{A}) = 0$.

2.8.3. If the operator ideal $\mathfrak{A}$ is endowed with an arbitrary quasi-norm, then it may happen that the n-th $\mathfrak{A}$-approximation number does not depend continuously on the operator. However, we have the following result, analogous to 2.2.3.

Lemma. Let $\mathfrak{A}$ be a p-Banach operator ideal. Then

$$|a_n(S \mid \mathfrak{A})^p - a_n(T \mid \mathfrak{A})^p| \leqq \|S - T \mid \mathfrak{A}\|^p \quad \text{for } S, T \in \mathfrak{A}(E, F).$$

2.8.4. Let $\mathfrak{A}$ be any quasi-Banach operator ideal. An operator $T \in \mathfrak{A}(E, F)$ is said to be of **$\mathfrak{A}$-approximation type** $l_{r,w}$ if $(a_n(T \mid \mathfrak{A})) \in l_{r,w}$. The set of these operators is denoted by $\mathfrak{A}^{(a)}_{r,w}(E, F)$.

For $T \in \mathfrak{A}^{(a)}_{r,w}(E, F)$ we define

$$\|T \mid \mathfrak{A}^{(a)}_{r,w}\| := \|(a_n(T \mid \mathfrak{A})) \mid l_{r,w}\|.$$

Remark. Note that the definition of $a_n(T \mid \mathfrak{A})$ depends on the underlying quasi-norm. However, if two quasi-norms are equivalent, then corresponding inequalities hold for the associated $\mathfrak{A}$-approximation numbers. This shows, in view of D.1.4, that $\mathfrak{A}^{(a)}_{r,w}$ is uniquely determined by $\mathfrak{A}$. Therefore, without loss of generality, we may occasionally assume that $\mathfrak{A}$ is equipped with a p-norm.

2.8.5. The following theorem can be obtained by straightforward modification of the proof given in 2.2.5.

Theorem. $\mathfrak{A}^{(a)}_{r,w}$ is a quasi-Banach operator ideal.

2.8.6. Proposition. Let $\mathfrak{A}$ and $\mathfrak{B}$ be quasi-Banach operator ideals with $\mathfrak{A} \subseteqq \mathfrak{B}$. Then $\mathfrak{A}^{(a)}_{r,w} \subseteqq \mathfrak{B}^{(a)}_{r,w}$.

Proof. By D.1.9, there exists a constant $c > 0$ such that

$$\|T \mid \mathfrak{B}\| \leqq c\|T \mid \mathfrak{A}\| \quad \text{for all } T \in \mathfrak{A}.$$

It easily follows that corresponding inequalities hold for the associated approximation numbers:

$$a_n(T \mid \mathfrak{B}) \leqq c a_n(T \mid \mathfrak{A}) \quad \text{for all } T \in \mathfrak{A}.$$

This implies the desired inclusion.

2.8.7. We now establish various multiplication formulas.

Proposition. Let $\mathfrak{A}$ and $\mathfrak{B}$ be quasi-Banach operator ideals. Then

$$\mathfrak{A} \circ \mathfrak{B}^{(a)}_{r,w} \subseteqq (\mathfrak{A} \circ \mathfrak{B})^{(a)}_{r.w} \quad \text{and} \quad \mathfrak{A}^{(a)}_{r,w} \circ \mathfrak{B} \subseteqq (\mathfrak{A} \circ \mathfrak{B})^{(a)}_{r,w}.$$

Moreover,

$$\mathfrak{A}^{(a)}_{p,u} \circ \mathfrak{B}^{(a)}_{q,v} \subseteqq (\mathfrak{A} \circ \mathfrak{B})^{(a)}_{r,w}$$

provided that $1/p + 1/q = 1/r$ and $1/u + 1/v = 1/w$.

Proof. Note that, for $T \in \mathfrak{B}(E, F)$ and $S \in \mathfrak{A}(F,G)$,

$$a_{m+n-1}(ST \mid \mathfrak{A} \circ \mathfrak{B}) \leqq a_m(S \mid \mathfrak{A})\, a_n(T \mid \mathfrak{B}).$$

Modifying the estimates in the proof of 2.2.9, we obtain the third inclusion. The other ones can be checked by putting $m := 1$ or $n := 1$ in the inequality stated above.

2.8.8. Next we formulate a generalization of 2.3.8 which can be proved in the same way.

Representation theorem (A. Pietsch 1981:a). Let $q = 2, 3, \ldots$ An operator $T \in \mathfrak{A}(E, F)$ is of $\mathfrak{A}$-approximation type $l_{r,w}$ if and only if it can be written in the form

$$T = \sum_{k=0}^{\infty} T_k$$

such that $T_k \in \mathfrak{F}(E, F)$, rank $(T_k) \leqq q^k$ and $(q^{k/r}\|T_k \mid \mathfrak{A}\|) \in l_w$. Moreover,

$$\|T \mid \mathfrak{A}_{r,w}^{(a)}\|_q^{rep} := \inf \|(q^{k/r}\|T_k \mid \mathfrak{A}\|) \mid l_w\|,$$

where the infimum is taken over all possible representations, defines an equivalent quasi-norm on $\mathfrak{A}_{r,w}^{(a)}$.

2.8.9. The following statement is analogous to 2.3.9.

Proposition. If $0 < w < \infty$, then the quasi-Banach operator ideal $\mathfrak{A}_{r,w}^{(a)}$ is approximative.

2.8.10. In the setting of generalized approximation numbers the inclusion stated in 2.3.10 can be refined significantly.

Embedding theorem (A. Pietsch 1981:a). Let $0 < r, s < \infty$ and $1/p = 1/r + 1/s$. Assume that $\mathfrak{A}$ and $\mathfrak{B}$ are quasi-Banach operator ideals such that

$$\|T \mid \mathfrak{A}\| \leqq cn^{1/r}\|T \mid \mathfrak{B}\| \quad \text{whenever} \quad \text{rank}\,(T) \leqq n,$$

where $c > 0$ is a constant. Then

$$\mathfrak{B}_{p,w}^{(a)} \subseteq \mathfrak{A}_{s,w}^{(a)}.$$

Moreover, $\mathfrak{B}_{r,w}^{(a)} \subseteq \mathfrak{A}$ provided that $\mathfrak{A}$ is a w-Banach operator ideal $(0 < w \leqq 1)$.

Proof. Given $T \in \mathfrak{B}_{p,w}^{(a)}(E, F)$, we choose a representation

$$T = \sum_{k=0}^{\infty} T_k$$

such that $T_k \in \mathfrak{F}(E, F)$, rank $(T_k) \leqq 2^k$ and $(2^{k/p}\|T_k \mid \mathfrak{B}\|) \in l_w$. Then

$$\|T_k \mid \mathfrak{A}\| \leqq c\, 2^{k/r}\|T_k \mid \mathfrak{B}\|$$

implies that $(2^{k/s}\|T_k \mid \mathfrak{A}\|) \in l_w$. Hence $T \in \mathfrak{A}_{s,w}^{(a)}(E, F)$.
The proof of $\mathfrak{B}_{r,w}^{(a)} \subseteq \mathfrak{A}$ can be adapted from 2.3.10.

2.8.11. We now provide some auxiliary inequalities.

Lemma. There exists a constant $c > 0$ such that

$$\|T \mid \mathfrak{A}_{r,w}^{(a)}\| \leqq cn^{1/r}\|T \mid \mathfrak{A}\| \quad \text{whenever} \quad \text{rank}\,(T) \leqq n.$$

Proof. By G.3.2, we can find $c > 0$ such that

$$\left(\sum_{k=1}^{n} k^{w/r-1}\right)^{1/w} \leqq cn^{1/r}.$$

Hence

$$\|T \mid \mathfrak{A}_{r,w}^{(a)}\| = \left(\sum_{k=1}^{n} [k^{1/r-1/w} a_k(T \mid \mathfrak{A})]^w\right)^{1/w} \leqq cn^{1/r}\|T \mid \mathfrak{A}\|.$$

2.8.12. Lemma. $n^{1/r} a_{2n-1}(T \mid \mathfrak{A}) \leqq a_n(T \mid \mathfrak{A}_{r,\infty}^{(a)})$ for $T \in \mathfrak{A}_{r,\infty}^{(a)}(E, F)$.

Proof. Given $\varepsilon > 0$, we choose $X, Y \in \mathfrak{F}(E, F)$ such that

$$\|T - X \mid \mathfrak{A}_{r,\infty}^{(a)}\| \leqq (1 + \varepsilon)\, a_n(T \mid \mathfrak{A}_{r,\infty}^{(a)}) \quad \text{and} \quad \operatorname{rank}(X) < n,$$

$$\|T - X - Y \mid \mathfrak{A}\| \leqq (1 + \varepsilon)\, a_n(T - X \mid \mathfrak{A}) \quad \text{and} \quad \operatorname{rank}(Y) < n.$$

Since rank $(X + Y) < 2n - 1$, it follows that

$$n^{1/r} a_{2n-1}(T \mid \mathfrak{A}) \leqq n^{1/r}\|T - X - Y \mid \mathfrak{A}\| \leqq (1 + \varepsilon)\, n^{1/r} a_n(T - X \mid \mathfrak{A})$$
$$\leqq (1 + \varepsilon)\, \|T - X \mid \mathfrak{A}_{r,\infty}^{(a)}\| \leqq (1 + \varepsilon)^2\, a_n(T \mid \mathfrak{A}_{r,\infty}^{(a)}).$$

Letting $\varepsilon \to 0$ yields the desired inequality.

2.8.13. We are now in a position to describe what happens when the procedure $\mathfrak{A} \to \mathfrak{A}_{r,w}^{(a)}$ is applied repeatedly.

Reiteration theorem (A. Pietsch 1981:a). Let $0 < r,\ s < \infty$ and $1/p = 1/r + 1/s$. Then

$$(\mathfrak{A}_{r,u}^{(a)})_{s,w}^{(a)} = \mathfrak{A}_{p,w}^{(a)}.$$

Proof. Assume that $T \in \mathfrak{L}(E, F)$ is of $\mathfrak{A}_{r,\infty}^{(a)}$-approximation type $l_{s,w}$. Then, in view of 2.1.9, we deduce from

$$n^{1/r} a_{2n-1}(T \mid \mathfrak{A}) \leqq a_n(T \mid \mathfrak{A}_{r,\infty}^{(a)})$$

that $T \in \mathfrak{A}_{p,w}^{(a)}(E, F)$. Consequently,

$$(\mathfrak{A}_{r,u}^{(a)})_{s,w}^{(a)} \subseteq (\mathfrak{A}_{r,\infty}^{(a)})_{s,w}^{(a)} \subseteq \mathfrak{A}_{p,w}^{(a)}.$$

The reverse inclusion follows immediately from the embedding theorem and Lemma 2.8.11.

2.8.14. Lemma. $n^{1/r} x_{2n-1}(T) \leqq a_n(T \mid \mathfrak{L}_{r,\infty}^{(x)})$ for $T \in \mathfrak{L}_{r,\infty}^{(x)}(E, F)$.

Proof. Recall that

$$a_n(TX) = x_n(TX) \quad \text{for all } X \in \mathfrak{L}(l_2, E).$$

Hence, by 2.8.12, we have

$$n^{1/r} a_{2n-1}(TX) \leqq a_n(TX \mid \mathfrak{L}_{r,\infty}^{(a)}) = a_n(TX \mid \mathfrak{L}_{r,\infty}^{(x)}) \leqq a_n(T \mid \mathfrak{L}_{r,\infty}^{(x)})\, \|X\|,$$

and the definition of Weyl numbers yields the inequality we are looking for.

2.8.15. The following inclusion is an immediate consequence of the preceding lemma.

Theorem (A. Pietsch 1980:c). Let $0 < r, s < \infty$ and $1/p = 1/r + 1/s$. Then $(\mathfrak{L}_{r,\infty}^{(x)})_{s,w}^{(a)} \subseteq \mathfrak{L}_{p,w}^{(x)}$.

2.8.16. For later use, we prove a further inequality which could also be obtained as a corollary of 2.7.4 and 2.8.14.

Lemma. $n^{1/r} x_{2n-1}(T) \leqq a_n(T \mid \mathfrak{P}_{r,2})$ for $T \in \mathfrak{P}_{r,2}(E, F)$.

Proof. Given $\varepsilon > 0$, we choose $L \in \mathfrak{F}(E, F)$ such that

$$\|T - L \mid \mathfrak{P}_{r,2}\| \leqq (1 + \varepsilon)\, a_n(T \mid \mathfrak{P}_{r,2}) \quad \text{and} \quad \operatorname{rank}(T) < n.$$

Then it follows from 2.7.3 that

$$n^{1/r} x_{2n-1}(T) \leqq n^{1/r} x_n(T - L) \leqq \|T - L \mid \mathfrak{P}_{r,2}\| \leqq (1 + \varepsilon)\, a_n(T \mid \mathfrak{P}_{r,2}).$$

Letting $\varepsilon \to 0$ yields the result.

2.8.17. In the remaining part of this section we deal with operators of $\mathfrak{P}_2$-approximation type.

Proposition (A. Pietsch 1981:b). The quasi-Banach operator ideal $(\mathfrak{P}_2)^{(a)}_{r,w}$ is injective.

Proof. The assertion follows from the fact that, given any metric injection $J \in \mathfrak{L}(F, F_0)$, we have

$$a_n(T \mid \mathfrak{P}_2) = a_n(JT \mid \mathfrak{P}_2) \quad \text{for all } T \in \mathfrak{P}_2(E, F).$$

To verify this formula, we choose $L_0 \in \mathfrak{F}(E, F_0)$ such that

$$\|JT - L_0 \mid \mathfrak{P}_2\| \leqq (1 + \varepsilon)\, a_n(JT \mid \mathfrak{P}_2) \quad \text{and} \quad \operatorname{rank}(L_0) < n.$$

By 1.5.1, there exists a factorization

$$JT - L_0 = Y_0 A,$$

where $A \in \mathfrak{P}_2(E, H_0)$, $Y_0 \in \mathfrak{L}(H_0, F_0)$ and $\|Y_0\|\, \|A \mid \mathfrak{P}_2\| = \|JT - L_0 \mid \mathfrak{P}_2\|$. Let H be the inverse image of $M(J)$ with respect to Y_0, and denote by P the orthogonal projection from H_0 onto H. Then

$$Yf := J^{-1} Y_0 P f$$

defines an operator from H_0 into F such that $JY = Y_0 P$ and $\|Y\| \leqq \|Y_0\|$. Put

$$L := T - YA.$$

If $x \in N(L_0)$, then it follows from $Y_0 A x = JTx$ that $Ax \in H$. Therefore

$$JLx = JTx - JYAx = JTx - Y_0 PAx = JTx - Y_0 Ax = L_0 x = o.$$

Thus we have $x \in N(L)$. This proves that $N(L_0) \subseteqq N(L)$. Hence

$$\operatorname{rank}(L) \leqq \operatorname{rank}(L_0) < n.$$

We now obtain

$$\begin{aligned} a_n(T \mid \mathfrak{P}_2) &\leqq \|T - L \mid \mathfrak{P}_2\| = \|YA \mid \mathfrak{P}_2\| \leqq \|Y\|\, \|A \mid \mathfrak{P}_2\| \\ &\leqq \|Y_0\|\, \|A \mid \mathfrak{P}_2\| = \|JT - L_0 \mid \mathfrak{P}_2\| \leqq (1 + \varepsilon)\, \|JT \mid \mathfrak{P}_2\|. \end{aligned}$$

Letting $\varepsilon \to 0$ yields $a_n(T \mid \mathfrak{P}_2) \leqq a_n(JT \mid \mathfrak{P}_2)$.
The reverse inequality is obvious.

2.8.18. Proposition (A. Pietsch 1981:b). Let $0 < s < \infty$ and $1/r = 1/s + 1/2$.

Then

$$\mathfrak{L}^{(c)}_{r,w} \subseteq (\mathfrak{P}_2)^{(a)}_{s,w} \subseteq \mathfrak{L}^{(x)}_{r,w}.$$

Proof. From 1.2.8 we know that

$$\|T \mid \mathfrak{P}_2\| \leq n^{1/2} \|T\| \quad \text{whenever} \quad \operatorname{rank}(T) \leq n.$$

Hence the embedding theorem 2.8.10 yields

$$\mathfrak{L}^{(a)}_{r,w} \subseteq (\mathfrak{P}_2)^{(a)}_{s,w}.$$

We now assume that $T \in \mathfrak{L}^{(c)}_{r,w}(E, F)$. Take any metric injection $J \in \mathfrak{L}(F, F_0)$, where F_0 has the metric extension property. It follows from 2.4.7 that $JT \in \mathfrak{L}^{(a)}_{r,w}(E, F_0)$. Hence $JT \in (\mathfrak{P}_2)^{(a)}_{s,w}(E, F_0)$. Finally, the injectivity of the operator ideal $(\mathfrak{P}_2)^{(a)}_{s,w}$ implies that $T \in (\mathfrak{P}_2)^{(a)}_{s,w}(E, F)$. This completes the proof of the left-hand inclusion.

Furthermore, by 2.7.4 and 2.8.15, we have

$$(\mathfrak{P}_2)^{(a)}_{s,w} \subseteq (\mathfrak{L}^{(x)}_{2,\infty})^{(a)}_{s,w} \subseteq \mathfrak{L}^{(x)}_{r,w}.$$

Remark. By considering diagonal operators from l_∞ into l_2, it can be shown that the left-hand inclusion is strict. We conjecture that the right-hand inclusion is strict, as well.

2.8.19. Proposition. $(\mathfrak{P}_2)^{(a)}_{2/m,2/(m+2)} \subseteq (\mathfrak{P}_2)^{m+1} \subseteq (\mathfrak{P}_2)^{(a)}_{2/m,2/m}$.

Proof. In view of 1.2.8,

$$\|T \mid (\mathfrak{P}_2)^{m+1}\| \leq n^{m/2} \|T \mid \mathfrak{P}_2\| \quad \text{whenever} \quad \operatorname{rank}(T) \leq n.$$

Furthermore, we know from 1.5.7 that $(\mathfrak{P}_2)^{m+1}$ is $2/(m+2)$-normed. Thus 2.8.10 yields

$$(\mathfrak{P}_2)^{(a)}_{2/m,2/(m+2)} \subseteq (\mathfrak{P}_2)^{m+1}.$$

We now assume that $T \in (\mathfrak{P}_2)^{m+1}(E, F)$. Then, by 1.5.1, there exists a factorization

$$T = YT_m \dots T_1 A$$

such that $A \in \mathfrak{P}_2(E, H_1)$, $T_k \in \mathfrak{P}_2(H_k, H_{k+1})$ and $Y \in \mathfrak{L}(H_{m+1}, F)$, where $H_1, \dots, H_{m+1}$ are Hilbert spaces. In view of 2.7.2, we have $T_k \in \mathfrak{L}^{(a)}_2(H_k, H_{k+1})$. Therefore

$$T \in (\mathfrak{L}^{(a)}_2)^m \circ \mathfrak{P}_2(E, F) \subseteq (\mathfrak{P}_2)^{(a)}_{2/m}(E, F),$$

by 2.8.7. This proves that

$$(\mathfrak{P}_2)^{m+1} \subseteq (\mathfrak{P}_2)^{(a)}_{2/m}.$$

Remark. The following examples show that the parameters in the above inclusions are the best possible. In particular, it turns out that both inclusions are strict.

From 1.6.8 and 2.9.14 we know that

$$D_t \in (\mathfrak{P}_2)^{m+1}(l_2, l_1) \quad \text{if and only if} \quad t \in l_{2/(m+2),2/(m+2)}$$

and

$$D_t \in (\mathfrak{P}_2)^{(a)}_{2/m,w}(l_2, l_1) \quad \text{if and only if} \quad t \in l_{2/(m+2),w}.$$

Therefore

$$(\mathfrak{P}_2)^{(a)}_{2/m,w}(l_2, l_1) \nsubseteq (\mathfrak{P}_2)^{m+1}(l_2, l_1) \quad \text{whenever } w > 2/(m+2).$$

Furthermore, 1.6.10 and 2.9.16 state that

$$D_t \in (\mathfrak{P}_2)^{m+1}(l_1, l_2) \quad \text{if and only if} \quad t \in l_{2/m,2/m}$$

and

$$D_t \in (\mathfrak{P}_2)^{(a)}_{2/m,w}(l_1, l_2) \quad \text{if and only if} \quad t \in l_{2/m,w}.$$

Hence

$$(\mathfrak{P}_2)^{m+1}(l_1, l_2) \nsubseteq (\mathfrak{P}_2)^{(a)}_{2/m,w} \quad \text{whenever } w < 2/m.$$

2.8.20. The next result is an immediate consequence of 1.7.10 and 2.9.15.

Proposition. Let $0 < r < 2$ and $1/s = 1/r - 1/2$. Then

$$\mathfrak{N}_{r,2} \subseteq (\mathfrak{P}_2)^{(a)}_{s,r}.$$

Remark. By 2.9.14, we have

$$D_t \in (\mathfrak{P}_2)^{(a)}_{s,r}(l_2, l_1) \quad \text{if and only if} \quad t \in l_{p,r},$$

where $1/p = 1/s + 1$. Moreover, it can be shown that

$$D_t \in \mathfrak{N}_{r,2}(l_2, l_1) \quad \text{if and only if} \quad t \in l_p,$$

where $1/p = 1/s + 1/2$. This proves that the inclusion

$$\mathfrak{N}_{r,2}(l_2, l_1) \subseteq (\mathfrak{P}_2)^{(a)}_{s,r}(l_2, l_1)$$

is strict.

2.8.21. Finally, for later applications, we establish a multiplication formula which is analogous to 2.3.13.

Proposition (A. Pietsch 1981:b). If $1/p + 1/q = 1/r$ and $1/u + 1/v = 1/w$, then

$$\mathfrak{L}^{(a)}_{p,u} \circ \mathfrak{H} \circ (\mathfrak{P}_2)^{(a)}_{q,v} = (\mathfrak{P}_2)^{(a)}_{r,w}.$$

Proof. The inclusion

$$\mathfrak{L}^{(a)}_{p,u} \circ \mathfrak{H} \circ (\mathfrak{P}_2)^{(a)}_{q,v} \subseteq \mathfrak{L}^{(a)}_{p,v} \circ (\mathfrak{P}_2)^{(a)}_{q,o} \subseteq (\mathfrak{P}_2)^{(a)}_{r,w}$$

follows immediately from 2.8.7.

The converse can be checked by modifying the proof of 2.3.13. Given $T \in (\mathfrak{P}_2)^{(a)}_{r,w}(E, F)$, we consider a representation

$$T = \sum_{k=0}^{\infty} T_k$$

such that $T_k \in \mathfrak{F}(E, F)$, rank $(T_k) \leq 2^k$, and $(2^{k/r}\|T_k \mid \mathfrak{P}_2\|) \in l_w$. In view of 1.5.1, there exist factorizations $T_k = Y_k A_k$ with $A_k \in \mathfrak{L}(E, l_2(2^k))$, $Y_k \in \mathfrak{L}(l_2(2^k), F)$ and $\|Y_k\| \, \|A_k \mid \mathfrak{P}_2\| = \|T_k \mid \mathfrak{P}_2\|$. If $0 < w < \infty$, it can be arranged that

$$2^{k/p}\|Y_k\| = (2^{k/r}\|T_k \mid \mathfrak{P}_2\|)^{w/u} \quad \text{and} \quad 2^{k/q}\|A_k \mid \mathfrak{P}_2\| = (2^{k/r}\|T_k \, \mathfrak{P}_2\|)^{w/v}.$$

Form the Hilbert space $H := [l_2, l_2(2^k)]$ as described in C.2.1. Let $J_k \in \mathfrak{L}(l_2(2^k), H)$ and $Q_k \in \mathfrak{L}(H, l_2(2^k))$ denote the canonical injections and surjections, respectively. Observe that

$$(2^{k/p}\|Y_k Q_k\|) \in l_u \quad \text{and} \quad (2^{k/q}\|J_k A_k \mid \mathfrak{P}_2\|) \in l_v.$$

Therefore the operators

$$Y := \sum_{k=0}^{\infty} Y_k Q_k \quad \text{and} \quad A := \sum_{k=0}^{\infty} J_k A_k$$

are of approximation type $l_{p,u}$ and of $\mathfrak{P}_2$-approximation type $l_{q,v}$, respectively. Thus $T = YA$ is the factorization we are looking for.

In the case $w = \infty$ the proof requires the same modification as in 2.3.13.

2.9. Diagonal operators

Throughout this section we assume that $1 \leqq p, q \leqq \infty$.

2.9.1.* In the following we are concerned with **diagonal operators**

$$D_t : (\xi_k) \to (\tau_k \xi_k)$$

induced by scalar sequences $t = (\tau_k)$ and acting from l_p into l_q. To simplify the formulation of the results, it is always assumed that $\tau_1 \geqq \tau_2 \geqq \ldots \geqq 0$.

2.9.2.* For later use, we define the operators

$$J_m : (\xi_1, \ldots, \xi_m) \to (\xi_1, \ldots, \xi_m, 0, 0, \ldots)$$

and

$$Q_m : (\xi_1, \ldots, \xi_m, \xi_{m+1}, \ldots) \to (\xi_1, \ldots, \xi_m).$$

Note that

$$\|J_m : l_p(m) \to l_p\| = 1 \quad \text{and} \quad \|Q_m : l_p \to l_p(m)\| = 1.$$

Moreover, $P_m := J_m Q_m$ is the projection given by

$$P_m : (\xi_1, \ldots, \xi_m, \xi_{m+1}, \ldots) \to (\xi_1, \ldots, \xi_m, 0, 0, \ldots).$$

2.9.3.* **Lemma.** Let s be any s-scale. Then

$$\tau_m s_n(I : l_p(m) \to l_q(m)) \leqq s_n(D_t : l_p \to l_q).$$

Proof. We may assume that $\tau_m > 0$. Then $D_m := Q_m D_t J_m$ is invertible, and we have

$$\|D_m^{-1} : l_p(m) \to l_p(m)\| = \tau_m^{-1}.$$

Therefore

$$\begin{aligned} s_n(I : l_p(m) \to l_q(m)) &\leqq \|D_m^{-1} : l_p(m) \to l_p(m)\| \, s_n(D_m : l_p(m) \to l_q(m)) \\ &\leqq \tau_m^{-1} s_n(D_t : l_p \to l_q), \end{aligned}$$

which implies the desired inequality.

2.9.4.* We now show that, for any diagonal operator in l_2, all the s-numbers coincide.

Proposition (A. Pietsch 1974:a). Let s be any s-scale. Then

$$s_n(D_t : l_2 \to l_2) = \tau_n.$$

Proof. Since

$$\|D_t - D_t P_{n-1} : l_2 \to l_2\| = \tau_n \quad \text{and} \quad \text{rank}(P_{n-1}) < n,$$

we have

$$s_n(D_t : l_2 \to l_2) \leqq \|D_t - D_t P_{n-1} : l_2 \to l_2\| + s_n(D_t P_{n-1} : l_2 \to l_2) = \tau_n .$$

On the other hand, applying Lemma 2.9.3 with $m = n$, we deduce from (S_5) that

$$\tau_n = \tau_n s_n(I : l_2(n) \to l_2(n)) \leqq s_n(D_t : l_2 \to l_2).$$

2.9.5.* In the case of approximation numbers the preceding formula remains true in all sequence spaces l_p.

Proposition. $a_n(D_t : l_p \to l_p) = \tau_n$.

Remark. The same formula holds for the Gel'fand and Kolmogorov numbers.

2.9.6. To prove the next proposition, we need some preliminaries.

Lemma (V. D. Milman 1970). Let N be any subspace of $l_\infty(m)$ such that codim $(N) < n$. Then there exists a normalized vector $e = (\varepsilon_k) \in N$ with card $\{k : |\varepsilon_k| = 1\} \geqq m - n + 1$.

Proof. By the Krejn-Milman theorem (DUN, V.8.4), the convex compact subset

$$U_N := \{x = (\xi_k) \in N : \|x \mid l_\infty(m)\| \leqq 1\}$$

contains an extremal vector $e = (\varepsilon_k)$. Let $K := \{k : |\varepsilon_k| = 1\}$, and define the subspace

$$M := \{x = (\xi_k) \in l_\infty(m) : \xi_k = 0 \text{ for } k \in K\}.$$

Obviously, card (K) + dim $(M) = m$. Assume that card $(K) <$ dim (N). Then we can find a normalized vector $x \in M \cap N$. Note that

$$\delta := 1 - \max\{|\varepsilon_k| : k \notin K\} > 0.$$

Hence it follows from $e \pm \delta x \in U_N$ that e cannot be extremal. This contradiction shows that card $(K) \geqq$ dim $(N) \geqq m - n + 1$.

2.9.7. **Lemma.** Let $1 \leqq q < p < \infty$ and $|\varepsilon_{n+1}| \leqq \min(|\varepsilon_1|, \ldots, |\varepsilon_n|)$. Then

$$\frac{\left(\sum_{k=1}^{n+1} |\varepsilon_k|^q\right)^{1/q}}{\left(\sum_{k=1}^{n+1} |\varepsilon_k|^p\right)^{1/p}} \geqq \frac{\left(\sum_{k=1}^{n} |\varepsilon_k|^q\right)^{1/q}}{\left(\sum_{k=1}^{n} |\varepsilon_k|^p\right)^{1/p}} .$$

Proof. Put

$$\alpha := \left(\sum_{k=1}^{n} |\varepsilon_k|^p\right)^{1/p} \quad \text{and} \quad \beta := \left(\sum_{k=1}^{n} |\varepsilon_k|^q\right)^{1/q} .$$

Of course, we may assume that $\varepsilon_{n+1} \neq 0$. Then

$$|\varepsilon_k/\varepsilon_{n+1}| \geqq 1 \quad \text{implies} \quad |\varepsilon_k/\varepsilon_{n+1}|^p \geqq |\varepsilon_k/\varepsilon_{n+1}|^q$$

for $k = 1, \ldots, n$. Hence

$$|\alpha/\varepsilon_{n+1}|^p \geqq |\beta/\varepsilon_{n+1}|^q .$$

We now obtain

$$\frac{(\beta^q + |\varepsilon_{n+1}|^q)^{1/q}}{(\alpha^p + |\varepsilon_{n+1}|^p)^{1/p}} = \frac{\beta}{\alpha} \frac{(1 + |\varepsilon_{n+1}/\beta|^q)^{1/q}}{(1 + |\varepsilon_{n+1}/\alpha|^p)^{1/p}} \geqq \frac{\beta}{\alpha}.$$

This completes the proof.

2.9.8. We are now prepared to compute the approximation numbers of certain identity operators.

Proposition (A. Pietsch 1974:a). Let $1 \leqq q < p \leqq \infty$. Then

$$a_n(I: l_p(m) \to l_q(m)) = (m - n + 1)^{1/q-1/p} \quad \text{for } n = 1, \dots, m.$$

Proof. Define

$$P_{n-1}: (\xi_1, \dots, \xi_{n-1}, \xi_n, \dots, \xi_m) \to (\xi_1, \dots, \xi_{n-1}, 0, \dots, 0).$$

It follows from rank $(P_{n-1}) < n$ that

$$a_n(I: l_p(m) \to l_q(m)) \leqq \|I - P_{n-1}: l_p(m) \to l_q(m)\| = (m - n + 1)^{1/q-1/p}.$$

To verify the converse, we assume that

$$L \in \mathfrak{L}(l_p(m), l_q(m)) \quad \text{and} \quad \text{rank}(L) < n.$$

Let N be the null space of this operator. Since codim $(N) =$ rank $(L) < n$, we may choose a vector $e = (\varepsilon_k) \in N$ according to Lemma 2.9.6. Then

$$\|e \mid l_q(m)\| = \|(I - L)\, e \mid l_q(m)\| \leqq \|I - L: l_p(m) \to l_q(m)\| \, \|e \mid l_p(m)\|.$$

Without loss of generality, it can be arranged that $|\varepsilon_1| \geqq \dots \geqq |\varepsilon_m|$. Repeated application of Lemma 2.9.7 now yields

$$\|I - L \mid l_p(m) \to l_q(m)\| \geqq \frac{\|e \mid l_q(m)\|}{\|e \mid l_p(m)\|} = \frac{\left(\sum_{k=1}^{m} |\varepsilon_k|^q\right)^{1/q}}{\left(\sum_{k=1}^{m} |\varepsilon_k|^p\right)^{1/p}} \geqq \frac{\left(\sum_{k=1}^{m-n+1} |\varepsilon_k|^q\right)^{1/q}}{\left(\sum_{k=1}^{m-n+1} |\varepsilon_k|^p\right)^{1/p}} = (m - n + 1)^{1/q-1/p}$$

whenever $1 \leqq q < p < \infty$. Clearly, the same estimate holds for $p = \infty$. This proves that

$$a_n(I: l_p(m) \to l_q(m)) \geqq (m - n + 1)^{1/q-1/p}.$$

Remark. The Gel'fand and Kolmogorov numbers of the identity operator from $l_p(m)$ onto $l_q(m)$ are given by the same formula; see (PIE, 11.11.4).

2.9.9. The next lemma is basic for the subsequent considerations.

Lemma. Let $\mathfrak{A}$ be a quasi-Banach operator ideal such that

$$\|I: l_p(m) \to l_q(m) \mid \mathfrak{A}\| \prec m^\alpha,$$

where $\alpha \geqq 0$. If $0 < r < \infty$ and $1/s = 1/r + \alpha$, then

$$t \in l_{s,w} \text{ implies } D_t \in \mathfrak{A}^{(a)}_{r,w}(l_p, l_q).$$

Proof. Given $t = (\tau_n)$, we define $t_k = (\tau_{kn})$ by

$$\tau_{kn} := \begin{cases} \tau_n & \text{if} \quad 2^k \leqq n < 2^{k+1}, \\ 0 & \text{otherwise.} \end{cases}$$

Note that

$$t = \sum_{k=0}^{\infty} t_k \quad \text{and} \quad \text{card}\,(t_k) \leqq 2^k.$$

Let D_k denote the diagonal operator induced by t_k. Then

$$D_t = \sum_{k=0}^{\infty} D_k \quad \text{and} \quad \text{rank}\,(D_k) \leqq 2^k.$$

Moreover,

$$\|D_k : l_p \to l_q \mid \mathfrak{A}\| \leqq \tau_{2^k} \|I : l_p(2^k) \to l_q(2^k) \mid \mathfrak{A}\| \prec 2^{k\alpha}\tau_{2^k}.$$

By 2.1.10, we have $(2^{k/s}\tau_{2^k}) \in l_w$. Hence

$$(2^{k/r}\|D_k : l_p \to l_q \mid \mathfrak{A}\|) \in l_w.$$

Thus, in view of 2.8.8, we obtain $D_t \in \mathfrak{A}^{(a)}_{r,w}(l_p, l_q)$.

2.9.10. We are now in a position to establish one of the main results of this section.

Example. Let $0 < r < \infty$, $1 \leqq q \leqq p \leqq \infty$ and $1/s = 1/r - 1/p + 1/q$. Then

$$D_t \in \mathfrak{L}^{(a)}_{r,w}(l_p, l_q) \quad \text{if and only if} \quad t \in l_{s,w}.$$

Proof. Since

$$\|I : l_p(m) \to l_q(m)\| = m^{1/q - 1/p},$$

we see from 2.9.9 that

$$t \in l_{s,w} \quad \text{implies} \quad D_t \in \mathfrak{L}^{(a)}_{r,w}(l_p, l_q).$$

In view of 2.9.3 and 2.9.8, we have

$$(m - n + 1)^{1/q - 1/p}\tau_m = \tau_m a_n(I : l_p(m) \to l_q(m)) \leqq a_n(D_t : l_p \to l_q).$$

In particular, taking $m := 2n - 1$ yields

$$n^{1/q - 1/p}\tau_{2n-1} \leqq a_n(D_t : l_p \to l_q).$$

Hence, by 2.1.9,

$$D_t \in \mathfrak{L}^{(a)}_{r,w}(l_p, l_q) \quad \text{implies} \quad t \in l_{s,w}.$$

2.9.11. Next we state a counterpart of 2.9.8.

Proposition (S. B. Stechkin 1954).

$$a_n(I : l_1(m) \to l_2(m)) = \left(\frac{m - n + 1}{m}\right)^{1/2} \quad \text{for } n = 1, \ldots, m.$$

Proof. Consider any operator

$$L \in \mathfrak{L}(l_1(m), l_2(m)) \quad \text{with rank}\,(L) < n.$$

Let $P \in \mathfrak{L}(l_2(m))$ be the orthogonal projection from $l_2(m)$ onto the orthogonal complement of $M(L)$. This means that $N(P) = M(L)$. Then

$$\sum_{k=1}^{m} \|Pe_k \mid l_2(m)\|^2 = \|P \mid \mathfrak{S}\|^2 = \text{rank}\,(P) = \text{codim}\,(N(P)) = \text{codim}\,(M(L)) \geqq m - n + 1.$$

Therefore

$$\|I - L\| \geqq \|P(I - L)\| = \|PI\| = \max\{\|Pe_k \mid l_2(m)\| : k = 1, \ldots, m\} \geqq \left(\frac{m - n + 1}{m}\right)^{1/2}.$$

This implies that

$$a_n(I : l_1(m) \to l_2(m)) \geqq \left(\frac{m - n + 1}{m}\right)^{1/2}.$$

We even have equality. For a proof of this fact, which is not used later, we refer to (PIE, 11.11.8).

2.9.12. Example. Let $0 < r < \infty$. Then

$$D_t \in \mathfrak{L}_{r,w}^{(a)}(l_1, l_2) \quad \text{if and only if} \quad t \in l_{r,w}.$$

Proof. We know from 2.9.4 that

$$t \in l_{r,w} \quad \text{implies} \quad D_t \in \mathfrak{L}_{r,w}^{(a)}(l_2, l_2).$$

Hence $D_t \in \mathfrak{L}_{r,w}^{(a)}(l_1, l_2)$.

In view of 2.9.3 and 2.9.11, we have

$$\left(\frac{m - n + 1}{m}\right)^{1/2} \tau_m = \tau_m a_n(I : l_1(m) \to l_2(m)) \leqq a_n(D_t : l_1 \to l_2).$$

In particular, taking $m := 2n - 1$, it follows that

$$\tau_{2n-1} \leqq 2^{1/2} a_n(D_t : l_1 \to l_2).$$

Therefore, by 2.1.9,

$$D_t \in \mathfrak{L}_{r,w}^{(a)}(l_1, l_2) \quad \text{implies} \quad t \in l_{r,w}.$$

2.9.13. The following counterpart of the preceding example shows that the component $\mathfrak{L}_{r,w}^{(c)}(l_1, l_2)$ is considerably larger than $\mathfrak{L}_{r,w}^{(a)}(l_1, l_2)$.

Example. Let $0 < r < 2$ and $1/s = 1/r - 1/2$. Then

$$D_t \in \mathfrak{L}_{r,w}^{(c)}(l_1, l_2) \quad \text{if and only if} \quad t \in l_{s,w}.$$

Remark. The proof of this result is based on an asymptotic estimate of the Gel'fand numbers $c_n(I: l_1(m) \to l_2(m))$ due to B. S. Kashin (1977).

2.9.14. Example. Let $0 < r < \infty$ and $1/s = 1/r + 1$. Then

$$D_t \in (\mathfrak{P}_2)_{r,w}^{(a)}(l_2, l_1) \quad \text{if and only if} \quad t \in l_{s,w}.$$

Proof. By 1.2.8,

$$\|I : l_2(m) \to l_1(m) \mid \mathfrak{P}_2\| \leqq m^{1/2}\|I : l_2(m) \to l_1(m)\| = m.$$

Applying 2.9.9, we see that

$$t \in l_{s,w} \quad \text{implies} \quad D_t \in (\mathfrak{P}_2)^{(a)}_{r,w}(l_2, l_1).$$

Let $1/r_0 := 1/r + 1/2$. Then, by 2.8.18,

$$D_t \in (\mathfrak{P}_2)^{(a)}_{r,w}(l_2, l_1) \quad \text{implies} \quad D_t \in \mathfrak{L}^{(x)}_{r_0,w}(l_2, l_1).$$

As shown in 1.6.4, the embedding operator from l_1 into l_2 is absolutely 2-summing. Hence $I \in \mathfrak{L}^{(x)}_{2,\infty}(l_1, l_2)$, by 2.7.4. Since $1/r = 1/r_0 + 1/2$, it follows from 2.4.18 that $D_t \in \mathfrak{L}^{(x)}_{s,w}(l_2, l_2)$. Finally, 2.9.4 yields $t \in l_{s,w}$.

2.9.15. Example. Let $0 < r < \infty$ and $1/s = 1/r + 1/2$. Then

$$D_t \in (\mathfrak{P}_2)^{(a)}_{r,w}(l_\infty, l_2) \quad \text{if and only if} \quad t \in l_{s,w}.$$

Proof. We may deduce from 1.3.8 that

$$\|I : l_\infty(m) \to l_2(m) \mid \mathfrak{P}_2\| = m^{1/2}.$$

Applying 2.9.9, we see that

$$t \in l_{s,w} \quad \text{implies} \quad D_t \in (\mathfrak{P}_2)^{(a)}_{r,w}(l_\infty, l_2).$$

Conversely,

$$D_t \in (\mathfrak{P}_2)^{(a)}_{r,w}(l_\infty, l_2) \quad \text{implies} \quad D_t \in \mathfrak{L}^{(a)}_{r,w}(l_\infty, l_2).$$

Therefore it follows from 2.9.10 that $t \in l_{s,w}$.

2.9.16. Example. Let $0 < r < \infty$. Then

$$D_t \in (\mathfrak{P}_2)^{(a)}_{r,w}(l_1, l_2) \quad \text{if and only if} \quad t \in l_{r,w}.$$

Proof. By 2.9.4,

$$t \in l_{r,w} \quad \text{implies} \quad D_t \in \mathfrak{L}^{(a)}_{r,w}(l_2, l_2).$$

Hence it follows from 1.6.4 and 2.8.7 that $D_t \in (\mathfrak{P}_2)^{(a)}_{r,w}(l_1, l_2)$.
Conversely,

$$D_t \in (\mathfrak{P}_2)^{(a)}_{r,w}(l_1, l_2) \quad \text{implies} \quad D_t \in \mathfrak{L}^{(a)}_{r,w}(l_1, l_2).$$

Therefore $t \in l_{r,w}$, by 2.9.12.

2.9.17.* The "if" part of the following example is basic for applications in the theory of eigenvalue distributions of infinite matrices.

Example (C. Lubitz 1982). Let $0 < r < \max(q, 2)$ and

$$1/s = 1/r + \begin{cases} 1/q - 1/2 & \text{if } 1 \leqq q \leqq 2, \\ 0 & \text{if } 2 \leqq q \leqq \infty. \end{cases}$$

Then

$$D_t \in \mathfrak{L}^{(x)}_{r,w}(l_\infty, l_q) \quad \text{if and only if} \quad t \in l_{s,w}.$$

Proof. From 1.3.8 we know that

$$\|I : l_\infty(m) \to l_q(m) \mid \mathfrak{P}_q\| = m^{1/q}.$$

Define p by $1/s = 1/p + 1/q$. Then it follows from 2.9.9 that

$$t \in l_{s,w} \quad \text{implies} \quad D_t \in (\mathfrak{P}_q)^{(a)}_{p,w}(l_\infty, l_q).$$

Moreover, by 1.3.3, 2.7.4 and 2.8.15, we have

$$(\mathfrak{P}_q)^{(a)}_{p,w} \subseteq (\mathfrak{L}^{(x)}_{2,\infty})^{(a)}_{p,w} \subseteq \mathfrak{L}^{(x)}_{r,w} \quad \text{if } 1 \leqq q \leqq 2$$

and

$$(\mathfrak{P}_q)^{(a)}_{p,w} \subseteq (\mathfrak{L}^{(x)}_{q,\infty})^{(a)}_{p,w} \subseteq \mathfrak{L}^{(x)}_{r,w} \quad \text{if } 2 \leqq q \leqq \infty.$$

This proves that $D_t \in \mathfrak{L}^{(x)}_{r,w}(l_\infty, l_q)$.

For the convenience of the reader, we present one more proof of the preceding implication which avoids the use of generalized approximation numbers.

Since $1/s = 1/p + 1/q$, given any sequence $t \in l_s$, we may choose a factorization $t = ab$ such that $a \in l_p$ and $b \in l_q$. Then it follows from 2.9.5 and 1.6.3 that

$$D_a \in \mathfrak{L}^{(a)}_p(l_\infty, l_\infty) \quad \text{and} \quad D_b \in \mathfrak{P}_q(l_\infty, l_q).$$

Next, setting $q_0 := \max(q, 2)$, we conclude from 1.2.4 and 2.7.4 that

$$D_t \in \mathfrak{L}^{(x)}_{q_0,\infty} \circ \mathfrak{L}^{(a)}_p(l_\infty, l_q).$$

Thus, in view of the multiplication theorem 2.4.18,

$$t \in l_s \quad \text{implies} \quad D_t \in \mathfrak{L}^{(x)}_{r,p}(l_\infty, l_q).$$

The general case when $t \in l_{s,w}$ can now be treated by an obvious interpolation argument; see 2.1.14 and 2.2.10.

Conversely, by 2.4.20,

$$D_t \in \mathfrak{L}^{(x)}_{r,w}(l_\infty, l_q) \quad \text{implies} \quad D_t \in \mathfrak{L}^{(a)}_{r,w}(l_2, l_q).$$

If $1 \leqq q \leqq 2$, then it follows from 2.9.10 that $t \in l_{s,w}$ with $1/s = 1/r - 1/2 + 1/q$. If $2 \leqq q \leqq \infty$, then, turning to the dual operator, we conclude from $D_t \in \mathfrak{L}^{(a)}_{r,w}(l_2, l_\infty)$ that $D_t \in \mathfrak{L}^{(a)}_{r,w}(l_1, l_2)$. Hence 2.9.12 yields $t \in l_{r,w}$.

2.9.18. Example (C. Lubitz 1982). Let $0 < r < r_0$ and $1/s = 1/r - 1/r_0$, where

$$1/r_0 := \begin{cases} 1/p - 1/q & \text{if } 1 \leqq p \leqq q \leqq 2, \\ 1/p - 1/2 & \text{if } 1 \leqq p \leqq 2 \leqq q \leqq \infty, \\ 0 & \text{if } 2 \leqq p \leqq q \leqq \infty. \end{cases}$$

Then

$$D_t \in \mathfrak{L}^{(x)}_{r,w}(l_p, l_q) \quad \text{if and only if} \quad t \in l_{s,w}.$$

Proof. It follows from 2.9.5 that

$$t \in l_{s,w} \quad \text{implies} \quad D_t \in \mathfrak{L}^{(x)}_{s,w}(l_p, l_p).$$

Furthermore, by 1.6.7 and 2.7.4, we have

$$I \in \mathfrak{P}_{r_0,2}(l_p, l_q) \subseteq \mathfrak{L}^{(x)}_{r_0,\infty}(l_p, l_q).$$

Thus, in view of the multiplication theorem 2.4.18,

$$t \in l_{s,w} \quad \text{implies} \quad D_t \in \mathfrak{L}^{(x)}_{r,w}(l_p, l_q).$$

Next we prove the converse in the case $1 \leqq p \leqq q \leqq 2$. By 2.9.8, we have

$$\begin{aligned}(m-n+1)^{1/2} &= a_n(I : l_2(m) \to l_1(m)) = x_n(I : l_2(m) \to l_1(m)) \\ &\leqq \|I : l_2(m) \to l_p(m)\| \, x_n(I : l_p(m) \to l_q(m)) \, \|I : l_q(m) \to l_1(m)\| \\ &\leqq m^{(1/p-1/2)+(1-1/q)} x_n(I : l_p(m) \to l_q(m)).\end{aligned}$$

Applying 2.9.3 with $m = 2n - 1$ now yields

$$n^{1/2}\tau_{2n-1} \leqq (2n-1)^{1/p-1/q+1/2} x_n(D_t : l_p \to l_q).$$

Hence, by 2.1.9,

$$D_t \in \mathfrak{L}^{(x)}_{r,w}(l_p, l_q) \quad \text{implies} \quad t \in l_{s,w}.$$

To treat the remaining cases, we deduce from 2.9.11 that

$$\begin{aligned}\left(\frac{m-n+1}{m}\right)^{1/2} &= a_n(I : l_1(m) \to l_2(m)) = a_n(I : l_2(m) \to l_\infty(m)) \\ &= x_n(I : l_2(m) \to l_\infty(m)) \\ &\leqq \|I : l_2(m) \to l_p(m)\| \, x_n(I : l_p(m) \to l_q(m)) \, \|I : l_q(m) \to l_\infty(m)\|.\end{aligned}$$

Applying 2.9.3 with $m := 2n - 1$ yields

$$n^{1/2}\tau_{2n-1} \leqq (2n-1)^{1/p} \, x_n(D_t : l_p \to l_q) \quad \text{if } 1 \leqq p \leqq 2$$

and

$$n^{1/2}\tau_{2n-1} \leqq (2n-1)^{1/2} \, x_n(D_t : l_p \to l_q) \quad \text{if } 2 \leqq p \leqq \infty.$$

Hence, by 2.1.9

$$D_t \in \mathfrak{L}^{(x)}_{r,w}(l_p, l_q) \quad \text{implies} \quad t \in l_{s,w}.$$

2.9.19. We conclude this section with two special results about Hilbert numbers.

Proposition (B. Carl / A. Pietsch 1978).

$$h_n(I : l_1 \to l_1) \asymp n^{-1/2}.$$

Proof. According to the Grothendieck theorem, mentioned in 1.6.4 (Remark), there exists a constant $c > 0$ such that

$$\|Y \mid \mathfrak{P}_2\| \leqq c\|Y\| \quad \text{for} \quad Y \in \mathfrak{L}(l_1, l_2).$$

Let $X \in \mathfrak{L}(l_2, l_1)$. Then, by 2.7.2, we obtain

$$n^{1/2} a_n(YIX) \leqq \|YIX \mid \mathfrak{P}_2\| \leqq \|Y \mid \mathfrak{P}_2\| \, \|X\| \leqq c\|Y\| \, \|X\|.$$

This proves that

$$h_n(I : l_1 \to l_1) \leqq cn^{-1/2}.$$

Conversely, we have

$$\begin{aligned}1 &= h_n(I : l_2(n) \to l_2(n)) \\ &\leqq \|I : l_2(n) \to l_1(n)\| \, h_n(I : l_1(n) \to l_1(n)) \, \|I : l_1(n) \to l_2(n)\| \\ &= n^{1/2} h_n(I : l_1(n) \to l_1(n)) \leqq n^{1/2} h_n(I : l_1 \to l_1).\end{aligned}$$

Remark. The above result shows that the Hilbert numbers fail to be multiplicative.

2.9.20. Example. Let $0 < r < 2$ and $1/s = 1/r - 1/2$. Then

$$D_t \in \mathfrak{L}_{r,w}^{(h)}(l_1, l_1) \quad \text{if and only if} \quad t \in l_{s,w}.$$

Proof. It follows from $h_n(I : l_1 \to l_1) \leqq cn^{-1/2}$ that $I \in \mathfrak{L}_{2,\infty}^{(h)}(l_1, l_1)$. Furthermore, by 2.9.18,

$$t \in l_{s,w} \quad \text{implies} \quad D_t \in \mathfrak{L}_{s,w}^{(x)}(l_1, l_1).$$

Thus, in view of 2.6.7, we obtain $D_t \in \mathfrak{L}_{r,w}^{(h)}(l_1, l_1)$.

As shown at the end of the preceding paragraph, we have

$$n^{-1/2} \leqq h_n(I : l_1(n) \to l_1(n)).$$

Applying 2.9.3 with $m := n$ now yields

$$n^{-1/2}\tau_n \leqq h_n(D_t : l_1 \to l_1).$$

Hence

$$D_t \in \mathfrak{L}_{r,w}^{(h)}(l_1, l_1) \quad \text{implies} \quad t \in l_{s,w}.$$

2.10. Relationships between various s-numbers

2.10.1. In this section the approximation numbers, the Gel'fand, Kolmogorov, Weyl, Chang and Hilbert numbers are compared with each other. To begin with, we summarize some results already proved in 2.3.4, 2.4.19, 2.5.13 and 2.6.3.

Theorem. Let $T \in \mathfrak{L}(E, F)$. Then

$$h_n(T) \leqq x_n(T) \leqq c_n(T) \leqq a_n(T) \quad \text{and} \quad h_n(T) \leqq y_n(T) \leqq d_n(T) \leqq a_n(T).$$

2.10.2. We now turn to the much more complicated problem of estimating the larger s-numbers by the smaller ones.

Proposition (A. Pietsch 1974:a, C. V. Hutton / J. S. Morrell / J. R. Retherford 1976).

$$a_n(T) \leqq 2n^{1/2} c_n(T) \quad \text{for all } T \in \mathfrak{L}(E, F).$$

Proof. Fix any natural number n. Given $\varepsilon > 0$, we choose a subspace M of E such that

$$\|TJ_M^E\| \leqq (1 + \varepsilon)\, c_n(T) \quad \text{and} \quad \operatorname{codim}(M) < n.$$

By 1.7.17, there exists a projection $P \in \mathfrak{L}(E)$ with $M = N(P)$ and $\|P\| \leqq n^{1/2}$. Define $L := TP$. Then it follows from

$$\operatorname{rank}(L) \leqq \operatorname{rank}(P) = \operatorname{codim}(M) < n$$

that

$$\begin{aligned} a_n(T) &\leqq \|T - L\| = \|T(I - P)\| \leqq \|TJ_M^E\|\, \|I - P\| \\ &\leqq (1 + \varepsilon)(1 + n^{1/2})\, c_n(T). \end{aligned}$$

Letting $\varepsilon \to 0$ yields the inequality we are looking for.

2.10.3.* The following lemma is a counterpart of 2.7.1.

Lemma. Let $T \in \mathfrak{L}(E, F)$. If $c_n(T) > 0$, then for every $\varepsilon > 0$ there exist families $(x_1, \dots, x_n)$ in E and $(b_1, \dots, b_n)$ in F' such that $\|x_k\| = 1$, $\|b_k\| = 1$,

$$c_k(T) \leqq (1 + \varepsilon) |\langle Tx_k, b_k\rangle| \quad \text{for } k = 1, \dots, n$$

and

$$\langle Tx_i, b_j\rangle = 0 \quad \text{whenever } 1 \leqq j < i \leqq n.$$

Proof. The required families can be constructed by induction. If $x_1, \dots, x_{n-1} \in E$ and $b_1, \dots, b_{n-1} \in F'$ have already been found, then we define the subspace

$$M_n := \{x \in E : \langle Tx, b_k\rangle = 0 \text{ for } k = 1, \dots, n-1\}.$$

Since codim $(M_n) < n$, it follows that

$$c_n(T) \leqq \|TJ_{M_n}^E\|.$$

Hence there exists $x_n \in M_n$ such that

$$c_n(T) \leqq (1 + \varepsilon) \|Tx_n\| \quad \text{and} \quad \|x_n\| = 1.$$

Next we choose $b_n \in F'$ with $|\langle Tx_n, b_n\rangle| = \|Tx_n\|$ and $\|b_n\| = 1$. This completes the proof.

2.10.4.* Proposition (A. Pietsch 1980:c).

$$c_{2n-1}(T) \leqq 2en^{1/2} \left[\prod_{k=1}^{n} x_k(T)\right]^{1/n} \quad \text{for all } T \in \mathfrak{L}(E, F).$$

Proof. Given $\varepsilon > 0$, we choose

$$x_1, \dots, x_{2n-1} \in E \quad \text{and} \quad b_1, \dots, b_{2n-1} \in F'$$

according to the preceding lemma. Define $X_{2n-1} \in \mathfrak{L}(l_2, E)$ and $B_{2n-1} \in \mathfrak{L}(F, l_2)$ by

$$X_{2n-1} := \sum_{k=1}^{2n-1} e_k \otimes x_k \quad \text{and} \quad B_{2n-1} := \sum_{k=1}^{2n-1} b_k \otimes e_k.$$

Then

$$\|X_{2n-1}\| \leqq (2n-1)^{1/2} \quad \text{and} \quad \|B_{2n-1} \mid \mathfrak{P}_2\| \leqq (2n-1)^{1/2}.$$

Since the matrix

$$(\langle Tx_i, b_j\rangle) = ((B_{2n-1}TX_{2n-1}e_i, e_j))$$

has upper triangular form, it follows from 2.7.8 and 2.11.13 that

$$\begin{aligned}
\frac{1}{1+\varepsilon} c_{2n-1}(T) &\leqq \left[\prod_{k=1}^{2n-1} |\langle Tx_k, b_k\rangle|\right]^{1/(2n-1)} \\
&= |\det((B_{2n-1}TX_{2n-1}e_i, e_j))|^{1/(2n-1)} \\
&\leqq \left[\prod_{k=1}^{2n-1} x_k(B_{2n-1}TX_{2n-1})\right]^{1/(2n-1)} \\
&\leqq \exp\left(\frac{1}{2n-1} \|B_{2n-1} \mid \mathfrak{P}_2\|^2\right) \left[\prod_{k=1}^{n} x_k(TX_{2n-1})\right]^{1/n} \\
&\leqq 2en^{1/2} \left[\prod_{k=1}^{n} x_k(T)\right]^{1/n}.
\end{aligned}$$

Letting $\varepsilon \to 0$ completes the proof.

2.10.5. The next lemma is closely related to 2.7.1 and 2.10.3.

Lemma. Let $T \in \mathfrak{L}(H, F)$. If $a_{2n-1}(T) > 0$, then for every $\varepsilon > 0$ there exist an orthonormal family $(x_1, \ldots, x_n)$ in H and a family $(b_1, \ldots, b_n)$ in F' such that $\|b_k\| = 1$,

$$a_{2k-1}(T) \leqq (1 + \varepsilon)\, |\langle Tx_k, b_k\rangle| \quad \text{for } k = 1, \ldots, n$$

and

$$\langle Tx_i, b_j\rangle = 0 \quad \text{whenever } 1 \leqq j < i \leqq n.$$

Proof. The required families can be constructed by induction. If $x_1, \ldots, x_{n-1} \in H$ and $b_1, \ldots, b_{n-1} \in F'$ have already been found, then we define the subspace

$$M_n := \{x \in H : (x, x_k) = 0 \text{ and } \langle Tx, b_k\rangle = 0 \text{ for } k = 1, \ldots, n-1\}.$$

Since codim $(M_n) < 2n - 1$, it follows from 2.4.6 that

$$a_{2n-1}(T) = c_{2n-1}(T) \leqq \|TJ^H_{M_n}\|.$$

Hence there exists $x_n \in M_n$ such that

$$a_{2n-1}(T) \leqq (1 + \varepsilon)\, \|Tx_n\| \quad \text{and} \quad \|x_n\| = 1.$$

Next we choose $b_n \in F'$ with $|\langle Tx_n, b_n\rangle| = \|Tx_n\|$ and $\|b_n\| = 1$. This completes the proof.

2.10.6. Proposition.

$$x_{2n-1}(T) \leqq n^{1/2} \left[\prod_{k=1}^{n} h_k(T)\right]^{1/n} \quad \text{for all } T \in \mathfrak{L}(E, F).$$

Proof. Let $X \in \mathfrak{L}(l_2, E)$. Applying the preceding lemma to the operator $TX \in \mathfrak{L}(l_2, F)$, given $\varepsilon > 0$, we find

$$u_1, \ldots, u_n \in l_2 \quad \text{and} \quad b_1, \ldots, b_n \in F'.$$

Define $U_n \in \mathfrak{L}(l_2, l_2)$ and $B_n \in \mathfrak{L}(F, l_2)$ by

$$U_n := \sum_{k=1}^{n} e_k \otimes u_k \quad \text{and} \quad B_n := \sum_{k=1}^{n} b_k \otimes e_k.$$

Then

$$\|U_n\| = 1 \quad \text{and} \quad \|B_n\| \leqq n^{1/2}.$$

Since the matrix

$$(\langle TXu_i, b_j\rangle) = ((B_n TXU_n e_i, e_j))$$

has upper triangular form, it follows from 2.11.13 that

$$\frac{1}{1+\varepsilon}\, a_{2n-1}(TX) \leqq \left[\prod_{k=1}^{n} |\langle TXu_k, b_k\rangle|\right]^{1/n} = |\det((B_n TXU_n e_i, e_j))|^{1/n}$$

$$\leqq \left[\prod_{k=1}^{n} a_k(B_n TXU_n)\right]^{1/n} \leqq n^{1/2} \left[\prod_{k=1}^{n} h_k(T)\right]^{1/n} \|X\|.$$

Hence

$$\frac{1}{1+\varepsilon}\, x_{2n-1}(T) \leqq n^{1/2} \left[\prod_{k=1}^{n} h_k(T)\right]^{1/n}.$$

Letting $\varepsilon \to 0$ yields the inequality we are looking for.

2.10.7. From 2.10.1 we immediately obtain the following diagram, where the arrows point from the smaller operator ideals to the larger ones:

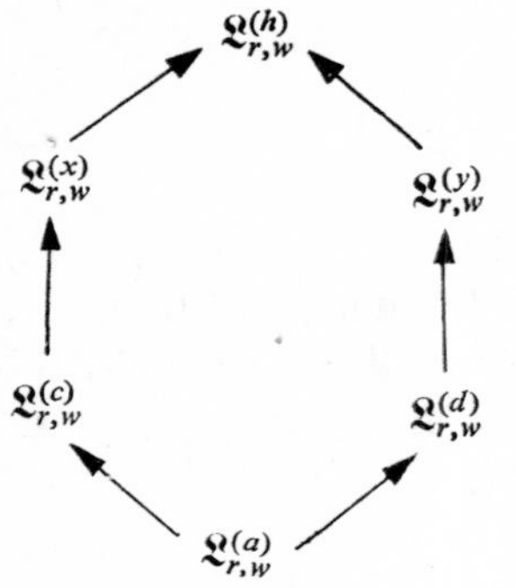

Furthermore, in view of 2.1.8 and 2.1.9, the inequalities proved in 2.10.2, 2.10.4 and 2.10.6 imply certain reverse inclusions. See also 2.3.16, 2.5.6, 2.5.12 and 2.6.9.

Theorem. Let $0 < p < 2$ and $1/q = 1/p - 1/2$. Then

$$\mathfrak{L}_{p,w}^{(h)} \subseteq \mathfrak{L}_{q,w}^{(x)}, \quad \mathfrak{L}_{p,w}^{(h)} \subseteq \mathfrak{L}_{q,w}^{(y)},$$
$$\mathfrak{L}_{p,w}^{(x)} \subseteq \mathfrak{L}_{q,w}^{(c)}, \quad \mathfrak{L}_{p,w}^{(y)} \subseteq \mathfrak{L}_{q,w}^{(d)},$$
$$\mathfrak{L}_{p,w}^{(c)} \subseteq \mathfrak{L}_{q,w}^{(a)}, \quad \mathfrak{L}_{p,w}^{(d)} \subseteq \mathfrak{L}_{q,w}^{(a)}.$$

Remark. Combining the above inclusions, we obtain

$$\mathfrak{L}_{p,w}^{(h)} \subseteq \mathfrak{L}_{q,w}^{(a)} \quad \text{if} \quad 0 < p < 2/3 \quad \text{and} \quad 1/q = 1/p - 3/2.$$

However, it seems unlikely that this result is the best possible. Since

$$a_n(I: l_1 \to l_\infty) = 1 \quad \text{and} \quad h_n(I: l_1 \to l_\infty) \asymp n^{-1},$$

the condition $1/q \leqq 1/p - 1$ is certainly necessary, and we conjecture that it is also sufficient.

2.10.8.* Finally, we state an easy consequence of 2.10.4 and 2.4.11.

Theorem (A. Pietsch 1980:c). Every operator $T \in \mathfrak{L}(E, F)$ with $(n^{1/2} x_n(T)) \in c_0$ is compact.

Remark. We see from $x_n(I: l_1 \to l_2) = n^{-1/2}$ that there are non-compact operators T with $(n^{1/2} x_n(T)) \in l_\infty$.

2.11. Schatten—von Neumann operators

Throughout this section we assume that H and K are complex Hilbert spaces.

2.11.1.* In a finite dimensional Hilbert space every orthonormal family (x_i) is of course finite. In this case, we can obtain an infinite sequence by defining the missing elements by $x_i := o$. This idea leads to the following concept:

We call (x_i) an **extended orthonormal sequence** provided that

$$(x_i, x_j) = 0 \text{ if } i \neq j \text{ and either } \|x_i\| = 1 \quad \text{or} \quad \|x_i\| = 0.$$

If not otherwise stated, in the following (x_i) and (y_i) denote extended orthonormal sequences in H and K, respectively.

Furthermore, (e_i) denotes the standard basis of l_2.

2.11.2.* The following result is an immediate consequence of Bessel's inequality.

Lemma. Let (x_i) and (y_i) be extended orthonormal sequences. Then

$$U := \sum_{i=1}^{\infty} x_i^* \otimes y_i$$

defines an operator from H into K with $\|U\| \leqq 1$.

Remark. The right-hand series converges with respect to the so-called strong operator topology induced by the semi-norms

$$p_M(T) := \sup \{\|Tx\| : x \in M\},$$

where M ranges over all finite subsets of H.

2.11.3.* A positive number τ is said to be a **singular value** of the operator $T \in \mathfrak{L}(H, K)$ if there exist normalized elements $x \in H$ and $y \in K$ such that

$$Tx = \tau y \quad \text{and} \quad T^* y = \tau x.$$

This is equivalent to say that τ^2 is an eigenvalue of TT^* and T^*T.

Note that T and T^* have the same singular values.

2.11.4.* We call

$$T = \sum_{i=1}^{\infty} \tau_i x_i^* \otimes y_i$$

a **Schmidt representation** of the operator $T \in \mathfrak{L}(H, K)$ if the following conditions are satisfied:

(1) $(\tau_i) \in c_0$.

(2) (x_i) and (y_i) are extended orthonormal sequences.

(3) $Tx = \sum_{i=1}^{\infty} \tau_i (x, x_i) y_i$ for all $x \in H$.

If $x_i \neq o$ and $y_i \neq o$, then it follows that $(Tx_i, y_i) = \tau_i$. In the opposite case, where at least one of the elements x_i and y_i is zero, the value of the corresponding coefficient τ_i does not have any effect. It is therefore natural to assume that $\tau_i = 0$. In this way we obtain

(4) $(Tx_i, y_i) = \tau_i$ for $i = 1, 2, \ldots$

The formula

$$\left\| T - \sum_{i=1}^{n} \tau_i x_i^* \otimes y_i \right\| = \left\| \sum_{i=n+1}^{\infty} \tau_i x_i^* \otimes y_i \right\| = \sup_{i>n} |\tau_i|$$

implies that the partial sums converge to T with respect to the operator norm.

A Schmidt representation is said to be **monotonic** if $\tau_1 \geqq \tau_2 \geqq \ldots \geqq 0$. Note that all positive coefficients τ_i are singular values of T.

2.11.5.* Diagonalization theorem. Let

$$T = \sum_{i=1}^{\infty} \tau_i x_i^* \otimes y_i$$

be a Schmidt representation of $T \in \mathfrak{L}(H, K)$. If

$$X := \sum_{i=1}^{\infty} e_i^* \otimes x_i \quad \text{and} \quad Y := \sum_{i=1}^{\infty} e_i^* \otimes y_i,$$

then

$$\begin{array}{ccc} H & \xrightarrow{\quad T \quad} & K \\ X^* \downarrow\uparrow X & & Y \uparrow\downarrow Y^* \\ l_2 & \xrightarrow[\quad D_t \quad]{} & l_2 \end{array},$$

where D_t is the diagonal operator induced by $t = (\tau_i)$. This diagram means that

$$T = Y D_t X^* \quad \text{and} \quad D_t = Y^* T X.$$

2.11.6.* Next we show that the coefficients of a monotonic Schmidt representation are uniquely determined by the underlying operator.

Proposition (D. Eh. Allakhverdiev 1957). Let

$$T = \sum_{i=1}^{\infty} \tau_i x_i^* \otimes y_i$$

be a monotonic Schmidt representation of $T \in \mathfrak{L}(H, K)$. Then $\tau_n = a_n(T)$.

Proof. From the above theorem we know that

$$T = Y D_t X^* \quad \text{and} \quad D_t = Y^* T X.$$

Hence

$$a_n(T) \leqq \|Y\|\, a_n(D_t)\, \|X^*\| \leqq a_n(D_t) \leqq \|Y^*\|\, a_n(T)\, \|X\| \leqq a_n(T).$$

Thus, by 2.9.4, we have $a_n(T) = a_n(D_t) = \tau_n$.

2.11.7.* Every operator admitting a Schmidt representation is of course approximable. In order to show the converse, we need the following lemma which guarantees the existence of singular values.

Lemma. Let $T \in \mathfrak{K}(H, K)$. If $T \neq O$, then $\tau := \|T\|$ is a singular value of T.

Proof. Choose any sequence (x_k) with

$$\lim_k \|Tx_k\| = \tau \quad \text{and} \quad \|x_k\| = 1.$$

Set $y_k := \tau^{-1} T x_k$. Since T is compact, we may suppose that (y_k) converges to some $y \in K$. Note that

$$\|y\| = \lim_k \|y_k\| = \tau^{-1} \lim_k \|Tx_k\| = 1.$$

It follows from

$$\|T^* y_k - \tau x_k\|^2 = \|T^* y_k\|^2 - (T^* y_k, \tau x_k) - (\tau x_k, T^* y_k) + \tau^2 \leqq 2\tau^2 - 2\|Tx_k\|^2$$

that (x_k) tends to $x := \tau^{-1}T^*y$. Finally, passing to the limit, we obtain from $\tau y_k = Tx_k$ and $\|x_k\| = 1$ that $\tau y = Tx$ and $\|x\| = 1$.

2.11.8.* We are now in a position to establish the most important tool of the theory of approximable operators on Hilbert spaces.

Representation theorem (E. Schmidt 1907:a). Every approximable operator $T \in \mathfrak{L}(H, K)$ admits a monotonic Schmidt representation.

Proof. We are going to construct the desired representation by induction.

If $\tau_1 := \|T\| > 0$, then by the preceding lemma there exist normalized elements $x_1 \in H$ and $y_1 \in K$ such that

$$Tx_1 = \tau_1 y_1 \quad \text{and} \quad T^*y_1 = \tau_1 x_1.$$

Assume that we have already found a scalar family $(\tau_1, \ldots, \tau_{n-1})$ and orthonormal families $(x_1, \ldots, x_{n-1})$ and $(y_1, \ldots, y_{n-1})$ satisfying the following properties:

$$\tau_i > 0, \quad Tx_i = \tau_i y_i \quad \text{and} \quad T^*y_i = \tau_i x_i.$$

Put

$$\tau_n := \left\| T - \sum_{i=1}^{n-1} \tau_i x_i^* \otimes y_i \right\|.$$

If $\tau_n = 0$, then

$$T = \sum_{i=1}^{n-1} \tau_i x_i^* \otimes y_i.$$

Therefore, setting $\tau_i := 0$, $x_i := o$ and $y_i := o$ for $i \geqq n$, we arrive at the desired representation.

If $\tau_n > 0$, then Lemma 2.11.7, applied to the operator

$$T_n := T - \sum_{i=1}^{n-1} \tau_i x_i^* \otimes y_i,$$

yields normalized elements $x_n \in H$ and $y_n \in K$ such that

$$T_n x_n = \tau_n y_n \quad \text{and} \quad T_n^* y_n = \tau_n x_n.$$

For $i = 1, \ldots, n-1$ we have

$$\tau_n(x_i, x_n) = (x_i, T_n^* y_n) = (T_n x_i, y_n) = (Tx_i - \tau_i y_i, y_n) = 0$$

and

$$\tau_n(y_i, y_n) = (y_i, T_n x_n) = (T_n^* y_i, x_n) = (T^*y_i - \tau_i x_i, x_n) = 0.$$

Hence the enlarged families $(x_1, \ldots, x_{n-1}, x_n)$ and $(y_1, \ldots, y_{n-1}, y_n)$ are orthogonal, as well. This implies that

$$Tx_n = T_n x_n \quad \text{and} \quad T^*y_n = T_n^* y_n.$$

Thus we have

$$Tx_n = \tau_n y_n \quad \text{and} \quad T^*y_n = \tau_n x_n.$$

In the next step we put

$$\tau_{n+1} := \left\| T - \sum_{i=1}^{n} \tau_i x_i^* \otimes y_i \right\|.$$

It follows from

$$\left(Tx - \sum_{i=1}^{n} \tau_i(x, x_i)\, y_i, y_n\right) = (x, T^* y_n) - \tau_n(x, x_n) = 0 \quad \text{for } x \in H$$

that the elements

$$Tx - \sum_{i=1}^{n} \tau_i(x, x_i)\, y_i \quad \text{and} \quad \tau_n(x, x_n)\, y_n$$

are orthogonal. Applying Pythagoras's formula, we obtain

$$\left\| Tx - \sum_{i=1}^{n} \tau_i(x, x_i)\, y_i \right\|^2 + \|\tau_n(x, x_n)\, y_n\|^2 = \left\| Tx - \sum_{i=1}^{n-1} \tau_i(x, x_i)\, y_i \right\|^2.$$

This proves that $\tau_{n+1} \leqq \tau_n$.

If this process, starting with τ_1, x_1 and y_1, does not end after a finite number of steps, then we obtain infinite sequences (τ_i), (x_i) and (y_i). In particular, it follows that $\tau_1 \geqq \tau_2 \geqq \ldots > 0$. Since (Tx_i) contains a Cauchy subsequence,

$$\|Tx_i - Tx_j\|^2 = \tau_i^2 + \tau_j^2 \quad \text{whenever } i \neq j$$

implies that (τ_i) tends to zero. This means that

$$\lim_n \left\| T - \sum_{i=1}^{n} \tau_i x_i^* \otimes y_i \right\| = 0.$$

Therefore

$$T = \sum_{i=1}^{\infty} \tau_i x_i^* \otimes y_i$$

is the Schmidt representation we are looking for.

2.11.9.* Next we show that the s-numbers of any operator between Hilbert spaces are uniquely determined by their axiomatic properties.

Theorem (A. Pietsch 1974:a). There exists one and only one s-scale on the class of all operators acting between Hilbert spaces.

Proof. We first treat the case when $T \in \mathfrak{L}(H, K)$ is approximable. Then, by the preceding theorem, there exists a monotonic Schmidt representation

$$T = \sum_{i=1}^{\infty} \tau_i x_i^* \otimes y_i.$$

Reasoning as in the proof of 2.11.6, we see that $s_n(T) = s_n(D_t) = \tau_n$ for any s-scale s.

We now consider an arbitrary operator $T \in \mathfrak{L}(H, K)$. Then

$$c_n(TJ_M^H) = s_n(TJ_M^H) \leqq s_n(T)$$

for all finite dimensional subspaces M. Hence, by 2.3.4, 2.4.6 and 2.4.12,

$$a_n(T) = c_n(T) = \sup\{c_n(TJ_M^H) : \dim(M) < \infty\} \leqq s_n(T) \leqq a_n(T).$$

2.11.10.* We now provide an auxiliary result; see 2.7.1, 2.10.3 and 2.10.5.

Lemma. Let $T \in \mathfrak{L}(H, K)$. If $a_{2n-1}(T) > 0$, then for every $\varepsilon > 0$ there exists an orthonormal family $(x_1, \ldots, x_n)$ in H such that

$$a_{2k-1}(T) \leqq (1 + \varepsilon)\, \|Tx_k\| \quad \text{for } k = 1, \ldots, n,$$

and $Tx_1, \ldots, Tx_n$ are orthogonal.

Proof. The required family can be constructed by induction. If $x_1, \ldots, x_{n-1}$ have already been found, then we define the subspace

$$M_n := \{x \in H : (x, x_k) = 0 \quad \text{and} \quad (Tx, Tx_k) = 0 \text{ for } k = 1, \ldots, n-1\}.$$

Since codim $(M_n) < 2n - 1$, it follows from 2.4.6 that

$$a_{2n-1}(T) = c_{2n-1}(T) \leqq \|TJ_{M_n}^H\|.$$

Hence there exists $x_n \in M_n$ such that

$$a_{2n-1}(T) \leqq (1 + \varepsilon)\|Tx_n\| \quad \text{and} \quad \|x_n\| = 1.$$

This completes the proof.

2.11.11.* We are now able to establish an extremely important result.

Theorem (J. W. Calkin 1941). Let $\mathfrak{A}$ be an operator ideal which does not contain the identity operator of l_2. Then all components $\mathfrak{A}(H, K)$ consist of approximable operators exclusively.

Proof. Assume that $\mathfrak{A}(H, K)$ contains a non-approximable operator T. Then

$$\alpha := \lim_n a_n(T) > 0.$$

According to the proof of the preceding lemma, we can find an orthonormal sequence (x_k) such that

$$a_{2k-1}(T) \leqq 2\|Tx_k\| \quad \text{for } k = 1, 2, \ldots$$

Moreover, the sequence (y_k) given by $y_k := Tx_k/\|Tx_k\|$ is orthonormal, as well. Define $X \in \mathfrak{L}(l_2, H)$ and $Y \in \mathfrak{L}(l_2, K)$ by

$$X := \sum_{k=1}^{\infty} e_k^* \otimes x_k \quad \text{and} \quad Y := \sum_{k=1}^{\infty} e_k^* \otimes y_k.$$

Then $S := Y^*TX \in \mathfrak{A}(l_2)$ is induced by the diagonal matrix $((Tx_i, y_j))$. Moreover, we see from

$$2(Tx_k, y_k) = 2\|Tx_k\| \geqq a_{2k-1}(T) \geqq \alpha > 0$$

that S is invertible. Therefore we have $I = S^{-1}S \in \mathfrak{A}(l_2)$. This contradiction completes the proof.

2.11.12.* Lemma (Ky Fan 1951). Let $T \in \mathfrak{L}(H, K)$ and $1 \leqq r < \infty$. Then

$$\sum_{k=1}^{n} |(Tx_k, y_k)|^r \leqq \sum_{k=1}^{n} a_k(T)^r$$

for all orthonormal families $(x_1, \ldots, x_n)$ and $(y_1, \ldots, y_n)$.

Proof. Define $X \in \mathfrak{L}(l_2, H)$ and $Y \in \mathfrak{L}(l_2, K)$ by

$$X := \sum_{k=1}^{n} e_k^* \otimes x_k \quad \text{and} \quad Y := \sum_{k=1}^{n} e_k^* \otimes y_k.$$

Take a monotonic Schmidt representation

$$Y^*TX = \sum_{h=1}^{\infty} \sigma_h u_h^* \otimes v_h,$$

where (u_h) and (v_h) are extended orthonormal sequences in l_2. From 2.11.6 we conclude that

$$\sigma_h = a_h(Y^*TX) \leqq a_h(T).$$

In particular, we have $\sigma_h = 0$ for $h > n$. It follows that

$$\begin{aligned}\sum_{k=1}^n |(Tx_k, y_k)| &= \sum_{k=1}^n |(Y^*TXe_k, e_k)| \\ &= \sum_{k=1}^n \left|\sum_{h=1}^n \sigma_h(e_k, u_h)\,(v_h, e_k)\right| \\ &\leqq \sum_{h=1}^n \sigma_h \left(\sum_{k=1}^n |(e_k, u_h)|^2\right)^{1/2} \left(\sum_{k=1}^n |(v_h, e_k)|^2\right)^{1/2} \\ &\leqq \sum_{h=1}^n \sigma_h\|u_h\|\,\|v_h\| \leqq \sum_{h=1}^n \sigma_h \leqq \sum_{h=1}^n a_h(T).\end{aligned}$$

This proves the assertion for $r = 1$.

Next we treat the case $1 < r < \infty$. By Hölder's inequality,

$$\begin{aligned}\sum_{h=1}^n \sigma_h|(e_k, u_h)|^2 &= \sum_{h=1}^n \sigma_h|(e_k, u_h)|^{2/r}\,|(e_k, u_h)|^{2/r'} \\ &\leqq \left(\sum_{h=1}^n \sigma_h^r|(e_k, u_h)|^2\right)^{1/r} \left(\sum_{h=1}^n |(e_k, u_h)|^2\right)^{1/r'}.\end{aligned}$$

Hence

$$\left(\sum_{h=1}^n \sigma_h|(e_k, u_h)|^2\right)^{r/2} \leqq \left(\sum_{h=1}^n \sigma_h^r|(e_k, u_h)|^2\right)^{1/2}.$$

Analogously, we obtain

$$\left(\sum_{h=1}^n \sigma_h|(v_h, e_k)|^2\right)^{r/2} \leqq \left(\sum_{h=1}^n \sigma_h^r|(v_h, e_k)|^2\right)^{1/2}.$$

Moreover,

$$\begin{aligned}|(Tx_k, y_k)| &= \left|\sum_{h=1}^n \sigma_h^{1/2}(e_k, u_h)\,\sigma_h^{1/2}(v_h, e_k)\right| \\ &\leqq \left(\sum_{h=1}^n \sigma_h|(e_k, u_h)|^2\right)^{1/2} \left(\sum_{h=1}^n \sigma_h|(v_h, e_k)|^2\right)^{1/2}.\end{aligned}$$

Combining the previous inequalities, we finally see that

$$\begin{aligned}&\sum_{k=1}^n |(Tx_k, y_k)|^r \leqq \\ &\leqq \sum_{k=1}^n \left(\sum_{h=1}^n \sigma_h^r|(e_k, u_h)|^2\right)^{1/2} \left(\sum_{h=1}^n \sigma_h^r|(v_h, e_k)|^2\right)^{1/2} \\ &\leqq \left(\sum_{k=1}^n \sum_{h=1}^n \sigma_h^r|(e_k, u_h)|^2\right)^{1/2} \left(\sum_{k=1}^n \sum_{h=1}^n \sigma_h^r|(v_h, e_k)|^2\right)^{1/2} \\ &\leqq \left(\sum_{h=1}^n \sigma_h^r\|u_h\|^2\right)^{1/2} \left(\sum_{h=1}^n \sigma_h^r\|v_h\|^2\right)^{1/2} \leqq \sum_{h=1}^n a_h(T)^r.\end{aligned}$$

Remark. The above inequality fails for $0 < r < 1$.

2.11.13.* The following inequality is closely related to the preceding one.

Lemma (Ky Fan 1949/50). Let $T \in \mathfrak{L}(H, K)$. Then

$$|\det((Tx_i, y_j))| \leqq \prod_{k=1}^{n} a_k(T)$$

for all orthonormal families $(x_1, \ldots, x_n)$ and $(y_1, \ldots, y_n)$.

Proof. Adopting the notation from 2.11.12, we have

$$(Tx_i, y_j) = (Y^*TXe_i, e_j) = \sum_{h=1}^{n} \sigma_h(e_i, u_h)(v_h, e_j).$$

This implies that

$$\det((Tx_i, y_j)) = \det((e_i, u_h)) \left(\prod_{k=1}^{n} \sigma_h\right) \det((v_h, e_j)).$$

By Hadamard's inequality A.4.5,

$$|\det((e_i, u_h))| \leqq 1 \quad \text{and} \quad |\det((v_h, e_j))| \leqq 1.$$

Hence

$$|\det((Tx_i, y_j))| \leqq \prod_{h=1}^{n} \sigma_h \leqq \prod_{h=1}^{n} a_h(T).$$

2.11.14.* The following characterization is taken from (RIN, p. 58).

Theorem. An operator $T \in \mathfrak{L}(H, K)$ is approximable if and only if $((Tx_k, y_k)) \in c_0$ for all extended orthonormal sequences (x_k) and (y_k).

Proof. If T is non-approximable, then the orthonormal sequences (x_k) and (y_k) constructed in the proof of 2.11.11 have the property that

$$|(Tx_k, y_k)| \geqq \alpha/2 \quad \text{for } k = 1, 2, \ldots$$

Hence the above condition fails.

Let $T \in \mathfrak{G}(H, K)$, and assume that there exist extended orthonormal sequences (x_k) and (y_k) for which $((Tx_k, y_k))$ does not tend to zero. Then, passing to subsequences (u_k) and (v_k), we may find some $\varepsilon > 0$ such that

$$|(Tu_k, v_k)| \geqq \varepsilon \quad \text{for } k = 1, 2, \ldots$$

Now it follows from Lemma 2.11.12 that

$$\frac{1}{n}\sum_{k=1}^{n} a_k(T) \geqq \frac{1}{n}\sum_{k=1}^{n} |(Tu_k, v_k)| \geqq \varepsilon.$$

This is a contradiction, since the arithmetic means of the approximation numbers converge to zero.

2.11.15.* All operator ideals $\mathfrak{L}_{r,w}^{(s)}$ determined by arbitrary s-scales coincide on the class of Hilbert spaces. We denote this common restriction by $\mathfrak{S}_{r,w}$. The operators belonging to this class are said to be of **Schatten-von Neumann type** $l_{r,w}$.

For $T \in \mathfrak{S}_{r,w}(H, K)$ we define

$$\|T \mid \mathfrak{S}_{r,w}\| := \|T \mid \mathfrak{L}_{r,w}^{(s)}\|.$$

To simplify notation, the index w is omitted whenever $w = r$.

2.11.16.* The following criterion is an immediate consequence of 2.11.6 and 2.11.8.

Theorem. An operator $T \in \mathfrak{L}(H, K)$ is of Schatten-von Neumann type $l_{r,w}$ if and only if there exists a Schmidt representation

$$T = \sum_{i=1}^{\infty} \tau_i x_i^* \otimes y_i$$

such that $t = (\tau_i) \in l_{r,w}$. When this is so, then

$$\|T \mid \mathfrak{S}_{r,w}\| = \|t \mid l_{r,w}\|.$$

2.11.17.* The next result is straightforward, as well.

Proposition. An operator $T \in \mathfrak{L}(H, K)$ is of Schatten-von Neumann type l_2 if and only if it is Hilbert-Schmidt. When this is so, then $\|T \mid \mathfrak{S}_2\| = \|T \mid \mathfrak{S}\|$.

2.11.18.* We now establish a characterization of $\mathfrak{S}_{r,w}$ which is analogous to 2.11.14.

Theorem (A. Pietsch 1971). Let $1 < r < \infty$. An operator $T \in \mathfrak{L}(H, K)$ is of Schatten-von Neumann type $l_{r,w}$ if and only if $((Tx_k, y_k)) \in l_{r,w}$ for all extended orthonormal sequence (x_k) and (y_k). Moreover,

$$\|T \mid \mathfrak{S}_{r,w}\|_0 := \sup \{\|((Tx_k, y_k)) \mid l_{r,w}\| : (x_k), (y_k)\}$$

defines an equivalent quasi-norm on $\mathfrak{S}_{r,w}$.

Proof. Assume that the above condition is satisfied. Then 2.11.14 implies that T is approximable. Take any Schmidt representation

$$T = \sum_{i=1}^{\infty} \tau_i x_i^* \otimes y_i.$$

Since $(Tx_k, y_k) = \tau_k$, we have $t = (\tau_k) \in l_{r,w}$. Hence, by 2.11.16,

$$T \in \mathfrak{S}_{r,w}(H, K) \quad \text{and} \quad \|T \mid \mathfrak{S}_{r,w}\| = \|t \mid l_{r,w}\| \leqq \|T \mid \mathfrak{S}_{r,w}\|_0.$$

To verify the converse, let $T \in \mathfrak{S}_{r,w}(H, K)$. Given extended orthonormal sequences (x_k) and (y_k), we may assume that

$$|(Tx_1, y_1)| \geqq |(Tx_2, y_2)| \geqq \dots \geqq 0.$$

In view of 2.11.12, it follows that

$$\left\| \left(\frac{1}{n} \sum_{k=1}^{n} |(Tx_k, y_k)| \right) \mid l_{r,w} \right\| \leqq \left\| \left(\frac{1}{n} \sum_{k=1}^{n} a_k(T) \right) \mid l_{r,w} \right\|.$$

Therefore 2.1.7 yields

$$((Tx_k, y_k)) \in l_{r,w} \quad \text{and} \quad \|((Tx_k, y_k)) \mid l_{r,w}\| \leqq c\|T \mid \mathfrak{S}_{r,w}\|.$$

The last result can also be obtained by interpolation techniques. Given extended orthonormal sequences (x_k) and (y_k), we consider the operator

$$D : \mathfrak{L}(H, K) \to l_\infty$$

which assigns to every operator T the sequence $((Tx_k, y_k))$. We see from 2.11.12 that

$$D : \mathfrak{S}_1(H, K) \to l_1.$$

Hence, by 2.3.14 and 2.1.14, the interpolation property yields

$$D: \mathfrak{S}_{r,w}(H, K) \to l_{r,w}.$$

2.11.19.* In the special case $1 \leqq r = w < \infty$ a refined version of the preceding characterization can be obtained via 2.11.12.

Proposition. Let $1 \leqq r < \infty$. An operator $T \in \mathfrak{L}(H, K)$ is of Schatten-von Neumann type l_r if and only if $((Tx_k, y_k)) \in l_r$ for all extended orthonormal sequences (x_k) and (y_k). Moreover,

$$\|T \mid \mathfrak{S}_r\| = \sup \{\|((Tx_k, y_k)) \mid l_r\| : (x_k), (y_k)\}.$$

2.11.20. The description of $\|. \mid \mathfrak{S}_r\|$ just given implies the following result.

Proposition (R. Schatten / J. von Neumann 1948). If $1 \leqq r < \infty$, then $\mathfrak{S}_r$ is a Banach operator ideal.

Remark. It follows from 1.7.2 and 2.11.26 that $\mathfrak{S}_r$ is an r-Banach operator ideal for $0 < r < 1$; see also S. J. Rotfeld (1967).

2.11.21. Next we formulate an elementary observation; see 1.4.10.

Proposition. The following are equivalent for $T \in \mathfrak{L}(H, K)$:

$$T \in \mathfrak{S}_{r,w}(H, K), \quad T^* \in \mathfrak{S}_{r,w}(K, H), \quad T' \in \mathfrak{S}_{r,w}(K', H').$$

The corresponding quasi-norms coincide.

2.11.22. In the Hilbert space setting 2.3.15 can be improved.

Proposition. The operator ideal $\mathfrak{S}_r$ is stable with respect to the tensor norm σ. More precisely,

$$\|S \tilde{\otimes}_\sigma T \mid \mathfrak{S}_r\| = \|S \mid \mathfrak{S}_r\| \, \|T \mid \mathfrak{S}_r\| \quad \text{for} \quad S \in \mathfrak{S}_r(H, H_0) \quad \text{and} \quad T \in \mathfrak{S}_r(K, K_0).$$

Proof. Let

$$S = \sum_{i=1}^{\infty} \sigma_i x_i^* \otimes u_i \quad \text{and} \quad T = \sum_{j=1}^{\infty} \tau_j y_j^* \otimes v_j$$

be Schmidt representations. Note that $(x_i \otimes y_j)$ and $(u_i \otimes v_j)$ are extended orthonormal sequences in $H \tilde{\otimes}_\sigma K$ and $H_0 \tilde{\otimes}_\sigma K_0$, respectively. Since

$$S \tilde{\otimes}_\sigma T = \sum_{i=1}^{\infty} \sum_{j=1}^{\infty} \sigma_i \tau_j (x_i \otimes y_j)^* \otimes (u_i \otimes v_j)$$

and

$$(S \tilde{\otimes}_\sigma T)(x_i \otimes y_j) = \sigma_i \tau_j (u_i \otimes v_j)$$

it follows from 2.11.16 that

$$S \tilde{\otimes}_\sigma T \in \mathfrak{S}_r(H \tilde{\otimes}_\sigma K, H_0 \tilde{\otimes}_\sigma K_0)$$

and

$$\begin{aligned} \|S \tilde{\otimes}_\sigma T \mid \mathfrak{S}_r\| &= \|(\sigma_i \tau_j) \mid l_r(\mathbb{N} \times \mathbb{N})\| = \|(\sigma_i) \mid l_r\| \, \|(\tau_j) \mid l_r\| \\ &= \|S \mid \mathfrak{S}_r\| \, \|T \mid \mathfrak{S}_r\|. \end{aligned}$$

2.11.23. We now need three complex Hilbert spaces denoted by H, K and L.

Proposition (A. Horn 1950). Let $1/p + 1/q = 1/r$. Then

$$\|ST \mid \mathfrak{S}_r\| \leqq \|S \mid \mathfrak{S}_p\| \, \|T \mid \mathfrak{S}_q\| \quad \text{for} \quad T \in \mathfrak{S}_q(H, K) \quad \text{and} \quad S \in \mathfrak{S}_p(K, L).$$

Proof. From 2.2.9 we know that

$$(*) \qquad \|ST \mid \mathfrak{S}_r\| \leqq c\|S \mid \mathfrak{S}_p\| \, \|T \mid \mathfrak{S}_q\|,$$

where $c \geqq 1$ is a constant. Note that this inequality holds for all operators of Schatten-von Neumann type l_r acting between arbitrary Hilbert spaces. Consequently,

$$\|(S \tilde{\otimes}_\sigma S)(T \tilde{\otimes}_\sigma T) \mid \mathfrak{S}_r\| \leqq c\|S \tilde{\otimes}_\sigma S \mid \mathfrak{S}_p\| \, \|T \tilde{\otimes}_\sigma T \mid \mathfrak{S}_p\|.$$

Obviously,

$$(S \tilde{\otimes}_\sigma S)(T \tilde{\otimes}_\sigma T) = ST \tilde{\otimes}_\sigma ST.$$

Applying the preceding proposition, we now obtain

$$\|ST \mid \mathfrak{S}_r\|^2 \leqq c\|S \mid \mathfrak{S}_p\|^2 \, \|T \mid \mathfrak{S}_q\|^2.$$

This implies that (*) holds with the constant $c^{1/2}$, as well. Hence, if $c \geqq 1$ is choosen as small as possible, it follows that $c \leqq c^{1/2}$. Thus $c = 1$.

2.11.24. In the following we investigate the restriction of some operator ideals to the class of Hilbert spaces. On the other hand, we try to find extensions of the Schatten-von Neumann ideals to the class of all Banach spaces; see D.3.

The first result along this line can be obtained from 1.4.5 and 2.11.17.

Proposition. The Banach operator ideal $\mathfrak{P}_2$ is a metric extension of $\mathfrak{S}_2$.

Remark. We stress the fact that, by 1.3.16, all Banach ideals $\mathfrak{P}_r$ with $1 \leqq r < \infty$ are extensions of $\mathfrak{S}_2$. However, their norms coincide with $\|\cdot \mid \mathfrak{S}_2\|$ only when $r = 2$.

2.11.25. Proposition (A. F. Ruston 1951: a). The Banach operator ideal $\mathfrak{N}$ is a metric extension of $\mathfrak{S}_1$.

Proof. Using a Schmidt representation

$$T = \sum_{i=1}^{\infty} \tau_i x_i^* \otimes y_i$$

of $T \in \mathfrak{S}_1(H, K)$, we deduce from

$$\sum_{i=1}^{\infty} |\tau_i| \, \|x_i\| \, \|y_i\| \leqq \sum_{i=1}^{\infty} |\tau_i| = \|T \mid \mathfrak{S}_1\|$$

that

$$T \in \mathfrak{N}(H, K) \quad \text{and} \quad \|T \mid \mathfrak{N}\| \leqq \|T \mid \mathfrak{S}_1\|.$$

We now assume that $T \in \mathfrak{N}(H, K)$. Since T is approximable, there exists a monotonic Schmidt representation

$$T = \sum_{i=1}^{\infty} \tau_i x_i^* \otimes y_i.$$

Furthermore, given $\varepsilon > 0$, we choose a representation

$$T = \sum_{j=1}^{\infty} u_j^* \otimes v_j$$

such that

$$\sum_{j=1}^{\infty} \|u_j\| \, \|v_j\| \leqq (1 + \varepsilon) \, \|T \mid \mathfrak{N}\| .$$

Then

$$\begin{aligned} \sum_{i=1}^{\infty} a_i(T) = \sum_{i=1}^{\infty} \tau_i &= \sum_{i=1}^{\infty} |(Tx_i, y_i)| \leqq \sum_{i=1}^{\infty} \sum_{j=1}^{\infty} |(x_i, u_j)\,(v_j, y_i)| \\ &\leqq \sum_{j=1}^{\infty} \left(\sum_{i=1}^{\infty} |(x_i, u_j)|^2 \right)^{1/2} \left(\sum_{i=1}^{\infty} |(v_j, y_i)|^2 \right)^{1/2} \\ &\leqq \sum_{j=1}^{\infty} \|u_j\| \, \|y_j\| \leqq (1 + \varepsilon) \, \|T \mid \mathfrak{N}\| . \end{aligned}$$

Thus we have

$$T \in \mathfrak{S}_1(H, K) \quad \text{and} \quad \|T \mid \mathfrak{S}_1\| \leqq \|T \mid \mathfrak{N}\| .$$

2.11.26. The preceding result extends as follows.

Proposition (R. Oloff 1969, 1972). The quasi-Banach operator ideal $\mathfrak{N}_p$ is a metric extension of $\mathfrak{S}_p$ for $0 < p \leqq 1$.

Proof. Reasoning as in the previous proof, it follows from $T \in \mathfrak{S}_p(H, K)$ that

$$T \in \mathfrak{N}_p(H, K) \quad \text{and} \quad \|T \mid \mathfrak{N}_p\| \leqq \|T \mid \mathfrak{S}_p\| .$$

We now assume that $T \in \mathfrak{N}_p(H, K)$. Let $1/r := 1/p - 1$. Given $\varepsilon > 0$, by 1.7.3, there exists a factorization

$$\begin{array}{ccc} H & \xrightarrow{\quad T \quad} & K \\ {\scriptstyle A}\big\downarrow & & \big\uparrow{\scriptstyle Y} \\ l_\infty & \xrightarrow[\quad D_t \quad]{} & l_1 \\ {\scriptstyle D_a}\big\downarrow & & \big\uparrow{\scriptstyle D_y} \\ l_2 & \xrightarrow[\quad D_s \quad]{} & l_2 \end{array}$$

such that

$$\|Y\| \, \|y \mid l_2\| \, \|s \mid l_r\| \, \|a \mid l_2\| \, \|A\| \leqq (1 + \varepsilon) \, \|T \mid \mathfrak{N}_p\| .$$

In view of 1.6.3 (Proof), 2.11.21, 2.11.23 and 2.11.24, we now obtain

$$T = (YD_y) \, D_s(D_a A) \in \mathfrak{S}_2 \circ \mathfrak{S}_r \circ \mathfrak{S}_2(H, K) \subseteq \mathfrak{S}_p(H, K)$$

and

$$\begin{aligned} \|T \mid \mathfrak{S}_p\| &\leqq \|YD_y \mid \mathfrak{S}_2\| \, \|D_s \mid \mathfrak{S}_r\| \, \|D_a A \mid \mathfrak{S}_2\| \\ &\leqq \|Y\| \, \|y \mid l_2\| \, \|s \mid l_r\| \, \|a \mid l_2\| \, \|A\| \leqq (1 + \varepsilon) \, \|T \mid \mathfrak{N}_p\| . \end{aligned}$$

Letting $\varepsilon \to 0$ completes the proof.

2.11.27. Proposition. The quasi-Banach operator ideal $\mathfrak{N}_{r,2}$ is a metric extension of $\mathfrak{S}_r$ for $0 < r \leqq 2$.

Proof. Using a Schmidt representation of $T \in \mathfrak{S}_r(H, K)$, it follows immediately that

$$T \in \mathfrak{N}_{r,2}(H, K) \quad \text{and} \quad \|T \mid \mathfrak{N}_{r,2}\| \leqq \|T \mid \mathfrak{S}_r\| .$$

We now assume that $T \in \mathfrak{N}_{r,2}(H, K)$. Let $1/p := 1/r - 1/2$. Given $\varepsilon > 0$, by 1.7.10, there exists a factorization

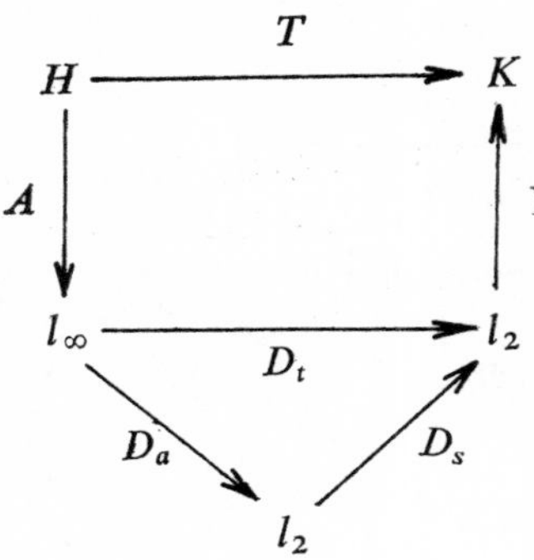

such that

$$\|Y\| \, \|s \mid l_p\| \, \|a \mid l_2\| \, \|A\| \leqq (1 + \varepsilon) \, \|T \mid \mathfrak{N}_{r,2}\|.$$

In view of 1.6.3 (Proof), 2.11.23 and 2.11.24, we now obtain

$$T = YD_s(D_aA) \in \mathfrak{S}_p \circ \mathfrak{S}_2(H, K) \subseteq \mathfrak{S}_r(H, K)$$

and

$$\|T \mid \mathfrak{S}_r\| \leqq \|Y\| \, \|D_s \mid \mathfrak{S}_p\| \, \|D_aA \mid \mathfrak{S}_2\| \leqq \|Y\| \, \|s \mid l_p\| \, \|a \mid l_2\| \, \|A\|$$
$$\leqq (1 + \varepsilon) \, \|T \mid \mathfrak{N}_{r,2}\|.$$

Letting $\varepsilon \to 0$ completes the proof.

2.11.28. According to S. Kwapień (1968) the next result, which is a special case of 2.7.6, was first established by B. S. Mityagin (unpublished). Here we give a direct proof.

Proposition. The Banach operator ideal $\mathfrak{P}_{r,2}$ is a metric extension of $\mathfrak{S}_r$ for $2 \leqq r < \infty$.

Proof. Since the identity operator of l_2 fails to be absolutely $(r, 2)$-summing, we conclude from 2.11.11 that all operators $T \in \mathfrak{P}_{r,2}(H, K)$ are approximable. Thus there exists a Schmidt representation

$$T = \sum_{i=1}^{\infty} \tau_i x_i^* \otimes y_i.$$

It follows from

$$\|(Tx_i) \mid l_r\| \leqq \|T \mid \mathfrak{P}_{r,2}\| \, \|(x_i) \mid w_2\|$$

as well as

$$|\tau_i| = |(Tx_i, y_i)| \leqq \|Tx_i\| \quad \text{and} \quad \|(x_i) \mid w_2\| \leqq 1$$

that

$$\|(\tau_i) \mid l_r\| \leqq \|T \mid \mathfrak{P}_{r,2}\|.$$

Hence

$$T \in \mathfrak{S}_r(H, K) \quad \text{and} \quad \|T \mid \mathfrak{S}_r\| \leqq \|T \mid \mathfrak{P}_{r,2}\|.$$

Conversely, assume that $T \in \mathfrak{S}_r(H, K)$. Given $x_1, \ldots, x_n \in H$, we define the operator $X \in \mathfrak{L}(l_2, H)$ with $\|X\| = \|(x_i) \mid w_2\|$ by

$$X := \sum_{i=1}^{n} e_i^* \otimes x_i .$$

Let

$$TX = \sum_{j=1}^{\infty} \sigma_j u_j^* \otimes y_j$$

be a monotonic Schmidt representation. Applying Hölder's inequality with respect to the exponents $p := r/2$ and $p' := r/(r-2)$ yields

$$\begin{aligned}\|Tx_i\|^2 &= \|TXe_i\|^2 = \sum_{j=1}^{\infty} \sigma_j^2 |(e_i, u_j)|^2 \\ &= \sum_{j=1}^{\infty} \sigma_j^2 |(e_i, u_j)|^{2/p} \, |(e_i, u_j)|^{2/p'} \\ &\leqq \Big(\sum_{j=1}^{\infty} \sigma_j^{2p} |(e_i, u_j)|^2\Big)^{1/p} \Big(\sum_{j=1}^{\infty} |(e_i, u_j)|^2\Big)^{1/p'} \\ &\leqq \Big(\sum_{j=1}^{\infty} \sigma_j^{r} |(e_i, u_j)|^2\Big)^{2/r} .\end{aligned}$$

Since $\sigma_j = a_j(TX)$, we obtain

$$\begin{aligned}\|(Tx_i) \mid l_r\| &= \Big(\sum_{i=1}^{n} \|Tx_i\|^r\Big)^{1/r} \leqq \Big(\sum_{i=1}^{n} \sum_{j=1}^{\infty} \sigma_j^r |(e_i . u_j)|^2\Big)^{1/r} \\ &\leqq \Big(\sum_{j=1}^{\infty} \sigma_j^r\Big)^{1/r} = \Big(\sum_{j=1}^{\infty} a_j(TX)^r\Big)^{1/r} \\ &= \|TX \mid \mathfrak{S}_r\| \leqq \|T \mid \mathfrak{S}_r\| \, \|X\| \leqq \|T \mid \mathfrak{S}_r\| \, \|(x_i) \mid w_2\| .\end{aligned}$$

This proves that

$$T \in \mathfrak{P}_{r,2}(H, K) \quad \text{and} \quad \|T \mid \mathfrak{P}_{r,2}\| \leqq \|T \mid \mathfrak{S}_r\| .$$

2.11.29. Proposition (H. König 1979, 1980: c). Let $2 \leqq p < \infty$, $0 < q < \infty$ and $1/r = 1/p + 1/q$. Then the operator ideal $\mathfrak{P}_{p,2} \circ \mathfrak{L}_{q,w}^{(a)}$ is an extension of $\mathfrak{S}_{r,w}$.

Proof. It follows from 2.7.4, 2.7.6 and 2.4.18 that

$$\mathfrak{P}_{p,2} \circ \mathfrak{L}_{q,w}^{(a)}(H, F) \subseteq \mathfrak{L}_{p,\infty}^{(x)} \circ \mathfrak{L}_{q,w}^{(x)}(H, F) \subseteq \mathfrak{L}_{r,w}^{(x)}(H, F) = \mathfrak{L}_{r,w}^{(a)}(H, F) .$$

To verify the converse, given $T \in \mathfrak{S}_{r,w}(H, K)$, we consider the factorization

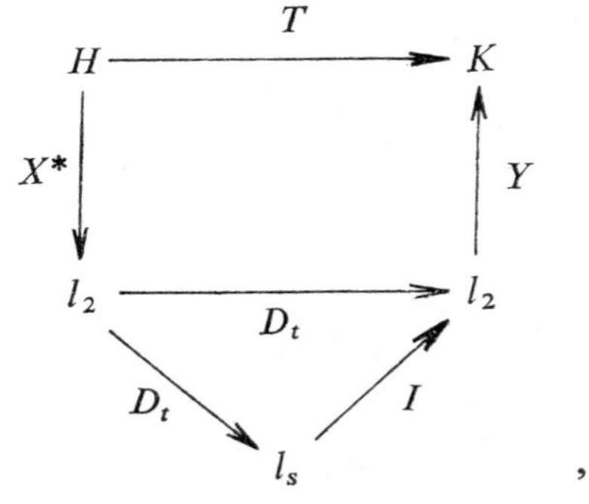

,

where $1/s := 1/p + 1/2$. By 1.6.7 and 2.9.10,

$$D_t \in \mathfrak{L}_{q,w}^{(a)}(l_2, l_s) \quad \text{and} \quad I \in \mathfrak{P}_{p,2}(l_s, l_2).$$

This proves that $T \in \mathfrak{P}_{p,2} \circ \mathfrak{L}_{q,w}^{(a)}(H, K)$.

2.11.30. We now formulate an immediate consequence of 2.7.6 and the preceding result.

Proposition (H. König 1980: c). Let $2 \leqq p < \infty$, $2 < q < \infty$ and $1/r = 1/p + 1/q$. Then the operator ideal $\mathfrak{P}_{p,2} \circ \mathfrak{P}_{q,2}$ is an extension of $\mathfrak{S}_{r,q}$.

2.11.31. An operator $T \in \mathfrak{L}(E, F)$ is said to be of **weak Schatten-von Neumann type** $l_{r,w}$ if

$$YTX \in \mathfrak{S}_{r,w}(l_2) \quad \text{for all } X \in \mathfrak{L}(l_2, E) \quad \text{and} \quad Y \in \mathfrak{L}(F, l_2).$$

The set of these operators is denoted by $\mathfrak{S}_{r,w}^{\text{weak}}(E, F)$.

For $T \in \mathfrak{S}_{r,w}^{\text{weak}}(E, F)$ we define

$$\|T \mid \mathfrak{S}_{r,w}^{\text{weak}}\| := \sup\left\{\|YTX \mid \mathfrak{S}_{r,w}\| : \begin{array}{l} X \in \mathfrak{L}(l_2, E),\ \|X\| \leqq 1 \\ Y \in \mathfrak{L}(F, l_2),\ \|Y\| \leqq 1 \end{array}\right\}.$$

To simplify notation, the index w is omitted whenever $w = r$.

2.11.32. Theorem (A. Pietsch 1970, 1976). $\mathfrak{S}_{r,w}^{\text{weak}}$ is a quasi-Banach operator ideal.

Proof. The ideal properties can be checked by standard techniques. Therefore we only show that the quasi-norm defined above is finite. Assume the contrary. Then there exist $X_k \in \mathfrak{L}(l_2, H)$ and $Y_k \in \mathfrak{L}(F, l_2)$ such that

$$\|X_k\| \leqq 1, \quad \|Y_k\| \leqq 1 \quad \text{and} \quad \|Y_k T X_k \mid \mathfrak{S}_{r,w}\| \geqq 4^k k.$$

Form the Hilbert space $H := [l_2, l_2]$ as described in C.2.1. Define

$$X := \sum_{k=1}^{\infty} 2^{-k} X_k Q_k \quad \text{and} \quad Y := \sum_{k=1}^{\infty} 2^{-k} J_k Y_k,$$

where $J_k \in \mathfrak{L}(l_2, H)$ and $Q_k \in \mathfrak{L}(H, l_2)$ denote the canonical injections and surjections, respectively. Since H can be identified with l_2, it follows that $YTX \in \mathfrak{S}_{r,w}(H)$. We now deduce from

$$XJ_k = 2^{-k} X_k \quad \text{and} \quad Q_k Y = 2^{-k} Y_k$$

that

$$k \leqq \|4^{-k} Y_k T X_k \mid \mathfrak{S}_{r,w}\| = \|Q_k YTX J_k\| \, |\mathfrak{S}_{r,w}|| \leqq \|YTX \mid \mathfrak{S}_{r,w}\|$$

for $k = 1, 2, \ldots$ This contradiction completes the proof.

2.11.33. Proposition (A. Pietsch 1970, 1976). The operator ideal $\mathfrak{S}_{r,w}^{\text{weak}}$ is the largest extension of $\mathfrak{S}_{r,w}$.

Proof. It follows from 2.11.11 that every operator $T \in \mathfrak{S}_{r,w}^{\text{weak}}(H, K)$ is approximable. Hence, by 2.11.5 and 2.11.8, there exist $t \in c_0$, $X \in \mathfrak{L}(l_2, H)$ and $Y \in \mathfrak{L}(l_2, K)$ such that

$$D_t = Y^* T X \quad \text{and} \quad T = Y D_t X^*.$$

From $T \in \mathfrak{S}_{r,w}^{\text{weak}}(H, K)$ we conclude that $D_t \in \mathfrak{S}_{r,w}(l_2)$. Therefore $T \in \mathfrak{S}_{r,w}(H, K)$. This proves that

$$\mathfrak{S}_{r,w}^{\text{weak}}(H, K) \subseteq \mathfrak{S}_{r,w}(H, K).$$

The reverse inclusion is obvious. Thus $\mathfrak{S}_{r,w}^{\text{weak}}$ is indeed an extension of $\mathfrak{S}_{r,w}$.
Finally, let $\mathfrak{A}$ be any extension of $\mathfrak{S}_{r,w}$. If $T \in \mathfrak{A}(E, F)$, then

$$YTX \in \mathfrak{A}(l_2) = \mathfrak{S}_{r,w}(l_2) \text{ for all } X \in \mathfrak{L}(l_2, E) \text{ and } Y \in \mathfrak{L}(F, l_2).$$

Hence $T \in \mathfrak{S}_{r,w}^{\text{weak}}(E, F)$. Thus we have $\mathfrak{A} \subseteq \mathfrak{S}_{r,w}^{\text{weak}}$.

2.11.34. Proposition. $\mathfrak{S}_{r,\infty}^{\text{weak}} = \mathfrak{L}_{r,\infty}^{(h)}$.

Proof. Let $T \in \mathfrak{S}_{r,\infty}^{\text{weak}}(E, F)$. Then

$$n^{1/r} a_n(YTX) \leqq \|YTX \mid \mathfrak{S}_{r,\infty}\| \leqq \|Y\| \, \|T \mid \mathfrak{S}_{r,\infty}^{\text{weak}}\| \, \|X\|$$

for all $X \in \mathfrak{L}(l_2, E)$ and $Y \in \mathfrak{L}(F, l_2)$. Hence

$$n^{1/r} h_n(T) \leqq \|T \mid \mathfrak{S}_{r,\infty}^{\text{weak}}\|.$$

This proves that

$$\mathfrak{S}_{r,\infty}^{\text{weak}} \subseteq \mathfrak{L}_{r,\infty}^{(h)}.$$

The reverse inclusion follows immediately from 2.11.33 and the fact that $\mathfrak{L}_{r,\infty}^{(h)}$ is an extension of $\mathfrak{S}_{r,\infty}$.

2.11.35*. Proposition (H. König 1979). Let $1 \leqq r < 2$ and $1/s = 1/r - 1/2$. Then

$$\mathfrak{S}_r^{\text{weak}} \subseteq \mathfrak{P}_{s,2}.$$

Proof. Assume that $T \in \mathfrak{S}_r^{\text{weak}}(E, F)$. Given $(x_i) \in [w_2, E]$, we choose $b_i \in F'$ such that

$$\langle Tx_i, b_i \rangle = \|Tx_i\| \text{ and } \|b_i\| \leqq 1.$$

Define $X \in \mathfrak{L}(l_2, E)$ and $B \in \mathfrak{L}(F, l_\infty)$ by

$$X := \sum_{i=1}^{\infty} e_i \otimes x_i \quad \text{and} \quad B := \sum_{i=1}^{\infty} b_i \otimes e_i.$$

Moreover, let $D_s \in \mathfrak{L}(l_\infty, l_2)$ be the diagonal operator determined by an arbitrary sequence $s = (\sigma_i) \in l_2$. Since

$$\sigma_i \|Tx_i\| = \sigma_i \langle Tx_i, b_i \rangle = (D_s BTXe_i, e_i),$$

it follows from 2.11.19 that $(\sigma_i \|Tx_i\|) \in l_r$. Hence $(Tx_i) \in [l_s, F]$. This proves that $T \in \mathfrak{P}_{s,2}(E, F)$, by 1.2.2.

2.11.36. In the case $r = 1$ the preceding result can essentially be improved. For a proof we refer to (PIE, 17.4.3 and 17.5.2).

Proposition (S. Kwapień 1972). $\mathfrak{S}_1^{\text{weak}} = \mathfrak{P}_2' \circ \mathfrak{P}_2$.

2.11.37. The following diagram illustrates the containment relationships between various extension of $\mathfrak{S}_1$. The arrows point from the smaller operator ideals to the larger ones. The results are taken from 1.7.16, 2.3.11, 2.8.18, 2.8.20, 2.11.33 and 2.11.36. Concerning the inner hexagon we refer to 2.10.7.

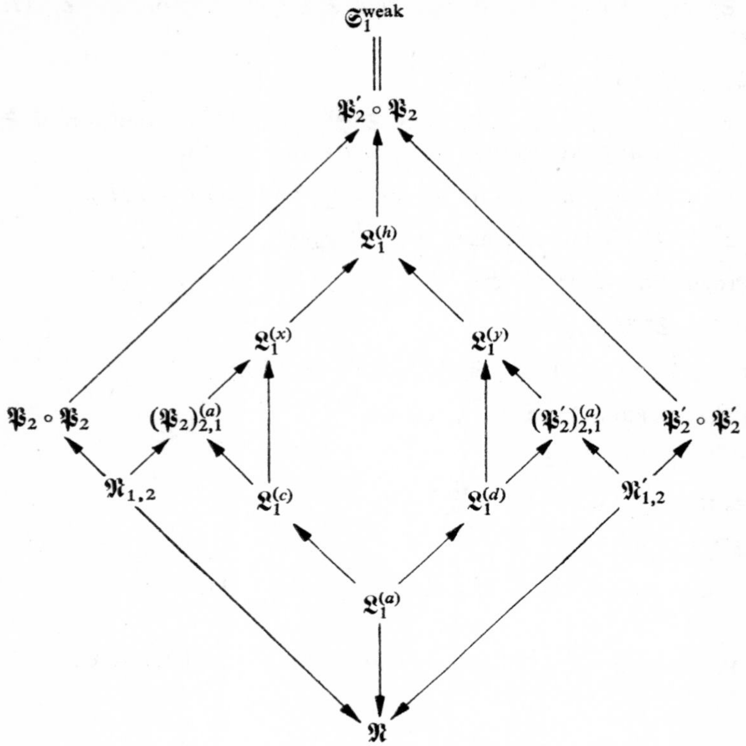

Further extensions can be obtained by forming finite sums of the ideals listed above.

This is indeed only a small part of the "ideal zoo" in the sense of B. Simon (SIM, p. 114), isn't it?

CHAPTER 3

Eigenvalues

In this chapter we investigate the asymptotic behaviour of the eigenvalues of operators belonging to a given ideal.

The background of this theory is the concept of a Riesz operator which can be introduced from very different points of view. We present here a geometric approach. It turns out that every Riesz operator $T \in \mathfrak{L}(E)$ possesses a countable set of eigenvalues. Apart from the case when this set is empty or finite, we get a sequence $(\lambda_n(T))$ tending to zero. It is our main goal to obtain information about the rate of this convergence which is measured in terms of Lorentz sequence spaces. Roughly speaking, an operator ideal $\mathfrak{A}$ is said to be of eigenvalue type $l_{r,w}$ if

$$(\lambda_n(T)) \in l_{r,w} \quad \text{for all operators } T \in \mathfrak{A}(E)$$

and arbitrary Banach spaces E.

By means of some effective tools such as

the principle of related operators,
the principle of iteration,
the principle of uniform boundedness,
the principle of tensor stability,

and more especially

the Weyl inequalities,

we determine the best possible eigenvalue type of several concrete operator ideals. A survey of these results is given in 7.4.6.

3.1. The Riesz decomposition

Throughout this section A denotes an arbitrary linear map acting on a linear space E. All considerations are purely algebraic.

3.1.1.* Let

$$N_k(A) := \{x \in E : A^k x = o\} \quad \text{and} \quad M_h(A) := \{A^h x : x \in E\}.$$

Obviously,

$$\{o\} = N_0(A) \subseteqq N_1(A) \subseteqq \ldots \quad \text{and} \quad \ldots \subseteqq M_1(A) \subseteqq M_0(A) = E.$$

Write

$$N_\infty(A) := \bigcup_{k=0}^{\infty} N_k(A) \qquad \text{and} \qquad M_\infty(A) := \bigcap_{h=0}^{\infty} M_h(A).$$

Furthermore, define

$$n(A) := \dim\,[N_\infty(A)] \qquad \text{and} \quad m(A) := \operatorname{codim}\,[M_\infty(A)],$$

where the right-hand quantities may or may not be finite.

Remark. To simplify notation, in the following proofs we write N_k and M_h instead of $N_k(A)$ and $M_h(A)$, respectively.

3.1.2.* If there exists an integer k such that

$$N_k(A) = N_{k+1}(A),$$

then A is said to have **finite ascent**. The smallest such k is denoted by $d_N(A)$.

3.1.3.* We now show that $N_k(A)$ is constant for $k \geqq d_N(A)$.

Lemma. $N_k(A) = N_{k+1}(A)$ implies $N_{k+1}(A) = N_{k+2}(A)$.

Proof. Let $x \in N_{k+2}$. Then $Ax \in N_{k+1}$, and the assumption says that $Ax \in N_k$. Hence $x \in N_{k+1}$.

3.1.4.* **Lemma.** If A has finite ascent, then

$$M_h(A) \cap N_k(A) = \{o\} \quad \text{for} \quad k = 0, 1, \ldots \quad \text{and} \quad h \geqq d_N(A).$$

Proof. Let $y \in M_h \cap N_k$. Because $y \in M_h$, there exists $x \in E$ with $y = A^h x$. We now conclude from $A^{h+k}x = A^k y = o$ that $x \in N_{h+k}$. Hence $x \in N_h$. This implies that $y = o$.

3.1.5.* **Lemma.** If A has finite ascent, then $n(A) \geqq d_N(A)$.

Proof. Note that

$$\dim (N_k/N_{k-1}) \geqq 1 \quad \text{for} \quad k = 1, \ldots, d_N(A).$$

Hence

$$n(A) = \dim [N_\infty(A)] = \sum_{k=1}^{\infty} \dim (N_k/N_{k-1}) \geqq d_N(A).$$

3.1.6.* If there exists an integer h such that

$$M_h(A) = M_{h+1}(A),$$

then A is said to have **finite descent**. The smallest such h is denoted by $d_M(A)$.

3.1.7.* We now show that $M_h(A)$ is constant for $h \geqq d_M(A)$.

Lemma. $M_h(A) = M_{h+1}(A)$ implies $M_{h+1}(A) = M_{h+2}(A)$.

Proof. Let $y \in M_{h+1}$. Then there exists $x \in E$ with $y = A^{h+1}x$. By assumption, we can find $x_0 \in E$ such that $A^h x = A^{h+1}x_0$. Hence $y = A^{h+1}x = A^{h+2}x_0$, which means that $y \in M_{h+2}$.

3.1.8.* **Lemma.** If A has finite descent, then

$$M_h(A) + N_k(A) = E \quad \text{for} \quad h = 0, 1, \ldots \quad \text{and} \quad k \geqq d_M(A).$$

Proof. Let $x \in E$. Because $A^k x \in M_k = M_{h+k}$, we can choose $x_0 \in E$ such that $A^k x = A^{h+k}x_0$. Set

$$x_M := A^h x_0 \quad \text{and} \quad x_N := x - A^h x_0.$$

Then $x_M \in M_h$ and $x_N \in N_k$. Thus $x = x_M + x_N$ is a decomposition we are looking for.

3.1.9.* **Lemma.** If A has finite descent, then $m(A) \geqq d_M(A)$.

Proof. Note that

$$\dim (M_{h-1}/M_h) \geqq 1 \quad \text{for} \quad h = 1, \ldots, d_M(A).$$

Hence

$$m(A) = \operatorname{codim} [M_\infty(A)] = \sum_{h=1}^{\infty} \dim (M_{h-1}/M_h) \geqq d_M(A).$$

3.1.10.* Proposition. If A has finite ascent and finite descent, then $d_N(A) = d_M(A)$.

Proof. Put $h := d_N(A)$ and $k := d_M(A)$. By 3.1.8, every element $x \in E$ possesses a decomposition

$$x = x_M + x_N \quad \text{with} \quad x_M \in M_{h+1} \quad \text{and} \quad x_N \in N_k.$$

We now assume that $x \in N_{k+1}$. Then it follows from

$$x_M = x - x_N \in N_{k+1} + N_k = N_{k+1}$$

that $x_M \in M_{h+1} \cap N_{k+1}$. Thus 3.1.4 implies that $x_M = o$. Hence $x = x_N \in N_k$. This proves that $N_k = N_{k+1}$. Consequently, $d_N(A) \leqq h := d_M(A)$.

Next we assume that $x \in M_h$. Then it follows from

$$x_N = x - x_M \in M_h + M_{h+1} = M_h$$

that $x_N \in M_h \cap N_k$. Thus 3.1.4 implies that $x_N = o$. Hence $x = x_M \in M_{h+1}$. This proves that $M_h = M_{h+1}$. Consequently, $d_M(A) \leqq h := d_N(A)$.

3.1.11.* For every linear map A having finite ascent and finite descent the common value of $d_N(A)$ and $d_M(A)$ is denoted by $d(A)$.

3.1.12.* A linear subset M of a linear space E is **invariant under a linear map** A if $x \in M$ implies $Ax \in M$. When this is so, then we speak of an **A-invariant** linear subset.

3.1.13.* A linear map A is **nilpotent** if there exists an exponent m with $T^m = O$. The smallest such m is called the **order of nilpotency**.

3.1.14.* We are now prepared to establish the main result of this section.

Decomposition theorem (F. Riesz 1918, A. E. Taylor 1966). If the linear map A has finite ascent and finite descent, then the linear space E is the direct sum of the A-invariant linear subsets $M_\infty(A)$ and $N_\infty(A)$. Moreover, the following holds:

(1) The restriction of A to $M_\infty(A)$ is invertible.

(2) The restriction of A to $N_\infty(A)$ is nilpotent of order $d(A)$.

Proof. Write $d := d(A)$. Then

$$M_\infty(A) = M_d \quad \text{and} \quad N_\infty(A) = N_d.$$

Now it follows from 3.1.4 and 3.1.8 that E is indeed the direct sum of $M_\infty(A)$ and $N_\infty(A)$. Obviously, both linear subsets are invariant under A.

Given $y \in M_\infty(A)$, there exists $x_0 \in E$ such that $y = A^{d+1}x_0$. Hence $y = Ax$, where $x := A^d x_0 \in M_d$. This proves that A maps $M_\infty(A)$ onto $M_\infty(A)$. Furthermore, we know from 3.1.4 that $M_d \cap N_1 = \{o\}$. Thus the restriction of A to $M_\infty(A)$ is one-to-one.

Lastly, we note that

$$A^d x = o \quad \text{for all } x \in N_\infty(A).$$

3.1.15.* Finally, we prove the result stated in A.2.5. under a very special assumption.

Lemma. If A is a nilpotent linear map on a finite dimensional linear space E, then there exists a basis $(u_1, \ldots, u_n)$ such that the representing matrix $M = (\mu_{ij})$ has upper triangular form with zeros on the principal diagonal. This means that $\mu_{ij} = 0$ whenever $i \geqq j$.

Proof. We begin by choosing a basis $(u_{11}, \ldots, u_{1n_1})$ of N_1. Then we extend this to a basis of N_2 by adding $(u_{21}, \ldots, u_{2n_2})$, and repeat the process. In this way we obtain a basis

$$(u_{11}, \ldots, u_{1n_1}; \ldots; u_{d1}, \ldots, u_{dn_d})$$

of E, where $d = d(A)$. The proof is finished if we number these elements according to their lexicographical order.

3.2. Riesz operators

Further information about Riesz operators is to be found in (BAS), (CAR) and (DOW).

3.2.1.* An operator $T \in \mathfrak{L}(E)$ is said to be **iteratively compact** if for every $\varepsilon > 0$ there exist an exponent n and elements $u_1, \ldots, u_k \in E$ such that

$$T^n(U) \subseteqq \bigcup_{h=1}^{k} \{u_h + \varepsilon U\},$$

where U denotes the closed unit ball of the underlying Banach space E.

Remark. It can be shown that an operator $T \in \mathfrak{L}(E)$ is iteratively compact if and only if we can find an exponent n and a compact operator $K \in \mathfrak{L}(E)$ such that $\|T^n - K\| < 1$; see K. Yosida (1939) and A. Pietsch (1961).

3.2.2.* We now establish a **pigeon-hole principle** which is basic for the considerations that follow.

Lemma. Suppose that $T \in \mathfrak{L}(E)$ is iteratively compact. Let (x_i) be any sequence in U. Then for every $\varepsilon > 0$ there exist an exponent n and an infinite subset I of $\mathbb{N}$ such that

$$\|T^n x_i - T^n x_j\| \leqq \varepsilon \quad \text{for all } i, j \in I.$$

Proof. Choose n and $u_1, \ldots, u_k \in E$ such that

$$T^n(U) \subseteqq \bigcup_{h=1}^{k} \left\{u_h + \frac{\varepsilon}{2} U\right\}.$$

Setting

$$I_h := \left\{i \in \mathbb{N} : T^n x_i \in u_h + \frac{\varepsilon}{2} U\right\} \quad \text{for } h = 1, \ldots, k,$$

we have

$$\|T^n x_i - T^n x_j\| \leqq \varepsilon \quad \text{for all } i, j \in I_h.$$

Furthermore, it follows from

$$\bigcup_{h=1}^{k} I_h = \mathbb{N}$$

that at least one of the sets $I_1, \ldots, I_k$ is infinite.

3.2.3.* Next we state the **Riesz lemma** which, in the following proofs, is the opponent of the preceding pigeon-hole principle.

Lemma (F. Riesz 1918). Let M be a proper subspace of the Banach space E. Then for every $\varepsilon > 0$ there exists an element $x_0 \in E$ such that $\|x_0\| = 1$ and

$$\|x_0 - x\| \geqq \frac{1}{1+\varepsilon} \quad \text{for all } x \in M.$$

Proof. Take any $y_0 \in E \backslash M$, and set

$$\varrho := \inf \{\|y_0 - x\| : x \in M\}.$$

Since M is closed, it follows that $\varrho > 0$. Next we choose $z_0 \in M$ such that

$$\varrho_0 := \|y_0 - z_0\| \leqq (1 + \varepsilon)\,\varrho.$$

Define $x_0 := \varrho_0^{-1}(y_0 - z_0)$. Then, for every $x \in M$, we have

$$\varrho_0\|x_0 - x\| = \|y_0 - (z_0 + \varrho_0 x)\| \geqq \varrho.$$

Therefore x_0 has the required properties.

3.2.4.* We now deal with operators of the form $I - T$, where I denotes the identity operator of the underlying Banach space.

Proposition. If $T \in \mathfrak{L}(E)$ is iteratively compact, then all null spaces $N_k(I - T)$ are finite dimensional.

Proof. Assume the contrary, and take the smallest k for which $N_k(I - T)$ is infinite dimensional. Applying the Riesz lemma with $\varepsilon = 1/3$, we can inductively choose elements $x_i \in N_k(I - T)$ such that $\|x_i\| = 1$ and

$$\|x_i - x\| \geqq 3/4 \quad \text{for all } x \in \operatorname{span}(x_1, \ldots, x_{i-1}) + N_{k-1}(I - T)$$

It follows from

$$x - T^n x = (I + T + \ldots + T^{n-1})(I - T)\,x$$

that

$$x - T^n x \in N_{k-1}(I - T) \quad \text{for all } x \in N_k(I - T) \quad \text{and} \quad n = 1, 2, \ldots$$

Hence

$$T^n x_i - T^n x_j \in x_i - x_j + N_{k-1}(I - T),$$

which implies that

$$\|T^n x_i - T^n x_j\| \geqq 3/4 \quad \text{whenever } i > j \quad \text{and} \quad n = 1, 2, \ldots$$

On the other hand, by the pigeon-hole principle, there exist an exponent n and different indices i and j such that

$$\|T^n x_i - T^n x_j\| \leqq 1/2.$$

This contradiction proves the assertion.

3.2.5.* **Proposition.** If $T \in \mathfrak{L}(E)$ is iteratively compact, then $I - T$ has finite ascent.

Proof. Assume the contrary. Applying the Riesz lemma with $\varepsilon = 1/3$, we can choose $x_k \in N_k(I - T)$ such that $\|x_k\| = 1$ and

$$\|x_k - x\| \geqq 3/4 \quad \text{for all } x \in N_{k-1}(I - T).$$

It follows from

$$x - T^n x = (I + T + \ldots + T^{n-1})(I - T)x$$

that

$$x_k - T^n x_k \in N_{k-1}(I - T) \quad \text{for } n = 1, 2, \ldots$$

Hence

$$T^n x_h - T^n x_k \in x_h - x_k + N_{h-1}(I - T) + N_{k-1}(I - T),$$

which implies that

$$\|T^n x_h - T^n x_k\| \geqq 3/4 \quad \text{whenever } h > k \quad \text{and} \quad n = 1, 2, \ldots$$

On the other hand, by the pigeon-hole principle, there exist an exponent n and different indices h and k such that

$$\|T^n x_h - T^n x_k\| \leqq 1/2.$$

This contradiction proves the assertion.

3.2.6.* Lemma. Let $T \in \mathfrak{L}(E)$ be iteratively compact. Then every bounded sequence (x_i) for which $((I - T)x_i)$ is convergent has a convergent subsequence.

Proof. Without loss of generality, we may assume that (x_i) is contained in U. Given $\varepsilon > 0$, by the pigoen-hole principle, there exist an exponent n and an infinite subset I such that

$$\|T^n x_i - T^n x_j\| \leqq \varepsilon \quad \text{for all } i, j \in I.$$

It follows from

$$x = T^n x + (I + T + \ldots + T^{n-1})(I - T)x$$

that

$$\begin{aligned} &\|x_i - x_j\| \leqq \\ &\leqq \|T^n x_i - T^n x_j\| + \|I + T + \ldots + T^{n-1}\| \, \|(I - T)x_i - (I-T)x_j\|. \end{aligned}$$

Thus, since $((I - T)x_i)$ is Cauchy, we can find an infinite subset I_0 of I such that

$$\|x_i - x_j\| \leqq 2\varepsilon \quad \text{for all } i, j \in I_0.$$

Let $(x_i^0) := (x_i)$ and $\varepsilon_m := 2^{-m-1}$ for $m = 1, 2, \ldots$ Applying the preceding construction infinitely many times, we obtain sequences (x_i^m) each of which is a subsequence of its predecessor (x_i^{m-1}) and such that

$$\|x_i^m - x_j^m\| \leqq 2\varepsilon_m \quad \text{for all } i \text{ and } j.$$

Then the diagonal (x_i^i) is the desired convergent subsequence, because

$$\|x_i^i - x_j^j\| \leqq 2^{-m} \quad \text{whenever } i, j \geqq m.$$

3.2.7.* Proposition. If $T \in \mathfrak{L}(E)$ is iteratively compact, then all ranges $M_h(I - T)$ are closed.

Proof. Let $y = \lim_i y_i$, where (y_i) is contained in $M_h(I - T)$. Set

$$\varrho_i := \inf\{\|x\| : (I - T)^h x = y_i\},$$

and choose $x_i \in E$ such that

$$(I - T)^h x_i = y_i \quad \text{and} \quad \|x_i\| \leqq 2\varrho_i.$$

Assuming that $\varrho_i \to \infty$, we put

$$u_i := \varrho_i^{-1} x_i \quad \text{and} \quad v_i := \varrho_i^{-1} y_i.$$

Then $\|u_i\| \leqq 2$, and (v_i) tends to zero. Next, h-fold application of 3.2.6 yields a subsequence of (u_i) which converges to some $u \in E$. Passing to the limit, we see that

$$(I - T)^h u_i = v_i \quad \text{implies} \quad (I - T)^h u = o.$$

Hence

$$(I - T)^h (x_i - \varrho_i u) = y_i.$$

Thus, by the definition of ϱ_i, we have

$$\|x_i - \varrho_i u\| \geqq \varrho_i \quad \text{or} \quad \|u_i - u\| \geqq 1.$$

This contradiction shows that (ϱ_i) has a bounded subsequence.

Applying 3.2.6 once again, we find a convergent subsequence of (x_i). If x is the corresponding limit, then $(I - T)^h x = y$. This proves that $y \in M_h(I - T)$.

3.2.8.* Proposition. If $T \in \mathfrak{L}(E)$ is iteratively compact, then $I - T$ has finite descent.

Proof. Assume the contrary. Applying the Riesz lemma with $\varepsilon = 1/3$, we can choose $y_h \in M_h(I - T)$ such that $\|y_h\| = 1$ and

$$\|y_h - y\| \geqq 3/4 \quad \text{for all } y \in M_{h+1}(I - T).$$

It follows from

$$y - T^n y = (I + T + \dots + T^{n-1})(I - T)\, y$$

that

$$y_h - T^n y_h \in M_{h+1}(I - T) \quad \text{for } n = 1, 2, \dots$$

Hence

$$T^n y_h - T^n y_k \in y_h - y_k + M_{h+1}(I - T) + M_{k+1}(I - T),$$

which implies that

$$\|T^n y_h - T^n y_k\| \geqq 3/4 \quad \text{whenever } h < k \quad \text{and} \quad n = 1, 2, \dots$$

On the other hand, by the pigeon-hole principle, there exist an exponent n and different indices h and k such that

$$\|T^n y_h - T^n y_k\| \leqq 1/2.$$

This contradiction proves the assertion.

3.2.9.* We are now in a position to establish the main result of this section.

Decomposition theorem (F. Riesz 1918, K. Yosida 1939, A. Pietsch 1961). Let $T \in \mathfrak{L}(E)$ be iteratively compact. Then the Banach space E is the direct sum of the T-invariant subspaces $M_\infty(I - T)$ and $N_\infty(I - T)$, the latter being finite dimensional. Moreover, the following holds:

(1) The restriction of $I - T$ to $M_\infty(I - T)$ is invertible.

(2) The restriction of $I - T$ to $N_\infty(I - T)$ is nilpotent of order $d(I - T)$.

Proof. According to 3.1.14 and the previous results of this section we need to show only that the restriction of $I - T$ to $M_\infty(I - T)$ is continuously invertible. Of course, since $M_\infty(I - T)$ is closed, this follows from the bounded inverse theorem. There is, however, a direct approach.

Suppose that the inverse fails to be continuous. Then we can find normalized elements $x_i \in M_\infty(I - T)$ such that $((I - T)x_i)$ tends to zero. Hence, by 3.2.6, there exists a subsequence converging to an element $x \in M_\infty(I - T)$. Passing to the limit, we see that $\|x_i\| = 1$ implies $\|x\| = 1$. Furthermore, $(I - T)x_i \to o$ yields $(I - T)x = o$. Hence $x \in M_\infty(I - T) \cap N_1(I - T)$. Thus, by 3.1.4, we have $x = o$. This contradiction proves the continuity.

3.2.10.* For every iteratively compact operator $T \in \mathfrak{L}(E)$ we let $P_\infty(I - T)$ denote the projection from E onto $N_\infty(I - T)$ along $M_\infty(I - T)$.

3.2.11.* The decomposition theorem 3.2.9 can be rephrased as follows.

Theorem. Let $T \in \mathfrak{L}(E)$ be iteratively compact. Then there exists a finite projection $P \in \mathfrak{L}(E)$ commuting with T such that the operators

$$T_M := (I - P)T \quad \text{and} \quad T_N := PT$$

have the following properties:

(1) $I - T_M$ is invertible.

(2) $P - T_N$ is nilpotent of order $d(I - T)$.

Proof. Define $P := P_\infty(I - T)$.

3.2.12.* An operator $T \in \mathfrak{L}(E)$ is said to be **Riesz** if for every $\varepsilon > 0$ there exist an exponent n and elements $u_1, \ldots, u_k \in E$ such that

$$T^n(U) \subseteq \bigcup_{h=1}^{k} \{u_h + \varepsilon^n U\}.$$

3.2.13.* The following criterion relates the concepts defined in 3.2.1 and 3.2.12 with each other.

Proposition. An operator $T \in \mathfrak{L}(E)$ is Riesz if and only if ζT is iteratively compact for all $\zeta \in \mathbb{C}$.

Proof. Assume that T is Riesz. Given $\zeta \in \mathbb{C}$ and $\varepsilon > 0$, we put

$$\delta := \frac{\min(\varepsilon, 1)}{1 + |\zeta|}.$$

Then there exist n and $u_1, \ldots, u_k \in E$ such that

$$T^n(U) \subseteq \bigcup_{h=1}^{k} \{u_h + \delta^n U\}.$$

It follows from $|\zeta|\,\delta \leqq \varepsilon$ and $|\zeta|\,\delta \leqq 1$ that $|\zeta|^n\,\delta^n \leqq \varepsilon$. Hence

$$(\zeta T)^n(U) \subseteqq \bigcup_{h=1}^{k} \{\zeta^n u_h + \varepsilon U\}.$$

Thus ζT is iteratively compact.

In order to prove the converse, we let $\varepsilon > 0$. Take

$$\zeta := \max\left(\frac{1}{\varepsilon}, 1\right).$$

Since ζT is assumed to be iteratively compact, we find n and $u_1, \ldots, u_k \in E$ such that

$$(\zeta T)^n\,(U) \subseteqq \bigcup_{h=1}^{k} \{u_h + \varepsilon U\}.$$

It follows from $\zeta\varepsilon \geqq 1$ and $\zeta \geqq 1$ that $\zeta^n\varepsilon^{n-1} \geqq 1$. Hence $\zeta^{-n}\varepsilon \leqq \varepsilon^n$, which implies that

$$T^n(U) \subseteqq \bigcup_{h=1}^{k} \{\zeta^{-n} u_h + \varepsilon^n U\}.$$

Thus T is Riesz.

3.2.14.* In view of the preceding criterion, Theorem 3.2.9 implies the following basic result.

Decomposition theorem. Let $T \in \mathfrak{L}(E)$ be a Riesz operator. Then, for every complex number ζ, the Banach space E is the direct sum of the T-invariant subspaces $M_\infty(I + \zeta T)$ and $N_\infty(I + \zeta T)$, the latter being finite dimensional. Moreover, the following holds:

(1) The restriction of $I + \zeta T$ to $M_\infty(I + \zeta T)$ is invertible.

(2) The restriction of $I + \zeta T$ to $N_\infty(I + \zeta T)$ is nilpotent of order $d(I + \zeta T)$.

Remark. We stress the fact that, conversely, an operator $T \in \mathfrak{L}(E)$ is Riesz if it has the above decomposition property for all $\zeta \in \mathbb{C}$; see (BAS, 0.3.5).

3.2.15.* A complex number λ_0 is an **eigenvalue** of the operator $T \in \mathfrak{L}(E)$ if there exists $x \in E$ such that

$$Tx = \lambda_0 x \quad \text{and} \quad x \neq o.$$

Every element $x \neq o$ such that $(\lambda_0 I - T)^k\, x = o$ for some exponent k is called a **principal element** of T associated with the eigenvalue λ_0. If $k = 1$, then we have $Tx = \lambda_0 x$, and x is said to be an **eigenelement**.

The **multiplicity** of an eigenvalue λ_0 is defined by

$$n(\lambda_0 I - T) := \dim\,[N_\infty(\lambda_0 I - T)],$$

where the right-hand quantity may or may not be finite.

Remark. In contrast to the *algebraic multiplicity* just defined, the dimension of the subspace $N(\lambda_0 I - T)$ is called the *geometric multiplicity* of an eigenvalue λ_0.

3.2.16.* First we establish a very elementary property.

Proposition. Eigenelements $x_1, \dots, x_m$ of any operator $T \in \mathfrak{L}(E)$ associated with distinct eigenvalues $\lambda_1, \dots, \lambda_m$ are linearly independent.

Proof. Suppose that $\alpha_1 x_1 + \dots + \alpha_m x_m = o$. Applying T^k to both sides of this equation, we obtain

$$\lambda_1^k \alpha_1 x_1 + \dots + \lambda_m^k \alpha_m x_m = o \quad \text{for } k = 0, 1, \dots, m-1.$$

Since Vandermonde's determinant

$$\det \begin{pmatrix} 1 & \dots & 1 \\ \lambda_1 & & \lambda_m \\ \vdots & & \vdots \\ \lambda_1^{m-1} & \dots & \lambda_m^{m-1} \end{pmatrix} = \prod_{i>j} (\lambda_i - \lambda_j)$$

does not vanish, we have $\alpha_j x_j = o$. It follows from $x_j \neq o$ that $\alpha_j = 0$ for $j = 1, \dots, m$.

3.2.17.* The next proposition contains the preceding result as a special case.

Proposition. Principal elements $x_1, \dots, x_m$ of any operator $T \in \mathfrak{L}(E)$ associated with distinct eigenvalues $\lambda_1, \dots, \lambda_m$ are linearly independent.

Proof. Suppose that $\alpha_1 x_1 + \dots + \alpha_m x_m = o$, and choose a natural number k such that $(\lambda_i I - T)^k x_i = o$ for $i = 1, \dots, m$. Fix any λ_j. Since the polynomials $(\lambda_j - \lambda)^k$ and $\prod_{i \neq j} (\lambda_i - \lambda)^k$ are relatively prime, there exist polynomials p_j and q_j which satisfy

$$p_j(\lambda)\,(\lambda_j - \lambda)^k + q_j(\lambda) \prod_{i \neq j} (\lambda_i - \lambda)^k = 1.$$

Hence

$$p_j(T)\,(\lambda_j I - T)^k + q_j(T) \prod_{i \neq j} (\lambda_i I - T)^k = I.$$

Obviously,

$$p_j(T)\,(\lambda_j I - T)^k\,(\alpha_j x_j) = o.$$

Moreover, it follows from

$$\alpha_j x_j = -\sum_{i \neq j} \alpha_i x_i$$

that

$$q_j(T) \prod_{i \neq j} (\lambda_i I - T)^k\,(\alpha_j x_j) = o.$$

This proves that $\alpha_j x_j = o$. Since $x_j \neq o$, we have $\alpha_j = 0$ for $j = 1, \dots, m$.

3.2.18.* Let $T \in \mathfrak{L}(E)$ be a Riesz operator. Then, as observed in 3.2.14, the operators $I + \zeta T$ with $\zeta \in \mathbb{C}$ enjoy many important properties. In connection with eigenvalues, however, it is more natural to consider the operators $\lambda I - T$ with $\lambda \in \mathbb{C}$. The point is that we have to exclude the value $\lambda = 0$. Then, setting $\zeta = -1/\lambda$, it follows that

$$N_k(\lambda I - T) = N_k(I + \zeta T) \quad \text{and} \quad M_h(\lambda I - T) = M_h(I + \zeta T).$$

In particular, by 3.2.4, the multiplicity of every eigenvalue $\lambda \neq 0$ is finite.

3.2.19.* Theorem (F. Riesz 1918). Let $T \in \mathfrak{L}(E)$ be a Riesz operator. Then, for every $\varrho > 0$, the set of all eigenvalues λ with $|\lambda| \geqq \varrho$ is finite.

Proof. Suppose that T possesses a sequence of distinct eigenvalues $\lambda_1, \lambda_2, \ldots$ such that $|\lambda_k| \geqq \varrho$. Take any sequence of associated eigenelements $u_1, u_2, \ldots \in E$. It follows from 3.2.16 that the subspace

$$E_k := \operatorname{span}(u_1, \ldots, u_k)$$

is k-dimensional, $E_0 := \{o\}$. Applying the Riesz lemma with $\varepsilon = 1/3$, we may choose elements $x_k \in E_k$ such that $\|x_k\| = 1$ and

$$\|x_k - x\| \geqq 3/4 \quad \text{for all } x \in E_{k-1}.$$

There exist coefficients α_k for which

$$x_k - \alpha_k u_k \in E_{k-1}.$$

Hence

$$T^n x_k - \alpha_k \lambda_k^n u_k \in E_{k-1} \quad \text{and} \quad \lambda_k^n x_k - \alpha_k \lambda_k^n u_k \in E_{k-1}.$$

This implies that

$$T^n x_k - \lambda_k^n x_k \in E_{k-1} \quad \text{for } n = 1, 2, \ldots$$

Consequently,

$$T^n x_h - T^n x_k \in \lambda_h^n x_h - \lambda_k^n x_k + E_{h-1} + E_{k-1},$$

and we obtain

$$\|T^n x_h - T^n x_k\| \geqq \tfrac{3}{4}\varrho^n \quad \text{whenever } h > k \quad \text{and} \quad n = 1, 2, \ldots$$

On the other hand, applying the pigeon-hole principle to the operator $\varrho^{-1} T$, we may choose an exponent n and different indices h and k such that

$$\|T^n x_h - T^n x_k\| \leqq \tfrac{1}{2}\varrho^n.$$

This contradiction completes the proof.

3.2.20.* With every Riesz operator $T \in \mathfrak{L}(E)$ we associate the **eigenvalue sequence** $(\lambda_n(T))$ defined in the following way:

(1) Every eigenvalue $\lambda_0 \neq 0$ is counted according to its multiplicity. This means that it occurs $n(\lambda_0 I - T)$-times, one after the other.

(2) The eigenvalues are arranged in order of non-increasing magnitude:

$$|\lambda_1(T)| \geqq |\lambda_2(T)| \geqq \ldots \geqq 0.$$

In case there are distinct eigenvalues having the same modulus these can be written in any order we please.

(3) If T possesses less than n eigenvalues $\lambda \neq 0$, then $\lambda_n(T) := 0$. Thus, even for finite operators, $(\lambda_n(T))$ is an infinite sequence.

3.2.21.* In order to prove the following theorem, we require an auxiliary result.

Lemma. Let $T \in \mathfrak{L}(E)$ and $\lambda \neq \mu$. Then $N_\infty(\lambda I - T) \subseteq M_\infty(\mu I - T)$.

Proof. Given $x \in N_\infty(\lambda I - T)$, there exists an exponent k such that $(\lambda I - T)^k x = o$. Applying the binomial formula to

$$[(\lambda - \mu) I + (\mu I - T)]^k x = o$$

and dividing by $(\lambda - \mu)^k \neq 0$, we obtain $x = (\mu I - T)\,Sx$, where the operator S commutes with T. Hence

$$x = (\mu I - T)^h S^h x \quad \text{for } h = 1, 2, \ldots$$

This proves that $x \in M_\infty(\mu I - T)$.

3.2.22.* We are now able to establish a generalization of the Jordan decomposition, well-known for linear maps on finite dimensional linear spaces; see A.2.6.

Theorem. Let $T \in \mathfrak{L}(E)$ be a Riesz operator. Then, given any finite set of distinct eigenvalues $\lambda_1, \ldots, \lambda_m \neq 0$, the Banach space E admits the decomposition

$$E = \bigcap_{i=1}^{m} M_\infty(\lambda_i I - T) \oplus N_\infty(\lambda_1 I - T) \oplus \ldots \oplus N_\infty(\lambda_m I - T).$$

Proof. Applying the Riesz decomposition to the operators $\lambda_i I - T$ with $i = 1, \ldots, m$, we may write every element $x \in E$ in the form

$$x = x_{M,i} + x_{N,i} \text{ with } x_{M,i} \in M_\infty(\lambda_i I - T) \quad \text{and} \quad x_{N,i} \in N_\infty(\lambda_i I - T).$$

The preceding lemma implies that

$$x_M := x - \sum_{i=1}^{m} x_{N,i} = x_{M,j} - \sum_{i \neq j} x_{N,i} \in M_\infty(\lambda_j I - T)$$

for $j = 1, \ldots, m$. Hence

$$x = x_M + \sum_{i=1}^{m} x_{N,i}$$

is the desired decomposition. The uniqueness follows from 3.2.14 and 3.2.21.

Remark. Note that every linear map T on a finite dimensional linear space E has at least one eigenvalue. Hence, if $\{\lambda_1, \ldots, \lambda_m\}$ denotes the set of all distinct eigenvalues (possibly including $\lambda = 0$), then we have the classical Jordan decomposition

$$E = N_\infty(\lambda_1 I - T) \oplus \ldots \oplus N_\infty(\lambda_m I - T).$$

In the infinite dimensional case the situation is completely different, since then there are Riesz operators without any eigenvalue. As an important example we mention the integration operator

$$S : g(\eta) \to f(\xi) := \int_0^\xi g(\eta)\,\mathrm{d}\eta$$

defined on $C(0, 1)$ or $L_p(0, 1)$.

3.2.23.* We now provide a lemma which is an elementary but basic tool in the theory of eigenvalue distributions.

Lemma. Let $T \in \mathfrak{L}(E)$ be a Riesz operator. If $\lambda_n(T) \neq 0$, then there exists an n-dimensional T-invariant subspace E_n such that the operator $T_n \in \mathfrak{L}(E_n)$ induced by T has precisely $\lambda_1(T), \ldots, \lambda_n(T)$ as its eigenvalues.

Proof. Denote by $\{\lambda_1, \ldots, \lambda_m\}$ the set of distinct complex numbers appearing in $\{\lambda_1(T), \ldots, \lambda_n(T)\}$. In particular, we let $\lambda_m := \lambda_n(T)$. Since $\lambda_m I - T$ is nilpotent on

$N_\infty(\lambda_m I - T)$, we may choose a basis (x_h) of this subspace according to 3.1.15. Let

$$k := n - \sum_{i=1}^{m-1} n(\lambda_i I - T).$$

Then $1 \leqq k \leqq n(\lambda_m I - T)$, and

$$E_n := \sum_{i=1}^{m-1} N_\infty(\lambda_i I - T) + \operatorname{span}(x_1, \ldots, x_k)$$

is the T-invariant subspace we are looking for.

3.2.24.* We now prove a special but very precise version of the famous **spectral mapping theorem**; see (DUN, VII.3.11) and (TAY, V.3.4).

Theorem. An operator $T \in \mathfrak{L}(E)$ is Riesz if and only if T^m is Riesz for some (every) exponent m. In this case, the eigenvalue sequences of T and T^m can be arranged in such a way that

$$(\lambda_n(T)^m) = (\lambda_n(T^m)).$$

Proof. Given $\varepsilon > 0$, we put

$$\delta := \frac{1}{\|T^{m-1}\|}\,\varepsilon.$$

If T is Riesz, then there exist n and $u_1, \ldots, u_k \in E$ such that

$$T^n(U) \subseteqq \bigcup_{h=1}^{k} \{u_h + \delta^n U\}.$$

Since

$$T^{mn}(U) \subseteqq \|T^{m-1}\|^n\, T^n(U),$$

it follows that

$$(T^m)^n\,(U) \subseteqq \bigcup_{h=1}^{k} \{\|T^{m-1}\|^n\, u_h + \varepsilon^n U\}.$$

This proves that T^m is Riesz for all exponents m.

Conversely, assume that T^m is Riesz for some m. Given $\varepsilon > 0$, we let $\delta := \varepsilon^m$. Then there exist n and $u_1, \ldots, u_k \in E$ such that

$$(T^m)^n\,(U) \subseteqq \bigcup_{h=1}^{k} \{u_h + \delta^n U\}.$$

This means that

$$T^{mn}(U) \subseteqq \bigcup_{h=1}^{k} \{u_h + \varepsilon^{mn} U\}.$$

Hence T is Riesz, as well.

The remaining part of the assertion follows from

$$N_\infty(\mu I - T^m) = N_\infty(\lambda_1 I - T) \oplus \ldots \oplus N_\infty(\lambda_m I - T),$$

where $\lambda_1, \ldots, \lambda_m$ are the roots of $\lambda^m = \mu$, and $\mu \neq 0$. We first note that

$$(\mu I - T^m) = (\lambda_i^{m-1} I + \ldots + T^{m-1})\,(\lambda_i I - T)$$

implies

$$N_\infty(\lambda_i I - T) \subseteqq N_\infty(\mu I - T^m) \quad \text{for } i = 1, \ldots, m.$$

Next let λ be any eigenvalue of the operator induced by T on the invariant subspace $N_\infty(\mu I - T^m)$. Take any associated eigenelement x. Then we may choose k such that $(\mu I - T^m)^k x = o$. Hence $(\mu - \lambda^m)^k x = o$, which shows that λ must be a root of $\lambda^m = \mu$. Thus, in view of 3.2.22 (Remark), $N_\infty(\mu I - T^m)$ has the Jordan decomposition described above.

3.2.25. Next we generalize another classical result.

Proposition (J. Schauder 1930). An operator $T \in \mathfrak{L}(E)$ is iteratively compact if and only if T' is iteratively compact.

Proof. If T is supposed to be iteratively compact, then for every $\varepsilon > 0$ there exist n and $u_1, \ldots, u_h \in E$ such that

$$T^n(U) \subseteqq \bigcup_{i=1}^{h} \{u_i + \varepsilon U\}.$$

We may assume that $u_i = Tx_i$ with $x_i \in U$. Define $X \in \mathfrak{L}(l_1(h), E)$ by

$$X := \sum_{i=1}^{h} e_i \otimes x_i.$$

Since $X'(T')^n$ has finite rank, we can choose $a_1, \ldots, a_k \in U^0$ such that

$$X'(T')^n(U^0) \subseteqq \bigcup_{j=1}^{k} \{X'(T')^n a_j + \varepsilon U_\infty(h)\},$$

where U^0 and $U_\infty(h)$ denote the closed unit balls of E' and $l_\infty(h)$, respectively. Note that $X'a = (\langle x_i, a\rangle)$ for $a \in E'$. Given $a \in U^0$, there exists a_j with

$$|\langle x_i, (T')^n a - (T')^n a_j\rangle| \leqq \varepsilon \quad \text{for } i = 1, \ldots, h. \tag{1}$$

Next we pick some $x \in U$ such that

$$\|(T')^n a - (T')^n a_j\| \leqq 2|\langle x, (T')^n a - (T')^n a_j\rangle|. \tag{2}$$

Finally, we select x_i which satisfies

$$\|T^n x - T^n x_i\| \leqq \varepsilon. \tag{3}$$

Combining (1), (2) and (3), we obtain

$$\begin{aligned} \|(T')^n a - (T')^n a_j\| &\leqq 2|\langle x, (T')^n a - (T')^n a_j\rangle| = 2|\langle T^n x, a - a_j\rangle| \\ &\leqq 2|\langle T^n x_i, a - a_j\rangle| + 2|\langle T^n x - T^n x_i, a - a_j\rangle| \\ &\leqq 2|\langle x_i, (T')^n a - (T')^n a_j\rangle| + 4\|T^n x - T^n x_i\| \leqq 6\varepsilon. \end{aligned}$$

This proves that

$$(T')^n(U^0) \subseteqq \bigcup_{j=1}^{k} \{(T')^n a_j + 6\varepsilon U^0\}.$$

Thus T' is iteratively compact.

Conversely, if T' is iteratively compact, then so is T'', and it follows from $K_F T^n = (T'')^n K_E$ that T is iteratively compact, as well.

3.2.26. We conclude with an important theorem which is an easy corollary of the preceding proposition.

Theorem (T. T. West 1966). An operator $T \in \mathfrak{L}(E)$ is Riesz if and only if T' is Riesz. In this case, the eigenvalue sequences of T and T' can be arranged in such a way that

$$(\lambda_n(T)) = (\lambda_n(T')).$$

Proof. The first part of the assertion follows from 3.2.13 and 3.2.25.

To finish the proof, we observe that

$$N_\infty(\lambda I' - T') = \{a \in E' : \langle y, a \rangle = 0 \quad \text{for } y \in M_\infty(\lambda I - T)\}.$$

Hence

$$\dim [N_\infty(\lambda I' - T')] = \operatorname{codim} [M_\infty(\lambda I - T)] = \dim [N_\infty(\lambda I - T)].$$

This means that T and T' have the same eigenvalues $\lambda \neq 0$ with the same multiplicities.

3.3. Related operators

3.3.1.* We begin with some auxiliary results.

Lemma (A. Pietsch 1963: a). Let $A \in \mathfrak{L}(F, E)$ and $B \in \mathfrak{L}(E, F)$. Then

$$B[N_k(I_E - AB)] = N_k(I_F - BA) \quad \text{for } k = 1, 2, \ldots$$

Moreover, $n(I_E - AB) = n(I_F - BA)$.

Proof. We first treat the case $k = 1$. If $x \in N(I_E - AB)$, then $x = ABx$. Hence $Bx = BA(Bx)$, which means that $Bx \in N(I_F - BA)$. Therefore

$$B[N(I_E - AB)] \subseteq N(I_F - BA).$$

Analogously, it follows from $y \in N(I_F - BA)$ that $Ay \in N(I_E - AB)$. Since $y = BAy$, we have $y \in B[N(I_E - AB)]$. Consequently,

$$N(I_F - BA) \subseteq B[N(I_E - AB)].$$

If $k = 1, 2, \ldots$, then

$$(I_E - AB)^k = I_E - A_k B \quad \text{and} \quad (I_F - BA)^k = I_F - BA_k,$$

where

$$A_k := \sum_{i=1}^{k} \binom{k}{i} (-AB)^{i-1} A = \sum_{i=1}^{k} \binom{k}{i} A(-BA)^{i-1}.$$

Thus, applying the result already verified to the operators A_k and B, we obtain

$$B[N_k(I_E - AB)] = N_k(I_F - BA).$$

This implies that

$$\dim [N_k(I_E - AB)] \geq \dim [N_k(I_F - BA)].$$

Interchanging the roles of A and B, we see that in fact equality holds. Hence $n(I_E - AB) = n(I_F - BA)$.

3.3.2.* **Lemma.** Let $A \in \mathfrak{L}(F, E)$ and $B \in \mathfrak{L}(E, F)$. If $AB \in \mathfrak{L}(E)$ is iteratively compact, then so is $BA \in \mathfrak{L}(F)$.

Proof. Given $\varepsilon > 0$, we put

$$\delta := \frac{1}{\|A\| \, \|B\|} \, \varepsilon .$$

Then there exist n and $u_1, \ldots, u_h \in E$ such that

$$(AB)^n (U) \subseteqq \bigcup_{h=1}^{k} \{u_h + \delta U\}.$$

Note that

$$A(V) \subseteqq \|A\| \, U \quad \text{and} \quad B(U) \subseteqq \|B\| \, V.$$

We now obtain

$$\begin{aligned}(BA)^{n+1} (V) &= B(AB)^n A(V) \subseteqq \|A\| \, B(AB)^n (U) \\ &\subseteqq \bigcup_{h=1}^{k} \{\|A\| \, Bu_h + \delta \|A\| \, B(U)\} \subseteqq \bigcup_{h=1}^{k} \{\|A\| \, Bu_h + \varepsilon V\}.\end{aligned}$$

Hence BA is iteratively compact.

3.3.3.* Operators $S \in \mathfrak{L}(E)$ and $T \in \mathfrak{L}(F)$ are said to be **related** if there exist $A \in \mathfrak{L}(F, E)$ and $B \in \mathfrak{L}(E, F)$ such that $S = AB$ and $T = BA$:

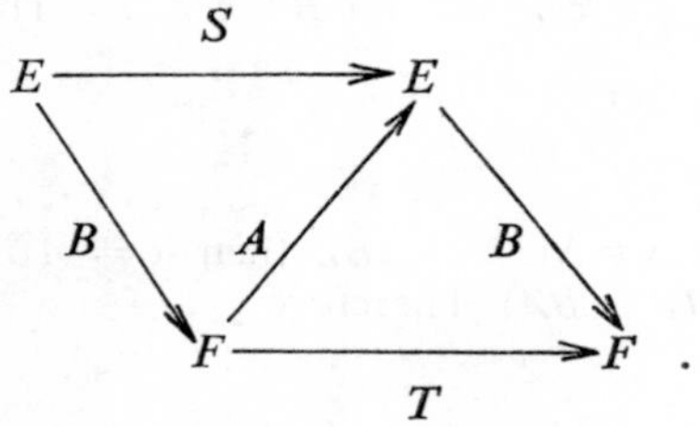

3.3.4.* Combining the preceding lemmas with 3.2.13, we obtain an elementary but extremely powerful result.

Principle of related operators (A. Pietsch 1963: a). Let $S \in \mathfrak{L}(E)$ and $T \in \mathfrak{L}(F)$ be related. If S is Riesz, then so is T. Moreover, both operators have the same non-zero eigenvalues with the same multiplicities.

3.3.5. Finally, we give the simplest example of related operators.

Proposition. Every finite operator $T \in \mathfrak{L}(E)$ of the form

$$T = \sum_{i=1}^{m} a_i \otimes x_i$$

with $a_1, \ldots, a_m \in E'$ and $x_1, \ldots, x_m \in E$ is related to the operator $M_{\text{op}} \in \mathfrak{L}(l_2(m))$ induced by the matrix $M := (\langle x_j, a_i \rangle)$.

Proof. Define $A \in \mathfrak{L}(E, l_2(m))$ and $X \in \mathfrak{L}(l_2(m), E)$ by

$$A := \sum_{i=1}^{m} a_i \otimes e_i \quad \text{and} \quad X := \sum_{j=1}^{m} e_j \otimes x_j .$$

Then we have $T = XA$ and $\langle AXe_j, e_i \rangle = \langle x_j, a_i \rangle$. The latter formula means that AX is indeed induced by the matrix $M := (\langle x_j, a_i \rangle)$.

3.4. The eigenvalue type of operator ideals

3.4.1.* An operator ideal $\mathfrak{A}$ is said to be of **eigenvalue type** $l_{r,w}$ if, for arbitrary Banach spaces E, all operators $T \in \mathfrak{A}(E)$ are Riesz and $(\lambda_n(T)) \in l_{r,w}$.

The collection of these operator ideals is denoted by $\mathbb{E}_{r,w}$. If $r = w$, then we simply write $\mathbb{E}_r$ instead of $\mathbb{E}_{r,r}$.

We say that an operator ideal $\mathfrak{A}$ has **optimum eigenvalue type** $l_{r,w}$ if $\mathfrak{A} \in \mathbb{E}_{r,w}$ but $\mathfrak{A} \notin \mathbb{E}_{r,w_0}$ whenever $0 < w_0 < w \leqq \infty$. Of course, such an "infimum" (with respect to the lexicographical order of Lorentz sequence spaces) must not exist. However, for numerous concrete operator ideals the best possible eigenvalue type is indeed attained.

3.4.2.* In this section we prove some general facts about the eigenvalue type of operator ideals. To begin with, we state an immediate consequence of 3.2.26.

Proposition. Let $\mathfrak{A}$ be an operator ideal. Then $\mathfrak{A}$ and $\mathfrak{A}'$ have the same eigenvalue type.

3.4.3.* **Principle of iteration** (A. Pietsch 1982: a). An operator ideal $\mathfrak{A}$ is of eigenvalue type $l_{r,w}$ if and only if $\mathfrak{A}^m$ is of eigenvalue type $l_{r/m,w/m}$ for some (every) exponent m.

Proof. Given $T \in \mathfrak{A}^m(E)$, we consider a factorization

$$T = T_m \dots T_1, \text{ where } T_k \in \mathfrak{A}(E_k, E_{k+1}) \quad \text{for } k = 1, \dots, m$$

and $E_1 = E_{m+1} = E$. Form the direct sum $F := [l_2(m), E_k]$ as described in C.2.1, and define

$$S := J_1 T_m Q_m + J_2 T_1 Q_1 + \dots + J_m T_{m-1} Q_{m-1},$$

where $J_k \in \mathfrak{L}(E_k, F)$ and $Q_k \in \mathfrak{L}(F, E_k)$ denote the canonical injections and surjections, respectively. Then $S \in \mathfrak{A}(F)$, and we have

$$S(x_1, \dots, x_{m-1}, x_m) = (T_m x_m, T_1 x_1, \dots, T_{m-1} x_{m-1})$$

for $x_1 \in E_1, \dots, x_{m-1} \in E_{m-1}$, $x_m \in E_m$. Observe that E can be identified with the subspace $M(J_1)$ of F which is invariant under S^m. Moreover, the restriction of S^m to $M(J_1)$ coincides with $T = T_m \dots T_1$. This shows that $(\lambda_n(T))$ is a subsequence of $(\lambda_n(S^m))$. By the spectral mapping theorem 3.2.24, we have $(\lambda_n(S^m)) = (\lambda_n(S)^m)$. Hence, if $(\lambda_n(S)) \in l_{r,w}$, then $(\lambda_n(T)) \in l_{r/m,w/m}$. This proves that

$$\mathfrak{A} \in \mathbb{E}_{r,w} \quad \text{implies} \quad \mathfrak{A}^m \in \mathbb{E}_{r/m,w/m}.$$

The converse is obvious, by 3.2.24.

3.4.4. In the following we look for upper bounds for all the operator ideals belonging to $\mathbb{E}_{r,w}$.

Proposition. If $\mathfrak{A}$ is of eigenvalue type $l_{r,w}$, then $\mathfrak{A} \subseteq \mathfrak{S}_{r,w}^{\text{weak}}$.

Proof. Let $T \in \mathfrak{A}(l_2)$. Since the identity operator of l_2 fails to be Riesz, it follows from Calkin's theorem 2.11.11 that T must be approximable. Consequently, by 2.11.5 and 2.11.8, there exist $X, Y \in \mathfrak{L}(l_2)$ such that $D_t = Y^* T X$ and $T = Y D_t X^*$, where the diagonal operator D_t is induced by a non-increasing sequence $t = (\tau_n) \in c_0$. Note that $\lambda_n(D_t) = \tau_n$. Since $T \in \mathfrak{A}(l_2)$, we have $D_t \in \mathfrak{A}(l_2)$ and therefore $t \in l_{r,w}$. This proves that $D_t \in \mathfrak{S}_{r,w}(l_2)$, which in turn yields $T \in \mathfrak{S}_{r,w}(l_2)$.

Using the fact that $\mathfrak{S}_{r,w}^{\text{weak}}$ is the largest extension of $\mathfrak{S}_{r,w}$, we finally obtain the desired result.

Remark. Note that $\mathbb{E}_{r,w}$ does not contain a largest operator ideal; see 3.9.1.

3.4.5. The next result is along the same line.

Proposition (A. Pietsch 1982: a). If $\mathfrak{A}$ is of eigenvalue type l_r, then $\mathfrak{A}^m \subseteq \mathfrak{P}_2$ whenever $m \geqq r$.

Proof. By 3.4.3, we have $\mathfrak{A}^m \in \mathbb{E}_1$. Hence it follows from 3.4.4 and 2.11.35 that $\mathfrak{A}^m \subseteq \mathfrak{S}_1^{\text{weak}} \subseteq \mathfrak{P}_2$.

3.4.6.* We now establish one of the most important tools of the theory of eigenvalue distributions.

Principle of uniform boundedness (A. Pietsch 1972: b). Let $\mathfrak{A}$ be a quasi-Banach operator ideal which is of eigenvalue type $l_{r,w}$. Then there exists a constant $c \geqq 1$ such that

$$\|(\lambda_n(T)) \mid l_{r,w}\| \leqq c\|T \mid \mathfrak{A}\| \quad \text{for all } T \in \mathfrak{A}(E)$$

and arbitrary Banach spaces E.

Proof. Assume that the assertion fails. Then we can find operators $T_k \in \mathfrak{A}(E_k)$ such that

$$\|(\lambda_n(T_k)) \mid l_{r,w}\| \geqq k \quad \text{and} \quad \|T_k \mid A\| \leqq (2c_{\mathfrak{A}})^{-k} \quad \text{for } k = 1, 2, \dots,$$

where $c_{\mathfrak{A}}$ denotes the constant in the quasi-triangle inequality of $\|\cdot \mid \mathfrak{A}\|$. Define $E := [l_2, E_k]$ as described in C.2.1. Then it follows from

$$\left\|\sum_{k=m+1}^{n} J_k T_k Q_k \mid \mathfrak{A}\right\| \leqq \sum_{k=1}^{\infty} c_{\mathfrak{A}}^k \|T_{m+k} \mid \mathfrak{A}\| \leqq (2c_{\mathfrak{A}})^{-m} \quad \text{for } n > m$$

that

$$T := \sum_{k=1}^{\infty} J_k T_k Q_k$$

belongs to $\mathfrak{A}(E)$. Because $T_k = Q_k T J_k$, we may identify $(\lambda_n(T_k))$ with a subsequence of $(\lambda_n(T))$. This implies that

$$\|(\lambda_n(T)) \mid l_{r,w}\| \geqq \|(\lambda_n(T_k)) \mid l_{r,w}\| \geqq k \quad \text{for } k = 1, 2, \dots$$

This contradiction proves the existence of the desired constant.

3.4.7.* In order to prove the next result, we need an auxiliary result of purely algebraic character.

Lemma. Let $S \in \mathfrak{L}(E)$ and $T \in \mathfrak{L}(F)$. Then

$$N_\infty[(\lambda I_E - S)] \otimes N_\infty[(\mu I_F - T)] \subseteq N_\infty(\lambda\mu I_E \otimes I_F - S \otimes T) \quad \text{for } \lambda, \mu \in \mathbb{C}$$

Proof. Given $x \in N_\infty(\lambda I_E - S)$ and $y \in N_\infty(\mu I_F - T)$, there exist h and k such that

$$(\lambda I_E - S)^h x = o \quad \text{and} \quad (\mu I_F - T)^k y = o.$$

Write $m := h + k - 1$. Then it follows from

$$(\lambda I_E - S)^i x = o \quad \text{for } i = h, \dots, m,$$
$$(\mu I_F - T)^{m-i} y = o \quad \text{for } i = 0, \dots, h - 1$$

and

$$\lambda\mu I_{E\otimes F} - S \otimes T = (\lambda I_E - S) \otimes (\mu I_F) + S \otimes (\mu I_F - T)$$

that

$$(\lambda\mu I_{E\otimes F} - S \otimes T)^m x \otimes y =$$

$$= \sum_{i=0}^{m} \binom{m}{i} (\lambda I_E - S)^i S^{m-i}x \otimes (\mu I_F)^i (\mu I_F - T)^{m-i} y = o.$$

Hence $x \otimes y \in N_\infty(\lambda\mu I_E \otimes I_F - S \otimes T)$.

3.4.8.* The following result is very surprising, since it connects two concepts which—at a first glance—have nothing in common with each other.

Principle of tensor stability (A. Pietsch 1986). Suppose that the quasi-Banach operator ideal $\mathfrak{A}$ is stable with respect to a tensor norm α such that

$$\|T \tilde{\otimes}_\alpha T \mid \mathfrak{A}\| \leqq c\|T \mid \mathfrak{A}\|^2 \quad \text{for all } T \in \mathfrak{A}(E)$$

and arbitrary Banach spaces E. If $\mathfrak{A}$ is of eigenvalue type l_r, then (with the above constant $c \geqq 1$) we have

$$\|(\lambda_n(T)) \mid l_r\| \leqq c\|T \mid \mathfrak{A}\| \quad \text{for all } T \in \mathfrak{A}(E).$$

Proof. The preceding lemma implies that the double sequence $(\lambda_h(T)\,\lambda_k(T))$ can be identified with a subsequence of $(\lambda_n(T \tilde{\otimes}_\alpha T))$.

By the principle of uniform boundedness, there exists a constant $c_r \geqq 1$ such that

$$\|(\lambda_n(S)) \mid l_r\| \leqq c_r\|S \mid \mathfrak{A}\| \quad \text{for all } S \in \mathfrak{A}(F)$$

and all Banach spaces F. Taking $F := E \tilde{\otimes}_\alpha E$ and $S := T \tilde{\otimes}_\alpha T$, we obtain

$$\left(\sum_{n=1}^{\infty} |\lambda_n(T)|^r\right)^{2/r} = \left(\sum_{h=1}^{\infty} \sum_{k=1}^{\infty} |\lambda_h(T)\,\lambda_k(T)|^r\right)^{1/r} \leqq \left(\sum_{n=1}^{\infty} |\lambda_n(T \tilde{\otimes}_\alpha T)|^r\right)^{1/r}$$

$$\leqq c_r\|T \tilde{\otimes}_\alpha T \mid \mathfrak{A}\| \leqq c_r c\|T \mid \mathfrak{A}\|^2.$$

This yields

$$\|(\lambda_n(T)) \mid l_r\| \leqq (c_r c)^{1/2}\, \|T \mid \mathfrak{A}\|.$$

If c_r is chosen as small as possible, it follows that $c_r \leqq (c_r c)^{1/2}$. Thus $c_r \leqq c$.

3.4.9.* Next we show that, roughly speaking, for tensor stable quasi-Banach operator ideals the set of eigenvalue types l_r is closed.

Theorem (A. Pietsch 1986). Suppose that the quasi-Banach operator ideal $\mathfrak{A}$ is stable with respect to a tensor norm α such that

$$\|T \tilde{\otimes}_\alpha T \mid \mathfrak{A}\| \leqq c\|T \mid \mathfrak{A}\|^2 \quad \text{for all } T \in \mathfrak{A}(E)$$

and arbitrary Banach spaces E. If $\mathfrak{A}$ is of eigenvalue type $l_{r+\varepsilon}$ for all $\varepsilon > 0$, then it is even of eigenvalue type l_r. Moreover, we have

$$\|(\lambda_n(T)) \mid l_r\| \leqq c\|T \mid \mathfrak{A}\| \quad \text{for all } T \in \mathfrak{A}(E).$$

Proof. The preceding theorem implies that

$$\|(\lambda_n(T)) \mid l_{r+\varepsilon}\| \leqq c\|T \mid \mathfrak{A}\| \quad \text{for all } T \in \mathfrak{A}(E).$$

Since the constant c does not depend on ε, we may pass to the limit as $\varepsilon \to 0$.

3.5. Eigenvalues of Schatten—von Neumann operators

Throughout this section we exclusively consider operators acting on a complex Hilbert space.

3.5.1.* First of all, we prove the **multiplicative Weyl inequality.**

Lemma (H. Weyl 1949). Let $T \in \mathfrak{S}(H)$. Then

$$\prod_{k=1}^{n} |\lambda_k(T)| \leqq \prod_{k=1}^{n} a_k(T) \quad \text{for } n = 1, 2, \ldots$$

Proof. Fix any natural number n for which $\lambda_n(T) \neq 0$. By 3.2.23, there exists an n-dimensional T-invariant subspace H_n such that the operator induced by T has precisely $\lambda_1(T), \ldots, \lambda_n(T)$ as its eigenvalues. Let $(x_1, \ldots, x_n)$ be any orthonormal basis of H_n. Then it follows from 2.11.13 that

$$\prod_{k=1}^{n} |\lambda_k(T)| = |\det((Tx_i, x_j))| \leqq \prod_{k=1}^{n} a_k(T).$$

3.5.2.* In order to establish the additive Weyl inequality, we need some purely analytic results.

Lemma. Let $\alpha_1 \geqq \alpha_2 \geqq \ldots \geqq -\infty$ and $\beta_1 \geqq \beta_2 \geqq \ldots \geqq -\infty$. Suppose that

$$\sum_{k=1}^{n} \beta_k \leqq \sum_{k=1}^{n} \alpha_k \quad \text{for } n = 1, 2, \ldots$$

Then

$$\sum_{k=1}^{n} \exp(\beta_k) \leqq \sum_{k=1}^{n} \exp(\alpha_k) \quad \text{for } n = 1, 2, \ldots$$

Proof. On the real line we define the functions

$$a_n(\tau) := \sum_{k=1}^{n} (\alpha_k - \tau)_+ \quad \text{and} \quad b_n(\tau) := \sum_{k=1}^{n} (\beta_k - \tau)_+ .$$

If $\beta_1 > \tau$, then we put $m := \max\{k \in \mathbb{Z}_n : \beta_k > \tau\}$. Now it follows that

$$b_n(\tau) = \sum_{k=1}^{m} (\beta_k - \tau) \leqq \sum_{k=1}^{m} (\alpha_k - \tau) \leqq a_n(\tau).$$

If $\beta_1 \leqq \tau$, then $b_n(\tau) = 0$. Thus we also have $b_n(\tau) \leqq a_n(\tau)$. Integration by parts yields

$$\exp(\xi) = \int_{-\infty}^{+\infty} (\xi - \tau)_+ \exp(\tau)\, d\tau .$$

Therefore

$$\sum_{k=1}^{n} \exp(\beta_k) = \int_{-\infty}^{+\infty} b_n(\tau) \exp(\tau)\, d\tau \leqq \int_{-\infty}^{+\infty} a_n(\tau) \exp(\tau)\, d\tau = \sum_{k=1}^{n} \exp(\alpha_k).$$

This completes the proof.

3.5.3.* **Lemma.** Let $\alpha_1 \geqq \alpha_2 \geqq \ldots \geqq 0$ and $\beta_1 \geqq \beta_2 \geqq \ldots \geqq 0$. Suppose that

$$\prod_{k=1}^{n} \beta_k \leqq \prod_{k=1}^{n} \alpha_k \quad \text{for } n = 1, 2, \ldots$$

If $0 < r < \infty$, then

$$\sum_{k=1}^{n} \beta_k^r \leqq \sum_{k=1}^{n} \alpha_k^r \quad \text{for } n = 1, 2, \ldots$$

Proof. Apply the preceding lemma to the sequences

$$(r \log(\alpha_k)) \quad \text{and} \quad (r \log(\beta_k)).$$

3.5.4.* We now establish the **additive Weyl inequality**.

Lemma (H. Weyl 1949). Let $T \in \mathfrak{G}(H)$ and $0 < r < \infty$. Then

$$\sum_{k=1}^{n} |\lambda_k(T)|^r \leqq \sum_{k=1}^{n} a_k(T)^r \text{ for } n = 1, 2, \ldots$$

Proof. Obviously, the assertion follows immediately from 3.5.1 and 3.5.3.

Since this result is the keystone for most of the further investigation, we give here another proof which works, however, only for the case $1 \leqq r < \infty$. Fix any natural number n for which $\lambda_n(T) \neq 0$. By 3.2.23, there exists an n-dimensional T-invariant subspace H_n such that the operator $T_n \in \mathfrak{L}(H_n)$ induced by T has precisely $\lambda_1(T), \ldots, \lambda_n(T)$ as its eigenvalues. Consider any Schur basis $(x_1, \ldots, x_n)$ as described in A.2.5. This means that $((Tx_i, x_j))$ is a triangular matrix. Hence, by A.5.5, $(Tx_1, x_1), \ldots, (Tx_n, x_n)$ are the eigenvalues of T_n. Now it follows from 2.11.12 that

$$\sum_{k=1}^{n} |\lambda_k(T)|^r = \sum_{k=1}^{n} |(Tx_k, x_k)|^r \leqq \sum_{k=1}^{n} a_k(T)^r.$$

3.5.5.* We are now in a position to establish the main result of this section which was the starting point of the theory of eigenvalue distribution of abstract operators.

Theorem (H. Weyl 1949). The operator ideal $\mathfrak{S}_r$ is of optimum eigenvalue type l_r. Moreover,

$$\|(\lambda_n(T)) \mid l_r\| \leqq \|T \mid \mathfrak{S}_r\| \quad \text{for all } T \in \mathfrak{S}_r(H).$$

Proof. The result follows from

$$\sum_{k=1}^{n} |\lambda_k(T)|^r \leqq \sum_{k=1}^{n} a_k(T)^r$$

by passing to the limit as $n \to \infty$.

Note that the second proof of the preceding lemma requires that $1 \leqq r < \infty$. In order to treat the case $0 < r < 1$ with this method, too, we choose a natural number m such that $m\,r \geqq 1$. Then $\mathfrak{S}_{mr}$ is of eigenvalue type l_{mr}, and the assertion can be inferred from the principle of iteration 3.4.3. Furthermore,

$$\|(\lambda_n(T))\| \, l_r\| \leqq \|T \mid \mathfrak{S}_r\| \quad \text{for all } T \in \mathfrak{S}_r(H)$$

follows by means of tensor stability arguments; see 2.11.22 and 3.4.8.

If $D_t \in \mathfrak{L}(l_2)$ is a diagonal operator induced by $t = (\tau_n)$ with $\tau_1 \geqq \tau_2 \geqq \ldots \geqq 0$, then $\lambda_n(D_t) = \tau_n = a_n(D_t)$. This shows that $\mathfrak{S}_r$ cannot have any eigenvalue type better than l_r.

3.5.6.* In view of 2.11.17, we obtain an important corollary of the preceding result.

Eigenvalue theorem for Hilbert-Schmidt operators (I. Schur 1909: a, T. Carleman 1921). The ideal of Hilbert-Schmidt operators is of optimum eigenvalue type l_2. Moreover,

$$\|(\lambda_n(T)) \mid l_2\| \leqq \|T \mid \mathfrak{S}\| \quad \text{for all } T \in \mathfrak{S}(H).$$

3.5.7.* The Weyl theorem 3.5.5 admits the following generalization.

Eigenvalue theorem for Schatten-von Neumann operators. The operator ideal $\mathfrak{S}_{r,w}$ is of optimum eigenvalue type $l_{r,w}$.

Proof. The assertion follows from 3.5.1 and 2.1.8, or 3.5.4 and 2.1.7.
Diagonal operators on l_2 show that this result is sharp.

3.6. Eigenvalues of s-type operators

3.6.1.* The following substitute for the multiplicative Weyl inequality is the key to almost all results about eigenvalue distributions of operators on Banach spaces.

Lemma (A. Pietsch 1980: c). Let $T \in \mathfrak{L}(E)$ be a Riesz operator. Then

$$|\lambda_{2n-1}(T)| \leqq \mathrm{e}\left[\prod_{k=1}^{n} x_k(T)\right]^{1/n} \quad \text{for } n = 1, 2, \ldots$$

Proof. Fix any natural number n for which $\lambda_{2n-1}(T) \neq 0$. By 3.2.23, we may choose a $(2n-1)$-dimensional T-invariant subspace E_n such that the operator $T_n \in \mathfrak{L}(E_n)$ induced by T has precisely $\lambda_1(T), \ldots, \lambda_{2n-1}(T)$ as its eigenvalues. In view of 1.2.8 and 1.5.1, the identity operator I_n of E_n admits a factorization $I_n = X_n A_n$, where

$$A_n \in \mathfrak{L}(E_n, l_2(2n-1)) \quad \text{and} \quad \|A_n \mid \mathfrak{P}_2\| \leqq (2n-1)^{1/2},$$
$$X_n \in \mathfrak{L}(l_2(2n-1), E_n) \quad \text{and} \quad \|X_n\| = 1.$$

Since $T_n = X_n A_n T_n$ and $S_n := A_n T_n X_n$ are related, both operators have the same eigenvalues. Therefore

$$|\lambda_{2n-1}(T)|^{2n-1} \leqq \prod_{k=1}^{2n-1} |\lambda_k(T)| = \prod_{k=1}^{2n-1} |\lambda_k(T_n)| = \prod_{k=1}^{2n-1} |\lambda_k(S_n)|.$$

By 2.4.20 or 2.11.9, the multiplicative Weyl inequality for $S_n \in \mathfrak{L}(l_2(2n-1))$ reads

$$\prod_{k=1}^{2n-1} |\lambda_k(S_n)| \leqq \prod_{k=1}^{2n-1} x_k(S_n).$$

Recall that the Weyl numbers are multiplicative. Hence, applying 2.7.8, we obtain

$$|\lambda_{2n-1}(T)| \leqq \exp\left(\frac{1}{2n-1}\|A_n \mid \mathfrak{P}_2\|^2\right)\left[\prod_{k=1}^{n} x_k(T_n X_n)\right]^{1/n}$$
$$\leqq \mathrm{e}\left[\prod_{k=1}^{n} x_k(T_n)\right]^{1/n} \leqq \mathrm{e}\left[\prod_{k=1}^{n} x_k(T)\right]^{1/n}.$$

3.6.2.* Within the theory of abstract operators on Banach spaces, the following generalization of the classical Weyl theorem 3.5.5 is the most important result of this monograph.

Eigenvalue theorem for Weyl operators (A. Pietsch 1980: c). The operator ideal $\mathfrak{L}_{r,w}^{(x)}$ is of optimum eigenvalue type $l_{r,w}$.

Proof. Choose any natural number m with $2m > r$. Then it follows from 2.4.18 and 2.2.7 that

$$T \in \mathfrak{L}_{r,w}^{(x)}(E) \quad \text{implies} \quad T^m \in \mathfrak{L}_{r/m,w/m}^{(x)}(E) \subseteq \mathfrak{L}_2^{(x)}(E).$$

Thus T^m is compact, by 2.10.8. This proves that T is a Riesz operator.

Finally, combining 3.6.1 with 2.1.8 and 2.1.9, we conclude that

$$(x_n(T)) \in l_{r,w} \quad \text{implies} \quad (\lambda_n(T)) \in l_{r,w}.$$

Diagonal operators on l_2 show that this result is sharp.

Remark. According to the principle of uniform boundedness, there exists a constant $c_r \geqq 1$ such that

$$\left(\sum_{n=1}^{\infty} |\lambda_n(T)|^r\right)^{1/r} \leqq c_r \left(\sum_{n=1}^{\infty} x_n(T)^r\right)^{1/r} \quad \text{for all } T \in \mathfrak{L}_r^{(x)}(E)$$

and arbitrary Banach spaces E. Suppose that c_r is chosen as small as possible. Then the following estimate is known from H. König (1984:b):

$$c_r \leqq \sqrt{(m+1)\,\mathrm{e}} \left(1 + \frac{1}{m}\right)^{1/r} \leqq \sqrt{(m+1)\,\mathrm{e}}\; \mathrm{e}^{1/mr} \quad \text{for } m = 1, 2, \ldots$$

However, it is an open problem whether or not $c_r \to \infty$ as $r \to 0$.

3.6.3. It seems to be worth while to formulate the following corollary of the preceding result.

Theorem (H. König 1978). The operator ideals $\mathfrak{L}_{r,w}^{(a)}$ and $\mathfrak{L}_{r,w}^{(c)}$ are of optimum eigenvalue type $l_{r,w}$.

3.6.4. Let $0 < w < \infty$. A quasi-Banach operator ideal $\mathfrak{A}$ is said to be of **uniform eigenvalue type** $l_{r,w}$ if it is of eigenvalue type $l_{r,w}$ such that the series

$$\sum_{n=1}^{\infty} [n^{1/r-1/w} \mid \lambda_n(T)|]^w$$

converge uniformly on every precompact subset $\mathfrak{C}$ of any component $\mathfrak{A}(E)$. This means that for every $\varepsilon > 0$ there exists n_0 such that

$$\left(\sum_{n \geqq n_0} [n^{1/r-1/w} \mid \lambda_n(T)|]^w\right)^{1/w} \leqq \varepsilon \quad \text{whenever } T \in \mathfrak{C}.$$

Remark. We conjecture that this condition is automatically satisfied for all quasi-Banach operator ideals $\mathfrak{A} \in \mathbb{E}_{r,w}$.

3.6.5. Using the concept just introduced, we now improve theorem 3.6.2.

Proposition (H. Leiterer / A. Pietsch 1982). If $0 < w < \infty$, then the quasi-Banach operator ideal $\mathfrak{L}_{r,w}^{(x)}$ is of uniform eigenvalue type $l_{r,w}$.

Proof. We know from 2.2.6 that

$$\|T \mid \mathfrak{L}_{r,w}^{(x)}\|_p^{\text{mean}} := \left(\sum_{n=1}^{\infty} \left[n^{1/r-1/w} \left(\frac{1}{n} \sum_{k=1}^{n} x_k(T)^p\right)^{1/p}\right]^w\right)^{1/w}$$

defines an equivalent quasi-norm on $\mathfrak{L}_{r,w}^{(x)}$ provided that $0 < p < r$.

Let $\mathfrak{C}$ be any precompact subset of $\mathfrak{L}_{r,w}^{(x)}(E)$. Given $\varepsilon > 0$, we choose a finite ε-net of operators $T_1, \ldots, T_m \in \mathfrak{L}_{r,w}^{(x)}(E)$. Then there exists n_0 such that

$$\left(\sum_{n \geqq n_0} \left[n^{1/r-1/w} \left(\frac{1}{n} \sum_{k=1}^{n} x_k(T_i)^p\right)^{1/p}\right]^w\right)^{1/w} \leqq \varepsilon \quad \text{for } i = 1, \ldots, m.$$

Next, for any $T \in \mathfrak{C}$, we pick T_i with $\|T - T_i \mid \mathfrak{L}_{r,w}^{(x)}\|_p^{\text{mean}} \leqq \varepsilon$. Note that

$$\left(\sum_{k=1}^{n} x_k(T)^p\right)^{1/p} \leqq 2^{1/p} c_p \left[\left(\sum_{k=1}^{n} x_k(T - T_i)^p\right)^{1/p} + \left(\sum_{k=1}^{n} x_k(T_i)^p\right)^{1/p}\right],$$

where $c_p := \max(2^{1/p-1}, 1)$ is the constant in the quasi-triangle inequality of $\|\cdot \mid l_p(n)\|$. Hence

$$\left(\sum_{n \geqq n_0} \left[n^{1/r-1/w} \left(\frac{1}{n} \sum_{k=1}^{n} x_k(T)^p\right)^{1/p}\right]^w\right)^{1/w} \leqq$$

$$\leqq 2^{1/p} c_p c_w \left[\|T - T_i \mid \mathfrak{L}_{r,w}^{(x)}\|_p^{\text{mean}} + \left(\sum_{n \geqq n_0} \left[n^{1/r-1/w} \left(\frac{1}{n} \sum_{k=1}^{n} x_k(T_i)^p\right)^{1/p}\right]^w\right)^{1/w}\right]$$

$$\leqq 2^{1+1/p} c_p c_w \varepsilon.$$

Applying the inequality of means, we deduce from 3.6.1 that

$$|\lambda_{2n-1}(T)| \leqq e \left(\frac{1}{n} \sum_{k=1}^{n} x_k(T)^p\right)^{1/p}.$$

Therefore

$$\left(\sum_{n \geqq n_0} [n^{1/r-1/w} \mid \lambda_{2n-1}(T)|]^w\right)^{1/w} \leqq 2^{1+1/p} c_p c_w e \varepsilon.$$

Finally, reasoning as in 2.1.9, we obtain

$$\left(\sum_{n \geqq 2n_0 - 1} [n^{1/r-1/w} \mid \lambda_n(T)|]^w\right)^{1/w} \leqq c\varepsilon \quad \text{for all } T \in \mathfrak{C},$$

where the constant $c > 0$ depends on p, r and w. This completes the proof.

3.6.6. Theorem. Let $0 < r < 2$ and $1/s = 1/r - 1/2$. Then the operator ideal $\mathfrak{L}_{r,w}^{(h)}$ is of optimum eigenvalue type $l_{s,w}$.

Proof. By 2.10.7, we have $\mathfrak{L}_{r,w}^{(h)} \subseteq \mathfrak{L}_{s,w}^{(x)}$. Thus $\mathfrak{L}_{r,w}^{(h)}$ indeed has eigenvalue type $l_{s,w}$. Furthermore, Example 2.9.20 shows that this type is the best possible.

3.6.7. Theorem (H. König 1979). Let $0 < r < 2$ and $1/s = 1/r - 1/2$. Then the operator ideal $\mathfrak{S}_{r,\infty}^{\text{weak}}$ is of optimum eigenvalue type $l_{s,\infty}$.

Proof. The result follows from 2.11.34 and 3.6.6.

3.7. Eigenvalues of absolutely summing operators

3.7.1*. To begin with, we establish the most elementary result about eigenvalue distributions of absolutely summing operators.

Theorem (A. Pietsch 1963: a, 1967 and 1972: b). If $1 \leqq r \leqq 2$, then the operator ideal $\mathfrak{P}_r$ is of optimum eigenvalue type l_2. Moreover,

$$\|(\lambda_n(T)) \mid l_2\| \leqq \|T \mid \mathfrak{P}_r\| \quad \text{for all } T \in \mathfrak{P}_r(E).$$

Proof. We know from 1.3.3 that $\mathfrak{P}_r \subseteq \mathfrak{P}_2$. Therefore it is enough to treat the operator ideal $\mathfrak{P}_2$.

Let $T \in \mathfrak{P}_2(E)$, and consider a factorization $T = XA$, where

$$A \in \mathfrak{P}_2(E, H), \quad X \in \mathfrak{L}(H, E) \quad \text{and} \quad \|X\| \, \|A \mid \mathfrak{P}_2\| = \|T \mid \mathfrak{P}_2\|.$$

It follows from 1.4.5 that $S := AX$ is a Hilbert-Schmidt operator. Hence the principle of related operators combined with 3.5.6 yields

$$\|(\lambda_n(T)) \mid l_2\| = \|(\lambda_n(S)) \mid l_2\| \leqq \|S \mid \mathfrak{S}\| \leqq \|A \mid \mathfrak{P}_2\| \, \|X\| = \|T \mid \mathfrak{P}_2\|.$$

By 1.3.16 and 1.4.7, we have

$$D_t \in \mathfrak{P}_r(l_2) \quad \text{if and only if} \quad t \in l_2.$$

Consequently, l_2 is indeed the best possible eigenvalue type of $\mathfrak{P}_r$.

3.7.2.* Next we establish a profound counterpart of the preceding result.

Eigenvalue theorem for absolutely *r*-summing operators (W. B. Johnson / H. König / B. Mauey / J. R. Retherford 1979). If $2 \leqq r < \infty$, then the operator ideal $\mathfrak{P}_r$ is of optimum eigenvalue type l_r. Moreover,

$$\|(\lambda_n(T)) \mid l_r\| \leqq \|T \mid \mathfrak{P}_r\| \quad \text{for all } T \in \mathfrak{P}_r(E).$$

Proof. By 1.2.4 and 2.7.4, we have $\mathfrak{P}_r \subseteq \mathfrak{P}_{r,2} \subseteq \mathfrak{L}^{(x)}_{r,\infty}$. Hence it follows from 3.6.2 that $\mathfrak{P}_r \in \mathbb{E}_{r,\infty}$. This implies that $\mathfrak{P}_r$ is of eigenvalue type $l_{r+\varepsilon}$ for all $\varepsilon > 0$.Furthermore, as stated in 1.3.11, the Banach operator ideal $\mathfrak{P}_r$ is stable with respect to the tensor norm ε. Therefore, in view of 3.4.9, we obtain

$$\|(\lambda_n(T)) \mid l_r\| \leqq \|T \mid \mathfrak{P}_r\| \quad \text{for all } T \in \mathfrak{P}_r(E).$$

It follows immediately from 1.6.3 that

$$D_t \in \mathfrak{P}_r(l_\infty) \quad \text{if and only if} \quad t \in l_r.$$

Thus $\mathfrak{P}_r$ cannot have any eigenvalue type better than l_r.

Remark. If $r = 2m$ is an even number, then 1.3.10 implies that $(\mathfrak{P}_{2m})^m \subseteq \mathfrak{P}_2$. Therefore the result can be deduced from 3.7.1 and the principle of iteration 3.4.3.

3.7.3. The previous theorem admits the following generalization.

Theorem (W. B. Johnson / H. König / B. Maurey / J. R. Retherford 1979, A. Pietsch 1986). Let $2 \leqq r_1, \ldots, r_m < \infty$ and $1/r = 1/r_1 + \ldots + 1/r_m$. Then the operator ideal $\mathfrak{P}_{r_1} \circ \ldots \circ \mathfrak{P}_{r_m}$ is of optimum eigenvalue type l_r. Moreover,

$$\|(\lambda_n(T)) \mid l_r\| \leqq \|T \mid \mathfrak{P}_{r_1} \circ \ldots \circ \mathfrak{P}_{r_m}\| \quad \text{for all } T \in \mathfrak{P}_{r_1} \circ \ldots \circ \mathfrak{P}_{r_m}(E).$$

Proof. Reasoning as in the preceding paragraph, the assertion follows from

$$\mathfrak{P}_{r_1} \circ \ldots \circ \mathfrak{P}_{r_m} \subseteq \mathfrak{L}^{(x)}_{r_1,\infty} \circ \ldots \circ \mathfrak{L}^{(x)}_{r_m,\infty} \subseteq \mathfrak{L}^{(x)}_{r,\infty};$$

see 2.4.18.

Remark. If $r_1 = \ldots = r_m = 2$, then the result can be deduced from 3.7.1 and the principle of iteration 3.4.3.

3.7.4. Theorem (A. Pietsch 1981: b). The operator ideal $(\mathfrak{P}_2)^{(a)}_{s,w}$ is of optimum eigenvalue type $l_{r,w}$, where $1/r = 1/s + 1/2$.

Proof. We know from 2.8.18 that $(\mathfrak{P}_2)^{(a)}_{s,w} \subseteq \mathfrak{L}^{(x)}_{r,w}$. Hence $(\mathfrak{P}_2)^{(a)}_{s,w}$ has eigenvalue type $l_{r,w}$.

Recall that $\mathfrak{P}_2(H) = \mathfrak{L}^{(a)}_2(H)$ for all Hilbert spaces H. Thus the reiteration theorem 2.8.13 yields

$$(\mathfrak{P}_2)^{(a)}_{s,w}(H) = \mathfrak{L}^{(a)}_{r,w}(H).$$

This implies that $l_{r,w}$ is indeed the best possible eigenvalue type of $(\mathfrak{P}_2)^{(a)}_{s,w}$.

The theorem can be proved by a completely different technique. For this purpose, we consider arbitrary decompositions

$$1/s = 1/p + 1/q \quad \text{and} \quad 1/w = 1/u + 1/v.$$

By 2.8.21, every operator $T \in (\mathfrak{P}_2)^{(a)}_{r,w}(E)$ admits a factorization $T = BA$ such that

$$A \in (\mathfrak{P}_2)^{(a)}_{p,u}(E, H) \quad \text{and} \quad B \in \mathfrak{L}^{(a)}_{q,v}(H, E),$$

where H is a Hilbert space. Applying 2.8.7, we obtain

$$S := AB \in (\mathfrak{P}_2)^{(a)}_{p,u} \circ \mathfrak{L}^{(a)}_{q,v}(H) \subseteq (\mathfrak{P}_2)^{(a)}_{s,w}(H) = \mathfrak{L}^{(a)}_{r,w}(H).$$

Since S and T are related, 3.3.4 and 3.5.7 imply $(\lambda_n(T)) \in l_{r,w}$.

3.7.5. Using the preceding result, we now give another proof of 3.6.3.

Theorem. The operator ideals $\mathfrak{L}^{(a)}_{r,w}$ and $\mathfrak{L}^{(c)}_{r,w}$ are of optimum eigenvalue type $l_{r,w}$.

Proof. Choose any natural number m such that $2m > r$. Then, in view of 2.4.9 and 2.8.18, we have

$$(\mathfrak{L}^{(c)}_{r,w})^m \subseteq \mathfrak{L}^{(c)}_{r/m,\, w/m} \subseteq (\mathfrak{P}_2)^{(a)}_{s,\, w/m},$$

where $1/s := m/r - 1/2$. Hence $(\mathfrak{L}^{(c)}_{r,w})^m \in \mathbb{E}_{r/m, w/m}$, and the principle of iteration implies that $\mathfrak{L}^{(c)}_{r,w} \in \mathbb{E}_{r,w}$.

3.7.6. Eigenvalue theorem for absolutely $(r,2)$-summing operators (H. König / J. R. Retherford / N. Tomczak-Jaegerman 1980). Let $2 < r < \infty$. Then the operator ideal $\mathfrak{P}_{r,2}$ is of optimum eigenvalue type $l_{r,\infty}$.

Proof. It follows from 2.7.4 that $\mathfrak{P}_{r,2} \subseteq \mathfrak{L}^{(x)}_{r,\infty}$. Therefore $\mathfrak{P}_{r,2}$ is of eigenvalue type $l_{r,\infty}$, by 3.6.2.

We now consider the function $f^{2\pi}_{r',0} \in L_{r',\infty}(2\pi)$ defined in 6.5.8. Then, as shown in the proof of 6.5.9, the corresponding convolution operator is absolutely $(r, 2)$-summing on $L_\infty(2\pi)$. Furthermore, we have $(\gamma_k(f^{2\pi}_{r',0})) \notin l_{r,w}(\mathbb{Z})$ for $0 < w < \infty$. This proves that the above result is sharp.

3.8. Eigenvalues of nuclear operators

3.8.1. We begin with a classical result.

Eigenvalue theorem for nuclear operators (GRO, Chap. II, p. 17). The operator ideal $\mathfrak{N}$ is of optimum eigenvalue type l_2. Moreover,

$$\|(\lambda_n(T)) \mid l_2\| \leq \|T \mid \mathfrak{N}\| \quad \text{for all } T \in \mathfrak{N}(E).$$

Proof. Obviously, $\mathfrak{N} \in \mathbb{E}_2$ can be deduced immediately from $\mathfrak{N} \subseteq \mathfrak{P}_2$ and $\mathfrak{P}_2 \in \mathbb{E}_2$. We prefer, however, to give a more direct proof of this important fact.

Let $T \in \mathfrak{N}(E)$ and $\varepsilon > 0$. Then there exists a representation

$$T = \sum_{i=1}^{\infty} a_i \otimes x_i$$

such that

$$\sum_{i=1}^{\infty} \|a_i\| \, \|x_i\| \leqq (1 + \varepsilon) \, \|T \mid \mathfrak{N}\|.$$

Without loss of generality, we may assume that $\|x_i\| = \|a_i\|$. Define $A \in \mathfrak{L}(E, l_2)$ and $X \in \mathfrak{L}(l_2, E)$ by

$$A := \sum_{i=1}^{\infty} a_i \otimes e_i \quad \text{and} \quad X := \sum_{j=1}^{\infty} e_j \otimes x_j.$$

Note that

$$\left(\sum_{i=1}^{\infty} \sum_{j=1}^{\infty} |\langle x_j, a_i \rangle|^2 \right)^{1/2} \leqq \left(\sum_{j=1}^{\infty} \|x_j\|^2 \right)^{1/2} \left(\sum_{i=1}^{\infty} \|a_i\|^2 \right)^{1/2}.$$

Since $S := AX \in \mathfrak{L}(l_2)$ is induced by the matrix $M := (\langle x_j, a_i \rangle)$, it must be a Hilbert-Schmidt operator. Therefore the principle of related operators combined with 3.5.6 yields

$$\|(\lambda_n(T)) \mid l_2\| = \|(\lambda_n(S)) \mid l_2\| \leqq \|S \mid \mathfrak{S}\| \leqq (1 + \varepsilon) \, \|T \mid \mathfrak{N}\|.$$

The example given in the following paragraph shows that the operator ideal $\mathfrak{N}$ cannot have any better eigenvalue type than l_2; see also 6.5.11 and 6.6.12.

3.8.2. The following construction is analogous to that carried out in 3.4.3.

Example (R. J. Kaiser / J. R. Retherford 1983, 1984). Given $(\lambda_n) \in l_2$ with $\lambda_n \neq 0$, there exists an operator $T \in \mathfrak{N}(l_1 \oplus l_\infty)$ having precisely $\pm\lambda_1, \pm\lambda_2, \ldots$ as its eigenvalues.

Proof. By 1.7.4, we can find a factorization.

$$(\lambda_n) = (\alpha_n \beta_n) \quad \text{with} \quad (\alpha_n) \in c_0 \quad \text{and} \quad (\beta_n) \in l_2.$$

Let $s := (\alpha_n^2)$ and $t := (\beta_n^2)$. It follows from 1.7.3 and 1.7.19 that

$$D_s \in \mathfrak{N}(l_1, l_\infty) \quad \text{and} \quad D_t \in \mathfrak{N}(l_\infty, l_1).$$

Therefore

$$T := J_\infty D_s Q_1 + J_1 D_t Q_\infty$$

defines a nuclear operator on $l_1 \oplus l_\infty$. Note that

$$T(x, y) = (D_t y, D_s x) \quad \text{for} \quad x \in l_1 \quad \text{and} \quad y \in l_\infty.$$

In particular, we have

$$T(\beta_n e_n, \pm\alpha_n e_n) = \pm\lambda_n(\beta_n e_n, \pm\alpha_n e_n).$$

Thus $\pm\lambda_n$ is an eigenvalue of T.

Finally, an easy calculation shows that T cannot have any other eigenvalue than $\pm\lambda_1, \pm\lambda_2, \ldots$

Remark. Very recently G. Pisier (1983) constructed a sophisticated Banach space E on which all approximable operators are indeed nuclear. Using this fact, R. J. Kaiser / J. R. Retherford (1983) were able to find an operator $T \in \mathfrak{N}(E)$ having precisely the prescribed eigenvalues $\lambda_1, \lambda_2, \ldots$ provided that $(\lambda_n) \in l_2$ and $\lambda_n \neq 0$.

3.8.3. Since $\mathfrak{N}(H) = \mathfrak{S}_1(H)$, we see from 3.5.5 that

$$\|(\lambda_n(T)) \mid l_1\| \leqq \|T \mid \mathfrak{N}\| \quad \text{for all } T \in \mathfrak{N}(H).$$

This fact indicates that the preceding theorem can be improved for nuclear operators acting on special Banach spaces.

Proposition (W. B. Johnson / H. König / B. Maurey / J. R. Retherford 1979). Let $2 \leqq p < \infty$ and $1/r = 1/p + 1/2$. Then

$$\|(\lambda_n(T)) \mid l_r\| \leqq \|T \mid \mathfrak{N}\| \quad \text{for all } T \in \mathfrak{N}(L_p).$$

Proof. Given $\varepsilon > 0$, we consider a factorization

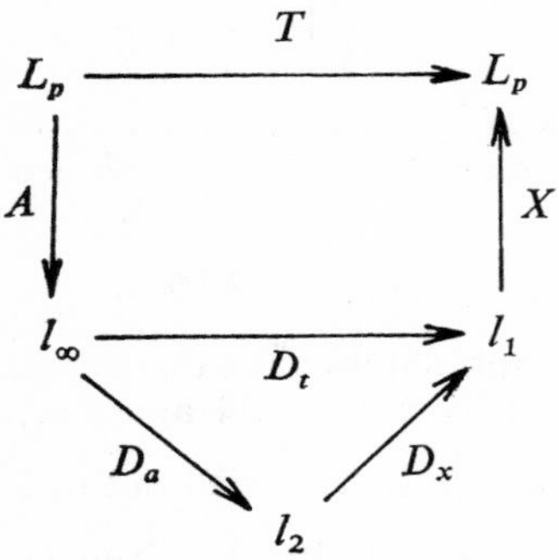

such that

$$\|X\| \, \|x \mid l_2\| \, \|a \mid l_2\| \, \|A\| \leqq (1 + \varepsilon) \, \|T \mid \mathfrak{N}\|.$$

Since $D_x \in \mathfrak{P}_2'(l_2, l_1)$, we have $XD_x \in \mathfrak{P}_p'(l_2, L_p)$. Therefore 1.3.13 implies that $XD_x \in \mathfrak{P}_p(l_2, L_p)$. Furthermore, we know from 1.6.3 that $D_a \in \mathfrak{P}_2(l_\infty, l_2)$. Hence

$$T = (XD_x)(D_a A) \in \mathfrak{P}_p \circ \mathfrak{P}_2(L_p).$$

For the corresponding quasi-norms we obtain the estimate

$$\begin{aligned} \|T \mid \mathfrak{P}_p \circ \mathfrak{P}_2\| &\leqq \|XD_x \mid \mathfrak{P}_p\| \, \|D_a A \mid \mathfrak{P}_2\| \leqq \|X\| \, \|x \mid l_2\| \, \|a \mid l_2\| \, \|A\| \\ &\leqq (1 + \varepsilon) \, \|T \mid \mathfrak{N}\|. \end{aligned}$$

Applying 3.7.3 and letting $\varepsilon \to 0$, we finally arrive at the desired result.

Remark. A Banach space E is said to be **θ-hilbertian**, $0 < \theta < 1$, if there exists a Banach interpolation couple (E_0, H), where H is a Hilbert space, such that E can be obtained by complex interpolation: $E = (E_0, H)_\theta$. For example, we have $L_p = (L_\infty, L_2)_\theta$ with $\theta := 2/p$. If E possesses this property and $1/r = (1 + \theta)/2$, then $T \in \mathfrak{N}(E)$ implies $(\lambda_n(T)) \in l_r$; see G. Pisier (1979).

3.8.4. We now state, without proof, a certain converse of the preceding result.

Proposition (W. B. Johnson / H. König / B. Maurey / J. R. Retherford 1979).

Every Banach space E with the property that

$$(\lambda_n(T)) \in l_1 \quad \text{for all } T \in \mathfrak{N}(E)$$

is isomorphic to a Hilbert space.

Remark. Let $1 < r < 2$. It would be interesting to characterize the class of those Banach spaces E such that

$$(\lambda_n(T)) \in l_r \quad \text{for all } T \in \mathfrak{N}(E).$$

Partial results were obtained by H. König / J. R. Retherford / N. Tomczak-Jaegermann (1980).

3.8.5. Eigenvalue theorem for p-nuclear operators (H. König 1977). Let $0 < p < 1$ and $1/r = 1/p - 1/2$. Then the operator ideal $\mathfrak{N}_p$ is of optimum eigenvalue type $l_{r,p}$.

Proof. In view of 1.7.3, every operator $T \in \mathfrak{N}_p(E)$ factors through a diagonal operator $D_t \in \mathfrak{L}(l_\infty, l_1)$ with $t \in l_p$. By 2.9.17, we have $D_t \in \mathfrak{L}^{(x)}_{r,p}(l_\infty, l_1)$. Hence $T \in \mathfrak{L}^{(x)}_{r,p}(E)$. This proves that $\mathfrak{N}_p$ is of eigenvalue type $l_{r,p}$.

Let $0 < w < p$. Then, for the weighted Littlewood-Walsh matrix $W_{r,\alpha}$ with $\alpha := 1/w$ defined in 5.2.3, we have

$$(W_{r,\alpha})_{\mathrm{op}} \in \mathfrak{N}_p(l_1) \quad \text{and} \quad (\lambda_n(W_{r,\alpha})) \notin l_{r,w}.$$

Thus $\mathfrak{N}_p \in \mathbb{E}_{r,p}$ is the best possible result.

3.8.6. Eigenvalue theorem for $(r,2)$-nuclear operators (P. Saphar 1966, B. Carl 1976). Let $0 < r \leqq 2$. Then the operator ideal $\mathfrak{N}_{r,2}$ is of optimum eigenvalue type l_r. Moreover,

$$\|(\lambda_n(T)) \mid l_r\| \leqq \|T \mid \mathfrak{N}_{r,2}\| \quad \text{for all } T \in \mathfrak{N}_{r,2}(E).$$

Proof. Define $1/p := 1/r - 1/2$. Given $\varepsilon > 0$, by 1.7.10, there exists a factorization

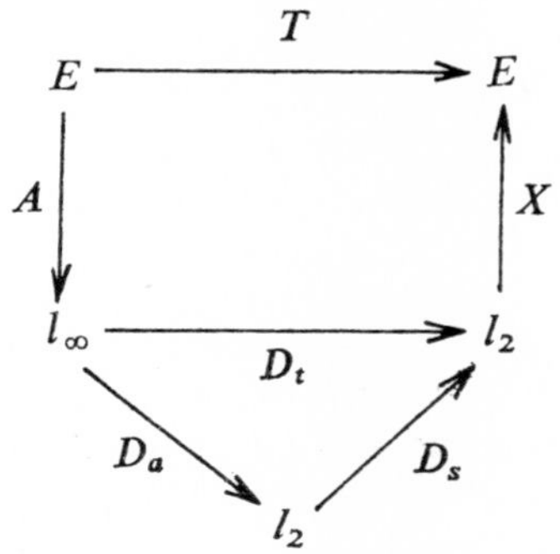

such that

$$\|X\| \, \|s \mid l_p\| \, \|a \mid l_2\| \, \|A\| \leqq (1 + \varepsilon) \, \|T \mid \mathfrak{N}_{r,2}\|.$$

By 1.6.3, we have $D_a \in \mathfrak{P}_2(l_\infty, l_2)$. Therefore it follows from 1.4.5 and 2.11.17 that $D_a AX \in \mathfrak{S}_2(l_2)$. Consequently, by 2.11.23,

$$S := D_a AXD_s \in \mathfrak{S}_2 \circ \mathfrak{S}_p(l_2) = \mathfrak{S}_r(l_2).$$

Furthermore,

$$\|S \mid \mathfrak{S}_r\| \leqq \|D_a AX \mid \mathfrak{S}_2\| \, \|D_s \mid \mathfrak{S}_p\| \leqq \|a \mid l_2\| \, \|A\| \, \|X\| \, \|s \mid l_2\|.$$

Next we deduce from the principle of related operators and 3.5.5 that

$$\|(\lambda_n(T)) \mid l_r\| = \|(\lambda_n(S)) \mid l_r\| \leqq \|S \mid \mathfrak{S}_r\| \leqq (1+\varepsilon)\, \|T \mid \mathfrak{N}_{r,2}\|.$$

Letting $\varepsilon \to 0$ now yields the inequality we are looking for. Hence $\mathfrak{N}_{r,2}$ is of eigenvalue type l_r.

Finally, since $\mathfrak{N}_{r,2}(H) = \mathfrak{S}_r(H)$, this is the best possible result.

3.9. The eigenvalue type of sums of operator ideals

3.9.1. Suppose that the operator ideals $\mathfrak{A}$ and $\mathfrak{B}$ have eigenvalue type $l_{r,w}$. Then we are interested to get some information about the eigenvalue type of $\mathfrak{A} + \mathfrak{B}$. A complete answer to this problem is still open. However, as the following examples show, the leading index r may change significantly for the worse. As a consequence of this observation we see that the class $\mathbb{E}_{r,w}$ does not contain a largest operator ideal.

3.9.2. Lemma. If $\mathfrak{A} + \mathfrak{B} \in \mathbb{E}_{r,w}$, then $\mathfrak{A} \circ \mathfrak{B} \in \mathbb{E}_{r/2,w/2}$.

Proof. In view of the principle of iteration 3.4.3, the assertion follows from

$$\mathfrak{A} \circ \mathfrak{B} \subseteqq (\mathfrak{A} + \mathfrak{B}) \circ (\mathfrak{A} + \mathfrak{B}).$$

3.9.3. We are now in a position to give the announced examples.

Theorem. The operator ideal $(\mathfrak{P}_2)^m + (\mathfrak{P}_2')^m$ is of optimum eigenvalue type $l_{4/(2m-1)}$.

Proof. We know from 1.5.10 that $(\mathfrak{P}_2')^m \subseteqq (\mathfrak{P}_2)^{m-1}$. Hence

$$[(\mathfrak{P}_2)^m + (\mathfrak{P}_2')^m]^2 \subseteqq (\mathfrak{P}_2)^{2m-1}.$$

By 3.4.3 and 3.7.3, this implies that $(\mathfrak{P}_2)^m + (\mathfrak{P}_2')^m$ indeed has the eigenvalue type $l_{4/(2m-1)}$.

Recall from 1.6.9 and 1.6.10 that

$$D_a \in (\mathfrak{P}_2)^m (l_1, l_2) \quad \text{if and only if} \quad a \in l_{2/(m-1)}$$

and

$$D_b \in (\mathfrak{P}_2)^m (l_\infty, l_2) \quad \text{if and only if} \quad b \in l_{2/m}.$$

Consequently,

$$D_t \in (\mathfrak{P}_2)^m \circ (\mathfrak{P}_2')^m (l_2) \quad \text{if and only if} \quad t \in l_{2/(2m-1)}.$$

Thus, in view of Lemma 3.9.2, the sum $(\mathfrak{P}_2)^m + (\mathfrak{P}_2')^m$ cannot have any eigenvalue type better than $l_{4/(2m-1)}$.

3.9.4. Theorem. Let $0 < r < 2$ and $1/r_0 = 1/r - 1/4$. Then the operator ideal $\mathfrak{L}^{(x)}_{r,w} + \mathfrak{L}^{(y)}_{r,w}$ is of optimum eigenvalue type $l_{r_0,w}$.

Proof. Define $1/s := 1/r - 1/2$. Then $2/r_0 = 1/r + 1/s$, and it follows from 2.4.18, 2.5.12 and 2.10.7 that

$$\begin{aligned}
\mathfrak{L}^{(x)}_{r,w} \circ \mathfrak{L}^{(x)}_{r,w} &\subseteqq \mathfrak{L}^{(x)}_{r/2,w/2} && \subseteqq \mathfrak{L}^{(x)}_{r_0/2,w/2},\\
\mathfrak{L}^{(x)}_{r,w} \circ \mathfrak{L}^{(y)}_{r,w} &\subseteqq \mathfrak{L}^{(x)}_{r,w} \circ \mathfrak{L}^{(x)}_{s,w} && \subseteqq \mathfrak{L}^{(x)}_{r_0/2,w/2},\\
\mathfrak{L}^{(y)}_{r,w} \circ \mathfrak{L}^{(x)}_{r,w} &\subseteqq \mathfrak{L}^{(x)}_{s,w} \circ \mathfrak{L}^{(x)}_{r,w} && \subseteqq \mathfrak{L}^{(x)}_{r_0/2,w/2},\\
\mathfrak{L}^{(y)}_{r,w} \circ \mathfrak{L}^{(y)}_{r,w} &\subseteqq \mathfrak{L}^{(y)}_{r/2,w/2} && \subseteqq \mathfrak{L}^{(x)}_{r_0/2,w/2}.
\end{aligned}$$

Hence

$$[\mathfrak{L}^{(x)}_{r,w} + \mathfrak{L}^{(y)}_{r,w}]^2 \subseteq \mathfrak{L}^{(x)}_{r_0/2,\, w/2}.$$

By 3.4.3 and 3.6.2, this implies that $\mathfrak{L}^{(x)}_{r,w} + \mathfrak{L}^{(y)}_{r,w}$ indeed has the eigenvalue type $l_{r_0,w}$.

Recall from 2.9.17 and 2.9.18 that

$$D_a \in \mathfrak{L}^{(x)}_{r,w}(l_1, l_2) \quad \text{if and only if} \quad a \in l_{s,w}$$

and

$$D_b \in \mathfrak{L}^{(x)}_{r,w}(l_\infty, l_2) \quad \text{if and only if} \quad b \in l_{r,w}.$$

Consequently,

$$D_t \in \mathfrak{L}^{(x)}_{r,w} \circ \mathfrak{L}^{(y)}_{r,w}(l_2) \quad \text{if and only if} \quad t \in l_{r_0/2,\, w/2}.$$

Thus, in view of Lemma 3.9.2, the sum $\mathfrak{L}^{(x)}_{r,w} + \mathfrak{L}^{(y)}_{r,w}$ cannot have any eigenvalue type better than $l_{r_0,w}$.

Remark. In the case $2 \leqq r < \infty$ the optimum eigenvalue type of $\mathfrak{L}^{(x)}_{r,w} + \mathfrak{L}^{(y)}_{r,w}$ is unknown.

3.9.5. The following example is analogous to the preceding one.

Theorem. Let $1/r_0 = 1/s + 1/4$. Then the operator ideal $(\mathfrak{P}_2)^{(a)}_{s,w} + (\mathfrak{P}_2')^{(a)}_{s,w}$ is of optimum eigenvalue type $l_{r_0,w}$.

Proof. Let $1/r := 1/s + 1/2$. Then $1/r_0 = 1/r - 1/4$, and we obtain from 2.8.18 and 2.5.12 that

$$(\mathfrak{P}_2)^{(a)}_{s,w} + (\mathfrak{P}_2')^{(a)}_{s,w} \subseteq \mathfrak{L}^{(x)}_{r,w} + \mathfrak{L}^{(y)}_{r,w}.$$

This proves that the operator ideal under consideration indeed has the eigenvalue type $l_{r_0,w}$.

Reasoning as in the proof of 3.9.4, we may infer from Examples 2.9.15 and 2.9.16 that this result is the best possible.

3.9.6. **Theorem.** Let $0 < r < 2$ and $1/r_0 = 1/r - 1/4$. Then the operator ideal $\mathfrak{N}_{r,2} + \mathfrak{N}'_{r,2}$ is of optimum eigenvalue type $l_{r_0,r}$.

Proof. We deduce from 1.7.10, 2.9.17 and 2.5.12 that

$$\mathfrak{N}_{r,2} + \mathfrak{N}'_{r,2} \subseteq \mathfrak{L}^{(x)}_{r,r} + \mathfrak{L}^{(y)}_{r,r}.$$

Hence $\mathfrak{N}_{r,2} + \mathfrak{N}'_{r,2}$ has the eigenvalue type $l_{r_0,r}$.

Define $1/s := 1/r - 1/2$. We state, without proof, that

$$D_a \in \mathfrak{N}_{r,2}(l_1, l_2) \quad \text{if and only if} \quad a \in l_{s,r}$$

and

$$D_b \in \mathfrak{N}_{r,2}(l_\infty, l_2) \quad \text{if and only if} \quad b \in l_{r,r}.$$

Consequently,

$$D_t \in \mathfrak{N}_{r,2} \circ \mathfrak{N}'_{r,2}(l_2) \quad \text{if and only if} \quad t \in l_{r_0/2,\, r/2}.$$

Thus, in view of Lemma 3.9.2, the sum $\mathfrak{N}_{r,2} + \mathfrak{N}'_{r,2}$ cannot have any eigenvalue type better than $l_{r_0,r}$.

3.9.7. We conclude with a general result due to B. Carl (unpublished).

Proposition. Let $\mathfrak{A}$ and $\mathfrak{B}$ be operator ideals. Then

$$\mathfrak{A}, \mathfrak{B} \in \mathbb{E}_r \quad \text{implies } \mathfrak{A} + \mathfrak{B} \in \mathbb{E}_{2r+\varepsilon} \text{ for } \varepsilon > 0.$$

Proof. By the principle of iteration, we have

$$\mathfrak{A}^m, \mathfrak{B}^m \in \mathbb{E}_{r/m}.$$

It now follows from 3.4.4 that

$$(\mathfrak{A} + \mathfrak{B})^{2m} \subseteqq \mathfrak{A}^m + \mathfrak{B}^m \subseteqq \mathfrak{S}_{r/m}^{\text{weak}}.$$

In view of 3.6.7, this implies that

$$(\mathfrak{A} + \mathfrak{B})^{2m} \in \mathbb{E}_{s/2m,\infty},$$

where s is defined by

$$2m/s = m/r - 1/2.$$

Choosing m so large that

$$1/s = 1/2r - 1/4m > 1/(2r + \varepsilon),$$

we finally obtain

$$\mathfrak{A} + \mathfrak{B} \in \mathbb{E}_{s,\infty} \subset \mathbb{E}_{2r+\varepsilon}.$$

Remark. We conjecture that

$$\mathfrak{A}, \mathfrak{B} \in \mathbb{E}_r \quad \text{implies } \mathfrak{A} + \mathfrak{B} \in \mathbb{E}_s,$$

where $1/s := \max(1/2r, 1/r - 1/4)$.

CHAPTER 4

Traces and determinants

In this chapter we develop an axiomatic theory of traces and determinants defined on certain operator ideals. Having in mind the classical formulas

$$\operatorname{trace}(M) = \sum_{i=1}^{n} \lambda_i \quad \text{and} \quad \det(\lambda I - M) = \prod_{i=1}^{n} (\lambda - \lambda_i),$$

which hold for every (n, n)-matrix M, our main goal is to investigate the relationship between traces, determinants and eigenvalues. When passing to the limit as $n \to \infty$, it turns out to be advantageous to replace the complex parameter λ by $\zeta = -1/\lambda$ (the minus sign is a matter of taste).

The historical starting point was the theory of nuclear operators. In view of the finite dimensional setting, it seems natural to define the trace of a nuclear operator

$$T = \sum_{i=1}^{\infty} a_i \otimes x_i$$

by

$$\tau(T) = \sum_{i=1}^{\infty} \langle x_i, a_i \rangle .$$

However, a counterexample constructed by P. Enflo (1973) shows that, in certain Banach spaces, this value depends on the underlying representation. In fact, the situation is even more confused than this. It turns out that on the Banach ideal of nuclear operators (formerly called trace class operators) there does not exist any continuous trace. Even when restricting our considerations to good-natured Banach spaces, including c_0 or l_1, we cannot avoid ill-natured operators. Thus the operator ideal $\mathfrak{N}$ is inappropriate for the purpose it was originally designed for. As a consequence we observe that a substantial trace theory holds only on smaller operator ideals such as $\mathfrak{L}_1^{(a)}$, $(\mathfrak{P}_2)_{2,1}^{(a)}$ and $(\mathfrak{P}_2)^2$. In particular, due to the fact that $\mathfrak{N}$ is the smallest Banach operator ideal, we are necessarily led to the context of quasi-Banach operator ideals.

We begin this chapter with some preliminaries concerning the Fredholm resolvent

$$F(\zeta, T) := T(I + \zeta T)^{-1}$$

of a Riesz operator T which is a meromorphic operator-valued function of the complex parameter ζ.

Next an axiomatic theory of traces is established. We first deal with the ideal of finite operators. Subsequently, the trace obtained in this elementary case is extended continuously to larger quasi-Banach operator ideals.

Every quasi-Banach operator ideal $\mathfrak{A}$ admitting a continuous trace τ is of eigenvalue type l_2. This means that such operator ideals are comparatively small. On the other hand, they are so large that the sum of the eigenvalue sequence $(\lambda_n(T))$ need not converge for all operators $T \in \mathfrak{A}(E)$. However, if $\mathfrak{A}$ is of eigenvalue type l_1, then it may happen that the trace formula holds:

$$\tau(T) = \sum_{n=1}^{\infty} \lambda_n(T).$$

We give several examples of quasi-Banach operator ideals having a continuous trace with this property.

By analogy with the theory of traces, we develop an axiomatic theory of determinants. Given any continuous determinant δ defined on a quasi-Banach operator ideal $\mathfrak{A}$, the associated Fredholm denominator

$$\delta(\zeta, T) := \delta(I + \zeta T) \quad \text{for all } \zeta \in \mathbb{C}$$

is an entire function whose zeros ζ_0 are related to the eigenvalues λ_0 of the operator $T \in \mathfrak{A}(E)$ via $\zeta_0 = -1/\lambda_0$. Moreover, there exists an entire operator-valued function $D(\zeta, T)$ such that the Fredholm resolvent of T admits the representation

$$F(\zeta, T) = \frac{D(\zeta, T)}{\delta(\zeta, T)} \quad \text{for all } \zeta \in \varrho(T).$$

Using a regularization process, Fredholm denominators can also be obtained for operators $T \in \mathfrak{L}(E)$ possessing a power $T^m \in \mathfrak{A}(E)$.

One of the basic results of this chapter is the fact that, on every quasi-Banach operator ideal, there exists a one-to-one correspondence between continuous traces and continuous determinants.

Finally, we present the strange phenomena which can occur in the trace theory of nuclear operators on Banach spaces. It is further shown that all the trouble disappears for the ideal of (1,2)-nuclear operators.

The last section is an appendix containing some basic results from the theory of entire functions.

4.1. Fredholm resolvents

4.1.1. The **Fredholm resolvent set** $\varrho(T)$ of an operator $T \in \mathfrak{L}(E)$ consists of all complex numbers ζ for which $I + \zeta T$ is invertible. Then

$$F(\zeta, T) := T(I + \zeta T)^{-1} \quad \text{for all } \zeta \in \varrho(T)$$

is called the **Fredholm resolvent.** Note that $F(\zeta, T)$ and T commute.

Remark. The advantage of $T(I + \zeta T)^{-1}$ over $(I + \zeta T)^{-1}$ is that, for any operator ideal $\mathfrak{A}$, we have $T(I + \zeta T)^{-1} \in \mathfrak{A}(E)$ whenever $T \in \mathfrak{A}(E)$. Furthermore, if $\mathfrak{A}$ is endowed with a quasi-norm, then the Taylor and Laurent expansions considered below converge in the corresponding topology.

4.1.2. First of all, we state an elementary criterion.

Lemma. A complex number ζ belongs to the Fredholm resolvent set of $T \in \mathfrak{L}(E)$ if and only if there exists $F(\zeta, T) \in \mathfrak{L}(E)$ commuting with T such that

$$(I + \zeta T) F(\zeta, T) = T.$$

When this is so, then

$$(I + \zeta T)^{-1} = I - \zeta F(\zeta, T) \quad \text{and} \quad T(I + \zeta T)^{-1} = F(\zeta, T).$$

4.1.3. The next statement is routine, as well.

Proposition (RIE, p. 117). For every operator $T \in \mathfrak{L}(E)$ the $\mathfrak{L}(E)$-valued function

$$\zeta \to F(\zeta, T)$$

is holomorphic on $\varrho(T)$. More precisely, given $\zeta_0 \in \varrho(T)$, there exists $\varepsilon > 0$ such that

$|\zeta - \zeta_0| < \varepsilon$ implies $\zeta \in \varrho(T)$, and we have the **Taylor expansion**

$$F(\zeta, T) = \sum_{n=0}^{\infty} (-1)^n F(\zeta_0, T)^{n+1} (\zeta - \zeta_0)^n.$$

4.1.4. Let $T \in \mathfrak{L}(E)$ be a Riesz operator. Then, according to the decomposition theorem 3.2.14, for every complex number we have

$$E = M_\infty(I + \zeta T) \oplus N_\infty(I + \zeta T).$$

The projection from E onto $N_\infty(I + \zeta T)$ along $M_\infty(I + \zeta T)$ is denoted by $P_\infty(I + \zeta T)$.

4.1.5. We now consider the case in which the above decomposition is non-trivial.

Proposition (F. Riesz 1918). Let $T \in \mathfrak{L}(E)$ be a Riesz operator. Then $\zeta_0 \notin \varrho(T)$ if and only if $\lambda_0 = -1/\zeta_0$ is an eigenvalue of T.

Proof. If λ_0 is not an eigenvalue, then $N_\infty(I + \zeta_0 T) = \{o\}$. Therefore $I + \zeta_0 T$ must be invertible on $E = M_\infty(I + \zeta_0 T)$.

On the other hand, if λ_0 is an eigenvalue, then $I + \zeta_0 T$ fails to be one-to-one.

4.1.6. We are now in a position to establish the main result of this section.

Theorem (A. F. Ruston 1954). For every Riesz operator $T \in \mathfrak{L}(E)$ the $\mathfrak{L}(E)$-valued function

$$\zeta \to F(\zeta, T)$$

is meromorphic on the entire complex plane with poles located on the complement of $\varrho(T)$. More precisely, given $\zeta_0 \notin \varrho(T)$, there exists $\varepsilon > 0$ such that $0 < |\zeta - \zeta_0| < \varepsilon$ implies $\zeta \in \varrho(T)$, and we have the **Laurent expansion**

$$F(\zeta, T) = \sum_{n=1}^{d} \frac{X_n(\zeta_0, T)}{(\zeta - \zeta_0)^n} + \sum_{n=0}^{\infty} R_n(\zeta_0, T)(\zeta - \zeta_0)^n,$$

where the coefficients $X_2(\zeta_0, T), \ldots, X_d(\zeta_0, T)$ are nilpotent finite operators and

$$X_1(\zeta_0, T) = P_\infty(I + \zeta_0 T).$$

Moreover, the order of the pole at ζ_0 is equal to $d(I + \zeta_0 T)$.

Proof. Applying 3.2.11 to the operator $-\zeta_0 T$, we write

$$M := M_\infty(I + \zeta_0 T), \quad N := N_\infty(I + \zeta_0 T) \quad \text{and} \quad P := P_\infty(I + \zeta_0 T).$$

Note that

$$T = T_M + T_N \quad \text{with } T_M := (I - P)T \quad \text{and} \quad T_N := PT.$$

We further know that the operator

$$S := P(I + \zeta_0 T)$$

is nilpotent of order $d := d(I + \zeta_0 T)$. Define

$$X_1 := P \quad \text{and} \quad X_n := [\zeta_0 S(I + S + \ldots + S^{d-1})]^{n-1} \quad \text{for } n = 2, 3, \ldots$$

Since $S^d = O$, we have

$$X_n^d = O \quad \text{for } n = 2, \ldots, d \quad \text{and} \quad X_n = O \quad \text{for } n = d+1, d+2, \ldots$$

Moreover, it follows from

$$I = I - S^d = (I - S)(I + S + \dots + S^{d-1}) \quad \text{and} \quad P(I - S) = -\zeta_0 T_N$$

that

$$P = -\zeta_0 T_N (I + S + \dots + S^{d-1}).$$

Hence

$$S = -T_N X_2.$$

This in turn implies the recurrence formula

$$T_N X_{n+1} = T_N X_2^n = -S X_2^{n-1} = -S X_n.$$

Put

$$F(\zeta, T_N) := \sum_{n=1}^{d} \frac{X_n}{(\zeta - \zeta_0)^n} \quad \text{for} \quad \zeta \neq \zeta_0.$$

Then

$$\begin{aligned}(I + \zeta T_N) F(\zeta, T_N) &= (P + \zeta T_N) F(\zeta, T_N) = [S + (\zeta - \zeta_0) T_N] F(\zeta, T_N) \\ &= T_N X_1 + \sum_{n=1}^{d-1} \frac{S X_n + T_N X_{n+1}}{(\zeta - \zeta_0)^n} + \frac{S X_d}{(\zeta - \zeta_0)^d} = T_N.\end{aligned}$$

This proves that

$$(I + \zeta T_N) F(\zeta, T_N) = T_N.$$

Since $F(\zeta, T_N)$ and T_N commute, it follows from 4.1.2 that $F(\zeta, T_N)$ is indeed the Fredholm resolvent of T_N at $\zeta \neq \zeta_0$.

On the other hand, we have $\zeta_0 \in \varrho(T_M)$. Therefore, by 4.1.3, there exists $\varepsilon > 0$ such that $|\zeta - \zeta_0| < \varepsilon$ implies $\zeta \in \varrho(T_M)$ and

$$F(\zeta, T_M) = \sum_{n=0}^{\infty} (-1)^n F(\zeta_0, T_M)^{n+1} (\zeta - \zeta_0)^n.$$

Since $T_M T_N = O$ and $T_N T_M = O$, we have $T_M F(\zeta, T_N) = O$ and $T_N F(\zeta, T_M) = O$. Hence

$$[I + \zeta(T_M + T_N)]\,[F(\zeta, T_M) + F(\zeta, T_N)] = T_M + T_N.$$

This proves that

$$F(\zeta, T) := F(\zeta, T_M) + F(\zeta, T_N)$$

is the Fredholm resolvent of T at ζ whenever $0 < |\zeta - \zeta_0| < \varepsilon$.

Remarks. Observe that the principal part in the above Laurent expansion is determined by T_N while the regular part depends only on T_M.

We stress the fact that, conversely, an operator $T \in \mathfrak{L}(E)$ is Riesz if it has a meromorphic Fredholm resolvent such that all residues are finite operators; see (CAR, 3.4) and (DOW, p. 78).

4.2. Traces

4.2.1. By a **trace** τ defined on an operator ideal $\mathfrak{A}$ we mean a function which assigns to every operator $T \in \mathfrak{A}(E)$ a complex number $\tau(T)$ such that the following conditions are satisfied:

(T$_1$) $\tau(a \otimes x) = \langle x, a \rangle$ for $a \in E'$ and $x \in E$.

(T$_2$) $\tau(XT) = \tau(TX)$ for $T \in \mathfrak{A}(E, F)$ and $X \in \mathfrak{L}(F, E)$.

(T$_3$) $\tau(S + T) = \tau(S) + \tau(T)$ for $S, T \in \mathfrak{A}(E)$.

(T$_4$) $\tau(\lambda T) = \lambda\tau(T)$ for $T \in \mathfrak{A}(E)$ and $\lambda \in \mathbb{C}$.

4.2.2. We now define the trace of finite operators.

Lemma. Let $T \in \mathfrak{F}(E)$. Then

$$\operatorname{trace}(T) := \sum_{i=1}^{m} \langle x_i, a_i \rangle$$

does not depend on the special choice of the representation

$$T = \sum_{i=1}^{m} a_i \otimes x_i.$$

Proof. Given any representation

$$T = \sum_{i=1}^{m} a_i \otimes x_i,$$

we choose linearly independent elements $y_1, \ldots, y_n \in E$ such that

$$x_1, \ldots, x_m \in \operatorname{span}(y_1, \ldots, y_n).$$

Then there exists a representation

$$T = \sum_{h=1}^{n} b_h \otimes y_h.$$

Writing

$$x_i = \sum_{h=1}^{n} \xi_{ih} y_h,$$

we conclude that the functionals $b_1, \ldots, b_n \in E'$ are uniquely determined by

$$b_h = \sum_{i=1}^{m} \xi_{ih} a_i.$$

Therefore

$$\sum_{i=1}^{m} \langle x_i, a_i \rangle = \sum_{i=1}^{m} \sum_{h=1}^{n} \xi_{ih} \langle y_h, a_i \rangle = \sum_{h=1}^{n} \langle y_h, b_h \rangle.$$

The final conclusion now follows from a simultaneous application of this construction to two arbitrary representations.

4.2.3. **Theorem** (A. Pietsch 1981: b). There exists a unique trace on the operator ideal $\mathfrak{F}$.

Proof. For every finite operator

$$T = \sum_{i=1}^{m} a_i \otimes x_i$$

we define

$$\operatorname{trace}(T) := \sum_{i=1}^{m} \langle x_i, a_i \rangle .$$

In the special case $T = a \otimes x$ we therefore have

(T₁) $\operatorname{trace}(a \otimes x) = \langle x, a \rangle .$

Let $T \in \mathfrak{F}(E, F)$ and $X \in \mathfrak{L}(F, E)$. If

$$T = \sum_{i=1}^{m} a_i \otimes y_i ,$$

then

$$XT = \sum_{i=1}^{m} a_i \otimes Xy_i \quad \text{and} \quad TX = \sum_{i=1}^{m} X'a_i \otimes y_i .$$

Hence

$$\operatorname{trace}(XT) = \sum_{i=1}^{m} \langle Xy_i, a_i \rangle \quad \text{and} \quad \operatorname{trace}(TX) = \sum_{i=1}^{m} \langle y_i, X'a_i \rangle .$$

This implies that

(T_2) $\operatorname{trace}(XT) = \operatorname{trace}(TX).$

Properties (T_3) and (T_4) are obvious.
Finally, given any trace τ, for every finite operator

$$T = \sum_{i=1}^{m} a_i \otimes x_i$$

we have

$$\tau(T) = \sum_{i=1}^{m} \tau(a_i \otimes x_i) = \sum_{i=1}^{m} \langle x_i, a_i \rangle .$$

This proves the uniqueness.

4.2.4. A trace τ defined on a quasi-Banach operator ideal $\mathfrak{A}$ is said to be **continuous** if the function $T \to \tau(T)$ has this property on all components $\mathfrak{A}(E)$. Then there exists a constant $c \geq 1$ such that

$$|\tau(T)| \leq c \, \|T \mid \mathfrak{A}\| \quad \text{for all } T \in \mathfrak{A}(E)$$

and arbitrary Banach spaces E.

Remark. By the technique employed in the proof of 3.4.6, it can easily be shown that a trace τ is continuous if and only if $X \to \tau(XT)$ yields a continuous function on $\mathfrak{L}(F, E)$ for every fixed $T \in \mathfrak{A}(E, F)$. This means that $\tau(XT)$ viewed as a bilinear form on $\mathfrak{L}(F, E) \times \mathfrak{A}(E, F)$ is separately continuous in the variable X. We stress the point that this property can be used to define the concept of continuity for traces given on arbitrary operator ideals.

4.2.5. Next we state an elementary but important result.

Trace extension theorem (First version). Let $\mathfrak{A}$ be an approximative quasi-Banach operator ideal such that

$$|\operatorname{trace}(T)| \leq c \|T \mid \mathfrak{A}\| \quad \text{for all } T \in \mathfrak{F}(E),$$

where $c \geq 1$ is a constant. Then $\mathfrak{A}$ admits a unique continuous trace.

Proof. As indicated in the name of the theorem the desired trace is obtained by continuous extension of the unique trace defined for finite operators.

Remark. Suppose that $\mathfrak{A}$ and $\mathfrak{B}$ are approximative quasi-Banach operator ideals with continuous traces α and β, respectively. Then it is unknown whether or not it follows that

$$\alpha(T) = \beta(T) \quad \text{for all } T \in \mathfrak{A}(E) \cap \mathfrak{B}(E).$$

4.2.6. We are now in a position to give some non-trivial examples.

Theorem (H. König 1980: b). There exists a unique continous trace on the quasi-Banach operator ideal $(\mathfrak{P}_2)^2$.

Proof. Let $T \in \mathfrak{F}(E)$. By 1.5.1, for every $\varepsilon > 0$ we can find a factorization

$$\begin{array}{ccc} E & \xrightarrow{\quad T \quad} & E \\ {\scriptstyle A}\downarrow & & \uparrow{\scriptstyle X} \\ H & \xrightarrow[\quad S \quad]{} & K \end{array}$$

such that

$$\|X\| \, \|S \mid \mathfrak{P}_2\| \, \|A \mid \mathfrak{P}_2\| \leqq (1 + \varepsilon) \, \|T \mid (\mathfrak{P}_2)^2\|,$$

H and K being Hilbert spaces. In view of B.4.6, it may be further arranged that S is finite.

Consider a Schmidt representation

$$S = \sum_{i=1}^{n} \sigma_i x_i^* \otimes y_i.$$

Then 1.4.5 implies that

$$\|S \mid \mathfrak{P}_2\| = \left(\sum_{i=1}^{n} |\sigma_i|^2\right)^{1/2}.$$

Therefore it follows from

$$\operatorname{trace}(T) = \operatorname{trace}(XSA) = \operatorname{trace}(AXS) = \sum_{i=1}^{n} \sigma_i (AXy_i, x_i)$$

that

$$\begin{aligned} |\operatorname{trace}(T)| &\leqq \left(\sum_{i=1}^{n} |\sigma_i|^2\right)^{1/2} \left(\sum_{i=1}^{n} \|AXy_i\|^2\right)^{1/2} \\ &\leqq \|S \mid \mathfrak{P}_2\| \, \|A \mid \mathfrak{P}_2\| \, \|X\| \leqq (1 + \varepsilon) \, \|T \mid (\mathfrak{P}_2)^2\|. \end{aligned}$$

Letting $\varepsilon \to 0$, we obtain

$$|\operatorname{trace}(T)| \leqq \|T \mid (\mathfrak{P}_2)^2\| \quad \text{for all } T \in \mathfrak{F}(E).$$

We know from 1.5.9 that $(\mathfrak{P}_2)^2$ is approximative. Hence the extension theorem applies.

Remark. The assertion could also be deduced from 3.7.3 (Remark) and 4.2.17. A stronger result is proved in 4.2.30.

4.2.7. Theorem (A. Pietsch 1981: b). There exists a unique continuous trace on the quasi-Banach operator ideal $(\mathfrak{P}_2)_{2,1}^{(a)}$.

Proof. By 2.8.8, every operator $T \in (\mathfrak{P}_2)_{2,1}^{(a)}(E)$ admits a representation

$$T = \sum_{k=0}^{\infty} T_k$$

such that $T_k \in \mathfrak{F}(E)$, rank $(T_k) \leqq 2^k$ and

$$\sum_{k=0}^{\infty} 2^{k/2} \|T_k \mid \mathfrak{P}_2\| \leqq c\|T \mid (\mathfrak{P}_2)_{2,1}^{(a)}\|.$$

Here $c \geqq 1$ is a constant. Applying 1.5.1, we can find factorizations $T_k = X_k A_k$, where

$$A_k \in \mathfrak{L}(E, l_2(2^k)), \quad X_k \in \mathfrak{L}(l_2(2^k), E) \quad \text{and} \quad \|X_k\| \, \|A_k \mid \mathfrak{P}_2\| = \|T_k \mid \mathfrak{P}_2\|.$$

It follows from

$$\operatorname{trace}(A_k X_k) = \sum_{i=1}^{2^k} (A_k X_k e_i, e_i)$$

that

$$\begin{aligned} |\operatorname{trace}(A_k X_k)| &\leqq \sum_{i=1}^{2^k} \|A_k X_k e_i\| \\ &\leqq 2^{k/2} \left(\sum_{i=1}^{2^k} \|A_k X_k e_i\|^2 \right)^{1/2} \leqq 2^{k/2} \|T_k \mid \mathfrak{P}_2\|. \end{aligned}$$

If $T \in \mathfrak{F}(E)$, then the above representation possesses only a finite number of summands; see 2.3.8 (Remark). Therefore

$$\operatorname{trace}(T) = \sum_{k=0}^{\infty} \operatorname{trace}(T_k) = \sum_{k=0}^{\infty} \operatorname{trace}(X_k A_k) = \sum_{k=0}^{\infty} \operatorname{trace}(A_k X_k).$$

Thus we obtain

$$|\operatorname{trace}(T)| \leqq c\|T \mid (\mathfrak{P}_2)_{2,1}^{(a)}\| \quad \text{for all } T \in \mathfrak{F}(E).$$

We know from 2.8.9 that $(\mathfrak{P}_2)_{2,1}^{(a)}$ is approximative. Hence the extension theorem applies.

Remark. The assertion could also be deduced from 3.7.4 and 4.2.17. A stronger result is proved in 4.2.31.

4.2.8. An operator $T \in \mathfrak{L}(E, F)$ is called **integral** if there exists a constant $c \geqq 0$ such that

$$|\operatorname{trace}(XT)| \leqq c\|X\| \quad \text{for all } X \in \mathfrak{F}(F, E).$$

The set of these operators is denoted by $\mathfrak{I}(E, F)$.

For $T \in \mathfrak{I}(E, F)$ we define

$$\|T \mid \mathfrak{I}\| := \inf c,$$

the infimum being taken over all constants $c \geqq 0$ for which the above inequality holds.

4.2.9. The following result is straightforward; (PIE, 6.4.2.).

Theorem (GRO, Chap. I, p. 128). $\mathfrak{J}$ is a Banach operator ideal.

4.2.10. Next we prove an inclusion which can be found in the work of A. Grothendieck, at least implicitly.

Proposition. $\mathfrak{J} \subseteq \mathfrak{P}$.

Proof. Let $T \in \mathfrak{J}(E, F)$. Given $x_1, \ldots, x_n \in E$, we choose $b_1, \ldots, b_n \in F'$ such that

$$\|Tx_i\| = \langle Tx_i, b_i \rangle \quad \text{and} \quad \|b_i\| \leqq 1 .$$

Define

$$X := \sum_{i=1}^{n} b_i \otimes x_i ,$$

and note that

$$\|X\| \leqq \sup \left\{ \sum_{i=1}^{n} |\langle x_i, a \rangle| : a \in U^0 \right\}.$$

Then it follows that

$$\sum_{i=1}^{n} \|Tx_i\| = \sum_{i=1}^{n} \langle Tx_i, b_i \rangle = \tau(TX) \leqq \|T \mid \mathfrak{J}\| \, \|X\| .$$

Therefore we have

$$T \in \mathfrak{P}(E, F) \quad \text{and} \quad \|T \mid \mathfrak{P}\| \leqq \|T \mid \mathfrak{J}\| .$$

Remark. The embedding operator from l_1 into l_2 shows that the inclusion $\mathfrak{J} \subseteq \mathfrak{P}$ is strict.

4.2.11. **Proposition.** The Banach operator ideal $\mathfrak{J}$ is a metric extension of $\mathfrak{S}_1$.

Proof. Define the projection $P_m \in \mathfrak{F}(l_2)$ by

$$P_m(\zeta_1, \ldots, \xi_m, \xi_{m+1}, \ldots) := (\xi_1, \ldots, \xi_m, 0, \ldots).$$

Then it follows from

$$\text{trace}\,(P_m) = m \quad \text{and} \quad \|P_m\| = 1$$

that the identity operator of l_2 fails to be integral. Hence, by Calkin's theorem 2.11.11, every operator $T \in \mathfrak{J}(H, K)$ is approximable. Thus there exists a monotonic Schmidt representation

$$T = \sum_{i=1}^{\infty} \tau_i x_i^* \otimes y_i .$$

Define

$$U_m := \sum_{i=1}^{m} y_i^* \otimes x_i .$$

We then have

$$\sum_{i=1}^{m} \tau_i = \text{trace}\,(U_m T) \leqq \|T \mid \mathfrak{J}\| \, \|U_m\| \leqq \|T \mid \mathfrak{J}\| .$$

Letting $m \to \infty$, we see that $T \in \mathfrak{I}(H, K)$ implies

$$T \in \mathfrak{S}_1(H, K) \quad \text{and} \quad \|T \mid \mathfrak{S}_1\| \leqq \|T \mid \mathfrak{I}\|.$$

Conversely, it easily follows from $T \in \mathfrak{S}_1(H, K)$ that

$$T \in \mathfrak{I}(H, K) \quad \text{and} \quad \|T \mid \mathfrak{I}\| \leqq \|T \mid \mathfrak{S}_1\|.$$

4.2.12. We now show that $\mathfrak{I}$ is an upper bound for all quasi-Banach operator ideals which admit a continuous trace. This proves that those ideals are comparatively small. We stress the fact that there does not exist any continuous trace on the Banach operator ideal $\mathfrak{I}$ itself; see 4.7.6.

Theorem (A. Pietsch 1981: b). Let $\mathfrak{A}$ be a quasi-Banach operator ideal with a continuous trace. Then $\mathfrak{A} \subseteq \mathfrak{I}$.

Proof. By hypothesis, there exists a trace τ such that

$$|\tau(S)| \leqq c\|S \mid \mathfrak{A}\| \quad \text{for all } S \in \mathfrak{A}(E).$$

Hence, whenever $T \in \mathfrak{A}(E, F)$ and $X \in \mathfrak{F}(F, E)$, we have

$$|\text{trace}\,(XT)| \leqq c\|XT \mid \mathfrak{A}\| \leqq c\|T \mid \mathfrak{A}\| \, \|X\|.$$

This proves that

$$T \in \mathfrak{I}(E, F) \quad \text{and} \quad \|T \mid \mathfrak{I}\| \leqq c\|T \mid \mathfrak{A}\|.$$

4.2.13. We have the following corollary of the previous results.

Theorem. Let $\mathfrak{A}$ be a quasi-Banach operator ideal with a continuous trace. Then $\mathfrak{A}$ is of eigenvalue type l_2. In particular, all operators $T \in \mathfrak{A}(E)$ are Riesz.

4.2.14. Suppose that the operator ideal $\mathfrak{A}$ is of eigenvalue type l_1. Then we may define the **spectral sum**

$$\lambda(T) := \sum_{n=1}^{\infty} \lambda_n(T) \quad \text{for all } T \in \mathfrak{A}(E).$$

Unfortunately, we do not know whether this expression always yields a trace. The critical point is additivity. If λ has this property, then we say that the operator ideal $\mathfrak{A}$ admits a **spectral trace**. Furthermore, a trace τ is called **spectral** if the **trace formula** holds:

$$\tau(T) = \sum_{n=1}^{\infty} \lambda_n(T) \quad \text{for all } T \in \mathfrak{A}(E).$$

4.2.15. We now treat the most elementary example.

Theorem. The operator ideal $\mathfrak{F}$ admits a spectral trace.

Proof. Given $T \in \mathfrak{F}(E)$, we consider a representation

$$T = \sum_{i=1}^{m} a_i \otimes x_i$$

such that the matrix $M := (\langle x_j, a_i \rangle)$ has triangular form; see B.4.5. We know from 3.3.5 that the operators T and M_{op} are related. Thus both have the same non-zero eigenvalues. Therefore

$$\operatorname{trace}(T) = \sum_{i=1}^{m} \langle x_i, a_i \rangle = \sum_{n=1}^{m} \lambda_n(M) = \sum_{n=1}^{\infty} \lambda_n(T).$$

This proves that the unique trace of $\mathfrak{F}$ is spectral.

4.2.16. The following result is an immediate consequence of the principle of uniform boundedness 3.4.6.

Lemma. Suppose that the quasi-Banach operator ideal $\mathfrak{A}$ is of eigenvalue type l_1. Then the spectral sum is continuous at the origin.

Proof. There exists a constant $c \geqq 1$ such that

$$|\lambda(T)| \leqq \sum_{n=1}^{\infty} |\lambda_n(T)| \leqq c\|T \mid \mathfrak{A}\| \quad \text{for all} \quad T \in \mathfrak{A}(E).$$

4.2.17. The next theorem is closely related to 4.2.5.

Trace extension theorem (Second version). Let $\mathfrak{A}$ be an approximative quasi-Banach operator ideal which is of eigenvalue type l_1. Then $\mathfrak{A}$ admits a unique continuous trace.

Proof. By 4.2.15 and 4.2.16, there exists a constant $c \geqq 1$ such that

$$|\operatorname{trace}(T)| = |\lambda(T)| \leqq c\|T \mid \mathfrak{A}\| \quad \text{for all} \quad T \in \mathfrak{F}(E).$$

Remark. It is unknown whether or not the property of being spectral survives the extension procedure from $\mathfrak{F}(E)$ to $\mathfrak{A}(E)$.

4.2.18. For the following considerations some preliminaries are required. First of all, we state a lemma which is proved in (DUN, XI.9.5).

Lemma. Let $T_1, T_2, \ldots \in \mathfrak{L}(E)$ and $T \in \mathfrak{L}(E)$ be Riesz operators such that (T_k) converges to T with respect to the operator norm. Then the eigenvalue sequences of these operators can be arranged in such a way that

$$\lim_k \lambda_n(T_k) = \lambda_n(T) \quad \text{for } n = 1, 2, \ldots$$

Remark. Readers who accept this assertion without proof may proceed immediately to 4.2.23. As an alternative some results from perturbation theory can be used. These are proved in the following paragraphs.

4.2.19. Lemma. Suppose that $I - T$ with $T \in \mathfrak{L}(E)$ is invertible. Put

$$\alpha := \|(I - T)^{-1}\| > 0.$$

If

$$S \in \mathfrak{L}(E) \quad \text{and} \quad \alpha\|S - T\| \leqq q < 1,$$

then $I - S$ is invertible, as well. Moreover, we have

$$\|S(I - S)^{-1} - T(I - T)^{-1}\| \leqq \frac{\alpha^2}{1 - q} \|S - T\|.$$

Proof. Elementary manipulations show that

$$S(I - S)^{-1} - T(I - T)^{-1} = (I - S)^{-1} - (I - T)^{-1} = (I - T)^{-1} Q \sum_{k=0}^{\infty} Q^k,$$

where

$$Q := (S - T)(I - T)^{-1} \quad \text{and} \quad \|Q\| \leqq \alpha \|S - T\| \leqq q < 1 .$$

Therefore $(I - S)^{-1}$ exists, and the required inequality is obvious.

4.2.20. A number $\varrho > 0$ is said to be an **admissible radius** of an operator $T \in \mathfrak{L}(E)$ if the circle $|\zeta| = \varrho$ belongs to the Fredholm resolvent set $\varrho(T)$. Then we may define (the projection)

$$P_\infty(\varrho, T) := \frac{1}{2\pi i} \oint_{|\zeta| = \varrho} T(I + \zeta T)^{-1} \, d\zeta ,$$

the contour integral being taken in the positive direction.

4.2.21. Lemma. Suppose that ϱ is an admissible radius of an operator $T \in \mathfrak{L}(E)$. Put

$$\alpha := \sup \{\|(I + \zeta T)^{-1}\| : |\zeta| = \varrho\} > 0 .$$

If

$$S \in \mathfrak{L}(E) \quad \text{and} \quad \alpha\varrho \|S - T\| \leqq q < 1 ,$$

then ϱ is an admissible radius of S, as well. Moreover, we have

$$\|P_\infty(\varrho, S) - P_\infty(\varrho, T)\| \leqq \frac{\alpha^2 \varrho}{1 - q} \|S - T\| .$$

Proof. Applying Lemma 4.2.19, we obtain

$$\begin{aligned} \|P_\infty(\varrho, S) - P_\infty(\varrho, T)\| &\leqq \varrho \sup \{\|S(I + \zeta S)^{-1} - T(I + \zeta T)^{-1}\| : |\zeta| = \varrho\} \\ &\leqq \frac{\alpha^2 \varrho}{1 - q} \|S - T\| . \end{aligned}$$

4.2.22. Lemma. Let $\mathfrak{A}$ be a quasi-Banach operator ideal. Suppose that ϱ is an admissible radius of an operator $T \in \mathfrak{A}(E)$. Put

$$\alpha := \sup \{\|(I + \zeta T)^{-1}\| : |\zeta| = \varrho\} > 0 .$$

If

$$S \in \mathfrak{A}(E) \quad \text{and} \quad \alpha\varrho \|S - T \mid \mathfrak{A}\| \leqq q < 1 ,$$

then

$$\|SP_\infty(\varrho, S) - TP_\infty(\varrho, T) \mid \mathfrak{A}\| \leqq c \|S - T \mid \mathfrak{A}\| ,$$

where $c > 0$ is a constant only depending on q, ϱ and T.

Proof. Note that

$$\|S \mid \mathfrak{A}\| \leqq c_{\mathfrak{A}}(\|S - T \mid \mathfrak{A}\| + \|T \mid \mathfrak{A}\|) \leqq c_{\mathfrak{A}} \left(\frac{q}{\alpha\varrho} + \|T \mid \mathfrak{A}\| \right) .$$

Hence

$$\begin{aligned} &\|SP_\infty(\varrho, S) - TP_\infty(\varrho, T) \mid \mathfrak{A}\| \leqq \\ &\leqq c_{\mathfrak{A}} [\|S \mid \mathfrak{A}\| \, \|P_\infty(\varrho, S) - P_\infty(\varrho, T)\| + \|S - T \mid \mathfrak{A}\| \, \|P_\infty(\varrho, T)\|] \\ &\leqq c_{\mathfrak{A}} \left[c_{\mathfrak{A}} \left(\frac{q}{\alpha\varrho} + \|T \mid \mathfrak{A}\| \right) \frac{\alpha^2 \varrho}{1 - q} + \|P_\infty(\varrho, T)\| \right] \|S - T \mid \mathfrak{A}\| . \end{aligned}$$

4.2.23. We are now prepared to establish a stronger version of Lemma 4.2.16.

Lemma. Suppose that the quasi-Banach operator ideal $\mathfrak{A}$ is of uniform eigenvalue type l_1. Then the spectral sum is continuous everywhere.

Proof. We consider a sequence of operators $T_1, T_2, \ldots, \in \mathfrak{A}(E)$ converging to $T \in \mathfrak{A}(E)$. By hypothesis, given $\varepsilon > 0$, there exists n_0 such that

$$\sum_{n>n_0} |\lambda_n(T)| \leqq \varepsilon \quad \text{and} \quad \sum_{n>n_0} |\lambda_n(T_k)| \leqq \varepsilon \quad \text{for } k = 1, 2, \ldots$$

Hence

$$|\lambda(T_k) - \lambda(T)| \leqq \sum_{n=1}^{n_0} |\lambda_n(T_k) - \lambda_n(T)| + 2\varepsilon.$$

Applying Lemma 4.2.18, we find k_0 such that

$$|\lambda(T_k) - \lambda(T)| \leqq 3\varepsilon \quad \text{whenever} \quad k \geqq k_0.$$

Thus λ is indeed continuous.

Another proof, based on Lemmas 4.2.21 and 4.2.22, goes as follows.

Fix any T-admissible radius ϱ with $\varepsilon\varrho \geqq n_0$. Let

$$\alpha := \sup\{\|(I + \zeta T)^{-1}\| : |\zeta| = \varrho\} \quad \text{and} \quad q := 1/2.$$

Choose k_0 such that

$$\alpha\varrho\|T_k - T \mid A\| \leqq q \quad \text{for} \quad k \geqq k_0.$$

Then Lemma 4.2.21 implies that ϱ is T_k-admissible, as well.

Since T is a Riesz operator, it follows from 4.1.6 and the residue theorem that

$$P_\infty(\varrho, T) = \sum_{|\zeta|<\varrho} P_\infty(I + \zeta T),$$

where the right-hand expression contains only a finite number of non-trivial summands. If $\lambda_0 = -1/\zeta_0$ is an eigenvalue, then we have

$$\lambda[TP_\infty(I + \zeta_0 T)] = n(\lambda_0 I - T)\,\lambda_0.$$

Therefore

$$\lambda[TP_\infty(\varrho, T)] = \sum_{|\lambda_n|\varrho>1} \lambda_n(T).$$

This implies that

$$|\lambda(T) - \lambda[TP_\infty(\varrho, T)]| \leqq \sum_{|\lambda_n|\varrho<1} |\lambda_n(T)|$$
$$\leqq \sum_{\substack{n \leqq n_0 \\ |\lambda_n|\varrho<1}} |\lambda_n(T)| + \sum_{n>n_0} |\lambda_n(T)| \leqq n_0/\varrho + \varepsilon.$$

Hence

$$|\lambda(T) - \lambda[TP_\infty(\varrho, T)]| \leqq 2\varepsilon. \tag{1}$$

In the same way we obtain

$$|\lambda(T_k) - \lambda[T_k P_\infty(\varrho, T_k)]| \leqq 2\varepsilon \quad \text{for } k \geqq k_0. \tag{2}$$

According to 4.2.16, there exists $c_0 \geqq 1$ such that

$$|\lambda(S)| \leqq c_0\|S \mid \mathfrak{A}\| \quad \text{for all } S \in \mathfrak{F}(E).$$

Note that λ is additive on $\mathfrak{F}(E)$. Hence we see from 4.2.22 that

$$|\lambda[T_k P_\infty(\varrho, T_k)] - \lambda[TP_\infty(\varrho, T)]| \leqq c\, c_0 \|T_k - T \mid \mathfrak{A}\| \quad \text{for} \quad k \geqq k_0.$$

Thus we may choose $k_1 \geqq k_0$ such that

$$|\lambda[T_k P_\infty(\varrho, T_k)] - \lambda[TP_\infty(\varrho, T)]| \leqq \varepsilon \quad \text{for} \quad k \geqq k_1. \tag{3}$$

Combining inequalities (1), (2) and (3) yields

$$|\lambda(T_k) - \lambda(T)| \leqq 5\varepsilon \quad \text{for} \quad k \geqq k_1.$$

This completes the proof for everyone.

Remark. We conjecture that the spectral sum is continuous on all quasi-Banach operator ideals which are of eigenvalue type l_1. This would mean that the uniformity assumption is either automatically satisfied or superfluous.

4.2.24. The main result of this section is now easily obtained.

Spectral trace theorem (H. Leiterer / A. Pietsch 1982). Let $\mathfrak{A}$ be an approximative quasi-Banach operator ideal which is of uniform eigenvalue type l_1. Then $\mathfrak{A}$ admits a spectral trace.

Proof. We have just proved that the spectral sum λ is continuous. On the other hand, by 4.2.17, there exists a unique continuous trace τ. Since both functionals λ and τ coincide on $\mathfrak{F}(E)$, which is dense in $\mathfrak{A}(E)$, they must coincide everywhere. Thus λ is indeed a trace.

4.2.25. Since it is unknown whether or not the finite operators are dense in the quasi-Banach operator ideal $\mathfrak{L}_1^{(x)}$, the preceding theorem only applies to its approximative kernel.

Theorem (H. Leiterer / A. Pietsch 1982). The operator ideal $(\mathfrak{L}_1^{(x)})^{(a)}$ admits a spectral trace.

4.2.26. We have the following corollary of the previous result.

Theorem (H. König 1980: a). The operator ideals $\mathfrak{L}_1^{(a)}$ and $\mathfrak{L}_1^{(c)}$ admit a spectral trace.

Proof. We know from 2.10.7, 2.8.18 and 2.8.9 that

$$\mathfrak{L}_1^{(a)} \subseteqq \mathfrak{L}_1^{(c)} \subseteqq (\mathfrak{P}_2)_{2,1}^{(a)} \subseteqq (\mathfrak{L}_1^{(x)})^{(a)}.$$

Therefore the spectral trace can be restricted from $(\mathfrak{L}_1^{(x)})^{(a)}$ to the smaller operator ideals.

Remark. Since $\mathfrak{L}_1^{(a)}$ is approximative, the spectral sum yields the unique continuous trace on this quasi-Banach operator ideal. On the other hand, we do not know whether there exists more than one continuous trace on $\mathfrak{L}_1^{(c)}$.

4.2.27. A trace τ defined on an operator ideal $\mathfrak{A}$ is said to be **nilpotent** if $\tau(T) = 0$ for every nilpotent operator $T \in \mathfrak{A}(E)$.

4.2.28. Proposition. The spectral trace is nilpotent.

Proof. Note that a nilpotent operator cannot possess any non-zero eigenvalue. Therefore the spectral sum vanishes.

4.2.29. We now describe a useful method for transferring a trace from certain operator ideals to related ones.

Trace transfer theorem (A. Pietsch 1981: b). Let $\mathfrak{A}$ and $\mathfrak{B}$ be quasi-Banach operator ideals. If σ is a nilpotent continuous trace on $\mathfrak{A} \circ \mathfrak{B}$, then there exists a nilpotent continuous trace τ on $\mathfrak{B} \circ \mathfrak{A}$ such that

$$\tau(BA) = \sigma(AB) \quad \text{for all} \quad A \in \mathfrak{A}(E, F) \quad \text{and} \quad B \in \mathfrak{B}(F, E).$$

In particular, if σ is spectral, then so is τ.

Proof. Given any operator $T \in \mathfrak{B} \circ \mathfrak{A}(E)$, we consider two factorizations $T = B_1 A_1$ and $T = B_2 A_2$, where

$$A_1 \in \mathfrak{A}(E, F_1), \quad B_1 \in \mathfrak{B}(F_1, E) \quad \text{and} \quad A_2 \in \mathfrak{A}(E, F_2), \quad B_2 \in \mathfrak{B}(F_2, E).$$

Form the direct sum $F := F_1 \oplus F_2$, and let J_1 and Q_1, J_2 and Q_2, denote the canonical injections and surjections, respectively. It follows from

$$(B_1 Q_1 + B_2 Q_2)(J_2 A_2 - J_1 A_1) = B_2 A_2 - B_1 A_1 = O$$

and

$$(B_1 Q_1 - B_2 Q_2)(J_1 A_1 + J_2 A_2) = B_1 A_1 - B_2 A_2 = O$$

that the operators

$$N_1 := \tfrac{1}{2}(J_2 A_2 - J_1 A_1)(B_1 Q_1 + B_2 Q_2)$$

and

$$N_2 := \tfrac{1}{2}(J_1 A_1 + J_2 A_2)(B_1 Q_1 - B_2 Q_2)$$

are nilpotent of order 2. Note that

$$J_1 A_1 B_1 Q_1 + N_1 = J_2 A_2 B_2 Q_2 + N_2.$$

Hence

$$\sigma(A_1 B_1) = \sigma(J_1 A_1 B_1 Q_1) = \sigma(J_2 A_2 B_2 Q_2) = \sigma(A_2 B_2).$$

This proves that every factorization of an operator $T \in \mathfrak{B} \circ \mathfrak{A}(E)$ in the form $T = BA$ with $A \in \mathfrak{A}(E, F)$ and $B \in \mathfrak{B}(F, E)$ yields the same value $\sigma(AB)$. Hence we may define $\tau(T) := \sigma(AB)$.

If $x \in E$ and $a \in E'$, then we have

$$a \otimes x = (1 \otimes x)(a \otimes 1) \quad \text{and} \quad (a \otimes 1)(1 \otimes x) = \langle x, a \rangle\, 1 \otimes 1.$$

Note that $1 \otimes 1$ is the identity operator of $\mathbb{C}$, and $\sigma(1 \otimes 1) = 1$. Therefore

(T$_1$) $\quad \tau(a \otimes x) = \langle x, a \rangle.$

Let $T \in \mathfrak{B} \circ \mathfrak{A}(E, F)$ and $X \in \mathfrak{L}(F, E)$. Choose any factorization $T = BA$ with $A \in \mathfrak{A}(E, G)$ and $B \in \mathfrak{B}(G, F)$. Then we have

$$\tau(XT) = \tau(XBA) = \sigma(AXB) \quad \text{and} \quad \tau(TX) = \tau(BAX) = \sigma(AXB).$$

This proves that

(T$_2$) $\quad \tau(XT) = \tau(TX).$

Given $T_1, T_2 \in \mathfrak{B} \circ \mathfrak{A}(E)$, we consider factorizations $T_1 = B_1 A_1$ and $T_2 = B_2 A_2$, where

$$A_1 \in \mathfrak{A}(E, F_1), \quad B_1 \in \mathfrak{B}(F_1, E) \quad \text{and} \quad A_2 \in \mathfrak{A}(E, F_2), \quad B_2 \in \mathfrak{B}(F_2, E).$$

Form the direct sum $F := F_1 \oplus F_2$ as described above. Then

$$T_1 + T_2 = (B_1 Q_1 + B_2 Q_2)(J_1 A_1 + J_2 A_2).$$

Observe that the operators $N_1 := J_1 A_1 B_2 Q_2$ and $N_2 := J_2 A_2 B_1 Q_1$ are nilpotent of order 2. Hence it follows from

$$(J_1 A_1 + J_2 A_2)(B_1 Q_1 + B_2 Q_2) = J_1 A_1 B_1 Q_1 + J_2 A_2 B_2 Q_2 + N_1 + N_2$$

that

$$\begin{aligned}\sigma[(J_1 A_1 + J_2 A_2)(B_1 Q_1 + B_2 Q_2)] &= \sigma(J_1 A_1 B_1 Q_1) + \sigma(J_2 A_2 B_2 Q_2)\\ &= \sigma(A_1 B_1) + \sigma(A_2 B_2).\end{aligned}$$

Thus we have

(T₃) $\quad \tau(T_1 + T_2) = \tau(T_1) + \tau(T_2)$.

Property (T_4) is obvious.

Since the trace σ is assumed to be continuous, there exists a constant $c \geqq 1$ such that

$$|\sigma(S)| \leqq c \| S \mid \mathfrak{A} \circ \mathfrak{B} \| \quad \text{for all} \quad S \in \mathfrak{A} \circ \mathfrak{B}(F)$$

and arbitrary Banach spaces F. Given $T \in \mathfrak{B} \circ \mathfrak{A}(E)$ and $\varepsilon > 0$, we choose a factorization $T = BA$ such that

$$A \in \mathfrak{A}(E, F), \quad B \in \mathfrak{B}(F, E) \quad \text{and} \quad \|B \mid \mathfrak{B}\| \, \|A \mid \mathfrak{A}\| \leqq (1 + \varepsilon) \, \|T \mid \mathfrak{B} \circ \mathfrak{A}\|.$$

Then it follows that

$$\begin{aligned}|\tau(T)| = |\sigma(AB)| &\leqq c \|AB \mid \mathfrak{A} \circ \mathfrak{B}\| \leqq c \|A \mid \mathfrak{A}\| \, \|B \mid \mathfrak{B}\|\\ &\leqq (1 + \varepsilon) \, c \|T \mid \mathfrak{B} \circ \mathfrak{A}\|.\end{aligned}$$

Letting $\varepsilon \to 0$ yields that

$$|\tau(T)| \leqq c \|T \mid \mathfrak{B} \circ \mathfrak{A}\| \quad \text{for all} \quad T \in \mathfrak{B} \circ \mathfrak{A}(E).$$

We now assume that $T \in \mathfrak{B} \circ \mathfrak{A}(E)$ is nilpotent. Considering any factorization $T = BA$ with $A \in \mathfrak{A}(E, F)$ and $B \in \mathfrak{B}(F, E)$, we conclude from $T^m = O$ that $(AB)^{m+1} = AT^m B = O$. Thus, since σ is nilpotent, it follows that $\tau(T) = \sigma(AB) = 0$.

Finally, the principle of related operators implies that

$$\tau(T) = \sigma(AB) = \sum_{n=1}^{\infty} \lambda_n(AB) = \sum_{n=1}^{\infty} \lambda_n(BA) = \sum_{n=1}^{\infty} \lambda_n(T)$$

provided that σ is spectral.

4.2.30. Applying the preceding theorem, we are now in a position to give some important examples.

Theorem (H. König 1980: b). The operator ideal $(\mathfrak{P}_2)^2$ admits a spectral trace.

Proof. By 1.5.3, we have

$$(\mathfrak{P}_2)^2 = \mathfrak{B} \circ \mathfrak{A} \quad \text{with} \quad \mathfrak{B} := \mathfrak{H} \circ \mathfrak{P}_2 \circ \mathfrak{H} \quad \text{and} \quad \mathfrak{A} := \mathfrak{P}_2 .$$

It follows from 2.7.2 and 2.3.13 that

$$\mathfrak{A} \circ \mathfrak{B} \subseteq \mathfrak{P}_2 \circ \mathfrak{H} \circ \mathfrak{P}_2 \circ \mathfrak{H} \subseteq \mathfrak{L}_2^{(a)} \circ \mathfrak{L}_2^{(a)} \subseteq \mathfrak{L}_1^{(a)} .$$

Thus the spectral trace existing on $\mathfrak{L}_1^{(a)}$ can be transferred to $(\mathfrak{P}_2)^2$.

Remark. Since the quasi-Banach operator ideal $(\mathfrak{P}_2)^2$ is approximative and of uniform eigenvalue type l_1 (H. König 1980: b), the assertion could also be deduced from the spectral trace theorem 4.2.24.

4.2.31. Theorem (A. Pietsch 1981: b). The operator ideal $(\mathfrak{P}_2)_{2,1}^{(a)}$ admits a spectral trace.

Proof. By 2.8.21, we have

$$(\mathfrak{P}_2)_{2,1}^{(a)} = \mathfrak{B} \circ \mathfrak{A} \quad \text{with} \quad \mathfrak{B} := \mathfrak{L}_{4,2}^{(a)} \circ \mathfrak{H} \quad \text{and} \quad \mathfrak{A} := (\mathfrak{P}_2)_{4,2}^{(a)} .$$

It follows from 2.8.7, 2.7.2 and 2.8.13 that

$$\mathfrak{A} \circ \mathfrak{B} = (\mathfrak{P}_2)_{4,2}^{(a)} \circ \mathfrak{L}_{4,2}^{(a)} \circ \mathfrak{H} \subseteq (\mathfrak{P}_2 \circ \mathfrak{H})_{2,1}^{(a)} \subseteq (\mathfrak{L}_2^{(a)})_{2,1}^{(a)} = \mathfrak{L}_1^{(a)} .$$

Thus the spectral trace existing on $\mathfrak{L}_1^{(a)}$ can be transferred to $(\mathfrak{P}_2)_{2,1}^{(a)}$.

Remark. Since the quasi-Banach operator ideal $(\mathfrak{P}_2)_{2,1}^{(a)}$ is approximative and of uniform eigenvalue type l_1, the assertion could also be deduced from the spectral trace theorem 4.2.24.

4.2.32. With the help of the following proposition we can produce more examples of quasi-Banach operator ideals admitting a continuous trace.

Proposition. Let τ be a continuous trace on the quasi-Banach operator ideal $\mathfrak{A}$. Then

$$\tau'(T) := \tau(T') \quad \text{for all} \quad T \in \mathfrak{A}'(E)$$

defines a continuous trace τ' on the dual quasi-Banach operator ideal $\mathfrak{A}'$. In particular, if τ is spectral, then so is τ'.

4.2.33. The next proposition yields examples of comparatively large quasi-Banach operator ideals admitting a trace.

Proposition. Let $\mathfrak{A}$ and $\mathfrak{B}$ be quasi-Banach operator ideals with the continuous traces α and β, respectively. If

$$\alpha(T) = \beta(T) \quad \text{for all} \quad T \in \mathfrak{A}(E) \cap \mathfrak{B}(E),$$

then there exists a continuous trace τ on $\mathfrak{A} + \mathfrak{B}$ such that

$$\tau(A + B) = \alpha(A) + \beta(B) \quad \text{for all} \quad A \in \mathfrak{A}(E) \quad \text{and} \quad B \in \mathfrak{B}(E).$$

Remark. The above assumption is satisfied in the special case when the traces α and β are spectral.

4.2.34. Summarizing the previous results, we see that all quasi-Banach operator ideals occuring in the following diagram admit a spectral trace:

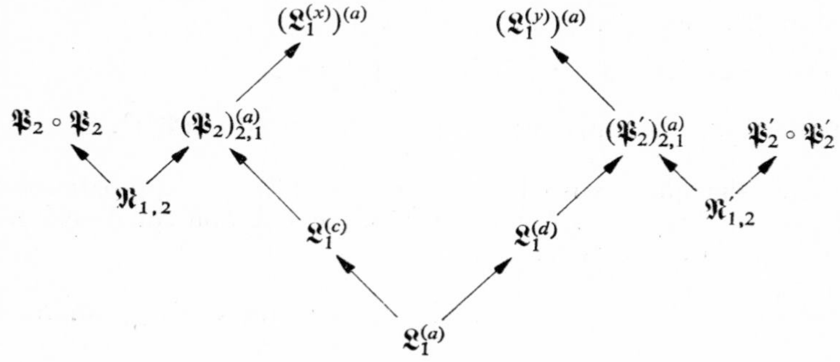

Furthermore, in view of 4.2.33, there exists a continuous trace on each of the following quasi-Banach operator ideals:

$$(\mathfrak{P}_2)^2 + (\mathfrak{P}_2')^2, \quad (\mathfrak{L}_1^{(x)})^{(a)} + (\mathfrak{L}_1^{(y)})^{(a)}, \quad (\mathfrak{P}_2)_{2,1}^{(a)} + (\mathfrak{P}_2')_{2,1}^{(a)}.$$

These traces, however, fail to be spectral.

4.2.35. Proposition. Every continuous trace defined on a quasi-Banach operator ideal $\mathfrak{A}$ is spectral on $(\mathfrak{H} \circ \mathfrak{A})^{(a)}$ and $(\mathfrak{A} \circ \mathfrak{H})^{(a)}$.

Proof. It follows from 4.2.11 and 4.2.12 that

$$(\mathfrak{H} \circ \mathfrak{A}) \circ \mathfrak{H} \subseteq \mathfrak{H} \circ \mathfrak{J} \circ \mathfrak{H} \subseteq \mathfrak{L}_1^{(a)}.$$

Thus the spectral trace existing on $\mathfrak{L}_1^{(a)}$ can be transferred to $\mathfrak{H} \circ \mathfrak{A} = \mathfrak{H} \circ (\mathfrak{H} \circ \mathfrak{A})$. Since the given trace and the spectral trace are continuous, they must coincide on the approximative kernel of $\mathfrak{H} \circ \mathfrak{A}$.

The quasi-Banach operator ideal $(\mathfrak{A} \circ \mathfrak{H})^{(a)}$ can be treated analogously.

4.2.36. Proposition (J. C. Engelbrecht / J. J. Grobler 1983). Every continuous trace defined on an approximative quasi-Banach operator ideal $\mathfrak{A}$ is spectral on $\mathfrak{A}^2$.

Proof. It follows from

$$\mathfrak{A} \subseteq \mathfrak{J} \subseteq \mathfrak{P}_2 \subseteq \mathfrak{H}$$

that

$$\mathfrak{A} \circ \mathfrak{A} \subseteq \mathfrak{H} \circ \mathfrak{A}^{(a)} \subseteq (\mathfrak{H} \circ \mathfrak{A})^{(a)}.$$

4.2.37. It is very likely that the above proposition remains true in the case of non-approximative quasi-Banach operator ideals, as well. At present, we know only the following result.

Proposition. Every continuous trace defined on a quasi-Banach operator ideal $\mathfrak{A}$ is spectral on $\mathfrak{A}^3$.

Proof. It follows from

$$\mathfrak{A} \subseteq \mathfrak{J} \subseteq \mathfrak{P}_2 \subseteq \mathfrak{H} \quad \text{and} \quad \mathfrak{P}_2 \circ \mathfrak{P}_2 \subseteq \mathfrak{H} \circ \mathfrak{P}_2 \circ \mathfrak{H} \subseteq \mathfrak{H}^{(a)}$$

that

$$\mathfrak{A} \circ \mathfrak{A} \circ \mathfrak{A} \subseteq \mathfrak{P}_2 \circ \mathfrak{P}_2 \circ \mathfrak{A} \subseteq \mathfrak{H}^{(a)} \circ \mathfrak{A} \subseteq (\mathfrak{H} \circ \mathfrak{A})^{(a)}.$$

4.2.38. All examples of quasi-Banach operator ideals, treated in this chapter, admit either none or exactly one continuous trace. Only recently, it was proved by N. J. Kalton that this is not so in general. For a proof of the following result we refer to the Appendix of this book.

Theorem. There exist quasi-Banach operator ideals admitting different continuous traces.

4.3. Determinants

4.3.1. By a **determinant** δ defined on an operator ideal $\mathfrak{A}$ we mean a function which assigns to every operator of the form $I_E + T$ with $T \in \mathfrak{A}(E)$ a complex number $\delta(I_E + T)$ such that the following conditions are satisfied:

(D_1) $\delta(I_E + a \otimes x) = 1 + \langle x, a\rangle$ for $a \in E'$ and $x \in E$.

(D_2) $\delta(I_E + XT) = \delta(I_F + TX)$ for $T \in \mathfrak{A}(E, F)$ and $X \in \mathfrak{L}(F, E)$.

(D_3) $\delta[(I_E + S)(I_E + T)] = \delta(I_E + S)\,\delta(I_E + T)$ for $S, T \in \mathfrak{A}(E)$.

(D_4) $\delta(I_E + \zeta T)$ is an entire function for fixed $T \in \mathfrak{A}(E)$.

Remark. If there is no risk of confusion, then all identity operators $I_E, I_F, \ldots$ are denoted by I.

4.3.2. In the sequel we need an elementary result from the classical theory of determinants.

Lemma. Let X be an (m, n)-matrix and Y an (n, m)-matrix. Then

$$\det(I(m) + XY) = \det(I(n) + YX).$$

Proof. It follows from

$$\begin{pmatrix} I(m) & X \\ -Y & I(n) \end{pmatrix} \begin{pmatrix} I(m) & O \\ Y & I(n) \end{pmatrix} = \begin{pmatrix} I(m) + XY & X \\ O & I(n) \end{pmatrix}$$

and

$$\begin{pmatrix} I(m) & O \\ Y & I(n) \end{pmatrix} \begin{pmatrix} I(m) & X \\ -Y & I(n) \end{pmatrix} = \begin{pmatrix} I(m) & X \\ O & I(n) + YX \end{pmatrix}$$

that the determinants of the right-hand $(m + n, m + n)$-matrices are equal. This implies the formula stated above, since

$$\det \begin{pmatrix} I(m) + XY & X \\ O & I(n) \end{pmatrix} = \det(I(m) + XY)$$

and

$$\det \begin{pmatrix} I(m) & X \\ O & I(n) + YX \end{pmatrix} = \det(I(n) + YX).$$

4.3.3. **Lemma.** Let $T \in \mathfrak{F}(E)$. Then

$$\det(I + T) := \det(\delta_{ij} + \langle x_i, a_j\rangle)$$

does not depend on the special choice of the representation

$$T = \sum_{i=1}^{m} a_i \otimes x_i .$$

Proof. Given any representation

$$T = \sum_{i=1}^{m} a_i \otimes x_i ,$$

we choose linearly independent elements $y_1, \ldots, y_n \in E$ such that

$$x_1, \ldots, x_m \in \operatorname{span}(y_1, \ldots, y_n).$$

Then there exists a representation

$$T = \sum_{h=1}^{n} b_h \otimes y_h .$$

Writing

$$x_i = \sum_{h=1}^{n} \xi_{ih} y_h ,$$

we conclude that the functionals $b_1, \ldots, b_n \in E'$ are uniquely determined by

$$b_h = \sum_{i=1}^{m} \xi_{ih} a_i .$$

Hence

$$\langle x_i, a_j \rangle = \sum_{h=1}^{n} \xi_{ih} \langle y_h, a_j \rangle \quad \text{and} \quad \langle y_h, b_k \rangle = \sum_{j=1}^{m} \langle y_h, a_j \rangle \, \xi_{jk} .$$

If $X := (\xi_{ih})$ and $Y := (\langle y_h, a_j \rangle)$, then we have

$$\det(\delta_{ij} + \langle x_i, a_j \rangle) = \det(I(m) + XY)$$

and

$$\det(\delta_{hk} + \langle y_h, b_k \rangle) = \det(I(n) + YX).$$

Therefore the preceding lemma implies that

$$\det(\delta_{ij} + \langle x_i, a_j \rangle) = \det(\delta_{hk} + \langle y_h, b_k \rangle).$$

The final conclusion now follows from a simultaneous application of this construction to two arbitrary representations.

4.3.4. We now establish a counterpart of 4.2.3.

Theorem. There exists a unique determinant on the operator ideal $\mathfrak{F}$.

Proof. For every finite operator

$$T = \sum_{i=1}^{m} a_i \otimes x_i$$

we define

$$\det(I + T) := \det(\delta_{ij} + \langle x_i, a_j \rangle).$$

In the special case $T = a \otimes x$ we therefore have

(D_1) $\det(I + a \otimes x) = 1 + \langle x, a \rangle .$

Let $T \in \mathfrak{F}(E, F)$ and $X \in \mathfrak{L}(F, E)$. If

$$T = \sum_{i=1}^{m} a_i \otimes y_i,$$

then

$$XT = \sum_{i=1}^{m} a_i \otimes Xy_i \quad \text{and} \quad TX = \sum_{i=1}^{m} X'a_i \otimes y_i.$$

Hence

$$\det(I + XT) = \det(\delta_{ij} + \langle Xy_i, a_j\rangle)$$

and

$$\det(I + TX) = \det(\delta_{ij} + \langle y_i, X'a_j\rangle).$$

This implies that

$$(\mathrm{D}_2) \qquad \det(I + XT) = \det(I + TX).$$

Given $S, T \in \mathfrak{F}(E)$, we choose representations

$$S = \sum_{i=1}^{m} a_i \otimes x_i \quad \text{and} \quad T = \sum_{h=1}^{n} b_h \otimes y_h.$$

Then

$$S + (I + S)\,T = \sum_{i=1}^{m} a_i \otimes x_i + \sum_{h=1}^{n} b_h \otimes \left(y_h + \sum_{p=1}^{m} \langle y_h, a_p\rangle\, x_p\right).$$

Thus

$$\det[(I + S)(I + T)] = \det[I + S + (I + S)\,T]$$

is defined to be the determinant of the matrix

$$\begin{pmatrix} \delta_{ij} + \langle x_i, a_j\rangle & , & \langle x_i, b_k\rangle \\ \langle y_h, a_j\rangle + \sum_{p=1}^{m} \langle y_h, a_p\rangle \langle x_p, a_j\rangle, & & \delta_{hk} + \langle y_h, b_k\rangle + \sum_{p=1}^{m} \langle y_h, a_p\rangle \langle x_p, b_k\rangle \end{pmatrix},$$

which can be written in the form

$$\begin{pmatrix} \delta_{ip} & , & 0 \\ \langle y_h, a_p\rangle, & & \delta_{hq} + \langle y_h, b_q\rangle \end{pmatrix} \begin{pmatrix} \delta_{pj} + \langle x_p, a_j\rangle, & & \langle x_p, b_k\rangle \\ 0 & , & \delta_{qk} \end{pmatrix}.$$

Next we observe that the determinants of the right-hand and left-hand factor are equal to $\det(I + S)$ and $\det(I + T)$, respectively. This completes the proof of the multiplication formula

$$(\mathrm{D}_3) \qquad \det[(I + S)(I + T)] = \det(I + S)\det(I + T).$$

Property (D_4) follows from the fact that $\det(I + \zeta T)$ is a polynomial in the complex variable ζ.

Finally, we know from B.4.5 that every operator $T \in \mathfrak{F}(E)$ admits a representation

$$T = \sum_{i=1}^{m} a_i \otimes x_i$$

such that the matrix $(\langle x_i, a_j\rangle)$ has upper triangular form:

$$\langle x_i, a_j\rangle = 0 \quad \text{if} \quad i > j.$$

Therefore

$$I + T = (I + a_1 \otimes x_1) \dots (I + a_m \otimes x_m).$$

Thus, given any determinant δ, we have

$$\delta(I + T) = \prod_{i=1}^{m} \delta(I + a_i \otimes x_i) = \prod_{i=1}^{m} (1 + \langle x_i, a_i \rangle) = \det(\delta_{ij} + \langle x_i, a_j \rangle).$$

This proves the uniqueness.

4.3.5. Let δ be any determinant on an operator ideal $\mathfrak{A}$. Then property (D_4) implies that the **Gâteaux derivative**

$$\delta^{\bullet}(T) := \lim_{\zeta \to 0} \frac{\delta(I + \zeta T) - 1}{\zeta}$$

exists for all $T \in \mathfrak{A}(E)$.

4.3.6. **Lemma.** Let δ be a determinant on an operator ideal $\mathfrak{A}$. Then

$$|\delta^{\bullet}(T)| \leqq \sup\{|\delta(I + \zeta T) - 1| : |\zeta| = 1\} \quad \text{for all} \quad T \in \mathfrak{A}(E).$$

Proof. Since $\delta(I + \zeta T)$ is an entire function, we have the Taylor expansion

$$\delta(I + \zeta T) = 1 + \sum_{n=1}^{\infty} \alpha_n(T)\, \zeta^n \quad \text{for all} \quad \zeta \in \mathbb{C}.$$

Hence the residue theorem implies that

$$\delta^{\bullet}(T) = \alpha_1(T) = \frac{1}{2\pi i} \oint_{|\zeta|=1} \frac{\delta(I + \zeta T) - 1}{\zeta^2}\, d\zeta,$$

the contour integral being taken in the positive direction. The desired inequality is now obvious.

4.3.7. A determinant δ defined on a quasi-Banach operator ideal $\mathfrak{A}$ is said to be **continuous** (at the origin) if the function $T \to \delta(I + T)$ has this property on all components $\mathfrak{A}(E)$.

4.3.8. Next we prove an analogue of 4.2.12.

Theorem. Let $\mathfrak{A}$ be a quasi-Banach operator ideal with a determinant which is continuous at the origin. Then $\mathfrak{A} \subseteq \mathfrak{T}$.

Proof. By hypothesis, there exist a determinant δ and $\varepsilon > 0$ such that

$$|\delta(I + S) - 1| \leqq 1 \quad \text{if} \quad \|S \mid \mathfrak{A}\| \leqq \varepsilon.$$

Hence, in view of 4.3.6,

$$|\delta^{\bullet}(S)| \leqq 1 \quad \text{if} \quad \|S \mid \mathfrak{A}\| \leqq \varepsilon.$$

Restricting attention to finite operators, we have $\delta^{\bullet}(S) = \operatorname{trace}(S)$. Hence

$$|\operatorname{trace}(S)| \leqq \varepsilon^{-1} \|S \mid \mathfrak{A}\| \quad \text{for all} \quad S \in \mathfrak{F}(E).$$

and, reasoning as in the proof of 4.2.12, we arrive at the desired conclusion.

4.3.9. We have the following corollary of the preceding result.

Theorem. Let $\mathfrak{A}$ be a quasi-Banach operator ideal with a determinant which is continuous at the origin. Then $\mathfrak{A}$ is of eigenvalue type l_2. In particular, all operators $T \in \mathfrak{A}(E)$ are Riesz.

4.3.10. Proposition. Let δ be a determinant on a quasi-Banach operator ideal $\mathfrak{A}$. If δ is continuous at the origin, then it is continuous everywhere.

Proof. Fix $S \in \mathfrak{A}(E)$. If $I + S$ is invertible, then the continuity at S follows from

$$\delta(I + S + T) = \delta(I + S)\,\delta(I + (I + S)^{-1}T) \quad \text{for all} \quad T \in \mathfrak{A}(E).$$

Otherwise, in view of the preceding result, there exists a Riesz decomposition $S = S_M + S_N$ such that $I + S_M$ is invertible and $S_N \in \mathfrak{F}(E)$. Then

$$\delta(I + S + T) = \delta(I + S_M + T)\,\delta(I + (I + S_M + T)^{-1}S_N)$$

for all $T \in \mathfrak{A}(E)$ provided that $\|T\|\;\|(I + S_M)^{-1}\| < 1$. As already shown $T \to O$ implies $\delta(I + S_M + T) \to \delta(I + S_M)$. Furthermore, we know from 4.3.4 that

$$\delta(I + (I + S_M + T)^{-1}S_N) = \det(\delta_{ij} + \langle (I + S_M + T)^{-1}x_i, a_j\rangle),$$

where

$$S_N = \sum_{i=1}^{n} a_i \otimes x_i$$

is any representation of S_N. Obviously, the right-hand determinant viewed as a function of T is continuous at $T = O$. This completes the proof.

4.3.11. Proposition. Let δ be a continuous determinant on a quasi-Banach operator deal $\mathfrak{A}$. Then $I + T$ with $T \in \mathfrak{A}(E)$ is invertible if and only if $\delta(I + T) \neq 0$.

Proof. Suppose that $I + T$ is invertible. Then the inverse can be written in the form $(I + T)^{-1} = I + S$, where $S \in \mathfrak{A}(E)$. We now obtain $\delta(I + S)\,\delta(I + T) = \delta(I) = 1$. Hence $\delta(I + T) \neq 0$.

In the case when $I + T$ fails to be invertible we consider a Riesz decomposition $T = T_M + T_N$. Since the restriction $I_N + Q_N T J_N$ of $I + T$ to $N := N_\infty(I + T)$ is nilpotent, it follows that

$$\delta(I + T_N) = \det(I + J_N Q_N T) = \det(I_N + Q_N T J_N) = 0.$$

Hence

$$\delta(I + T) = \delta[(I + T_M)(I + T_N)] = \delta(I + T_M)\,\delta(I + T_N) = 0.$$

4.3.12. For every operator $T \in \mathfrak{L}(E)$ we define

$$\exp(T) := I + \sum_{k=1}^{\infty} \frac{1}{k!} T^k.$$

Note that

$$\exp(T) = \lim_n \left(I + \frac{1}{n} T\right)^n.$$

The following result is obvious.

Lemma. Let $\mathfrak{A}$ be any quasi-Banach operator ideal. If $T \in \mathfrak{A}(E)$, then

$$\exp(T) - I \in \mathfrak{A}(E) \quad \text{and} \quad \|\exp(T) - I \mid \mathfrak{A}\| \leqq \|T \mid \mathfrak{A}\| \exp(\|T\|).$$

4.3.13. Proposition. Let δ be a continuous determinant on a quasi-Banach operator ideal $\mathfrak{A}$. Then

$$\delta[\exp(S + T)] = \delta[\exp(S)]\, \delta[\exp(T)] \quad \text{for all} \quad S, T \in \mathfrak{A}(E).$$

Proof. If the operators S and T commute, then the above formula follows immediately from $\exp(S + T) = \exp(S)\exp(T)$. The verification in the general case is more sophisticated.

Assume that $n \geqq 2\|S + T\|$. Then $I + \frac{1}{n}(S + T)$ is invertible, and we have

$$\left\| \left[I + \frac{1}{n}(S + T)\right]^{-1} \right\| \leqq 2.$$

Define U_n and R_n by the equations

$$\left[I + \frac{1}{n}(S + T)\right](I + U_n) = \left(I + \frac{1}{n}S\right)\left(I + \frac{1}{n}T\right)$$

and

$$I + R_n = (I + U_n)^n.$$

Since

$$n^2 \left[I + \frac{1}{n}(S + T)\right] U_n = ST,$$

we have

$$n^2 \|U_n \mid \mathfrak{A}\| \leqq 2\|ST \mid \mathfrak{A}\|.$$

Next we deduce from

$$R_n = U_n \sum_{k=1}^{n} \binom{n}{k} U_n^{k-1}$$

that $R_n \in \mathfrak{A}(E)$ and

$$\|R_n \mid \mathfrak{A}\| \leqq \|U_n \mid \mathfrak{A}\| \sum_{k=1}^{n} \frac{n^k}{k!} \|U_n\|^{k-1} \leqq n\|U_n \mid \mathfrak{A}\| \exp(n\|U_n\|).$$

Hence $\|R_n \mid \mathfrak{A}\| \to 0$ as $n \to \infty$. Note that

$$\delta\left[\left(I + \frac{1}{n}S\right)^n\right] \delta\left[\left(I + \frac{1}{n}T\right)^n\right] = \left[\delta\left(I + \frac{1}{n}S\right) \delta\left(I + \frac{1}{n}T\right)\right]^n$$
$$= \left(\delta\left[I + \frac{1}{n}(S + T)\right] \delta[I + U_n]\right)^n = \delta\left(\left[I + \frac{1}{n}(S + T)\right]^n\right) \delta(I + R_n).$$

Letting $n \to \infty$ yields the formula we are looking for.

4.3.14. Lemma. Let δ be a continuous determinant on a quasi-Banach operator ideal $\mathfrak{A}$. Then

$$\delta[\exp(T)] = \exp[\delta^{\cdot}(T)] \quad \text{for all} \quad T \in \mathfrak{A}(E).$$

Proof. We have

$$\delta\left(I + \frac{1}{n}T\right) = 1 + \frac{1}{n}\delta^{\cdot}(T) + \omega\left(\frac{1}{n}\right),$$

where $n\omega\left(\frac{1}{n}\right) \to 0$ as $n \to \infty$. Write

$$\varrho_n := \left[1 + \frac{1}{n}\delta^{\cdot}(T) + \omega\left(\frac{1}{n}\right)\right]^n - \left[1 + \frac{1}{n}\delta^{\cdot}(T)\right]^n.$$

Taking n so large that $n\left|\omega\left(\frac{1}{n}\right)\right| \leqq 1$, we get

$$\begin{aligned}|\varrho_n| &\leqq \sum_{k=1}^{n} \binom{n}{k} \left[1 + \frac{1}{n}\delta^{\cdot}(T)\right]^{n-k} \left|\omega\left(\frac{1}{n}\right)\right|^k \\ &\leqq \sum_{k=1}^{n} \frac{1}{k!} \left[1 + \frac{1}{n}|\delta^{\cdot}(T)|\right]^n \left|n\omega\left(\frac{1}{n}\right)\right|^k \\ &\leqq n\left|\omega\left(\frac{1}{n}\right)\right| \exp\left(1 + |\delta^{\cdot}(T)|\right).\end{aligned}$$

This proves that $\varrho_n \to 0$ as $n \to \infty$. Finally, the desired formula follows from

$$\begin{aligned}\delta\left[\left(I + \frac{1}{n}T\right)^n\right] &= \left[\delta\left(I + \frac{1}{n}T\right)\right]^n = \left[1 + \frac{1}{n}\delta^{\cdot}(T) + \omega\left(\frac{1}{n}\right)\right]^n \\ &= \left(1 + \frac{1}{n}\delta^{\cdot}(T)\right)^n + \varrho_n\end{aligned}$$

by passing to the limit as $n \to \infty$.

4.3.15. We are now in a position to establish the main result of this section.

Theorem. Let δ be a continuous determinant on a quasi-Banach operator ideal $\mathfrak{A}$. Then the Gâteaux derivative

$$\delta^{\cdot}(T) := \lim_{\zeta \to 0} \frac{\delta(I + \zeta T) - 1}{\zeta} \quad \text{for all} \quad T \in \mathfrak{A}(E)$$

defines a continuous trace.

Proof. If $a \in E'$ and $x \in E$, then it follows from

$$\delta(I + \zeta a \otimes x) = 1 + \zeta\langle x, a\rangle$$

that

(T₁) $\quad \delta^{\cdot}(a \otimes x) = \langle x, a\rangle.$

Let $T \in \mathfrak{A}(E, F)$ and $X \in \mathfrak{L}(F, E)$. Then

$$\delta(I_E + \zeta XT) = \delta(I_F + \zeta TX) \quad \text{for all} \quad \zeta \in \mathbb{C}.$$

This implies that

(T_2) $\quad \delta^{\cdot}(XT) = \delta^{\cdot}(TX).$

Given $T \in \mathfrak{A}(E)$ and $\lambda \neq 0$, we have

$$\lim_{\zeta \to 0} \frac{\delta(I + \zeta\lambda T) - 1}{\zeta} = \lambda \lim_{\zeta \to 0} \frac{\delta(I + \zeta\lambda T) - 1}{\zeta\lambda}.$$

Hence

(T_4) $\quad \delta^{\cdot}(\lambda T) = \lambda\delta^{\cdot}(T).$

Let $S, T \in \mathfrak{A}(E)$ and $\zeta \in \mathbb{C}$. Then, in view of 4.3.14, it follows from

$$\delta[\exp(\zeta S + \zeta T)] = \delta[\exp(\zeta S)]\,\delta[\exp(\zeta T)]$$

that

$$\exp[\zeta\delta^{\cdot}(S + T)] = \exp[\zeta\delta^{\cdot}(S)]\exp[\zeta\delta^{\cdot}(T)].$$

Since this formula holds for all $\zeta \in \mathbb{C}$, we may conclude that

(T_3) $\quad \delta^{\cdot}(S + T) = \delta^{\cdot}(S) + \delta^{\cdot}(T).$

The inequality 4.3.6 yields the continuity of $\delta^{\cdot}$.

4.3.16. For every operator $T \in \mathfrak{L}(E)$ with $\|T\| < 1$ we define

$$\log(I + T) := -\sum_{k=1}^{\infty} \frac{(-1)^k}{k} T^k.$$

Note that

$$\exp[\log(I + T)] = I + T.$$

The following result is obvious.

Lemma. Let $\mathfrak{A}$ be any quasi-Banach operator ideal. If $T \in \mathfrak{A}(E)$ and $\|T\| \leqq q < 1$, then

$$\log(I + T) \in \mathfrak{A}(E) \quad \text{and} \quad \|\log(I + T) \mid \mathfrak{A}\| \leqq \frac{1}{1 - q}\|T \mid \mathfrak{A}\|.$$

4.3.17. Using the logarithm, we may rephrase Lemma 4.3.14 as follows.

Lemma. Let δ be a continuous determinant on a quasi-Banach operator ideal $\mathfrak{A}$. Then

$$\delta(I + T) = \exp(\delta^{\cdot}[\log(I + T)]) \quad \text{for all} \quad T \in \mathfrak{A}(E)$$

provided that $\|T\| < 1$.

4.3.18. Lemma. Let δ be a continuous determinant on a quasi-Banach operator ideal $\mathfrak{A}$. If $0 < q < 1$, then there exists $c > 0$ such that

$$|\delta(I + T) - 1| \leqq c\|T \mid \mathfrak{A}\| \quad \text{for all} \quad T \in \mathfrak{A}(E)$$

provided that $\|T \mid \mathfrak{A}\| \leqq q$.

Proof. According to 4.3.15, we can find a constant $c_0 \geqq 1$ such that

$$|\delta^{\cdot}(T)| \leqq c_0\|T \mid \mathfrak{A}\| \quad \text{for all} \quad T \in \mathfrak{A}(E).$$

Recall from Lemma 4.3.16 that

$$\|\log(I + T) \mid \mathfrak{A}\| \leqq \frac{1}{1 - q}\|T \mid \mathfrak{A}\| \quad \text{if} \quad \|T \mid \mathfrak{A}\| \leqq q.$$

Hence

$$\|\delta^{\cdot}[\log(I + T)]\| \leqq \frac{1}{1 - q}\, c_0\|T \mid \mathfrak{A}\| \quad \text{if} \quad \|T \mid \mathfrak{A}\| \leqq q.$$

Note that

$$|\exp(\zeta) - 1| \leqq |\zeta| \exp(|\zeta|) \quad \text{for all} \quad \zeta \in \mathbb{C}.$$

Therefore, if $\|T \mid \mathfrak{A}\| \leqq q$, then we have

$$|\delta(I + T) - 1| = |\exp(\delta^{\cdot}[\log(I + T)]) - 1|$$
$$\leqq \left[\frac{1}{1-q} c_0 \exp\left(\frac{q}{1-q} c_0\right)\right] \|T \mid \mathfrak{A}\|.$$

4.3.19. In what follows the derivatives of functions $\varphi(\zeta)$ and $T(\zeta)$ are denoted by $\varphi^{\cdot}(\zeta)$ and $T^{\cdot}(\zeta)$, respectively.

Proposition. Let δ be a continuous determinant on a quasi-Banach operator ideal $\mathfrak{A}$. Suppose that the $\mathfrak{A}(E)$-valued function $T(\zeta)$ is defined on a domain of the complex plane. If $T(\zeta)$ is differentiable at a point ζ_0, then so is the complex-valued function $\varphi(\zeta) := \delta[I + T(\zeta)]$. In the particular case when $I + T(\zeta_0)$ is invertible the derivative is given by

$$\varphi^{\cdot}(\zeta_0) = \delta^{\cdot}[T^{\cdot}(\zeta_0)(I + T(\zeta_0))^{-1}]\,\delta[I + T(\zeta_0)].$$

Proof. By hypothesis, we have

$$T(\zeta) = T(\zeta_0) + (\zeta - \zeta_0)\,T^{\cdot}(\zeta_0) + O(\zeta, \zeta_0),$$

where

$$\lim_{\zeta \to \zeta_0} \frac{\|O(\zeta, \zeta_0) \mid \mathfrak{A}\|}{|\zeta - \zeta_0|} = 0.$$

First of all, we treat the case when $I + T(\zeta_0)$ is invertible. Then there exists

$$S(\zeta, \zeta_0) := (I + T(\zeta_0) + (\zeta - \zeta_0)\,T^{\cdot}(\zeta_0))^{-1}\,O(\zeta, \zeta_0)$$

provided that $|\zeta - \zeta_0|$ is sufficiently small. It now follows that

$$\delta[I + T(\zeta)] = \delta[I + T(\zeta_0) + (\zeta - \zeta_0)\,T^{\cdot}(\zeta_0)]\,\delta[I + S(\zeta, \zeta_0)].$$

Obviously, the function

$$\delta[I + T(\zeta_0) + (\zeta - \zeta_0)\,T^{\cdot}(\zeta_0)] =$$
$$= \delta[I + (\zeta - \zeta_0)\,T^{\cdot}(\zeta_0)(I + T(\zeta_0))^{-1}]\,\delta[I + T(\zeta_0)]$$

is differentiable at ζ_0. Furthermore, by the preceding lemma, we have

$$\frac{|\delta[I + S(\zeta, \zeta_0)] - 1|}{|\zeta - \zeta_0|} \leqq c\,\frac{\|S(\zeta, \zeta_0) \mid \mathfrak{A}\|}{|\zeta - \zeta_0|}.$$

This implies that

$$\lim_{\zeta \to \zeta_0} \frac{\delta[I + S(\zeta, \zeta_0)] - 1}{\zeta - \zeta_0} = 0.$$

Thus $\delta[I + T(\zeta)]$ is indeed differentiable at ζ_0, and the derivative can be obtained by applying the product formula.

In the general case we consider a Riesz decomposition $T(\zeta_0) = T_M(\zeta_0) + T_N(\zeta_0)$ such that $I + T_M(\zeta_0)$ is invertible and $T_N(\zeta_0) \in \mathfrak{F}(E)$. If $|\zeta - \zeta_0|$ is sufficiently small, then

$$\delta[I + T(\zeta)] = \delta[I + T(\zeta) - T_N(\zeta_0)]\,\delta[I + (I + T(\zeta) - T_N(\zeta_0))^{-1}\,T_N(\zeta_0)].$$

Using any representation

$$T_N(\zeta_0) = \sum_{i=1}^{m} a_i \otimes x_i,$$

by 4.3.4, we have

$$\delta[I + (I + T(\zeta) - T_N(\zeta_0))^{-1} T_N(\zeta_0)] =$$
$$= \det(\delta_{ij} + \langle (I + T(\zeta) - T_N(\zeta_0))^{-1} x_i, a_j \rangle).$$

This determinant is of course differentiable at ζ_0. The differentiability of $\delta[I + T(\zeta) - T_N(\zeta_0)]$ follows from the first part of the proof, since $I + T(\zeta_0) - T_N(\zeta_0)$ is invertible.

4.3.20. It is worth specializing the preceding result in the following way.

Proposition. Let δ be a continuous determinant on a quasi-Banach operator ideal $\mathfrak{A}$. Then the **Gâteaux derivative**

$$\delta^{\cdot}(I + T, S) := \lim_{\zeta \to 0} \frac{\delta(I + T + \zeta S) - \delta(I + T)}{\zeta}$$

exists for all operators $S, T \in \mathfrak{A}(E)$. In the particular case when $I + T$ is invertible we have **Fredholm's formula**

$$\frac{\delta^{\cdot}(I + T, S)}{\delta(I + T)} = \delta^{\cdot}[S(I + T)^{-1}].$$

4.3.21. Suppose that the operator ideal $\mathfrak{A}$ is of eigenvalue type l_1. Then we may define the **spectral product**

$$\pi(I + T) := \prod_{n=1}^{\infty} (1 + \lambda_n(T)) \quad \text{for all} \quad T \in \mathfrak{A}(E).$$

Unfortunately, we do not know whether this expression always yields a determinant. The critical point is multiplicativity. If π has this property, then we say that the operator ideal $\mathfrak{A}$ admits a **spectral determinant**. Furthermore, a determinant δ is called **spectral,** if the **determinant formula** holds:

$$\delta(I + T) = \prod_{n=1}^{\infty} (1 + \lambda_n(T)) \quad \text{for all} \quad T \in \mathfrak{A}(E).$$

4.3.22. We now treat the most elementary example.

Theorem. The operator ideal $\mathfrak{F}$ admits a spectral determinant.

Proof. Given $T \in \mathfrak{F}(E)$, we consider a representation

$$T = \sum_{i=1}^{m} a_i \otimes x_i$$

such that the matrix $M := (\langle x_j, a_i \rangle)$ has triangular form; see B.4.5. We know from 3.3.5 that the operators T and M_{op} are related. Thus both have the same non-zero eigenvalues. Therefore

$$\delta(I + T) = \det(\delta_{ij} + \langle x_i, a_j \rangle) = \prod_{i=1}^{m} (1 + \langle x_i, a_i \rangle)$$
$$= \prod_{n=1}^{m} (1 + \lambda_n(M)) = \prod_{n=1}^{\infty} (1 + \lambda_n(T)).$$

This proves that the unique determinant of $\mathfrak{F}$ is spectral.

4.3.23. Proposition. Let δ be a continuous determinant on a quasi-Banach operator ideal. If δ is spectral, then so is the associated continuous trace $\delta^{\cdot}$.

Proof. If the quasi-Banach operator ideal $\mathfrak{A}$ is of eigenvalue type l_1, then the spectral product is Gâteaux differentiable at the origin, and we have

$$\lim_{\zeta\to 0} \frac{\pi(I+\zeta T)-1}{\zeta} = \lambda(T) \quad \text{for all} \quad T\in\mathfrak{A}(E).$$

4.4. Fredholm denominators

4.4.1. Suppose that the Fredholm resolvent of a Riesz operator $T\in\mathfrak{L}(E)$ is written in the form

$$F(\zeta, T) = \frac{D(\zeta, T)}{\delta(\zeta, T)} \quad \text{for all} \quad \zeta\in\varrho(T),$$

where $\delta(\zeta, T)$ is an entire complex-valued function and $D(\zeta, T)$ is an entire $\mathfrak{L}(E)$-valued function. Then we call $\delta(\zeta, T)$ and $D(\zeta, T)$ a **Fredholm denominator** and **Fredholm numerator**, respectively. Obviously, both functions are uniquely related to each other.

Remark. Occasionally, $D(\zeta, T)$ is said to be a **Fredholm minor** of first order. Fredholm minors of higher order were defined by A. F. Ruston (1953, 1967).

4.4.2. Proposition. An entire function $\delta(\zeta, T)$ is a Fredholm denominator of a Riesz operator $T\in\mathfrak{L}(E)$ if and only if every point $\zeta_0\notin\varrho(T)$ is a zero of multiplicity greater or equal than $d(I+\zeta_0 T)$.

Proof. Let $\delta(\zeta, T)$ be any entire complex-valued function. Then

$$D(\zeta, T) := \delta(\zeta, T)\, F(\zeta, T)$$

defines a holomorphic $\mathfrak{L}(E)$-valued function on $\varrho(T)$. Of course, the isolated singularities of this function are removable if and only if the zeros of $\delta(\zeta, T)$ annihilate the poles of $F(\zeta, T)$.

4.4.3. Proposition. Let $T\in\mathfrak{L}(E)$ be a Riesz operator. Suppose that

$$F(\zeta, T) = \frac{D(\zeta, T)}{\delta(\zeta, T)} \quad \text{for all} \quad \zeta\in\varrho(T),$$

where

$$\delta(\zeta, T) = \sum_{n=0}^{\infty} \alpha_n(T)\,\zeta^n \quad \text{for all} \quad \zeta\in\mathbb{C}$$

and

$$D(\zeta, T) = \sum_{n=0}^{\infty} A_n(T)\,\zeta^n \quad \text{for all} \quad \zeta\in\mathbb{C}.$$

Then

$$A_n(T) = \sum_{h=0}^{n} (-1)^{n-h}\,\alpha_h(T)\,T^{n-h+1} \quad \text{for} \quad n = 0, 1, \ldots$$

Proof. Note that

$$\delta(\zeta, T)\,T = D(\zeta, T)\,(I+\zeta T) \quad \text{for all} \quad \zeta\in\varrho(T).$$

Substituting the power series and equating coefficients, we obtain

$$A_0(T) = \alpha_0(T)\,T \quad \text{and} \quad A_n(T) = \alpha_n(T)\,T - A_{n-1}(T)\,T \quad \text{for} \quad n = 1, 2, \ldots$$

This implies the formula we are looking for.

4.4.4. The existence of Fredholm denominators follows from the Weierstrass factorization theorem 4.8.4. This construction, however, is very unsatisfactory, since we need to know all the eigenvalues of the Riesz operator under consideration. For this reason, we are interested in finding a more direct approach.

Theorem. Let δ be a continuous determinant on a quasi-Banach operator ideal $\mathfrak{A}$. If $T \in \mathfrak{A}(E)$, then

$$\delta(\zeta, T) := \delta(I + \zeta T) \quad \text{for all} \quad \zeta \in \mathbb{C}$$

defines an entire function with zeros at all points $\zeta_0 \notin \varrho(T)$ and at these points only. Furthermore, the multiplicity of the zero ζ_0 is equal to the multiplicity of the eigenvalue $\lambda_0 = -1/\zeta_0$.

Proof. First of all, we note that $\delta(\zeta, T)$ is holomorphic, by property (D_4). Furthermore, it follows from 4.3.9 that T is a Riesz operator.

If $\zeta_0 \in \varrho(T)$, then $I + \zeta_0 T$ is invertible. Hence, by 4.3.11, we have $\delta(\zeta_0, T) \neq 0$.

If $\zeta_0 \notin \varrho(T)$, then we consider the Riesz decomposition

$$E = M_\infty(I + \zeta_0 T) \oplus N_\infty(I + \zeta_0 T)$$

as described in 3.2.14. Let $P := P_\infty(I + \zeta_0 T)$, and define

$$T_M := T(I - P) \quad \text{and} \quad T_N := TP.$$

It follows from $T = T_M + T_N$ and $T_M T_N = O$ that

$$(I + \zeta T) = (I + \zeta T_M)(I + \zeta T_N).$$

Hence

$$\delta(\zeta, T) = \delta(\zeta, T_M)\, \delta(\zeta, T_N).$$

Since the restriction of $I + \zeta_0 T$ to $N_\infty(I + \zeta_0 T)$ is nilpotent, we may choose a basis $(x_1, \ldots, x_n)$ according to 3.1.15. Next, taking the canonical extension of the corresponding coordinate functionals, we find a representation

$$P = \sum_{i=1}^{n} a_i \otimes x_i$$

with $a_1, \ldots, a_n \in E'$. By construction,

$$\langle x_i + \zeta_0 T x_i, a_j \rangle = 0 \quad \text{if} \quad i \geqq j.$$

Hence

$$-\zeta_0 \langle T x_i, a_j \rangle = \langle x_i, a_j \rangle = \delta_{ij} \quad \text{if} \quad i \geqq j.$$

This implies that the matrix $(\langle T x_i, a_j \rangle)$ has triangular form. Therefore

$$\delta(I + \zeta T_N) = \det(\delta_{ij} + \zeta \langle T x_i, a_j \rangle) = \prod_{i=1}^{n} (1 + \zeta \langle T x_i, a_i \rangle) = (1 - \zeta/\zeta_0)^n.$$

Consequently, we obtain

$$\delta(\zeta, T) = (1 - \zeta/\zeta_0)^n\, \delta(\zeta, T_M).$$

Since $I + \zeta_0 T_M$ is invertible, it follows that $\delta(\zeta_0, T_M) \neq 0$. Thus ζ_0 is indeed a zero of $\delta(\zeta, T)$ with multiplicity $n = n(I + \zeta_0 T)$.

4.4.5. The entire function

$$\delta(\zeta, T) := \delta(I + \zeta T) \quad \text{for all } \zeta \in \mathbb{C}$$

is said to be the **Fredholm denominator** of $T \in \mathfrak{A}(E)$ associated with the continuous determinant δ.

In what follows, viewing T as a fixed parameter, the derivative of $\delta(\zeta, T)$ with respect to the complex variable ζ is denoted by $\delta^{\cdot}(\zeta, T)$.

4.4.6. The next result shows that every continuous determinant δ can be reconstructed from the continuous trace $\delta^{\cdot}$ defined in 4.3.15. Hence the correspondence $\delta \to \delta^{\cdot}$ is one-to-one.

Proposition. Let δ be a continuous determinant on a quasi-Banach operator ideal $\mathfrak{A}$. If $T \in \mathfrak{A}(E)$, then the associated Fredholm denominator $\delta(\zeta, T)$ satisfies the differential equation

$$\frac{\delta^{\cdot}(\zeta, T)}{\delta(\zeta, T)} = \delta^{\cdot}[F(\zeta, T)] \quad \text{for all } \zeta \in \varrho(T)$$

under the initial condition $\delta(0, T) = 1$.

Proof. Apply 4.3.19 to the $\mathfrak{A}(E)$-valued function $T(\zeta) := \zeta T$.

4.4.7. The preceding result suggests the hypothesis that Fredholm denominators can also be obtained from continuous traces. This is indeed so.

Theorem (J. J. Grobler / H. Raubenheimer / P. van Eldik 1982). Let τ be a continuous trace on a quasi-Banach operator ideal $\mathfrak{A}$. If $T \in \mathfrak{A}(E,)$ then

$$\tau(\zeta, T) := \tau[F(\zeta, T)] \quad \text{for all} \quad \zeta \in \varrho(T)$$

defines a meromorphic function with simple poles at all points $\zeta_0 \notin \varrho(T)$ and at these points only. Furthermore, the residue of the pole ζ_0 is equal to the multiplicity of the eigenvalue $\lambda_0 = -1/\zeta_0$.

Proof. First of all, we know from 4.2.13 that T is a Riesz operator. Secondly, by 4.1.3, the function $\tau(\zeta, T)$ is holomorphic on $\varrho(T)$.

According to 4.1.6, about every point $\zeta_0 \notin \varrho(T)$ we have a Laurent expansion

$$F(\zeta, T) = \sum_{n=1}^{d} \frac{X_n(\zeta_0, T)}{(\zeta - \zeta_0)^n} + \sum_{n=0}^{\infty} R_n(\zeta_0, T)(\zeta - \zeta_0)^n \quad \text{if} \quad 0 < |\zeta - \zeta_0| < \varepsilon.$$

Note that all coefficients belong to $\mathfrak{A}(E)$. Since the operators $X_2(\zeta_0, T), \ldots, X_d(\zeta_0, T)$ are finite and nilpotent, it follows from 4.2.15 and 4.2.28 that their traces vanish. Moreover,

$$\tau[X_1(\zeta_0, T)] = \tau[P_\infty(I + \zeta_0 T)] = n(I + \zeta_0 T).$$

Combining these observations, we obtain

$$\tau(\zeta, T) = \frac{n(I + \zeta_0 T)}{\zeta - \zeta_0} + \sum_{n=0}^{\infty} \tau[R_n(\zeta_0, T)](\zeta - \zeta_0)^n \quad \text{if} \quad 0 < |\zeta - \zeta_0| < \varepsilon.$$

Thus ζ_0 is indeed a simple pole of $\tau(\zeta, T)$ with residue $n(I + \zeta_0 T)$.

4.4.8. We now recall some well-known facts from complex analysis.

Let δ be an entire function with $\delta(0) = 1$. If ζ_0 is any zero of multiplicity n, then we have

$$\delta(\zeta) = (\zeta - \zeta_0)^n \, \delta_0(\zeta),$$

where δ_0 is an entire function such that $\delta_0(\zeta_0) \neq 0$. By logarithmic differentation we obtain

$$\tau(\zeta) := \frac{\delta^{\cdot}(\zeta)}{\delta(\zeta)} = \frac{n}{\zeta - \zeta_0} + \frac{\delta_0^{\cdot}(\zeta)}{\delta_0(\zeta)}.$$

This shows that τ is a meromorphic function with simple poles. Furthermore, all residues are natural numbers.

Conversely, given any meromorphic function τ with these properties, we may solve the differential equation

$$\frac{\delta^{\cdot}(\zeta)}{\delta(\zeta)} = \tau(\zeta)$$

under the initial condition $\delta(0) = 1$ provided that τ is holomorphic at the origin. Then

$$\delta(\zeta) = \exp\left[\int_0^\zeta \tau(\gamma)\, d\gamma\right],$$

the integral being taken along any path from 0 to ζ avoiding the poles of τ. Of course, the value of the integral depends on the path chosen, but when we take the exponential all ambiguity disappears. If ζ_0 is a pole of τ, then we write

$$\tau(\zeta) = \frac{n}{\zeta - \zeta_0} + \tau_0(\zeta),$$

where τ_0 is holomorphic in a neighbourhood of ζ_0. We now see from

$$\delta(\zeta) = \left(\frac{\zeta - \zeta_0}{0 - \zeta_0}\right)^n \exp\left[\int_0^\zeta \tau_0(\gamma)\, d\gamma\right]$$

that the solution δ has indeed a zero of multiplicity n at ζ_0.

4.4.9. The entire function obtained by solving the differential equation

$$\frac{\delta^{\cdot}(\zeta, T)}{\delta(\zeta, T)} = \tau[F(\zeta, T)] \quad \text{for all} \quad \zeta \in \varrho(T)$$

under the initial condition $\delta(0, T) = 1$ is said to be the **Fredholm denominator** of $T \in \mathfrak{A}(E)$ associated with the continuous trace τ.

4.4.10. Next we put the preceding construction into an explicit form.

Theorem (J. Plemelj 1904, J. J. Grobler / H. Raubenheimer / P. van Eldik 1982). Let τ be a continuous trace on a quasi-Banach operator ideal $\mathfrak{A}$. If $T \in \mathfrak{A}(E)$, then the coefficients of the associated Fredholm denominator

$$\delta(\zeta, T) = \sum_{n=0}^{\infty} \alpha_n(T)\, \zeta^n \quad \text{for all} \quad \zeta \in \mathbb{C}$$

and of the associated Fredholm numerator

$$D(\zeta, T) = \sum_{n=0}^{\infty} A_n(T)\, \zeta^n \quad \text{for all} \quad \zeta \in \mathbb{C}$$

are determined in the following way:

For $n = 0$ we have

(0) $\alpha_0(T) = 1$ and $A_0(T) = T$.

For $n = 1, 2, \ldots$ we may use either the **recurrence formulas**

$$(R_0) \qquad \alpha_n(T) = \frac{1}{n}\tau[A_{n-1}(T)],$$

$$(R_1) \qquad A_n(T) = \sum_{h=0}^{n}(-1)^{n-h}\alpha_h(T)\,T^{n-h+1}$$

or **Plemelj's formulas**

$$(P_0) \qquad \alpha_n(T) := \frac{1}{n!}\det\begin{pmatrix} \tau(T) & 1 & & & 0 \\ \tau(T^2) & \tau(T) & 2 & & \\ \vdots & \vdots & & \ddots & \\ \tau(T^{n-1}) & \tau(T^{n-2}) & \ldots & \tau(T) & n-1 \\ \tau(T^n) & \tau(T^{n-1}) & \ldots & \tau(T^2) & \tau(T) \end{pmatrix},$$

$$(P_1) \qquad A_n(T) := \frac{1}{n!}\det\begin{pmatrix} \tau(T) & 1 & & & 0 \\ \tau(T^2) & \tau(T) & 2 & & \\ \vdots & \vdots & & \ddots & \\ \tau(T^n) & \tau(T^{n-1}) & \ldots & \tau(T) & n \\ T^{n+1} & T^n & \ldots & T^2 & T \end{pmatrix}.$$

Proof. Formula (R_1) is verified in 4.4.3. Moreover, it follows from $\delta(0, T) = 1$ that $\alpha_0(T) = 1$. Hence $A_0(T) = T$.

Observe that

$$\delta^{\cdot}(\zeta, T) = \delta(\zeta, T)\,\tau[F(\zeta, T)] = \tau[D(\zeta, T)].$$

Therefore

$$\sum_{n=1}^{\infty} n\alpha_n(T)\,\zeta^{n-1} = \sum_{n=0}^{\infty} \tau[A_n(T)]\,\zeta^n.$$

Equating coefficients, we obtain (R_0).

Suppose that $\alpha_n(T)$ and $A_n(T)$ are defined by the formulas (P_0) and (P_1), respectively. Expanding the determinants by the last rows yields

$$\alpha_n(T) = \frac{1}{n}\sum_{h=0}^{n-1}(-1)^{n-h-1}\,\alpha_h(T)\,\tau(T^{n-h})$$

and

$$A_n(T) = \sum_{h=0}^{n}(-1)^{n-h}\,\alpha_h(T)\,T^{n-h+1}.$$

Using the latter formula for $n - 1$ instead of n, we obtain

$$\tau[A_{n-1}(T)] = \sum_{h=0}^{n-1}(-1)^{n-h-1}\,\alpha_h(T)\,\tau(T^{n-h}).$$

Hence

$$\alpha_n(T) = \frac{1}{n}\tau[A_{n-1}(T)].$$

This proves that $\alpha_n(T)$ and $A_n(T)$ satisfy the recurrence formulas (R_0) and (R_1) which determine the coefficients of $\delta(\zeta, T)$ and $D(\zeta, T)$ uniquely.

Remark. For another proof of Plemelj's formula (P_0) we refer to 4.5.11.

4.4.11. Proposition (J. C. Engelbrecht / J. J. Grobler 1983). Let τ be the spectral trace defined on a quasi-Banach operator ideal $\mathfrak{A}$. If $T \in \mathfrak{A}(E)$, then the associated Fredholm denominator is given by the formula

$$\delta(\zeta, T) = \prod_{n=1}^{\infty} (1 + \zeta\lambda_n(T)) \quad \text{for all} \quad \zeta \in \mathbb{C}.$$

Proof. Let $|\zeta| \, \|T\| < 1$. Then it follows from

$$F(\zeta, T) = \sum_{k=1}^{\infty} (-1)^{k-1} \zeta^{k-1} T^k$$

that

$$\tau(\zeta, T) = \sum_{k=1}^{\infty} (-1)^{k-1} \zeta^{k-1} \tau(T^k).$$

Hence, in view of the spectral mapping theorem 3.2.24, we have

$$\begin{aligned}\tau(\zeta, T) &= \sum_{k=1}^{\infty} (-1)^{k-1} \zeta^{k-1} \sum_{n=1}^{\infty} \lambda_n(T^k) = \sum_{n=1}^{\infty} \sum_{k=1}^{\infty} (-1)^{k-1} \zeta^{k-1} \lambda_n(T)^k \\ &= \sum_{n=1}^{\infty} \lambda_n(T) (1 + \zeta\lambda_n(T))^{-1}.\end{aligned}$$

This implies that

$$\delta(\zeta, T) = \prod_{n=1}^{\infty} (1 + \zeta\lambda_n(T))$$

is the unique solution of the differential equation

$$\frac{\delta^{\cdot}(\zeta, T)}{\delta(\zeta, T)} = \tau(\zeta, T)$$

under the initial condition $\delta(0, T) = 1$.

4.5. Regularized Fredholm denominators

Throughout this section, when writing $T^m \in \mathfrak{A}(E)$, we tacitly assume that $T \in \mathfrak{L}(E)$.

4.5.1. The entire function ε_m is defined by

$$1 + \varepsilon_m(\zeta) := (1 + \zeta) \exp\left(\sum_{k=1}^{m-1} \frac{(-1)^k}{k} \zeta^k\right)$$

for $m = 2, 3, \ldots$ Furthermore, let $\varepsilon_1(\zeta) := \zeta$.

Lemma. There exists an entire function φ_m such that

$$\varepsilon_m(\zeta) = \zeta^m \varphi_m(\zeta) \quad \text{for all} \quad \zeta \in \mathbb{C}.$$

Proof. Assume that $|\zeta| < 1$. Then it follows from

$$\log(1 + \zeta) = -\sum_{k=1}^{\infty} \frac{(-1)^k}{k} \zeta^k$$

that

$$1 + \varepsilon_m(\zeta) = \exp\left(-\sum_{k=m}^{\infty} \frac{(-1)^k}{k}\zeta^k\right) = 1 - \frac{(-1)^m}{m}\zeta^m - \dots$$

This expansion shows that the entire function ε_m has a zero of multiplicity m at $\zeta = 0$. Therefore the holomorphic function

$$\varphi_m(\zeta) := \zeta^{-m}\varepsilon_m(\zeta)$$

defined for $\zeta \neq 0$ possesses a removable singularity at the origin.

4.5.2. We now define $\varepsilon_m(T)$ by substituting $T \in \mathfrak{L}(E)$ in place of $\zeta \in \mathbb{C}$ in the Taylor expansion

$$\varepsilon_m(\zeta) = \frac{(-1)^{m-1}}{m}\zeta^m + \dots$$

It turns out that $\varepsilon_m(T)$ behaves like T^m.

Lemma. Let $\mathfrak{A}$ be any operator ideal. Then

$$T^m \in \mathfrak{A}(E) \quad \text{implies} \quad \varepsilon_m(T) \in \mathfrak{A}(E).$$

Proof. Note that $\varepsilon_m(T) = T^m\varphi_m(T)$.

4.5.3. We are now in a position to carry out the basic construction of this section; see 4.4.4.

Theorem. Let δ be a continuous determinant on a quasi-Banach operator ideal $\mathfrak{A}$. If $T^m \in \mathfrak{A}(E)$, then

$$\delta_m(\zeta, T) := \delta\left[(I + \zeta T)\exp\left(\sum_{k=1}^{m-1} \frac{(-1)^k}{k}\zeta^k T^k\right)\right] \quad \text{for all} \quad \zeta \in \mathbb{C}$$

defines an entire function with zeros at all points $\zeta_0 \notin \varrho(T)$ and at these points only. Furthermore, the multiplicity of the zero ζ_0 is equal to the multiplicity of the eigenvalue $\lambda_0 = -1/\zeta_0$.

Proof. First of all, we observe that the definition

$$\delta_m(\zeta, T) := \delta[I + \varepsilon_m(\zeta T)]$$

makes sense, because $\varepsilon_m(\zeta T) \in \mathfrak{A}(E)$. Moreover, the composition $\delta[I + \varepsilon_m(\zeta T)]$ is holomorphic, by 4.3.19. Next we deduce from 4.3.9 and 3.2.24 that T is a Riesz operator.

If $\zeta_0 \in \varrho(T)$, then $I + \zeta_0 T$ is invertible, and so is

$$(I + \zeta_0 T)\exp\left(\sum_{k=1}^{m-1} \frac{(-1)^k}{k}\zeta_0^k T^k\right).$$

Hence, by 4.3.11, we have $\delta_m(\zeta_0, T) \neq 0$.

If $\zeta_0 \notin \varrho(T)$, then we consider the Riesz decomposition

$$E = M_\infty(I + \zeta_0 T) \oplus N_\infty(I + \zeta_0 T)$$

as described in 3.2.14. Let $P := P_\infty(I + \zeta_0 T)$, and define

$$T_M := T(I - P) \quad \text{and} \quad T_N := TP.$$

Using the Taylor expansion of ε_m, it easily follows from $T = T_M + T_N$ and $T_M T_N = T_N T_M = O$ that

$$\varepsilon_m(\zeta T) = \varepsilon_m(\zeta T_M) + \varepsilon_m(\zeta T_N) \quad \text{and} \quad \varepsilon_m(\zeta T_M)\,\varepsilon_m(\zeta T_N) = O.$$

Therefore

$$I + \varepsilon_m(\zeta T) = [I + \varepsilon_m(\zeta T_M)]\,[I + \varepsilon_m(\zeta T_N)].$$

This implies that

$$\delta_m(\zeta, T) = \delta_m(\zeta, T_M)\,\delta_m(\zeta, T_N).$$

As shown in the proof of 4.4.4, we have

$$\delta(I + \zeta T_N) = (1 - \zeta/\zeta_0)^n.$$

Hence

$$\delta_m(\zeta, T) = (1 - \zeta/\zeta_0)^n\,\delta\left[\exp\left(\sum_{k=1}^{m-1} \frac{(-1)^k}{k}\zeta^k T_N^k\right)\right]\delta_m(\zeta, T_M).$$

Since the operators

$$\exp\left(\sum_{k=1}^{m-1} \frac{(-1)^k}{k}\zeta_0^k T_N^k\right) \quad \text{and} \quad I + \zeta_0 T_M$$

are invertible, we obtain

$$\delta\left[\exp\left(\sum_{k=1}^{m-1} \frac{(-1)^k}{k}\zeta_0^k T_N^k\right)\right] \neq 0 \quad \text{and} \quad \delta_m(\zeta_0, T_M) \neq 0.$$

Thus ζ_0 is indeed a zero of $\delta_m(\zeta, T)$ with multiplicity $n(I + \zeta_0 T)$.

4.5.4. Let $m = 1, 2, \ldots$ If $T^m \in \mathfrak{A}(E)$, then the entire function

$$\delta_m(\zeta, T) := \delta\left[(I + \zeta T)\exp\left(\sum_{k=1}^{m-1} \frac{(-1)^k}{k}\zeta^k T^k\right)\right] \quad \text{for all} \quad \zeta \in \mathbb{C}$$

is said to be the **regularized Fredholm denominator** of order m associated with the continuous determinant δ.

4.5.5. Proposition. Let δ be the spectral determinant defined on a quasi-Banach operator ideal $\mathfrak{A}$. If $T^m \in \mathfrak{A}(E)$, then the associated regularized Fredholm denominator of order m is given by the formula

$$\delta_m(\zeta, T) = \prod_{n=1}^{\infty}\left[(1 + \zeta\lambda_n(T))\exp\left(\sum_{k=1}^{m-1} \frac{(-1)^k}{k}\zeta^k \lambda_n(T)^k\right)\right] \quad \text{for all} \quad \zeta \in \mathbb{C}.$$

Proof. Extending the spectral mapping theorem 3.2.24, we can show that the operator $\varepsilon_m(\zeta T)$ has the eigenvalues $\varepsilon_m(\zeta\lambda_n(T))$. This proves that

$$\delta_m(\zeta, T) = \delta[I + \varepsilon_m(\zeta T)] = \prod_{n=1}^{\infty}[1 + \varepsilon_m(\zeta\lambda_n(T))].$$

Remark. Another proof of this result follows from 4.5.12, 4.5.8 and 4.3.23.

4.5.6. Let $T \in \mathfrak{L}(E)$ and $m = 1, 2, \ldots$ Then

$$F_m(\zeta, T) := (-1)^{m-1}\zeta^{m-1} T^m (I + \zeta T)^{-1} \quad \text{for all} \quad \zeta \in \varrho(T)$$

is called the **regularized Fredholm resolvent** of order m. Note that $F_m(\zeta, T)$ and T commute.

4.5.7. The following formula shows that the functions $F(\zeta, T)$ and $F_m(\zeta, T)$ differ from each other only by a polynomial.

Lemma. Let $T \in \mathfrak{L}(E)$. Then

$$F_m(\zeta, T) = F(\zeta, T) + \sum_{k=1}^{m-1} (-1)^k \zeta^{k-1} T^k \quad \text{for all} \quad \zeta \in \varrho(T).$$

Proof. In view of 4.1.2, the above formula follows from

$$(I + \zeta T)\left[F_m(\zeta, T) + \sum_{k=1}^{m-1} (-\zeta)^{k-1} T^k\right] =$$

$$= \left[(-\zeta T)^{m-1} + (I + \zeta T) \sum_{k=1}^{m-1} (-\zeta T)^{k-1}\right] T = T.$$

4.5.8. The following result extends 4.4.6 to the context of regularized Fredholm denominators.

Proposition. Let δ be a continuous determinant on a quasi-Banach operator ideal $\mathfrak{A}$. If $T^m \in \mathfrak{A}(E)$, then the associated regularized Fredholm denominator $\delta_m(\zeta, T)$ satisfies the differential equation

$$\frac{\delta_m^{\cdot}(\zeta, T)}{\delta_m(\zeta, T)} = \delta^{\cdot}\,[F_m(\zeta, T)] \quad \text{for all} \quad \zeta \in \varrho(T)$$

under the initial condition $\delta_m(0, T) = 1$.

Proof. Apply 4.3.19 to the $\mathfrak{A}(E)$-valued function $T(\zeta) := \varepsilon_m(\zeta T)$ and use the fact that

$$\varepsilon_m^{\cdot}(\zeta T)\,(I + \varepsilon_m(\zeta T))^{-1} = F_m(\zeta, T) \quad \text{for all} \quad \zeta \in \varrho(T).$$

4.5.9. We now generalize the construction carried out in 4.4.7.

Theorem (J. J. Grobler / H. Raubenheimer / P. van Eldik 1982). Let τ be a continuous trace on a quasi-Banach operator ideal $\mathfrak{A}$. If $T^m \in \mathfrak{A}(E)$, then

$$\tau_m(\zeta, T) := \tau[F_m(\zeta, T)] \quad \text{for all} \quad \zeta \in \varrho(T)$$

defines a meromorphic function with simple poles at all points $\zeta_0 \notin \varrho(T)$ and at these points only. Furthermore, the residue of the pole ζ_0 is equal to the multiplicity of the eigenvalue $\lambda_0 = -1/\zeta_0$.

Proof. By Lemma 4.5.7, the meromorphic functions $F(\zeta, T)$ and $F_m(\zeta, T)$ have the same poles with the same principal parts. However, $F_m(\zeta, T)$ has the advantage that all of its values belong to $\mathfrak{A}(E)$. Therefore $\tau(F_m(\zeta, T))$ makes sense, and the proof can be completed by reasoning as in 4.4.7.

4.5.10. Let $m = 1, 2, \ldots$ If $T^m \in \mathfrak{A}(E)$, then the entire function obtained by solving the differential equation

$$\frac{\delta_m^{\cdot}(\zeta, T)}{\delta_m(\zeta, T)} = \tau[F_m(\zeta, T)] \quad \text{for all} \quad \zeta \in \varrho(T)$$

under the initial condition $\delta_m(0, T) = 1$ is said to be the **regularized Fredholm denominator** of order m associated with the continuous trace τ.

4.5.11. The next theorem is an analogue of 4.4.10.

Theorem (J. J. Grobler / H. Raubenheimer / P. van Eldik 1982). Let τ be a continuous trace on a quasi-Banach operator ideal $\mathfrak{A}$. If $T^m \in \mathfrak{A}(E)$, then the coefficients of the associated regularized Fredholm denominator

$$\delta_m(\zeta, T) = \sum_{n=0}^{\infty} \alpha_n^{(m)}(T)\,\zeta^n \quad \text{for all} \quad \zeta \in \mathbb{C}$$

are given by Plemelj's formula

$$(\mathrm{P}_0) \qquad \alpha_n^{(m)}(T) := \frac{1}{n!}\det\begin{pmatrix} \tau(T) & 1 & & & 0 \\ \tau(T^2) & \tau(T) & 2 & & \\ \vdots & \vdots & & \ddots & \\ \tau(T^{n-1}) & \tau(T^{n-2}) & \dots & \tau(T) & n-1 \\ \tau(T^n) & \tau(T^{n-1}) & \dots & \tau(T^2) & \tau(T) \end{pmatrix}$$

but with $\tau(T), \dots, \tau(T^{m-1})$ replaced by zeros. Note that $\alpha_0^{(m)}(T) = 1$.

Proof. Let $|\zeta|\,\|T\| < 1$. Then it follows from

$$F_m(\zeta, T) = \sum_{k=m}^{\infty} (-1)^{k-1}\zeta^{k-1}T^k$$

that

$$\tau_m(\zeta, T) = \sum_{k=m}^{\infty} (-1)^{k-1}\,\zeta^{k-1}\tau(T^k).$$

Above we made the convention that $\tau(T) = \dots = \tau(T^{m-1}) = 0$. Thus we may write

$$\tau_m(\zeta, T) = \sum_{k=1}^{\infty} (-1)^{k-1}\,\zeta^{k-1}\tau(T^k).$$

Hence, in view of the differential equation

$$\frac{\dot{\delta}_m(\zeta, T)}{\delta_m(\zeta, T)} = \tau_m(\zeta, T),$$

we have

$$\sum_{n=1}^{\infty} n\alpha_n^{(m)}(T)\,\zeta^{n-1} = \left[\sum_{h=0}^{\infty} \alpha_h^{(m)}(T)\,\zeta^h\right]\left[\sum_{k=1}^{\infty} (-1)^{k-1}\zeta^{k-1}\tau(T^k)\right].$$

Equating coefficients now yields

$$n\alpha_n^{(m)}(T) = \sum_{h=0}^{n-1} (-1)^{n-h-1}\,\alpha_h^{(m)}(T)\,\tau(T^{n-h}).$$

This implies that $\alpha_1^{(m)}(T), \dots, \alpha_n^{(m)}(T)$ can be obtained by solving the linear system

$$\begin{array}{llll} 1\cdot\alpha_1^{(m)}(T) & & & = \tau(T) \\ \tau(T)\cdot\alpha_1^{(m)}(T) - & 2\cdot\alpha_2^{(m)}(T) & & = \tau(T^2) \\ \vdots & \vdots & \ddots & \vdots \\ \tau(T^{n-1})\cdot\alpha_1^{(m)}(T) - & \tau(T^{n-2})\cdot\alpha_2^{(m)}(T) + & \dots + (-1)^{n-1}\,n\cdot\alpha_n^{(m)}(T) & = \tau(T^n). \end{array}$$

Therefore, applying Cramer's rule, we see that $\alpha_n^{(m)}(T)$ is given by Plemelj's formula (P_0). In particular, we have

$$\alpha_n^{(m)}(T) = 0 \quad \text{for} \quad n = 1, \ldots, m-1.$$

Remark. Replacing $\tau(T), \ldots, \tau(T^{m-1})$ and $T, \ldots, T^{m-1}$ by zeros, we may define the coefficients $A_n^{(m)}(T)$ via Plemelj's formula (P_1); see 4.4.10. Then

$$D_m(\zeta, T) := \sum_{n=0}^{\infty} A_n^{(m)}(T)\,\zeta^n \quad \text{for all } \zeta \in \mathbb{C}$$

is an entire $\mathfrak{A}(E)$-valued function, and we have

$$F_m(\zeta, T) = \frac{D_m(\zeta, T)}{\delta_m(\zeta, T)} \quad \text{for all} \quad \zeta \in \varrho(T).$$

4.5.12. Proposition (J. C. Engelbrecht / J. J. Grobler 1983). Let τ be the spectral trace defined on a quasi-Banach operator ideal $\mathfrak{A}$. If $T^m \in \mathfrak{A}(E)$, then the associated regularized Fredholm denominator of order m is given by the formula

$$\delta_m(\zeta, T) = \prod_{n=1}^{\infty} \left[(1 + \zeta\lambda_n(T)) \exp\left(\sum_{k=1}^{m-1} \frac{(-1)^k}{k} \zeta^k \lambda_n(T)^k\right)\right] \quad \text{for all} \quad \zeta \in \mathbb{C}.$$

Proof. Let $|\zeta|\,\|T\| < 1$. Then it follows from

$$F_m(\zeta, T) = \sum_{k=m}^{\infty} (-1)^{k-1} \zeta^{k-1} T^k$$

that

$$\tau_m(\zeta, T) = \sum_{k=m}^{\infty} (-1)^{k-1} \zeta^{k-1} \tau(T^k).$$

Hence, in view of the spectral mapping theorem 3.2.24, we have

$$\tau_m(\zeta, T) = \sum_{k=m}^{\infty} (-1)^{k-1} \zeta^{k-1} \sum_{n=1}^{\infty} \lambda_n(T^k) = \sum_{n=1}^{\infty} \sum_{k=m}^{\infty} (-1)^{k-1} \zeta^{k-1} \lambda_n(T)^k$$
$$= \sum_{n=1}^{\infty} (-1)^{m-1} \zeta^{m-1} \lambda_n(T)^m (1 + \zeta\lambda_n(T))^{-1}.$$

This implies that

$$\delta_m(\zeta, T) = \prod_{n=1}^{\infty} \left[(1 + \zeta\lambda_n(T)) \exp\left(\sum_{k=1}^{m-1} \frac{(-1)^k}{k} \zeta^k \lambda_n(T)^k\right)\right]$$

is the unique solution of the differential equation

$$\frac{\delta_m^{\cdot}(\zeta, T)}{\delta_m(\zeta, T)} = \tau_m(\zeta, T)$$

under the initial condition $\delta_m(0, T) = 1$.

4.5.13. From the preceding proposition we obtain an interesting corollary.

Theorem. Let τ be a continuous trace on an approximative quasi-Banach operator ideal $\mathfrak{A}$. If $T \in \mathfrak{A}(E)$, then

$$\delta_2(\zeta, T) = \prod_{n=1}^{\infty} [(1 + \zeta\lambda_n(T)) \exp(-\zeta\lambda_n(T))] \quad \text{for all} \quad \zeta \in \mathbb{C}.$$

Proof. As shown in 4.2.36, the restriction of τ to $\mathfrak{A}^2$ is spectral.

Remark. In view of 4.2.37, the approximability assumption can be dropped when we consider $\delta_m(\zeta, T)$ with $m \geqq 3$.

4.5.14. We now compare regularized Fredholm denominators of different orders with each other.

Proposition. Let τ be a continuous trace on a quasi-Banach operator ideal $\mathfrak{A}$. If $T^m \in \mathfrak{A}(E)$ and $m \leqq p \leqq q$, then

$$\delta_q(\zeta, T) = \delta_p(\zeta, T) \exp\left[\sum_{k=p}^{q-1} \frac{(-1)^k}{k} \zeta^k \tau(T^k)\right] \quad \text{for all} \quad \zeta \in \mathbb{C}.$$

Proof. Since $\delta_p(\zeta, T)$ and $\delta_q(\zeta, T)$ have the same zeros, there exists an entire function $\varepsilon(\zeta, T)$ such that

$$\delta_q(\zeta, T) = \delta_p(\zeta, T) \exp(\varepsilon(\zeta, T)).$$

In view of $\delta_p(0, T) = 1$ and $\delta_q(0, T) = 1$, we may assume that $\varepsilon(0, T) = 0$. By 4.5.10,

$$\frac{\delta_p^{\cdot}(\zeta, T)}{\delta_p(\zeta, T)} = \tau[F_p(\zeta, T)] \quad \text{and} \quad \frac{\delta_q^{\cdot}(\zeta, T)}{\delta_q(\zeta, T)} = \tau[F_q(\zeta, T)].$$

Furthermore, 4.5.7 implies that

$$F_q(\zeta, T) - F_p(\zeta, T) = \sum_{k=p}^{q-1} (-1)^k \zeta^{k-1} T^k.$$

Hence

$$\varepsilon^{\cdot}(\zeta, T) = \sum_{k=p}^{q-1} (-1)^k \zeta^{k-1} \tau(T^k),$$

and we obtain

$$\varepsilon(\zeta, T) = \sum_{k=p}^{q-1} \frac{(-1)^k}{k} \zeta^k \tau(T^k).$$

4.5.15. According to the previous results, regularized Fredholm denominators of order m are defined for all operators belonging to the following quasi-Banach operator ideals:

$\mathfrak{L}_r^{(a)}, \mathfrak{L}_r^{(c)}, \mathfrak{L}_r^{(d)}$	if $m \geqq r$,
$\mathfrak{L}_r^{(x)}, \mathfrak{L}_r^{(y)}$	if $m > r$ and m even,
$\mathfrak{P}_r, \mathfrak{P}_{r,2}$	if $m > r > 2$ and m even.

In the case when m is an even number, then $(\mathfrak{P}_m)^m$ admits a spectral trace. Consequently, the regularized Fredholm denominator $\delta_m(\zeta, T)$ exists for all operators $T \in \mathfrak{P}_m(E)$. We do not know whether this is so for m odd.

4.6. The relationship between traces and determinants

4.6.1. We know from Theorem 4.3.15 that every continuous determinant δ defined on a quasi-Banach operator ideal $\mathfrak{A}$ determines a continuous trace

$$\delta^{\cdot}(T) := \lim_{\zeta \to 0} \frac{\delta(I + \zeta T) - 1}{\zeta} \quad \text{for all} \quad T \in \mathfrak{A}(E).$$

Furthermore, by 4.4.6, the correspondence $\delta \to \delta^{\cdot}$ is one-to-one. Thus the question arises whether every continuous trace can be obtained in this way. In the following we give an affirmative answer.

Proposition. For every continuous trace τ defined on a quasi-Banach operator ideal $\mathfrak{A}$ there exists a continuous determinant δ such that $\tau = \delta^{\cdot}$.

Proof. Let $T \in \mathfrak{A}(E)$, and recall from 4.4.7 that $\tau[T(I + \zeta T)^{-1}]$ is a meromorphic function with simple poles, the residues being natural numbers. If $I + T$ is invertible, then we define

$$\delta(I + T) := \exp\left(\int_0^1 \tau[T(I + \gamma T)^{-1}]\mathrm{d}\gamma\right),$$

where the integral is taken along any path Γ from 0 to 1 avoiding the poles of the Fredholm resolvent. Of course, the value of the integral depends on the path chosen, but when we take the exponential all ambiguity disappears.

Next we show that the Gâteaux derivative

$$\delta^{\cdot}(I + T, S) := \lim_{\zeta \to 0} \frac{\delta(I + T + \zeta S) - \delta(I + T)}{\zeta}$$

exists for all $S \in \mathfrak{A}(E)$. To this end, let

$$\varrho := \sup\{\|\gamma(I + \gamma T)^{-1}\| : \gamma \in \Gamma\} > 0.$$

If $|\zeta|\,\varrho\|S\| < 1$, then

$$\delta(I + T + \zeta S) = \exp\left(\int_0^1 \tau[(T + \zeta S)\,(I + \gamma T + \zeta\gamma S)^{-1}]\,\mathrm{d}\gamma\right).$$

Easy manipulations yield the Taylor expansion

$$(T + \zeta S)\,(I + \gamma T + \zeta\gamma S)^{-1} = T(I + \gamma T)^{-1} + (I + \gamma T)^{-1}\,S(I + \gamma T)^{-1}\,\zeta + \ldots$$

converging uniformely on the disk $|\zeta|\,\varrho\|S\| < 1$. This implies that $\delta(I + T + \zeta S)$ is differentiable at $\zeta = 0$. Furthermore, we obtain

$$\frac{\delta^{\cdot}(I + T, S)}{\delta(I + T)} = \int_0^1 \tau[(I + \gamma T)^{-1}\,S(I + \gamma T)^{-1}]\,\mathrm{d}\gamma.$$

The obvious Taylor expansion

$$\gamma S(I + \gamma T)^{-1} = \gamma_0 S(I + \gamma_0 T)^{-1} + S(I + \gamma_0 T)^{-2}\,(\gamma - \gamma_0) + \ldots$$

shows that

$$\tau[\gamma S(I + \gamma T)^{-1}] \quad \text{is a primitive of } \tau[S(I + \gamma T)^{-2}].$$

Hence

$$\int_0^1 \tau[(I + \gamma T)^{-1}\,S(I + \gamma T)^{-1}]\,\mathrm{d}\gamma = \int_0^1 \tau[S(I + \gamma T)^{-2}]\,\mathrm{d}\gamma$$

$$= (\tau[\gamma S(I + \gamma T)^{-1}])_{\gamma=0}^{\gamma=1} = \tau[S(I + T)^{-1}].$$

Thus we have the **Fredholm formula**

$$\frac{\delta^{\cdot}(I + T, S)}{\delta(I + T)} = \tau[S(I + T)^{-1}].$$

In particular, if $T = O$, then it follows that $\delta^{\cdot}(S) = \tau(S)$.

In the case when $I + T$ is non-invertible we define $\delta(I + T) := 0$.

It remains to prove that the function $T \to \delta(I + T)$ is indeed a continuous determinant.

Write $\delta(\zeta, T) := \delta(I + \zeta T)$. Then Fredholm's formula implies that

$$\frac{\delta^{\cdot}(\zeta, T)}{\delta(\zeta, T)} = \tau[F(\zeta, T)] \quad \text{for all} \quad \zeta \in \varrho(T).$$

Thus $\delta(\zeta, T)$ is the Fredholm denominator associated with the continuous trace τ. This means that $\delta(I + \zeta T)$ is an entire function with zeros located on the complement of $\varrho(T)$. Hence condition (D_4) is satisfied.

In the case $T = a \otimes x$ we have

$$T(I + \gamma T)^{-1} = \frac{a \otimes x}{1 + \gamma\langle x, a\rangle} \quad \text{if} \quad 1 + \gamma\langle x, a\rangle \neq 0.$$

Therefore it follows from

$$\int_0^1 \frac{\tau(a \otimes x)}{1 + \gamma\langle x, a\rangle} d\gamma = \int_0^1 \frac{\langle x, a\rangle}{1 + \gamma\langle x, a\rangle} d\gamma = \log(1 + \langle x, a\rangle)$$

that

(D_1) $\quad \delta(I + a \otimes x) = 1 + \langle x, a\rangle.$

Next let $T \in \mathfrak{A}(E, F)$ and $X \in \mathfrak{L}(F, E)$. Note that $I_E + \gamma XT$ and $I_F + \gamma TX$ are simultaneously invertible or not. Since

$$XT(I_E + \gamma XT)^{-1} = X(I_F + \gamma TX)^{-1}T,$$

we have

$$\tau[XT(I_E + \gamma XT)^{-1}] = \tau[TX(I_F + \gamma TX)^{-1}].$$

Hence

(D_2) $\quad \delta(I_E + XT) = \delta(I_F + TX).$

Assume that $S, T \in \mathfrak{A}(E)$. If $I + T$ is non-invertible, then $N(I + T) \neq \{o\}$. Hence $(I + S)(I + T)$ is non-invertible, as well. Thus the multiplication formula

$$\delta[(I + S)(I + T)] = \delta(I + S)\,\delta(I + T)$$

becomes trivial. Next we treat the case when $I + T$ is invertible. Write

$$\delta(\zeta, S, T) := \delta[(I + \zeta S)(I + T)] = \delta[I + T + \zeta S(I + T)].$$

Applying Fredholm's formula, we obtain

$$\frac{\delta^{\cdot}(\zeta, S, T)}{\delta(\zeta, S, T)} = \tau[S(I + T)((I + \zeta S)(I + T))^{-1}] = \tau[F(\zeta, S)].$$

Hence the entire functions $\delta(\zeta, S, T)$ and $\delta(\zeta, S)$ satisfy one and the same differential equation. This implies that

$$\delta(\zeta, S, T) = c_0 \delta(\zeta, S),$$

where c_0 is an integration constant. Putting $\zeta = 0$, we conclude that

$$c_0 = \delta(0, S, T) = \delta(I + T).$$

Finally, letting $\zeta = 1$ yields

$$(\mathrm{D}_3) \qquad \delta[(I + S)(I + T)] = \delta(I + S)\,\delta(I + T).$$

Since the trace τ is assumed to be continuous, there exists a constant $c \geqq 1$ such that

$$|\tau(S)| \leqq c\|S \mid \mathfrak{A}\| \quad \text{for all } S \in \mathfrak{A}(E).$$

Let $2c\|T \mid \mathfrak{A}\| < 1$. Then

$$\|(I + \gamma T)^{-1}\| \leqq \frac{1}{1 - \|T\|} \leqq 2 \quad \text{for} \quad 0 \leqq \gamma \leqq 1.$$

Hence, integrating along the straight line from 0 to 1, we obtain

$$\left|\int_0^1 \tau[T(I + \gamma T)^{-1}]\,d\gamma\right| \leqq c\|T \mid \mathfrak{A}\| \int_0^1 \|(I + \gamma T)^{-1}\|\,d\gamma \leqq 2c\|T \mid \mathfrak{A}\| \leqq 1.$$

Since

$$|\exp(\zeta) - 1| \leqq |\zeta| \exp(|\zeta|) \quad \text{for all} \quad \zeta \in \mathbb{C}.$$

it follows that

$$|\delta(I + T) - 1| = \left|\exp\left(\int_0^1 \tau[T(I + \gamma T)^{-1}]\,d\gamma\right) - 1\right| \leqq 2ce\|T \mid \mathfrak{A}\|.$$

Thus the determinant δ is continuous, by 4.3.10.

Remark. Given any continuous trace τ, the associated continuous determinant δ could be defined by

$$\delta(I + T) = 1 + \sum_{n=1}^{\infty} \alpha_n(T) \quad \text{for all} \quad T \in \mathfrak{A}(E).$$

where the coefficients $\alpha_1(T), \alpha_2(T), \ldots$ are computed either recurrently or by Plemelj's formula; see 4.4.10. However, we are unaware of any direct proof of the convergence of the right-hand series. To obtain this, it would be necessary to verify that

$$\lim_n |\alpha_n(T)|^{1/n} = 0.$$

4.6.2. Let $T \in \mathfrak{L}(E)$ be a Riesz operator such that $I + T$ is invertible. Then we define

$$\log(I + T) := \int_0^1 T(I + \gamma T)^{-1}\,d\gamma,$$

where the contour integral is taken along any path from 0 to 1 avoiding the poles of the Fredholm resolvent. Determinations of $\log(I + T)$ obtained from different paths differ from each other by summands of the form

$$2\pi i \sum_{j=1}^{k} n_j P_\infty(I + \zeta_j T),$$

where $n_1, \ldots, n_k$ are integers.

In the case $\|T\| < 1$ we get the principal branch of the logarithm by integrating along the straight line from 0 to 1. Then, as in 4.3.16, we have

$$\log (I + T) = - \sum_{k=1}^{\infty} \frac{(-1)^k}{k} T^k.$$

Using the logarithm just introduced, the preceding proposition can be rephrased as follows.

Proposition. Let τ be a continuous trace on a quasi-Banach operator ideal $\mathfrak{A}$. Then

$$\delta(I + T) := \begin{cases} \exp(\tau[\log (I + T)]) & \text{if } I + T \text{ is invertible;} \\ 0 & \text{otherwise} \end{cases}$$

defines a continuous determinant on $\mathfrak{A}$.

Remark. Note that $\tau[\log(I + T)]$ is determined up to a summand $2\pi i n$, where n is an integer. Hence $\exp(\tau[\log(I + T)])$ is well-defined.

4.6.3. We now summarize the results stated in 4.3.15 and 4.6.1.

Trace-determinant theorem. There exists a one-to-one correspondence between continuous traces and continuous determinants on every quasi-Banach operator ideal.

4.6.4. Combining 4.3.23 and 4.4.11, we arrive at the following result.

Theorem. A quasi-Banach operator ideal admits a spectral determinant if and only if it admits a spectral trace.

4.6.5. In view of the preceding theorems, all quasi-Banach operator ideals listed in 4.2.34 admit a continuous determinant or even a spectral determinant.

4.7. Traces and determinants of nuclear operators

4.7.1. We begin with an analogue of Lemma 4.2.2.

Lemma (GRO, Chap. I, p. 165). Let $T \in \mathfrak{N}(E)$. If the Banach space E has the approximation property, then

$$\tau(T) := \sum_{i=1}^{\infty} \langle x_i, a_i \rangle$$

does not depend on the special choice of the nuclear representation

$$T = \sum_{i=1}^{\infty} a_i \otimes x_i,$$

and we have

$$|\tau(T)| \leqq \|T \mid \mathfrak{N}\|.$$

Proof. First of all, we show that

$$\operatorname{trace}(TL) = \sum_{i=1}^{\infty} \langle Lx_i, a_i \rangle \quad \text{for} \quad \text{all} \quad L \in \mathfrak{F}(E).$$

To this end, let

$$L = \sum_{j=1}^{n} b_j \otimes y_j$$

be any finite representation. Then it follows from

$$TL = \sum_{j=1}^{n} b_j \otimes Ty_j$$

that

$$\operatorname{trace}(TL) = \sum_{j=1}^{n} \langle Ty_j, b_j \rangle = \sum_{j=1}^{n} \sum_{i=1}^{\infty} \langle y_j, a_i \rangle \langle x_i, b_j \rangle$$
$$= \sum_{i=1}^{\infty} \sum_{j=1}^{n} \langle x_i, b_j \rangle \langle y_j, a_i \rangle = \sum_{i=1}^{\infty} \langle Lx_i, a_i \rangle.$$

Since E has the approximation property, given any precompact subset K, we can find $L_K \in \mathfrak{F}(E)$ such that

$$\|x - L_K x\| \leqq 1 \quad \text{for all} \quad x \in K.$$

By Lemma 1.7.4, there exists $(\alpha_i) \in c_0$ such that

$$\sum_{i=1}^{\infty} \alpha_i^{-1} \|x_i\| \, \|a_i\| < \infty \quad \text{and} \quad \alpha_i > 0.$$

Obviously,

$$K_0 := \left\{ \frac{\alpha_i}{\|x_i\|} x_i : i = 1, 2, \ldots \right\}$$

is precompact. If $\varepsilon > 0$, then for every precompact subset K with $\varepsilon K \supseteqq K_0$ and $i = 1, 2, \ldots$ it follows that

$$\|x_i - L_K x_i\| \leqq \varepsilon \alpha_i^{-1} \|x_i\|.$$

Consequently,

$$\left| \sum_{i=1}^{\infty} \langle x_i, a_i \rangle - \operatorname{trace}(TL_K) \right| = \left| \sum_{i=1}^{\infty} \langle x_i - L_K x_i, a_i \rangle \right| \leqq \varepsilon \sum_{i=1}^{\infty} \alpha_i^{-1} \|x_i\| \, \|a_i\|.$$

Note that the collection of all precompact subsets is directed upwards with respect to the set-theoretical inclusion. In this way $(\operatorname{trace}(TL_K))$ may be viewed as a generalized sequence. Letting $\varepsilon \to 0$, the preceding estimate implies that

$$\sum_{i=1}^{\infty} \langle x_i, a_i \rangle = \lim_{K} \operatorname{trace}(TL_K).$$

Since the right-hand expression does not depend on the underlying nuclear representation, the same holds for

$$\tau(T) := \sum_{i=1}^{\infty} \langle x_i, a_i \rangle .$$

Finally, we conclude from

$$|\tau(T)| \leqq \sum_{i=1}^{\infty} \|x_i\| \, \|a_i\|$$

that

$$|\tau(T)| \leqq \|T \mid \mathfrak{N}\| .$$

Remark. Conversely, it can be shown that a Banach space E must have the approximation property if $\tau(T)$ is well-defined for all operators $T \in \mathfrak{N}(E)$; see (DIL, pp. 239–240) and (GRO, Chap. I, p. 165).

4.7.2. Theorem. There exists a unique continuous trace on the Banach operator ideal $\mathfrak{N}$ restricted to the class of all Banach spaces with the approximation property.

Proof. It can easily be seen that the function $T \to \tau(T)$ defined in the preceding lemma is indeed a continuous trace. Furthermore, we know from 1.7.5 that the finite operators are dense in $\mathfrak{N}(E)$. This proves the uniqueness.

4.7.3. Next we provide some auxiliary results.

Lemma (A. Grothendieck 1956: a). An operator $T \in \mathfrak{L}(E, l_1)$ is nuclear if and only if the sequence $(T'e_i)$ is absolutely summable. When this is so, then

$$\|T \mid \mathfrak{N}\| = \sum_{i=1}^{\infty} \|T'e_i\| .$$

Proof. If $(T'e_i) \in [l_1, E']$, then T admits the nuclear representation

$$T = \sum_{i=1}^{\infty} T'e_i \otimes e_i ,$$

and we have

$$\|T \mid \mathfrak{N}\| \leqq \sum_{i=1}^{\infty} \|T'e_i\| .$$

Conversely, assume that $T \in \mathfrak{N}(E, l_1)$. Given any nuclear representation

$$T = \sum_{j=1}^{\infty} a_j \otimes y_j ,$$

it follows that

$$T'e_i = \sum_{j=1}^{\infty} a_j \langle y_j, e_i \rangle \quad \text{for} \quad i = 1, 2, \ldots$$

and

$$\|y_j\| = \sum_{i=1}^{\infty} |\langle y_j, e_i \rangle| \quad \text{for} \quad j = 1, 2, \ldots$$

Hence

$$\sum_{i=1}^{\infty} \|T'e_i\| \leq \sum_{i=1}^{\infty} \sum_{j=1}^{\infty} \|a_j\| \, |\langle y_j, e_i\rangle| \leq \sum_{j=1}^{\infty} \|a_j\| \sum_{i=1}^{\infty} |\langle y_j, e_i\rangle| = \sum_{j=1}^{\infty} \|a_j\| \, \|y_j\| .$$

This implies that

$$\sum_{i=1}^{\infty} \|T'e_i\| \leq \|T \mid \mathfrak{N}\| .$$

4.7.4. The ideal of nuclear operators fails to be surjective. It has, however, the following property.

Lemma. Let $T \in \mathfrak{L}(E, l_1)$. If there exists a surjection $Q \in \mathfrak{L}(E_0, E)$ such that $TQ \in \mathfrak{N}(E_0, l_1)$, then $T \in \mathfrak{N}(E, l_1)$.

Proof. By the preceding criterion, $TQ \in \mathfrak{N}(E_0, l_1)$ implies that $(Q'T'e_i) \in [l_1, E_0']$. Since Q' is an injection, we have $(T'e_i) \in [l_1, E']$. Hence $T \in \mathfrak{N}(E, l_1)$.

4.7.5. Since the Banach space l_1 has the approximation property, every operator $T \in \mathfrak{N}(l_1)$ possesses a well-defined trace. However, the following example shows that this trace behaves rather strangely. It was originally constructed by P. Enflo (1973). Simplified approaches are to be found in (LIN, Vol. I, 2.d.3) and (PIE, 10.4.5).

Theorem. There exists an operator $S \in \mathfrak{N}(l_1)$ such that

$$\tau(S) = 1 \quad \text{and} \quad S^2 = O .$$

Remark. The nilpotent operator S cannot possess any eigenvalue $\lambda_0 \neq 0$. Thus it is impossible to compute the trace $\tau(S) = 1$ from the trivial eigenvalue sequence $(0, 0, \ldots)$.

4.7.6. As an easy consequence of the preceding example we now establish the most unpleasant result of this monograph.

Theorem (A. Pietsch 1981: b). There does not exist a continuous trace on the Banach operator ideal $\mathfrak{N}$ considered on the class of all Banach spaces.

Proof. Consider an operator $S \in \mathfrak{N}(l_1)$ such that $\tau(S) = 1$ and $S^2 = O$. Denote by Q the canonical surjection from l_1 onto the quotient space $E := l_1/N(S)$. Furthermore, let $T \in \mathfrak{L}(E, l_1)$ be the operator induced by S. Then $S = TQ$, and we have the diagram

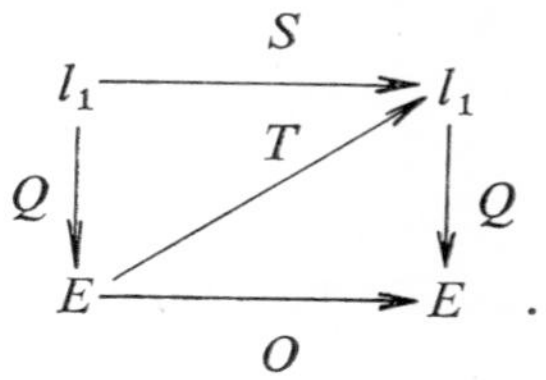

It follows from $S^2 = O$ that $M(S) \subseteq N(S)$. Therefore $QT = O$. Moreover, Lemma 4.7.4 shows that $T \in \mathfrak{N}(E, l_1)$.

We now assume that there exists a continuous trace σ defined on all components of the operator ideal $\mathfrak{N}$. Then

$$1 = \tau(S) = \sigma(S) = \sigma(TQ) = \sigma(QT) = \sigma(O) = 0 .$$

This contradiction completes the proof.

Remark. As a by-product of the preceding construction we see that the quotient space $l_1/N(S)$ cannot possess the approximation property.

4.7.7. Next we prove a counterpart of Lemma 4.7.1 which holds in arbitrary Banach spaces.

Lemma. Let $T \in \mathfrak{N}_{1,2}(E)$. Then

$$\tau(T) := \sum_{i=1}^{\infty} \langle x_i, a_i \rangle$$

does not depend on the special choice of the (1, 2)-nuclear representation,

$$T = \sum_{i=1}^{\infty} a_i \otimes x_i$$

and we have

$$|\tau(T)| \leqq \|T \mid \mathfrak{N}_{1,2}\|.$$

Proof. Let

$$\sum_{i=1}^{\infty} a_i \otimes x_i$$

be any (1, 2)-nuclear representation of the zero operator. Then we define the operators $A_m \in \mathfrak{N}(E, l_2)$ and $X_m \in \mathfrak{L}(l_2, E)$ by

$$A_m := \sum_{i=m+1}^{\infty} a_i \otimes e_i \quad \text{and} \quad X_m := \sum_{i=m+1}^{\infty} e_i \otimes x_i.$$

Note that the finite operators

$$T_m := \sum_{i=1}^{m} a_i \otimes x_i = - \sum_{i=m+1}^{\infty} a_i \otimes x_i$$

admit the factorizations $T_m = -X_m A_m$. Furthermore, by B.4.6, there exist orthogonal projections $P_m \in \mathfrak{F}(l_2)$ such that $X_m A_m = X_m P_m A_m$. We now obtain

$$\left|\sum_{i=1}^{m} \langle x_i, a_i \rangle\right| = |\operatorname{trace}(T_m)| = |\operatorname{trace}(X_m P_m A_m)| = |\operatorname{trace}(P_m A_m X_m)|$$

$$\leqq \|P_m A_m X_m \mid \mathfrak{N}\| \leqq \left(\sum_{i=m+1}^{\infty} \|a_i\|\right) \sup\left\{\left(\sum_{i=1}^{\infty} |\langle x_i, a\rangle|^2\right)^{1/2} : a \in U^0\right\}.$$

Letting $m \to \infty$ yields

$$\sum_{i=1}^{\infty} \langle x_i, a_i \rangle = 0.$$

This proves that $\tau(T)$ is well-defined for all operators $T \in \mathfrak{N}_{1,2}(E)$. The inequality $|\tau(T)| \leqq \|T \mid \mathfrak{N}_{1,2}\|$ is obvious.

4.7.8. The following statement is an immediate consequence of the preceding considerations.

Theorem. There exists a unique continuous trace on the quasi-Banach operator ideal $\mathfrak{N}_{1,2}$.

Remark. We know from 1.7.16 and 2.8.20 that

$$\mathfrak{N}_{1,2} \subseteqq (\mathfrak{P}_2)^2 \quad \text{and} \quad \mathfrak{N}_{1,2} \subseteqq (\mathfrak{P}_2)^{(a)}_{2,1}.$$

Therefore the assertion could also be obtained from 4.2.6 or 4.2.7. In view of 4.2.30 and 4.2.31 the continuous trace defined on $\mathfrak{N}_{1,2}$ is indeed spectral. For a completely different proof of this fact we refer to 4.7.15.

4.7.9. According to 4.6.1, every continuous trace induces a continuous determinant. This result remains true if the underlying quasi-Banach operator ideal is considered on any subclass of Banach spaces. Hence we have the following corollary of 4.7.2.

Theorem (A. F. Ruston 1951: a, A. Grothendieck 1956: a). There exists a unique continuous determinant on the Banach operator ideal $\mathfrak{N}$ restricted to the class of all Banach spaces with the approximation property.

4.7.10. The continuous determinant defined on the restricted Banach operator ideal $\mathfrak{N}$ can be described explicitely. To this end, we use the Taylor coefficients of the associated Fredholm denominator defined in 4.4.10.

Lemma. Let $T \in \mathfrak{N}(E)$. If the Banach space E has the approximation property, then

$$\alpha_n(T) := \frac{1}{n!} \sum_{i_1=1}^{\infty} \dots \sum_{i_n=1}^{\infty} \det \begin{pmatrix} \langle x_{i_1}, a_{i_1}\rangle & \dots & \langle x_{i_1}, a_{i_n}\rangle \\ \vdots & & \vdots \\ \langle x_{i_n}, a_{i_1}\rangle & \dots & \langle x_{i_n}, a_{i_n}\rangle \end{pmatrix}$$

does not depend on the special choice of the nuclear representation

$$T = \sum_{i=1}^{\infty} a_i \otimes x_i,$$

and we have

$$|\alpha_n(T)| \leqq \frac{n^{n/2}}{n!} \|T \mid \mathfrak{N}\|^n.$$

Proof. We consider the finite operators

$$T_m := \sum_{i=1}^{m} a_i \otimes x_i.$$

Then

$$\delta(I + T_m) = \det(\delta_{ij} + \langle x_i, a_j\rangle).$$

According to a well-known formula from multilinear algebra, the right-hand determinant can be written in the form

$$1 + \sum_{n=1}^{m} \alpha_n(T_m),$$

where

$$\alpha_n(T_m) := \frac{1}{n!} \sum_{i_1=1}^{m} \dots \sum_{i_n=1}^{m} \det \begin{pmatrix} \langle x_{i_1}, a_{i_1}\rangle & \dots & \langle x_{i_1}, a_{i_n}\rangle \\ \vdots & & \vdots \\ \langle x_{i_n}, a_{i_1}\rangle & \dots & \langle x_{i_n}, a_{i_n}\rangle \end{pmatrix}.$$

In view of 4.4.10, the coefficients $\alpha_n(T)$ are continuous functions of T. Thus, letting $m \to \infty$ yields the formula we are looking for.

Given $\varepsilon > 0$, we choose a nuclear representation

$$T = \sum_{i=1}^{\infty} a_i \otimes x_i$$

such that

$$\sum_{i=1}^{\infty} \|a_i\| \leqq (1+\varepsilon)\, \|T \mid \mathfrak{N}\| \quad \text{and} \quad \|x_i\| \leqq 1 \quad \text{for} \quad i = 1, 2, \ldots$$

Then Hadamard's inequality A.4.5 implies that

$$\left| \det \begin{pmatrix} \langle x_{i_1}, a_{i_1}\rangle & \ldots & \langle x_{i_1}, a_{i_n}\rangle \\ \vdots & & \vdots \\ \langle x_{i_n}, a_{i_1}\rangle & \ldots & \langle x_{i_n}, a_{i_n}\rangle \end{pmatrix} \right| \leqq n^{n/2} \|a_{i_1}\| \ldots \|a_{i_n}\|.$$

Hence

$$|\alpha_n(T)| \leqq \frac{n^{n/2}}{n!} \sum_{i_1=1}^{\infty} \ldots \sum_{i_n=1}^{\infty} \|a_{i_1}\| \ldots \|a_{i_n}\|$$

$$= \frac{n^{n/2}}{n!} \left(\sum_{i=1}^{\infty} \|a_i\| \right)^n \leqq (1+\varepsilon)^n \frac{n^{n/2}}{n!} \|T \mid \mathfrak{N}\|^n.$$

Letting $\varepsilon \to 0$, we obtain

$$|\alpha_n(T)| \leqq \frac{n^{n/2}}{n!} \|T \mid \mathfrak{N}\|^n.$$

4.7.11. The next result follows immediately from 4.7.8 and 4.6.1.

Theorem. There exists a unique continuous determinant on the quasi-Banach operator ideal $\mathfrak{N}_{1,2}$.

4.7.12. We now prove a counterpart of Lemma 4.7.10 which holds in arbitrary Banach spaces.

Lemma. Let $T \in \mathfrak{N}_{1,2}(E)$. Then

$$\alpha_n(T) := \frac{1}{n!} \sum_{i_1=1}^{\infty} \ldots \sum_{i_n=1}^{\infty} \det \begin{pmatrix} \langle x_{i_1}, a_{i_1}\rangle & \ldots & \langle x_{i_1}, a_{i_n}\rangle \\ \vdots & & \vdots \\ \langle x_{i_n}, a_{i_1}\rangle & \ldots & \langle x_{i_n}, a_{i_n}\rangle \end{pmatrix}$$

does not depend on the special choice of the (1,2)-nuclear representation

$$T = \sum_{i=1}^{\infty} a_i \otimes x_i,$$

and we have

$$|\alpha_n(T)| \leqq \frac{1}{n!} \|T \mid \mathfrak{N}_{1,2}\|^n.$$

Proof. The formula for the coefficients $\alpha_n(T)$ can be obtained in the same way as in 4.7.10.

In order to verify the estimate, we choose a (1,2)-nuclear representation

$$T = \sum_{i=1}^{\infty} a_i \otimes x_i$$

such that

$$\sum_{i=1}^{\infty} \|a_i\| \leqq (1+\varepsilon) \|T \mid \mathfrak{N}_{1,2}\| \quad \text{and} \quad \sum_{i=1}^{\infty} |\langle x_i, a\rangle|^2 \leqq 1 \quad \text{for} \quad a \in U^0.$$

Then Hadamard's inequality A.4.5 implies that

$$\left|\det \begin{pmatrix} \langle x_{i_1}, a_{i_1}\rangle & \dots & \langle x_{i_1}, a_{i_n}\rangle \\ \vdots & & \vdots \\ \langle x_{i_n}, a_{i_1}\rangle & \dots & \langle x_{i_n}, a_{i_n}\rangle \end{pmatrix}\right| \leqq \prod_{q=1}^{n} \left(\sum_{p=1}^{n} |\langle x_{i_p}, a_{i_q}\rangle|^2\right)^{1/2}$$
$$\leqq \|a_{i_1}\| \dots \|a_{i_n}\|.$$

Hence

$$|\alpha_n(T)| \leqq \frac{1}{n!} \sum_{i_1=1}^{\infty} \dots \sum_{i_n=1}^{\infty} \|a_{i_1}\| \dots \|a_{i_n}\|$$
$$= \frac{1}{n!} \left(\sum_{i=1}^{\infty} \|a_i\|\right)^n \leqq (1+\varepsilon)^n \frac{1}{n!} \|T \mid \mathfrak{N}_{1,2}\|^n.$$

Letting $\varepsilon \to 0$, we obtain

$$|\alpha_n(T)| \leqq \frac{1}{n!} \|T \mid \mathfrak{N}_{1,2}\|^n.$$

4.7.13. We now investigate the eigenvalue distribution of (1,2)-nuclear operators by the classical method based on the theory of determinants and entire functions.

Theorem. The operator ideal $\mathfrak{N}_{1,2}$ is of eigenvalue type l_1. Moreover,

$$\|(\lambda_n(T)) \mid l_1\| \leqq \|T \mid \mathfrak{N}_{1,2}\| \quad \text{for all } T \in \mathfrak{N}_{1,2}(E).$$

Proof. We consider the Fredholm denominator

$$\delta(\zeta, T) := 1 + \sum_{n=1}^{\infty} \alpha_n(T)\, \zeta^n \quad \text{for all} \quad \zeta \in \mathbb{C}.$$

In view of the estimate

$$|\alpha_n(T)| \leqq \frac{1}{n!} \|T \mid \mathfrak{N}_{1,2}\|^n,$$

it follows that

$$|\delta(\zeta, T)| \leqq 1 + \sum_{n=1}^{\infty} \frac{1}{n!} \|T \mid \mathfrak{N}_{1,2}\|^n |\zeta|^n \leqq \exp(\|T \mid \mathfrak{N}_{1,2}\| \, |\zeta|).$$

Since the zeros of $\delta(\zeta, T)$ and the eigenvalue of T are related by the formula $\zeta_n = -1/\lambda_n(T)$, we see from 4.8.6 that $(\lambda_n(T)) \in l_{1,\infty}$. Hence $(\lambda_n(T)) \in l_{1+\varepsilon}$ for all $\varepsilon > 0$. At this point the old-fashioned method terminates. However, using the recently discovered principle of tensor stability, we are able to deduce from 1.7.12 that

$$\|(\lambda_n(T)) \mid l_1\| \leqq \|T \mid \mathfrak{N}_{1,2}\| \quad \text{for all} \quad T \in \mathfrak{N}_{1,2}(E).$$

4.7.14. We now prove the main result of this section.

Theorem (V. S. Lidskij 1959). The operator ideal $\mathfrak{N}_{1,2}$ admits a spectral determinant.

Proof. Let $T \in \mathfrak{N}_{1,2}(E)$, and consider any (1,2)-nuclear representation

$$T = \sum_{i=1}^{\infty} a_i \otimes x_i$$

such that

$$\sum_{i=1}^{\infty} \|a_i\| < \infty \quad \text{and} \quad \sum_{i=1}^{\infty} |\langle x_i, a\rangle|^2 \leqq 1 \quad \text{for} \quad a \in U^0.$$

It is known from the proof in 4.7.12 that

$$\left| \det \begin{pmatrix} \langle x_{i_1}, a_{i_1}\rangle & \dots & \langle x_{i_1}, a_{i_n}\rangle \\ \vdots & & \vdots \\ \langle x_{i_n}, a_{i_1}\rangle & \dots & \langle x_{i_n}, a_{i_n}\rangle \end{pmatrix} \right| \leqq \|a_{i_1}\| \dots \|a_{i_n}\|.$$

Hence, writing the coefficients $\alpha_n(T)$ in the form

$$\alpha_n(T) = \sum_{i_1 < \dots < i_n} \det \begin{pmatrix} \langle x_{i_1}, a_{i_1}\rangle & \dots & \langle x_{i_1}, a_{i_n}\rangle \\ \vdots & & \vdots \\ \langle x_{i_n}, a_{i_1}\rangle & \dots & \langle x_{i_n}, a_{i_n}\rangle \end{pmatrix},$$

we obtain the estimate

$$|\delta(\zeta, T)| \leqq 1 + \sum_{n=1}^{\infty} |\alpha_n(T)|\, |\zeta|^n$$
$$\leqq 1 + \sum_{n=1}^{\infty} \sum_{i_1 < \dots < i_n} \|a_{i_1}\| \dots \|a_{i_n}\|\, |\zeta|^n = \prod_{i=1}^{\infty} (1 + \|a_i\|\, |\zeta|).$$

Given $\mu > 0$, we choose a natural number k and $M > 0$ such that

$$\sum_{i=k+1}^{\infty} \|a_i\| \leqq \tfrac{1}{2}\mu \quad \text{and} \quad \prod_{i=1}^{k} (1 + \|a_i\|\, |\zeta|) \leqq M \exp(\tfrac{1}{2}\mu|\zeta|).$$

Then

$$|\delta(\zeta, T)| \leqq \prod_{i=1}^{k} (1 + \|a_i\|\, |\zeta|) \prod_{i=k+1}^{\infty} (1 + \|a_i\|\, |\zeta|)$$
$$\leqq M \exp(\tfrac{1}{2}\mu|\zeta|) \exp\left(\sum_{i=k+1}^{\infty} \|a_i\|\, |\zeta|\right) \leqq M \exp(\mu\, |\zeta|).$$

This proves that $\delta(\zeta, T)$ satisfies condition (1) of 4.8.9. From 4.7.13 we know that condition (2) is fulfilled, as well. Therefore the Fredholm denominator $\delta(\zeta, T)$ admits the representation

$$\delta(\zeta, T) = \prod_{n=1}^{\infty} (1 - \zeta/\zeta_n) = \prod_{n=1}^{\infty} (1 + \zeta\lambda_n(T)) \quad \text{for all} \quad \zeta \in \mathbb{C}.$$

4.7.15. We are now in a position to give the classical proof of the trace formula which especially applies to nuclear operators on a Hilbert space; see 4.7.8 (Remark).

Theorem (V. B. Lidskij 1959). The operator ideal $\mathfrak{N}_{1,2}$ admits a spectral trace.

Proof. Let $T \in \mathfrak{N}_{1,2}(E)$. Comparing the coefficients of ζ in

$$\delta(\zeta, T) = 1 + \left(\sum_{i=1}^{\infty} \langle x_i, a_i\rangle\right) \zeta + \dots$$

and

$$\delta(\zeta, T) = 1 + \left(\sum_{n=1}^{\infty} \lambda_n(T)\right) \zeta + \dots,$$

we obtain

$$\tau(T) := \sum_{i=1}^{\infty} \langle x_i, a_i \rangle = \sum_{n=1}^{\infty} \lambda_n(T).$$

4.7.16. In order to prove the next theorem, an auxiliary result is required.

Lemma. Let $0 < r \leqq 1$. Then

$$1 + \varrho \leqq \exp\left(\frac{1}{r} \varrho^r\right) \quad \text{for all} \quad \varrho \geqq 0.$$

Proof. We have

$$\log(1 + \varrho) = \int_0^{\varrho} \frac{1}{1 + \xi} \, d\xi \leqq \int_0^{\varrho} \xi^{r-1} \, d\xi = \frac{1}{r} \varrho^r.$$

4.7.17. We are now prepared to estimate the growth of the Fredholm denominator of $(r, 2)$-nuclear operators.

Proposition. Let $T \in \mathfrak{N}_{r,2}(E)$ with $0 < r \leqq 1$. Then

$$|\delta(\zeta, T)| \leqq \exp\left(\frac{1}{r} \|T \mid \mathfrak{N}_{r,2}\|^r |\zeta|^r\right) \quad \text{for all} \quad \zeta \in \mathbb{C}.$$

Proof. Given $\varepsilon > 0$, we consider an $(r, 2)$-nuclear representation

$$T = \sum_{i=1}^{\infty} a_i \otimes x_i$$

such that

$$\left(\sum_{i=1}^{\infty} \|a_i\|^r\right)^{1/r} \leqq (1 + \varepsilon) \|T \mid \mathfrak{N}_{r,2}\| \quad \text{and} \quad \sum_{i=1}^{\infty} |\langle x_i, a \rangle|^2 \leqq 1 \quad \text{for} \quad a \in U^0.$$

As in the proof of 4.7.14 we obtain

$$|\delta(\zeta, T)| \leqq \prod_{i=1}^{\infty} (1 + \|a_i\| \, |\zeta|).$$

Therefore, by the preceding lemma, it follows that

$$|\delta(\zeta, T)| \leqq \exp\left(\frac{1}{r} \sum_{i=1}^{\infty} \|a_i\|^r |\zeta|^r\right) \leqq \exp\left(\frac{(1 + \varepsilon)^r}{r} \|T \mid \mathfrak{N}_{r,2}\|^r |\zeta|^r\right).$$

Letting $\varepsilon \to 0$ yields the estimate we are looking for.

4.7.18. Next we establish a counterpart of Lemma 4.7.16.

Lemma. Let $1 \leqq r \leqq 2$. Then

$$(|1 + \zeta|^2 + \varrho^2)^{1/2} \, |\exp(-\zeta)| \leqq \exp\left[\frac{4}{r 2^r} (|\zeta|^2 + \varrho^2)^{r/2}\right]$$

for all $\zeta \in \mathbb{C}$ and $\varrho > 0$.

Proof. Write $\zeta = \xi + i\eta$ with $\xi, \eta \in \mathbb{R}$. Then

$$(|1 + \zeta|^2 + \varrho^2)^{1/2} |\exp(-\zeta)| = (1 + 2\xi + \sigma^2)^{1/2} \exp(-\xi),$$

where $\sigma^2 := \xi^2 + \eta^2 + \varrho^2$ and $-\sigma \leqq \xi \leqq +\sigma$. Note that the function

$$f(\xi) := (1 + 2\xi + \sigma^2)^{1/2} \exp(-\xi)$$

defined for all $\xi \geqq -\frac{1}{2}(1 + \sigma^2)$ attains its maximum at $\xi_0 := -\frac{1}{2}\sigma^2$. Hence

$$f(\xi) \leqq \exp(\tfrac{1}{2}\sigma^2).$$

If $\sigma \leqq 2$, then $\sigma^{2-r} \leqq 2^{2-r}$. Therefore we have

$$f(\xi) \leqq \exp\left(\frac{1}{2}\sigma^2\right) \leqq \exp\left(\frac{2}{2^r}\sigma^r\right) \leqq \exp\left(\frac{4}{r2^r}\sigma^r\right).$$

If $\sigma \geqq 2$, then $\xi_0 \leqq -\sigma$. Thus the function $f(\xi)$ is decreasing on the interval $[-\sigma, \infty)$. Hence

$$f(\xi) \leqq f(-\sigma) = (\sigma - 1)\exp(\sigma) \leqq \exp(2\sigma - 2) \leqq \exp\left(\frac{4}{r2^r}\sigma^r\right)$$

for all $\xi \geqq -\sigma$. This completes the proof.

4.7.19. If $1 < r \leqq 2$, then there does not exist any continuous trace on the operator ideal $\mathfrak{N}_{r,2}$. However, given $T \in \mathfrak{N}_{r,2}(E)$, we may consider the regularized Fredholm denominator

$$\delta_2(\zeta, T) := \delta[(I + \zeta T)\exp(-\zeta T)] \quad \text{for all } \zeta \in \mathbb{C},$$

where δ is the spectral determinant defined on $(\mathfrak{N}_{r,2})^2$.

In the case of Hilbert-Schmidt operators the following inequality goes back to T. Carleman (1921).

Proposition. Let $T \in \mathfrak{N}_{r,2}(E)$ with $1 < r \leqq 2$. Then

$$|\delta_2(\zeta, T)| \leqq \exp\left(\frac{4}{r2^r} \|T \mid \mathfrak{N}_{r,2}\|^r |\zeta|^r\right) \quad \text{for all } \zeta \in \mathbb{C}.$$

Proof. Given $\varepsilon > 0$, we consider an $(r, 2)$-nuclear representation

$$T = \sum_{i=1}^{\infty} a_i \otimes x_i$$

such that

$$\left(\sum_{i=1}^{\infty} \|a_i\|^r\right)^{1/r} \leqq (1 + \varepsilon)\|T \mid \mathfrak{N}_{r,2}\| \quad \text{and} \quad \sum_{i=1}^{\infty} |\langle x_i, a\rangle|^2 \leqq 1 \quad \text{for } a \in U^0.$$

Write

$$T_m := \sum_{i=1}^{m} a_i \otimes x_i.$$

Then

$$\begin{aligned}\delta_2(\zeta, T_m) &= \delta[(I + \zeta T_m)\exp(-\zeta T_m)] = \delta[I + \zeta T_m]\exp[-\zeta\tau(T_m)] \\ &= \det(\delta_{ij} + \zeta\langle x_i, a_j\rangle)\exp\left(-\sum_{j=1}^{m} \zeta\langle x_j, a_j\rangle\right).\end{aligned}$$

It now follows from Hadamard's inequality and the preceding lemma that

$$|\delta_2(\zeta, T_m)| \leqq \prod_{j=1}^{m} \left(\sum_{i=1}^{m} |\delta_{ij} + \zeta\langle x_i, a_j\rangle|^2 \right)^{1/2} \prod_{j=1}^{m} |\exp(-\zeta\langle x_j, a_j\rangle)|$$
$$= \prod_{j=1}^{m} \left[\left(|1 + \zeta\langle x_j, a_j\rangle|^2 + \sum_{i \neq j} |\xi\langle x_i, a_j\rangle|^2 \right)^{1/2} |\exp(-\zeta\langle x_j, a_j\rangle)| \right]$$
$$\leqq \prod_{j=1}^{m} \exp\left[\frac{4}{r2^r} \left(\sum_{i=1}^{m} |\langle x_i, a_j\rangle|^2 \right)^{r/2} |\zeta|^r \right]$$
$$\leqq \prod_{j=1}^{m} \exp\left[\frac{4}{r2^r} \|a_j\|^r |\zeta|^r \right] = \exp\left[\frac{4}{r2^r} \sum_{j=1}^{m} \|a_j\|^r |\zeta|^r \right]$$
$$\leqq \exp\left[(1 + \varepsilon)^r \frac{4}{r2^r} \|T \mid \mathfrak{N}_{r,2}\|^r |\zeta|^r \right].$$

Passing to the limit as $m \to \infty$ and $\varepsilon \to 0$, we now obtain the desired inequality.

4.7.20. Finally, we investigate the eigenvalue distribution of $(r, 2)$-nuclear operators by the classical method already used in 4.7.13.

Theorem. The operator ideal $\mathfrak{N}_{r,2}$ is of eigenvalue type l_r. Moreover,

$$\|(\lambda_n(T)) \mid l_r\| \leqq \|T \mid \mathfrak{N}_{r,2}\| \quad \text{for all} \quad T \in \mathfrak{N}_{r,2}(E).$$

Proof. By 4.7.17 and 4.7.19 we have

$$|\delta_1(\zeta, T)| \leqq \exp(\mu\|T \mid \mathfrak{N}_{r,2}\|^r |\zeta|^r] \quad \text{if} \quad 0 < r \leqq 1$$

and

$$|\delta_2(\zeta, T)| \leqq \exp(\mu\|T \mid \mathfrak{N}_{r,2}\|^r |\zeta|^r) \quad \text{if} \quad 1 \leqq r \leqq 2,$$

where the constant $\mu > 0$ depends on the exponent r. Since the zeros of $\delta_m(\zeta, T)$ and the eigenvalues of T are related by the formula $\zeta_n = -1/\lambda_n(T)$, we see from 4.8.6 that $(\lambda_n(T)) \in l_{r,\infty}$. Hence $(\lambda_n(T)) \in l_{r+\varepsilon}$ for all $\varepsilon > 0$. Thus, in view of 1.7.12 and 3.4.9, we have

$$\|(\lambda_n(T)) \mid l_r\| \leqq \|T \mid \mathfrak{N}_{r,2}\| \quad \text{for all } T \in \mathfrak{N}_{r,2}(E).$$

4.8. Entire functions

For the convenience of the reader we now prove some basic results from the theory of entire functions which are used throughout this chapter. Complete treatments are to be found in (BOA), (LEV) and (VAL), for example. The classical textbook of E. C. Titchmarsh (TIT) can be recommended, as well.

4.8.1. An **entire function** φ is a complex-valued function defined and holomorphic in the complex plane. This means that about every point ζ_0 there exists a **Taylor expansion**

$$\varphi(\zeta) = \sum_{n=0}^{\infty} \alpha_n(\zeta_0)(\zeta - \zeta_0)^n.$$

Here and in the following, if not otherwise specified, ζ denotes an arbitrary complex number. For simplicity, we always suppose that $\varphi(0) = 1$.

4.8.2. A complex number ζ_0 is said to be a **zero** of the entire function φ if $\varphi(\zeta_0) = 0$, and its **multiplicity** is defined by

$$n(\zeta_0) := \min\{n : \alpha_n(\zeta_0) \neq 0\}.$$

The set of all zeros does not possess any finite cluster point. Therefore it may be arranged as a sequence (ζ_n) such that the following conditions are satisfied:

(1) Every zero ζ_0 is counted according to its multiplicity. This means that it occurs $n(\zeta_0)$-times, one after the other.

(2) The zeros are enumerated in order of non-decreasing magnitude:

$|\zeta_1| \leqq |\zeta_2| \leqq \ldots \leqq \infty$.

In case there are distinct zeros having the same modulus these can be written in any order we please.

(3) If φ has less than n zeros, then $\zeta_n := \infty$.

4.8.3. We repeat the definition of the **primary factors**

$$1 + \varepsilon_m(\zeta) := (1 + \zeta) \exp\left(\sum_{k=1}^{m-1} \frac{(-1)^k}{k} \zeta^k\right)$$

already given in 4.5.1. Note that $\varepsilon_1(\zeta) := \zeta$.

Lemma. Let $0 < q < 1$. Then there exists a constant $c > 0$ such that

$$|\varepsilon_m(\zeta)| \leqq \frac{c}{m} |\zeta|^m \quad \text{if} \quad |\zeta| \leqq q.$$

Proof. Put

$$\alpha := -\sum_{k=m}^{\infty} \frac{(-1)^k}{k} \zeta^k.$$

It follows from

$$|\alpha| \leqq \frac{1}{m} \frac{1}{1-q} |\zeta|^m \leqq \frac{1}{1-q}$$

that

$$|\varepsilon_m(\zeta)| = |\exp(\alpha) - 1| \leqq |\alpha| \exp(|\alpha|) \leqq \frac{1}{m} \frac{1}{1-q} \exp\left(\frac{1}{1-q}\right) |\zeta|^m.$$

4.8.4. The following classical result is basic for the theory of entire functions.

Theorem (K. Weierstrass 1876). For every sequence of complex numbers $\zeta_n \neq 0$ such that $\zeta_n \to \infty$ there exist entire functions with zeros at these points and at these points only.

A function having this property is given by the formula

$$\text{(W)} \qquad \varphi(\zeta) := \prod_{n=1}^{\infty} \left[(1 - \zeta/\zeta_n) \exp\left(\sum_{k=1}^{m_n - 1} \frac{1}{k} (\zeta/\zeta_n)^k\right)\right],$$

where the non-negative integers $m_1, m_2, \ldots$ are chosen such that

$$\sum_{n=1}^{\infty} |\zeta/\zeta_n|^{m_n} < \infty \quad \text{for all} \quad \zeta \in \mathbb{C}.$$

In view of Lemma 4.8.3, this condition guarantees the convergence of the above infinite product.

Of special interest is the case in which we can find a fixed exponent m such that

$$\sum_{n=1}^{\infty} |\zeta_n|^{-m} < \infty .$$

Then (W) is called **canonical Weierstrass product**. In particular, if

$$\sum_{n=1}^{\infty} |\zeta_n|^{-1} < \infty ,$$

then we have

$$\varphi(\zeta) := \prod_{n=1}^{\infty} (1 - \zeta/\zeta_n).$$

4.8.5. We now establish a fundamental inequality.

Proposition (J. L. W. V. Jensen 1899). Let $\zeta_1, \ldots, \zeta_m$ be the zeros of φ with modulus less than $\varrho > 0$. Then

$$\frac{\varrho^m}{|\zeta_1| \ldots |\zeta_m|} \leqq \sup\{|\varphi(\zeta)| : |\zeta| = \varrho\}.$$

Proof. Define the entire function ψ by

$$\psi(\zeta) := \varphi(\zeta)\, \varrho^{-m} \prod_{n=1}^{m} \frac{\varrho^2 - \zeta\zeta_n^*}{\zeta_n - \zeta}.$$

Then we have

$$\psi(0) = \frac{\varrho^m}{\zeta_1 \ldots \zeta_m}.$$

Observe that

$$\left| \frac{\varrho^2 - \zeta\zeta_n^*}{\zeta_n - \zeta} \right| = \varrho \quad \text{if} \quad |\zeta| = \varrho .$$

Applying the principle of maximum modulus, we obtain

$$|\psi(0)| \leqq \sup\{|\psi(\zeta)| : |\zeta| = \varrho\} = \sup\{|\varphi(\zeta)| : |\zeta| = \varrho\}.$$

This completes the proof.

4.8.6. If an entire function satisfies an estimate as stated in the following proposition, then the infimum of all admissible exponents r is said to be its **order.**

Theorem (J. Hadamard 1893: b, J. L. W. V. Jensen 1899). Suppose that

$$|\varphi(\zeta)| \leqq M \exp(\mu|\zeta|^r) \quad \text{for all} \quad \zeta \in \mathbb{C},$$

where M, μ and r are positive constants. Then $(\zeta_n^{-1}) \in l_{r,\infty}$.

Proof. Fix any natural number n. Put

$$\varrho := |e\zeta_n| \quad \text{and} \quad m := \max\{k : |\zeta_k| < \varrho\}.$$

Then Jensen's inequality 4.8.5 yields

$$e^n = \frac{|e\zeta_n|^m}{|\zeta_n|^n\, |e\zeta_n|^{m-n}} \leqq \frac{|e\zeta_n|^m}{|\zeta_1| \ldots |\zeta_m|} \leqq M \exp(\mu |e\zeta_n|^r).$$

Hence

$$n \leqq \log M + \mu|e\zeta_n|^r.$$

This proves that

$$n^{1/r}|\zeta_n|^{-1} \leqq (2\mu)^{1/r}e \quad \text{whenever } n \geqq 2\log M.$$

4.8.7. We now express the condition assumed in the preceding theorem in terms of Taylor coefficients.

Proposition (E. Lindelöf 1903, A. Pringsheim 1904). Let

$$\varphi(\zeta) = \sum_{n=0}^{\infty} \alpha_n \zeta^n \quad \text{for all} \quad \zeta \in \mathbb{C}.$$

Given any positive number r, the following are equivalent:

(1) There exist $M > 0$ and $\mu > 0$ such that

$$|\varphi(\zeta)| \leqq M \exp(\mu|\zeta|^r) \quad \text{for all} \quad \zeta \in \mathbb{C}.$$

(2) There exist $A > 0$ and $\alpha > 0$ such that

$$|\alpha_n| \leqq A(\alpha/n)^{n/r} \quad \text{for all } n = 0, 1, \ldots$$

More precisely, it follows from (1) that (2) holds with $\alpha := \mu r e$. Conversely, (2) implies (1) for all $\mu > \alpha/re$.

Proof. By Cauchy's formula we have

$$\alpha_n = \frac{1}{2\pi i} \oint_{|\zeta|=\varrho} \frac{\varphi(\zeta)}{\zeta^{n+1}} d\zeta.$$

Hence

$$|\alpha_n| \leqq M \exp(\mu\varrho^r)\, \varrho^{-n}.$$

Taking $\varrho := (n/\mu r)^{1/r}$, we get

$$|\alpha_n| \leqq M(\mu r e/n)^{n/r}.$$

Therefore (1) implies (2).

To prove the converse, we fix ζ. Put $\varrho := |\zeta|$, and choose the index m such that $m \leqq 2^r\varrho^r\alpha < m + 1$. Note that the maximum of the function

$$f(\xi) := (\alpha\varrho^r/\xi)^{\xi/r}$$

with $0 < \xi < \infty$ occurs at $\xi_0 := \alpha\varrho^r/e$. Since $f(\xi_0) = \exp(\alpha\varrho^r/er)$, we have

$$\left|\sum_{n=0}^{m} \alpha_n \zeta^n\right| \leqq (m+1) A \exp(\alpha\varrho^r/er). \tag{1}$$

Moreover,

$$|\alpha_n\zeta^n| \leqq A(\alpha\varrho^r/n)^{n/r} \leqq A/2^n \quad \text{if } n \geqq 2^r\varrho^r\alpha$$

implies

$$\left|\sum_{n=m+1}^{\infty} \alpha_n \zeta^n\right| \leqq A. \tag{2}$$

Combining (1) and (2), we obtain

$$|\varphi(\zeta)| \leqq A[(2^r\varrho^r\alpha + 1)\exp(\alpha\varrho^r/er) + 1].$$

Hence, whenever $\mu > \alpha/er$, condition (1) holds for all M sufficiently large.

4.8.8. In order to prove the following theorem, we need an inequality which goes back to E. Borel (1897). Its final form is due to C. Carathéodory who communicated his version to E. Landau (LAN, p. 89) in 1905; see also F. Schottky (1904).

Proposition. Let $0 < \varrho_0 < \varrho_1 < \infty$. Suppose that the complex-valued function ψ with $\psi(0) = 0$ is defined and holomorphic on a domain containing the disk $|\zeta| \leqq \varrho_1$. Then

$$\sup\{|\psi(\zeta)| : |\zeta| = \varrho_0\} \leqq \frac{2\varrho_0}{\varrho_1 - \varrho_0}\sup\{\operatorname{Re}\psi(\zeta) : |\zeta| = \varrho_1\}.$$

Proof. Put

$$\alpha := \sup\{\operatorname{Re}\psi(\zeta) : |\zeta| = \varrho_1\}.$$

Applying the principle of maximum modulus to $\exp(\psi)$, we obtain

$$\operatorname{Re}\psi(\zeta) \leqq \alpha \quad \text{whenever } |\zeta| \leqq \varrho_1.$$

In particular, we have $0 = \operatorname{Re}\psi(0) \leqq \alpha$. Obviously, if $\alpha = 0$, then ψ vanishes everywhere. Therefore we may assume that $\alpha > 0$. Then the function

$$\varphi(\zeta) := \frac{\varrho_1\psi(\zeta)}{\zeta(2\alpha - \psi(\zeta))}$$

is holomorphic on a domain containing the disk $|\zeta| \leqq \varrho_1$. Let

$$\alpha(\zeta) := \operatorname{Re}\psi(\zeta) \quad \text{and} \quad \beta(\zeta) := \operatorname{Im}\psi(\zeta).$$

Then

$$|\varphi(\zeta)|^2 = \frac{\alpha(\zeta)^2 + \beta(\zeta)^2}{(2\alpha - \alpha(\zeta))^2 + \beta(\zeta)^2} \leqq 1 \quad \text{if} \quad |\zeta| = \varrho_1.$$

Hence the principle of maximum modulus implies that

$$|\varphi(\zeta)| \leqq 1 \quad \text{whenever } |\zeta| \leqq \varrho_1.$$

Since

$$\psi(\zeta) = \frac{2\alpha\zeta\varphi(\zeta)}{\varrho_1 + \zeta\varphi(\zeta)},$$

it follows that

$$|\psi(\zeta)| \leqq \frac{2\alpha\varrho_0}{\varrho_1 - \varrho_0} \quad \text{if} \quad |\zeta| = \varrho_0.$$

This completes the proof of the desired inequality.

4.8.9. The next result is closely related to **Hadamard's factorization theorem.**

Theorem (E. Lindelöf 1905, VAL, p. 90). Suppose that the entire function φ with $\varphi(0) = 1$ has the following properties:

(1) Given $\mu > 0$, there exists $M > 0$ such that

$$|\varphi(\zeta)| \leqq M\exp(\mu|\zeta|) \quad \text{for all} \quad \zeta \in \mathbb{C}.$$

(2) $(\zeta_n^{-1}) \in l_1$.

Then φ coincides with the canonical Weierstrass product

$$\pi(\zeta) = \prod_{n=1}^{\infty} (1 - \zeta/\zeta_n).$$

Proof. Since φ and π have the same zeros, there is an entire function ε such that

$$\varphi(\zeta) = \exp(\varepsilon(\zeta))\, \pi(\zeta).$$

Moreover, because $\varphi(0) = 1$ and $\pi(0) = 1$, we may assume that $\varepsilon(0) = 0$.

Fix $\varrho \geqq 2$, and let $\zeta_1, \ldots, \zeta_m$ be the zeros of φ with $|\zeta_n| \leqq \varrho$. Then

$$\varphi_m(\zeta) := \varphi(\zeta) / \prod_{n=1}^{m} (1 - \zeta/\zeta_n)$$

defines an entire function φ_m not vanishing for $|\zeta| \leqq \varrho$. Hence, in view of the monodromy theorem,

$$\psi_m(\zeta) := \log \varphi_m(\zeta) \quad \text{with} \quad \psi_m(0) = 0$$

is uniquely determined on some domain containing the disk $|\zeta| \leqq \varrho$. Note that

$$|1 - \zeta/\zeta_n| \geqq 1 \quad \text{if} \quad |\zeta| = 2\varrho \quad \text{and} \quad n = 1, \ldots, m.$$

Therefore, by the principle of maximum modulus,

$$\begin{aligned} \sup\{|\varphi_m(\zeta)| : |\zeta| = \varrho\} &\leqq \sup\{|\varphi_m(\zeta)| : |\zeta| = 2\varrho\} \\ &\leqq \sup\{|\varphi(\zeta)| : |\zeta| = 2\varrho\} \leqq M \exp(2\mu\varrho). \end{aligned}$$

This implies that

$$\sup\{\operatorname{Re} \psi_m(\zeta) : |\zeta| = \varrho\} = \sup\{\log |\varphi_m(\zeta)| : |\zeta| = \varrho\} \leqq 2\mu\varrho + \log M.$$

We now apply the Borel-Carathéodory inequality 4.8.8 to ψ_m, where $\varrho_0 := 1$ and $\varrho_1 := \varrho$. This yields

$$\sup\{|\psi_m(\zeta)| : |\zeta| = 1\} \leqq \frac{2}{\varrho - 1} (2\mu\varrho + \log M).$$

Obviously,

$$|\log(1 - \zeta/\zeta_n)| \leqq 2\,|\zeta/\zeta_n| \quad \text{if} \quad |\zeta| \leqq 1 \quad \text{and} \quad |\zeta_n| > \varrho \geqq 2.$$

Furthermore,

$$\varepsilon(\zeta) = \psi_m(\zeta) - \sum_{n=m+1}^{\infty} \log(1 - \zeta/\zeta_n) \quad \text{if} \quad |\zeta| \leqq 1.$$

Combining the preceding observations, we obtain

$$\sup\{|\varepsilon(\zeta)| : |\zeta| = 1\} \leqq \frac{2}{\varrho - 1} (2\mu\varrho + \log M) + 2 \sum_{n=m+1}^{\infty} |\zeta_n|^{-1}.$$

Letting $\varrho \to \infty$ yields

$$\sup\{|\varepsilon(\zeta)| : |\zeta| = 1\} \leqq 4\mu.$$

Since $\mu > 0$ is arbitrary, it follows that $\varepsilon(\zeta) = 0$ for $|\zeta| \leqq 1$ and so for all $\zeta \in \mathbb{C}$.

Remark. We know from 4.8.6 that property (1) implies $(\zeta_n^{-1}) \in l_{1,\infty}$. However, examples show that the slightly stronger assumption $(\zeta_n^{-1}) \in l_1$ is indeed necessary. See (BOA, 2.10.3) and (LEV, Chap. 1, § 11).

CHAPTER 5

Matrix operators

In what follows we illustrate the abstract theory of eigenvalue distributions by applying this machinery to matrix operators on sequence spaces. Although the discrete case is interesting in its own right, our main goal is to prepare the reader for a better understanding of integral operators, which are treated in the next chapter.

In order to show that the basic eigenvalue theorems are sharp, we first provide some examples of finite and infinite matrices.

Next we deal with infinite matrices $M = (\mu_{ij})$ such that

$$\left(\sum_{i=1}^{\infty}\left[\sum_{j=1}^{\infty}|\mu_{ij}|^q\right]^{p/q}\right)^{1/p} < \infty .$$

The main result of this chapter is the eigenvalue theorem for Besov matrices.

Finally, we investigate traces and determinants of infinite matrices. For this purpose, the Dixon-von Koch matrices are of special importance. These are defined by the condition

$$\sum_{i=1}^{\infty} \sup_j |\mu_{ij}| < \infty .$$

We also describe regularized Fredholm denominators of Hilbert-Schmidt matrices.

5.1. Examples of finite matrices

5.1.1. Let $M = (\mu_{ij})$ be any (n, n)-matrix. Then the set of eigenvalues corresponding to the operator M_{op} induced on $l(n)$ is denoted by $(\lambda_1(M), \ldots, \lambda_n(M))$.

5.1.2. We call

$$I(n) := \begin{pmatrix} 1 & & 0 \\ & \ddots & \\ 0 & & 1 \end{pmatrix}$$

the **unit matrix** of order n.

5.1.3. For every vector $t = (\tau_1, \ldots, \tau_n)$ we define the **diagonal matrix**

$$D_t(n) := \begin{pmatrix} \tau_1 & & 0 \\ & \ddots & \\ 0 & & \tau_n \end{pmatrix}.$$

Obviously, the eigenvalues of the induced operator are $\tau_1, \ldots, \tau_n$.

5.1.4. The **shift matrix**

$$S(n) := \begin{pmatrix} 0 & 0 & \ldots & 0 & 1 \\ 1 & 0 & \ldots & 0 & 0 \\ \vdots & \vdots & & \vdots & \vdots \\ 0 & 0 & \ldots & 0 & 0 \\ 0 & 0 & \ldots & 1 & 0 \end{pmatrix}$$

has ones in the upper right-hand corner and just below the principal diagonal. All the other coefficients are zero. Since

$$S(n)_{\mathrm{op}} : (\xi_1, \ldots, \xi_{n-1}, \xi_n) \to (\xi_n, \xi_1, \ldots, \xi_{n-1})$$

the induced operator is unitary on $l_2(n)$.

5.1.5. We call

$$S_t(n) := \begin{pmatrix} 0 & 0 & \ldots & 0 & \tau_n \\ \tau_1 & 0 & \ldots & 0 & 0 \\ \vdots & \vdots & & \vdots & \vdots \\ 0 & 0 & \ldots & 0 & 0 \\ 0 & 0 & \ldots & \tau_{n-1} & 0 \end{pmatrix}$$

the **weighted shift matrix** associated with the vector $t = (\tau_1, \ldots, \tau_n)$. Note that

$$\det(\lambda I(n) - S_t(n)) = \lambda^n - \tau_1 \ldots \tau_n .$$

Hence

$$|\lambda_k(S_t(n))| = |\tau_1 \ldots \tau_n|^{1/n} \quad \text{for} \quad k = 1, \ldots, n .$$

Furthermore, if $\tau_1 \geqq \ldots \geqq \tau_n \geqq 0$, then it follows from $S_t(n) = S(n)\, D_t(n)$ that

$$s_k(S_t(n)_{\mathrm{op}} : l_2(n) \to l_2(n)) = \tau_k \quad \text{for} \quad k = 1, \ldots, n .$$

This example shows that it is impossible to improve the relationship between eigenvalues and s-numbers of operators on Hilbert spaces which is given by the multiplicative Weyl inequality 3.5.1.

5.1.6. Next we consider the **Fourier matrix**

$$F(n) := n^{-1/2} \left(\exp\left(\frac{2\pi \mathrm{i}}{n} hk\right)\right),$$

where $h, k = 1, \ldots, n$. It follows from

$$F(n)^* \, F(n) = F(n) \, F(n)^* = I(n)$$

that $F(n)$ is unitary. Hence

$$|\lambda_k(F(n))| = 1 \quad \text{for} \quad k = 1, \ldots, n .$$

Furthermore, if $2 \leqq p \leqq \infty$, then we have

$$\|F(n)_{\mathrm{op}} : l_p(n) \to l_p(n) \mid \mathfrak{N}\| \leqq n^{1/2+1/p} .$$

This example shows that the inequality

$$\|(\lambda_k(T)) \mid l_r\| \leqq c \|T \mid \mathfrak{N}\| \quad \text{for all} \quad T \in \mathfrak{N}(l_p)$$

can only hold under the assumption $1/r \leqq 1/2 + 1/p$; see 3.8.3.

Remark. The matrix $F(n)$ determines the Fourier transform on the cyclic group $\mathbb{Z}_n := \{1, \ldots, n\}$.

5.1.7.* The normalized **Littlewood-Walsh matrices** $W(2^h)$ are inductively defined by

$$W(2^{h+1}) := \frac{1}{\sqrt{2}} \begin{pmatrix} +W(2^h), & +W(2^h) \\ +W(2^h), & -W(2^h) \end{pmatrix} \quad \text{and} \quad W(2) := \frac{1}{\sqrt{2}} \begin{pmatrix} +1, & +1 \\ +1, & -1 \end{pmatrix}.$$

Obviously, all coefficients take the values $\pm 2^{-h/2}$. It follows from

$$W(2^h)^* = W(2^h) \quad \text{and} \quad W(2^h)\,W(2^h) = I(2^h)$$

that $W(2^h)$ is hermitian and unitary. Next we note that

$$\operatorname{trace}(W(2^h)) = 0 \quad \text{implies} \quad \sum_{i=1}^{2^h} \lambda_i(W(2^h)) = 0.$$

Hence $W(2^h)$ has eigenvalues $+1$ and -1 each with multiplicity 2^{h-1}.

Finally, by interpolation and duality, we obtain from

$$\| W(2^h)_{\mathrm{op}} : l_2(2^h) \to l_2(2^h) \| = 1$$

and

$$\| W(2^h)_{\mathrm{op}} : l_1(2^h) \to l_1(2^h) \| = 2^{h/2}$$

that

$$\| W(2^h)_{\mathrm{op}} : l_p(2^h) \to l_p(2^h) \| \leqq 2^{h|1/p-1/2|}.$$

Remark. The matrices $W(2^h)$ were first considered by J. Hadamard (1893) : a), J. L. Walsh (1923) and J. E. Littlewood (1930).

5.2. Examples of infinite matrices

5.2.1.* Observe that

$$\pi : (h, i) \to m := 2^h + i - 1$$

defines a one-to-one correspondence between

$$\mathbb{P} := \{(h, i) : h = 0, 1, \ldots;\ i = 1, \ldots, 2^h\} \quad \text{and} \quad \mathbb{N} := \{1, 2, \ldots\}.$$

In this way every sequence $x = (\xi_m)$ can be viewed as a family of vectors $x_h = (\xi_{hi})$, where $\xi_{hi} := \xi_m$. Analogously, every infinite matrix $M = (\mu_{mn})$ splits into $(2^h, 2^k)$-blocks $M_{hk} = (\mu_{hi,kj})$. In what follows the converse procedure is applied in order to build infinite matrices from finite ones.

5.2.2.* First, for $0 < r < \infty$ and $\alpha \geqq 0$, we consider the **diagonal matrices**

$$I_{r,\alpha} := \begin{pmatrix} 1 & & & & 0 \\ & \ddots & & & \\ & & \dfrac{1}{2^{h/r} h + 1} I(2^h) & & \\ & & & \ddots & \\ 0 & & & & 1 \end{pmatrix}.$$

Obviously, $I_{r,\alpha}$ induces an approximable operator on l_p with $1 \leqq p \leqq \infty$. Furthermore, in view of 2.1.12, for the corresponding eigenvalue sequence we have

$$(\lambda_n(I_{r,\alpha})) \in l_{r,w} \quad \text{if and only if } \alpha w > 1$$

and

$$(\lambda_n(I_{r,0})) \in l_{r,\infty}.$$

5.2.3.* Next, for $0 < r < \infty$ and $\alpha \geqq 0$, we define the **weighted Littlewood-Walsh matrices**

$$W_{r,\alpha} := \begin{pmatrix} 1 & & & & 0 \\ & \ddots & & & \\ & & \frac{1}{2^{h/r}h+1} W(2^h) & & \\ & & & \ddots & \\ 0 & & & & 1 \end{pmatrix}.$$

It follows from

$$\| W(2^h)_{\mathrm{op}} : l_p(2^h) \to l_p(2^h) \| \leqq 2^{h|1/2-1/p|}$$

that $W_{r,\alpha}$ induces an approximable operator on l_p whenever $1/r > |1/2 - 1/p|$. Furthermore, in view of 5.1.7 and 2.1.12, for the corresponding eigenvalue sequence we have

$$(\lambda_n(W_{r,\alpha})) \in l_{r,w} \quad \text{if and only if} \quad \alpha w > 1$$

and

$$(\lambda_n(W_{r,0})) \in l_{r,\infty}.$$

Finally, letting $0 < r \leqq 2$, it can easily be seen that

$$(W_{r,\alpha})_{\mathrm{op}} \in \mathfrak{N}_p(l_1) \quad \text{if} \quad 1/p = 1/r + 1/2 \quad \text{and} \quad \alpha p > 1.$$

5.3. Hille-Tamarkin matrices

Throughout this section we assume that $1 \leqq p < \infty$ and $1 \leqq q \leqq \infty$.

5.3.1.* Recall from 1.1.3 that $[l_p, E]$ denotes the Banach space of all E-valued sequences such that $(\|x_i\|) \in l_p$ equipped with the norm

$$\|(x_i) \mid [l_p, E]\| := \left(\sum_{i=1}^{\infty} \|x_i\|^p \right)^{1/p}.$$

5.3.2.* The following result is an immediate consequence of 1.3.4.

Proposition. Let $(x_i) \in [l_p, E]$. Then

$$(x_i)_{\mathrm{op}} : a \to (\langle x_i, a \rangle)$$

defines an absolutely p-summing operator from E' into l_p. Moreover,

$$\|(x_i)_{\mathrm{op}} \mid \mathfrak{P}_p\| \leqq \|(x_i) \mid [l_p, E]\|.$$

5.3.3.* An infinite matrix $M = (\mu_{ij})$ is said to be of **Hille-Tamarkin type** $[l_p, l_q]$ if the sequence-valued sequence formed by

$$m_i := (\mu_{i1}, \mu_{i2}, \ldots)$$

belongs to $[l_p, l_q]$. For $1 \leqq q < \infty$, this means that

$$\|M \mid [l_p, l_q]\| := \left(\sum_{i=1}^{\infty} \left[\sum_{j=1}^{\infty} |\mu_{ij}|^q \right]^{p/q} \right)^{1/p}$$

is finite. In the limiting case $q = \infty$ the same holds for the expression

$$\|M \mid [l_p, l_\infty]\| := \left(\sum_{i=1}^{\infty} \sup_j |\mu_{ij}|^p \right)^{1/p}.$$

In the case $p = q = 2$ we get the famous Hilbert-Schmidt matrices defined in 1.4.6. Furthermore it is worth mentioning that the **Dixon-von Koch matrices** obtained for $p = 1$ and $q = \infty$ just represent the nuclear operators acting on l_1; see 5.5.2.

5.3.4.* Given $M \in [l_p, l_q]$, it follows from 5.3.2 that

$$M_{\text{op}} : (\eta_j) \to \left(\sum_{j=1}^{\infty} \mu_{ij} \eta_j \right)$$

defines an operator from $(l_q)'$ into l_p. If $1/p + 1/q \geqq 1$, then there exists

$$\langle x, y \rangle := \sum_{i=1}^{\infty} \xi_i \eta_i \quad \text{for} \quad x = (\xi_i) \in l_p \quad \text{and} \quad y = (\eta_i) \in l_q .$$

Using this duality, we can embed l_p into $(l_q)'$. This implies that M_{op} acts on every intermediate Banach space E:

$$E \xrightarrow{\ I\ } (l_q)' \xrightarrow{\ M_{\text{op}}\ } l_p \xrightarrow{\ I\ } E.$$

For example, we may take $E := l_r$ with $p \leqq r \leqq q'$. Since the associated **eigenvalue sequence** does not depend on the choice of E, it is simply denoted by $(\lambda_n(M))$.

5.3.5.* We now state the main result of this section.

Eigenvalue theorem for Hille-Tamarkin matrices (B. Carl 1982: b). Let $1/p + 1/q > 1$ and

$$1/r := \begin{cases} 1/p & \text{if} \quad 1 \leqq q \leqq 2, \\ 1/p + 1/q - 1/2 & \text{if} \quad 2 \leqq q \leqq \infty . \end{cases}$$

Then

$$M \in [l_p, l_q] \quad \text{implies} \quad (\lambda_n(M)) \in l_{r,p} .$$

This result is the best possible.

Proof. Write

$$\tau_i := \left(\sum_{j=1}^{\infty} |\mu_{ij}|^q \right)^{1/q} .$$

Then $t = (\tau_i) \in l_p$, and it follows from 2.9.17 that $D_t \in \mathfrak{L}_{r,p}^{(x)}(l_\infty, l_{q'})$. Moreover, we define the infinite matrix $A = (\alpha_{ij})$ by setting $\alpha_{ij} := \tau_i^{-1} \mu_{ij}$. Since $A_{\text{op}} \in \mathfrak{L}(l_{q'}, l_\infty)$, we obtain the factorization

$$M_{\text{op}} : l_{q'} \xrightarrow{\ A_{\text{op}}\ } l_\infty \xrightarrow{\ D_t\ } l_{q'} .$$

This proves that $M_{\text{op}} \in \mathfrak{L}_{r,p}^{(x)}(l_{q'})$. Hence the eigenvalue theorem 3.6.2 implies that $(\lambda_n(M)) \in l_{r,p}$.

Using the diagonal matrices defined in 5.2.2, we have

$$I_{p,\alpha} \in [l_p, l_q] \quad \text{if and only if} \quad \alpha p > 1$$

and

$$(\lambda_n(I_{p,\alpha})) \in l_{p,w} \quad \text{if and only if} \quad \alpha w > 1 .$$

Similarly, employing the weighted Littlewood-Walsh matrices introduced in 5.2.3, and letting $1/r := 1/p + 1/q - 1/2$, it follows that

$$W_{r,\alpha} \in [l_p, l_q] \quad \text{if and only if} \quad \alpha p > 1$$

and

$$(\lambda_n(W_{r,\alpha})) \in l_{r,w} \quad \text{if and only if} \quad \alpha w > 1.$$

These examples show that $(\lambda_n(M)) \in l_{r,p}$ is indeed the best possible result which can be obtained for all matrices of Hille-Tamarkin type $[l_p, l_q]$.

5.3.6.* Finally, we deal with the limiting case $1/p + 1/q = 1$.

Theorem (W. B. Johnson / H. König / B. Maurey / J. R. Retherford 1979).

$$M \in [l_p, l_{p'}] \quad \text{implies} \quad (\lambda_n(M)) \in l_{\max(p,2)}.$$

This result is the best possible.

Proof. We know from 5.3.2 that M induces an absolutely p-summing operator on l_p. Hence the assertion can be deduced from 3.7.1 and 3.7.2.

It follows from the theory of nuclear operators that there are matrices $M \in [l_1, l_\infty]$ such that $(\lambda_n(M)) \notin l_{2,w}$ if $0 < w < 2$. Passing from $M = (\mu_{ij})$ to $N = (\varrho_i^{-1}\mu_{ij}\varrho_j)$, we may obtain matrices $N \in [l_p, l_{p'}|]$ with $\lambda_n(N) = \lambda_n(M)$. Therefore the conclusion $(\lambda_n(M)) \in l_2$ cannot be improved for any exponent p with $1 \leqq p \leqq 2$. Furthermore, diagonal operators show that $(\lambda_n(M)) \in l_p$ is the best possible result for $2 < p < \infty$.

5.4. Besov matrices

Throughout this section we assume that $-\infty < \sigma, \tau < +\infty$, $1 \leqq p, q \leqq \infty$ and $1 \leqq u, v \leqq \infty$.

5.4.1.* We define the **Besov sequence space**

$$[b^\sigma_{p,u}, E] := [l_u, 2^{h\sigma}[l_p(2^h), E]].$$

This means that $[b^\sigma_{p,u}, E]$ consists of all E-valued sequences (x_{hi}) indexed by

$$\mathbb{P} := \{(h, i) : h = 0, 1, \ldots; i = 1, \ldots, 2^h\}$$

such that

$$\|(x_{hi}) \mid [b^\sigma_{p,u}, E]\| := \left(\sum_{h=0}^{\infty} \left[2^{h\sigma}\left(\sum_{i=1}^{2^h} \|x_{hi}\|^p\right)^{1/p}\right]^u\right)^{1/u}$$

is finite. In the cases $p = \infty$ or $u = \infty$ the usual modifications are required.

If E is the complex field, then we simply write $b^\sigma_{p,u}$.

5.4.2.* We first establish an elementary inclusion.

Proposition. Let $-\infty < \sigma_1 \leqq \sigma_0 < +\infty$, $1 \leqq p_1 \leqq p_0 \leqq \infty$ and $\sigma_0 + 1/p_0 = \sigma_1 + 1/p_1$. Then

$$[b^{\sigma_0}_{p_0,u}, E] \subseteq [b^{\sigma_1}_{p_1,u}, E].$$

Proof. Observe that

$$\left(\sum_{i=1}^{2^h} \|x_i\|^{p_1}\right)^{1/p_1} \leqq 2^{h(1/p_1 - 1/p_0)} \left(\sum_{i=1}^{2^h} \|x_i\|^{p_0}\right)^{1/p_0}.$$

5.4.3.* The next result follows immediately from (BER, 5.6.1) and (TRI, 1.18.2).

Interpolation proposition. Let $-\infty < \sigma_0 < \sigma_1 < \infty$, $1 \leqq u_0, u_1, u \leqq \infty$ and $0 < \theta < 1$. If $\sigma = (1-\theta)\,\sigma_0 + \theta\sigma_1$, then

$$([b^{\sigma_0}_{p,u_0}, E], [b^{\sigma_1}_{p,u_1}, E])_{\theta,u} = [b^{\sigma}_{p,u}, E].$$

The norms on both sides are equivalent.

5.4.4.* **Lemma.** Let $\sigma + \tau > (1 - 1/p - 1/q)_+$. Then there exists

$$\langle x, y\rangle := \sum_{h=0}^{\infty} \sum_{i=1}^{2^h} \xi_{hi}\eta_{hi} \quad \text{for} \quad x = (\xi_{hi}) \in b^{\sigma}_{p,u} \quad \text{and} \quad y = (\eta_{hi}) \in b^{\tau}_{q,v}.$$

Using this duality, we can embed $b^{\sigma}_{p,u}$ into $(b^{\tau}_{q,v})'$.

Proof. Define $1/r := (1 - 1/p - 1/q)_+$. Then Hölder's inequality implies that

$$\sum_{i=1}^{2^h} |\xi_{hi}\eta_{hi}| \leqq 2^{h/r} \left(\sum_{i=1}^{2^h} |\xi_{hi}|^p\right)^{1/p} \left(\sum_{i=1}^{2^h} |\eta_{hi}|^q\right)^{1/q}.$$

Hence

$$\begin{aligned}|\langle x, y\rangle| &\leqq \sum_{h=0}^{\infty} \sum_{i=1}^{2^h} |\xi_{hi}\eta_{hi}| \\ &\leqq \sum_{h=0}^{\infty} 2^{-h(\sigma+\tau-1/r)} \left[2^{h\sigma}\left(\sum_{i=1}^{2^h} |\xi_{hi}|^p\right)^{1/p}\right] \left[2^{h\tau}\left(\sum_{i=1}^{2^h} |\eta_{hi}|^q\right)^{1/q}\right] \\ &\leqq \sum_{h=0}^{\infty} 2^{-h(\sigma+\tau-1/r)} \, \|x \mid b^{\sigma}_{p,u}\| \, \|y \mid b^{\tau}_{q,v}\|.\end{aligned}$$

Thus $\langle x, y\rangle$ defines a bounded bilinear form.

5.4.5.* The following result is related to 2.9.9.

Lemma. Let $\mathfrak{A}$ be a quasi-Banach operator ideal such that

$$\|I : l_p(m) \to l_q(m)' \mid \mathfrak{A}\| \prec m^{\alpha},$$

where $\alpha \geqq 0$. If $\sigma + \tau > \alpha$, then

$$a_n(I : b^{\sigma}_{p,u} \to (b^{\tau}_{q,v})' \mid \mathfrak{A}) \prec n^{-(\sigma+\tau-\alpha)}.$$

Proof. Let

$$Q_k(\xi_{hi}) := (\xi_{k1}, \ldots, \xi_{k2^k})$$

be the surjection which maps every sequence (ξ_{hi}) into its k-th block. Then

$$\|Q_k : b^{\sigma}_{p,v} \to l_p(2^k)\| = 2^{-k\sigma} \quad \text{and} \quad \|Q_k : b^{\tau}_{q,v} \to l_q(2^k)\| = 2^{-k\tau}.$$

Denoting the identity operator from $l_p(2^k)$ onto $l_q(2^k)'$ by I_k, the embedding operator from $b^{\delta}_{p,u}$ into $(b^{\tau}_{q,v})'$ can be written in the form

$$I = \sum_{k=0}^{\infty} Q_k' I_k Q_k.$$

According to D.1.8, we may assume that $\mathfrak{A}$ is r-normed. If $2^h \leqq n < 2^{h+1}$, then it follows from

$$\operatorname{rank}\left(\sum_{k=0}^{h-1} Q_k' I_k Q_k\right) = \sum_{k=0}^{h-1} 2^k < 2^h$$

that

$$a_n(I: b^\sigma_{p,u} \to (b^\tau_{q,v})' \mid \mathfrak{A}) \leqq \|I - \sum_{k=0}^{h-1} Q'_k I_k Q_k \mid \mathfrak{A}\| \leqq \left(\sum_{k=h}^{\infty} \|Q'_k I_k Q_k \mid \mathfrak{A}\|^r\right)^{1/r}$$

$$\leqq \left(\sum_{k=h}^{\infty} 2^{-k(\sigma+\tau-\alpha)r} \| I: l_p(2^k) \to l_q(2^k)' \mid \mathfrak{A}\|^r\right)^{1/r}$$

$$\prec \left(\sum_{k=h}^{\infty} 2^{-k(\sigma+\tau-\alpha)r}\right)^{1/r} \prec 2^{-h(\sigma+\tau-\alpha)} \prec n^{-(\sigma+\tau-\alpha)}.$$

5.4.6.* The result just obtained can be used to estimate the Weyl numbers.

Proposition (A. Pietsch 1980: d). Let $\sigma + \tau > 0$ and $1/p + 1/q \geqq 1$. Then $x_n(I: b^\sigma_{p,u} \to (b^\tau_{q,v})') \prec n^{-\varrho}$,

where

$$\varrho := \sigma + \tau + \begin{cases} 1/p + 1/q - 1 & \text{if } 1 \leqq p \leqq q' \leqq 2, \\ 1/p - 1/2 & \text{if } 1 \leqq p \leqq 2 \leqq q' \leqq \infty, \\ 0 & \text{if } 2 \leqq p \leqq q' \leqq \infty. \end{cases}$$

Proof. Put

$$1/r := \begin{cases} 1/p + 1/q - 1 & \text{if } 1 \leqq p \leqq q' \leqq 2, \\ 1/p - 1/2 & \text{if } 1 \leqq p \leqq 2 \leqq q' \leqq \infty, \\ 0 & \text{if } 2 \leqq p \leqq q' \leqq \infty. \end{cases}$$

In view of 1.6.7, we have

$$\|I: l_p(m) \to l_q(m)' \mid \mathfrak{P}_{r,2}\| \leqq c := \|I: l_p \to l_q' \mid \mathfrak{P}_{r,2}\|.$$

Furthermore, 2.8.16 implies that

$$x_{2n-1}(I: b^\sigma_{p,u} \to (b^\tau_{q,v})') \leqq n^{-1/r} a_n(I: b^\sigma_{p,u} \to (b^\tau_{q,v})' \mid \mathfrak{P}_{r,2}).$$

Finally, taking $\mathfrak{A} := \mathfrak{P}_{r,2}$ and $\alpha := 0$, the preceding lemma yields the desired estimate.

5.4.7.* We now generalize 5.3.2.

Proposition (A. Pietsch 1980: d). Let $(x_{hi}) \in [b^\sigma_{p,u}, E]$ and $r := \max(p, u)$. Then

$$(x_{hi})_{\mathrm{op}}: a \to (\langle x_{hi}, a\rangle)$$

defines an absolutely r-summing operator from E' into $b^\sigma_{p,u}$. Moreover,

$$\|(x_{hi})_{\mathrm{op}} \mid \mathfrak{P}_r\| \leqq \|(x_{hi}) \mid [b^\sigma_{p,u}, E]\|.$$

Proof. Given $a_1, \dots, a_n \in E'$, repeated application of Jessen's inequality C.3.10 yields

$$\left(\sum_{j=1}^{n} \|(\langle x_{hi}, a_j\rangle) \mid b^\sigma_{p,u}\|^r\right)^{1/r} =$$

$$= \left\{\sum_{j=1}^{n} \left[\sum_{h=0}^{\infty} \left(\sum_{i=1}^{2^h} \langle 2^{h\sigma}|x_{hi}, a_j\rangle|^p\right)^{u/p}\right]^{r/u}\right\}^{1/r}$$

$$\leqq \left\{\sum_{h=0}^{\infty} \left[\sum_{i=1}^{2^h} \left(\sum_{j=1}^{n} \langle 2^{h\sigma}|x_{hi}, a_j\rangle|^r\right)^{p/r}\right]^{u/p}\right\}^{1/u}$$

$$\leqq \|(x_{hi}) \mid [b^\sigma_{p,u}, E]\| \sup\left\{\left(\sum_{j=1}^{n} |\langle x, a_j\rangle|^r\right)^{1/r} : x \in U\right\}.$$

This shows that $(x_{hi})_{\mathrm{op}}$ is absolutely r-summing.

5.4.8.* An infinite matrix $M = (\mu_{hi,kj})$ is said to be of **Besov type** $[b^{\sigma}_{p,u}, b^{\tau}_{q,v}]$ if the sequence-valued sequence formed by

$$m_{hi} := (\mu_{hi,01}; \mu_{hi,11}, \mu_{hi,12}; \ldots)$$

belongs to $[b^{\sigma}_{p,u}, b^{\tau}_{q,v}]$.

5.4.9.* Given $M \in [b^{\delta}_{p,u}, b^{\tau}_{q,v}]$, it follows from 5.4.7 that

$$M_{\text{op}} : (\eta_{kj}) \to \left(\sum_{k=0}^{\infty} \sum_{j=1}^{2^k} \mu_{hi,kj} \eta_{kj} \right)$$

defines an operator from $(b^{\sigma}_{q,v})'$ into $b^{\delta}_{p,u}$. If $\sigma + \tau > (1 - 1/p - 1/q)_+$, then $b^{\sigma}_{p,u}$ is embedded into $(b^{\tau}_{q,v})'$. This implies that M_{op} acts on every intermediate Banach space E. Since the associated **eigenvalue sequence** does not depend on the choice of E, it is simply denoted by $(\lambda_n(M))$.

5.4.10.* We are now in a position to establish the main result of this chapter.

Eigenvalue theorem for Besov matrices (A. Pietsch 1980: d). Let $\sigma + \tau > (1 - 1/p - 1/q)_+$ and

$$1/r := \sigma + \tau + \begin{cases} 1/p & \text{if } \ 1 \leqq q \leqq 2, \\ 1/p + 1/q - 1/2 & \text{if } \ 2 \leqq q \leqq \infty. \end{cases}$$

Then

$$M \in [b^{\sigma}_{p,u}, b^{\tau}_{q,v}] \quad \text{implies} \quad (\lambda_n(M)) \in l_{r,u}.$$

This result is the best possible.

Proof. First we assume that $p \leqq q'$ and $u = 1$. Define

$$\varrho := \sigma + \tau + \begin{cases} 1/p + 1/q - 1 & \text{if } \ 1 \leqq p \leqq q' \leqq 2, \\ 1/p - 1/2 & \text{if } \ 1 \leqq p \leqq 2 \leqq q' \leqq \infty, \\ 0 & \text{if } \ 2 \leqq p \leqq q' \leqq \infty, \end{cases}$$

and $s := \max(p, 2)$. We know from 5.4.6 that

$$x_n(I : b^{\sigma}_{p,1} \to (b^{\tau}_{q,\infty})') \prec n^{-\varrho}.$$

Hence

$$I \in \mathfrak{L}^{(x)}_{1/\varrho,\infty}(b^{\sigma}_{p,1}, (b^{\tau}_{q,\infty})').$$

Furthermore, by 5.4.7 and 2.7.4,

$$M_{\text{op}} \in \mathfrak{L}^{(x)}_{s,\infty}((b^{\tau}_{q,v})', b^{\sigma}_{p,1}).$$

Thus, in view of the factorization

$$M_{\text{op}} : (b^{\tau}_{q,\infty})' \xrightarrow{M_{\text{op}}} b^{\sigma}_{p,1} \xrightarrow{\ I\ } (b^{\tau}_{q,v})',$$

we conclude from the multiplication theorem 2.4.18 that

$$M_{\text{op}} \in \mathfrak{L}^{(x)}_{r,\infty}((b^{\tau}_{q,\infty})'),$$

where $1/r = \varrho + 1/s$ takes precisely the values given in the above theorem.

If $p > q'$ and $u = 1$, then we see from 5.4.2 that

$$M \in [b^{\sigma}_{p,1}, b^{\tau}_{q,v}] \quad \text{implies} \quad M \in [b^{\sigma+1/p-1/q'}_{q',1}, b^{\tau}_{q,v}].$$

In this way the second case is reduced to the first one.

So far we have shown that the map $M \to M_{\text{op}}$, which assigns to every matrix the corresponding operator, acts as follows:

$$\text{op} : [b^{\sigma}_{p,1}, b^{\tau}_{q,v}] \to \mathfrak{L}^{(x)}_{r,\infty}((b^{\tau}_{q,v})').$$

This result can be improved by interpolation. To this end, choose σ_0, σ_1 and θ such that

$$\sigma = (1 - \theta)\,\sigma_0 + \theta\sigma_1 \quad \text{and} \quad -\infty < \sigma_0 < \sigma_1 < +\infty.$$

Since $\sigma + \tau > (1 - 1/p - 1/q)_+$, it is further possible to satisfy the condition $\sigma_0 + \tau > (1 - 1/p - 1/q)_+$. We now apply the formula

$$([b^{\sigma_0}_{p,1}, E], [b^{\sigma_1}_{p,1}, E])_{\theta,u} = [b^{\sigma}_{p,u}, E]$$

for $E := b^{\tau}_{q,v}$. Next, replacing σ by σ_0 and σ_1, we define r_0 and r_1 in the same way as r. Then it follows from $1/r = (1 - \theta)/r_0 + \theta/r_1$ that

$$(\mathfrak{L}^{(x)}_{r_0,\infty}, \mathfrak{L}^{(x)}_{r_1,\infty})_{\theta,u} \subseteq \mathfrak{L}^{(x)}_{r,u}.$$

Hence the interpolation property yields

$$\text{op} : [b^{\sigma}_{p,u}, b^{\tau}_{q,v}] \to \mathfrak{L}^{(x)}_{r,u}((b^{\tau}_{q,v})').$$

By the eigenvalue theorem 3.6.2, we therefore obtain $(\lambda_n(M)) \in l_{r,u}$.

Using the diagonal matrices defined in 5.2.2, for $1/r = \sigma + \tau + 1/p$, we have

$$I_{r,\alpha} \in [b^{\sigma}_{p,u}, b^{\tau}_{q,v}] \quad \text{if and only if} \quad \alpha u > 1$$

and

$$(\lambda_n(I_{r,\alpha})) \in l_{r,w} \quad \text{if and only if} \quad \alpha w > 1.$$

Similarly, employing the weighted Littlewood-Walsh matrices introduced in 5.2.3, and letting $1/r = \sigma + \tau + 1/p + 1/q - 1/2$, it follows that

$$W_{r,\alpha} \in [b^{\sigma}_{p,u}, b^{\tau}_{q,v}] \quad \text{if and only if} \quad \alpha u > 1$$

and

$$(\lambda_n(W_{r,\alpha})) \in l_{r,w} \quad \text{if and only if} \quad \alpha w > 1.$$

These examples show that $(\lambda_n(M)) \in l_{r,u}$ is indeed the best possible result which can be obtained for all matrices of Besov type $[b^{\sigma}_{p,u}, b^{\tau}_{q,v}]$.

Remark. The eigenvalue theorem for Hille-Tamarkin matrices can be deduced from the preceding one by taking $\sigma = 0$, $\tau = 0$, $u = p$ and $v = q$.

5.5. Traces and determinants of matrices

5.5.1. To begin with, we restate 4.7.3 in the following form.

Proposition (A. Grothendieck 1956: a). Let $(a_i) \in [l_1, E']$. Then

$$(a_i)_{\text{op}} : x \to (\langle x, a_i \rangle)$$

defines a nuclear operator from E into l_1 with

$$\|(a_i)_{\text{op}} \mid \mathfrak{N}\| = \|(a_i) \mid [l_1, E']\|.$$

Furthermore, all operators $T \in \mathfrak{N}(E, l_1)$ can be obtained in this way.

5.5.2. Specializing the preceding result, we arrive at a characterization of the nuclear operators acting on l_1.

Proposition. Every infinite matrix $M \in [l_1, l_\infty]$ induces an operator

$$M_{\text{op}} \in \mathfrak{N}(l_1) \quad \text{with} \quad \|M_{\text{op}} \mid \mathfrak{N}\| = \|M \mid [l_1, l_\infty]\|.$$

Furthermore, all operators $T \in \mathfrak{N}(l_1)$ can be obtained in this way.

5.5.3. Since l_1 has the approximation property, every operator $M_{\text{op}} \in \mathfrak{N}(l_1)$ possesses a well-defined **trace** which is simply denoted by $\tau(M)$.

Theorem. Let $M = (\mu_{ij}) \in [l_1, l_\infty]$. Then

$$\tau(M) = \sum_{i=1}^{\infty} \mu_{ii} \quad \text{and} \quad |\tau(M)| \leqq \|M \mid [l_1, l_\infty]\|.$$

Proof. Using the bounded sequences

$$m_i := (\mu_{i1}, \mu_{i2}, \dots),$$

we have the canonical nuclear representation

$$M_{\text{op}} = \sum_{i=1}^{\infty} m_i \otimes e_i.$$

Therefore

$$\tau(M) = \sum_{i=1}^{\infty} \langle e_i, m_i \rangle = \sum_{i=1}^{\infty} \mu_{ii}.$$

5.5.4. Given any infinite matrix $M = (\mu_{ij})$, we write

$$M\begin{pmatrix} i_1, \dots, i_n \\ j_1, \dots, j_n \end{pmatrix} := \det \begin{pmatrix} \mu_{i_1, j_1} & \dots & \mu_{i_1, j_n} \\ \vdots & & \vdots \\ \mu_{i_n, j_1} & \dots & \mu_{i_n, j_n} \end{pmatrix}.$$

In particular, observe that $M\binom{i}{j} = \mu_{ij}$.

5.5.5. We now describe the canonical **Fredholm denominator**

$$\delta(\zeta, M) = 1 + \sum_{n=1}^{\infty} \alpha_n(M) \zeta^n$$

of an operator M_{op} induced by a Dixon-von Koch matrix M.

Theorem (H. von Koch 1901). Let $M = (\mu_{ij}) \in [l_1, l_\infty]$. Then

$$\alpha_n(M) = \frac{1}{n!} \sum_{i_1=1}^{\infty} \dots \sum_{i_n=1}^{\infty} M\begin{pmatrix} i_1, \dots, i_n \\ i_1, \dots, i_n \end{pmatrix}$$

and

$$|\alpha_n(M)| \leqq \frac{n^{n/2}}{n!} \|M \mid [l_1, l_\infty]\|^n.$$

Proof. The formula for the coefficients of the Fredholm denominator could be deduced from 4.7.10. We prefer, however, to present here a direct approach.

Setting $\tau_i := \sup_j |\mu_{ij}|$ and applying Hadamard's inequality A.4.5, we obtain

$$\left| M \begin{pmatrix} i_1, \dots, i_n \\ i_1, \dots, i_n \end{pmatrix} \right| \leqq n^{n/2} \tau_{i_1} \dots \tau_{i_n}.$$

This implies that the series

$$\alpha_n(M) := \frac{1}{n!} \sum_{i_1=1}^{\infty} \dots \sum_{i_n=1}^{\infty} M \begin{pmatrix} i_1, \dots, i_n \\ i_1, \dots, i_n \end{pmatrix}$$

is absolutely convergent. Furthermore,

$$|\alpha_n(M)| \leqq \frac{n^{n/2}}{n!} \left(\sum_{i=1}^{\infty} \tau_i \right)^n = \frac{n^{n/2}}{n!} \| M \mid [l_1, l_\infty] \|^n.$$

It remains to show that the numbers $\alpha_n(M)$ just defined are indeed the coefficients of the Fredholm denominator $\delta(\zeta, M)$. To this end, we introduce the infinite matrices $A_n(M) = (\alpha_{n,ij}(M))$ given by

$$\alpha_{n,ij}(M) := \frac{1}{n!} \sum_{i_1=1}^{\infty} \dots \sum_{i_n=1}^{\infty} M \begin{pmatrix} i, i_1, \dots, i_n \\ j, i_1, \dots, i_n \end{pmatrix}.$$

Obviously,

$$\alpha_n(M) = \frac{1}{n} \tau[A_{n-1}(M)].$$

Expanding the determinant

$$M \begin{pmatrix} i, i_1, \dots, i_n \\ j, i_1, \dots, i_n \end{pmatrix}$$

by its first column, we get the expression

$$M \begin{pmatrix} i_1, \dots, i_n \\ i_1, \dots, i_n \end{pmatrix} M \begin{pmatrix} i \\ j \end{pmatrix} + \sum_{s=1}^{n} (-1)^s M \begin{pmatrix} i, & i_1, \dots, i_{s-1}, & i_{s+1}, \dots, i_n \\ i_1, & i_2, \dots, i_s, & i_{s+1}, \dots, i_n \end{pmatrix} M \begin{pmatrix} i_s \\ j \end{pmatrix}.$$

Hence

$$\alpha_{n,ij}(M) = \alpha_n(M) \mu_{ij} + \sum_{s=1}^{n} \beta_{ns,ij}(M),$$

where

$$\beta_{ns,ij}(M) := \frac{(-1)^s}{n!} \sum_{i_1=1}^{\infty} \dots \sum_{i_n=1}^{\infty} M \begin{pmatrix} i, & i_1, \dots, i_{s-1}, & i_{s+1}, \dots, i_n \\ i_1, & i_2, \dots, i_s, & i_{s+1}, \dots, i_n \end{pmatrix} M \begin{pmatrix} i_s \\ j \end{pmatrix}.$$

Substituting $k, i_s, \dots, i_{n-1}$ in place of $i_s, i_{s+1}, \dots, i_n$ yields

$$\begin{aligned} \beta_{ns,ij}(M) &= -\frac{1}{n!} \sum_{k=1}^{\infty} \sum_{i_1=1}^{\infty} \dots \sum_{i_{n-1}=1}^{\infty} M \begin{pmatrix} i, i_1, \dots, i_{n-1} \\ k, i_1, \dots, i_{n-1} \end{pmatrix} M \begin{pmatrix} k \\ j \end{pmatrix} \\ &= -\frac{1}{n} \sum_{k=1}^{\infty} \alpha_{n-1,ik}(M) \mu_{kj}. \end{aligned}$$

In particular, it turns out that $\beta_{ns,ij}(M)$ does not depend on the index s. Thus we obtain

$$\alpha_{n,ij}(M) = \alpha_n(M) \mu_{ij} - \sum_{k=1}^{\infty} \alpha_{n-1,ik}(M) \mu_{kj}.$$

In terms of matrices, the preceding formula reads

$$A_n(M) = \alpha_n(M)M - A_{n-1}(M)M.$$

Next, by induction, we see that

$$A_n(M) = \sum_{h=0}^{n} (-1)^{n-h}\alpha_h(M)\, M^{n-h+1}.$$

This proves that $\alpha_n(M)$ and $A_n(M)$ satisfy the recurrence formulas (R_0) and (R_1) stated in 4.4.10. Hence $\alpha_n(M)$ is the n-th coefficient of the Fredholm denominator we are looking for. Furthermore, the operator induced by the matrix $A_n(M)$ is the n-th coefficient of the associated Fredholm numerator.

5.5.6. According to the principle of related operators, the following result shows that the coefficients $\alpha_n(M)$ obtained in the preceding theorem can be used to define Fredholm denominators for nuclear operators on arbitrary Banach spaces. Of course, if the approximation property fails, then this construction depends on the choice of the nuclear representation.

Theorem (A. Pietsch 1963: a). Every operator $T \in \mathfrak{N}(E)$ is related to an operator $M_{\mathrm{op}} \in \mathfrak{N}(l_1)$ induced by an infinite matrix $M \in [l_1, l_\infty]$.

Proof. We consider a factorization

$$\begin{array}{ccc} E & \xrightarrow{\quad T \quad} & E \\ {\scriptstyle A}\downarrow & & \uparrow{\scriptstyle X} \\ l_\infty & \xrightarrow[\quad D_t \quad]{} & l_1 \end{array}$$

as described in 1.7.3. Since $D_t \in \mathfrak{N}(l_\infty, l_1)$ is nuclear, we can find a Dixon-von Koch matrix M such that $D_tAX = M_{\mathrm{op}}$.

5.5.7. Next we state an immediate corollary of 5.5.1.

Proposition. Every infinite matrix $M \in [l_1, l_2]$ induces an operator

$$M_{\mathrm{op}} \in \mathfrak{N}(l_2) \quad \text{with} \quad \|M_{\mathrm{op}} \mid \mathfrak{N}\| \leqq \|M \mid [l_1, l_2]\|.$$

Remark. It easily follows from the results in Chapter 2 that an infinite matrix represents a nuclear operator on l_2 if and only if it can be decomposed as a product of two Hilbert-Schmidt matrices.

5.5.8. We now show in a special case how the estimates for the coefficients of the Fredholm denominator improve when stronger conditions are placed on the type of the underlying infinite matrix.

Proposition. Let $M = (\mu_{ij}) \in [l_1, l_q]$ with $2 \leqq q < \infty$. Then

$$|\alpha_n(M)| \leqq \frac{n^{n(1/2-1/q)}}{n!} \|M \mid [l_1, l_q]\|^n.$$

Proof. Write

$$\tau_i := \left(\sum_{j=1}^{\infty} |\mu_{ij}|^q \right)^{1/q}.$$

Given $i_1, \ldots, i_n$, Hölder's inequality yields

$$\left(\sum_{k=1}^{n} |\mu_{ii_k}|^2\right)^{1/2} \leqq n^{1/2-1/q} \left(\sum_{k=1}^{n} |\mu_{ii_k}|^q\right)^{1/q} \leqq n^{1/2-1/q}\tau_i.$$

Next, applying Hadamard's inequality A.4.5, we obtain

$$\left|M\begin{pmatrix} i_1, \ldots, i_n \\ i_1, \ldots, i_n \end{pmatrix}\right| \leqq n^{n(1/2-1/q)}\,\tau_{i_1} \ldots \tau_{i_n}.$$

Hence

$$|\alpha_n(M)| \leqq \frac{n^{n(1/2-1/q)}}{n!}\left(\sum_{i=1}^{\infty} \tau_i\right)^n = \frac{n^{n(1/2-1/q)}}{n!}\,\|M \mid [l_1, l_q]\|^n.$$

Remark. It follows from 4.8.7 and Stirling's formula that $\delta(\zeta, M)$ is an entire function of order r, where $1/r := 1/2 + 1/q$.

5.5.9. To give explicit formulas for the coefficients of the regularized Fredholm denominators of infinite matrices, the following concept is required.

For every (n, n)-matrix $A = (\alpha_{hk})$ we define the **modified determinant** of order m by

$$\det{}_m(A) := \sum_{\pi}^{(m)} \operatorname{sign}(\pi)\,\alpha_{1\pi(1)} \ldots \alpha_{n\pi(n)},$$

where the sum is taken only over those permutations π for which all invariant subsets have at least m points. For example, if $m = 2$, then the original definition of a determinant is modified by cancelling all summands corresponding to permutations with a fixed point. This means that $\det_2(A)$ is the ordinary determinant of that matrix obtained from A by writing zeros on the principal diagonal.

Given any infinite matrix $M = (\mu_{ij})$, we put

$$M_m\begin{pmatrix} i_1, \ldots, i_n \\ j_1, \ldots, j_n \end{pmatrix} := \det{}_m\begin{pmatrix} \mu_{i_1,j_1} & \cdots & \mu_{i_1,j_n} \\ \vdots & & \vdots \\ \mu_{i_n,j_1} & \cdots & \mu_{i_n,j_n} \end{pmatrix}.$$

5.5.10. Let $M \in [l_p, l_q]$. If $1/p + 1/q \geqq 1$, then the induced operator belongs to $\mathfrak{P}_p(E)$ for every Banach space E lying between l_p and $(l_q)'$. Consequently, there exist regularized Fredholm denominators

$$\delta_m(\zeta, M) = 1 + \sum_{n=1}^{\infty} \alpha_n^{(m)}(M)\,\zeta^n$$

of all orders m greater than or equal to p. By analogy with a result of H. Poincaré (1910) we conjecture that the coefficients are given by the formula

$$\alpha_n^{(m)}(M) = \frac{1}{n!}\sum_{i_1=1}^{\infty} \cdots \sum_{i_n=1}^{\infty} M_m\begin{pmatrix} i_1, \ldots, i_n \\ i_1, \ldots, i_n \end{pmatrix}.$$

Furthermore, it should be possible to establish estimates for the modulus of $\alpha_n^{(m)}(M)$ which imply, by 4.8.6 and 4.8.7, that

$$(\lambda_n(M)) \in l_{r,\infty}, \quad \text{where } 1/r := \begin{cases} 1/p & \text{if } 1 \leqq q \leqq 2, \\ 1/p + 1/q - 1/2 & \text{if } 2 \leqq q \leqq \infty. \end{cases}$$

This would be the classical way for obtaining the eigenvalue type of the operator ideal $\mathfrak{P}_p$.

5.5.11. At present, the program envisaged in the preceding paragraph has only been carried out for Hilbert-Schmidt matrices; see (SMI, Chap. VI) and (ZAN, 9.16 and 9.17).

Theorem. Let $M = (\mu_{ij}) \in [l_2, l_2]$. Then

$$\alpha_n^{(2)}(M) = \frac{1}{n!} \sum_{i_1=1}^{\infty} \cdots \sum_{i_n=1}^{\infty} M_2 \begin{pmatrix} i_1, \ldots, i_n \\ i_1, \ldots, i_n \end{pmatrix}$$

and

$$|\alpha_n^{(2)}(M)| \leqq \frac{e^{n/2}}{n^{n/2}} \|M \mid [l_2, l_2]\|^n.$$

Remark. We stress the fact that the above estimate follows from Carleman's inequality 4.7.19,

$$|\delta_2(\zeta, M)| \leqq \exp\left(\tfrac{1}{2}\|M \mid [l_2, l_2]\|^2 \, |\zeta|^2\right) \quad \text{for all} \quad \zeta \in \mathbb{C},$$

by reasoning as in the proof of 4.8.7. No direct approach seems to be known.

CHAPTER 6

Integral operators

It is our aim to demonstrate the great importance of the methods of abstract operator theory to applications within the theory of integral operators.

First of all, we prove the classical Schur theorem on the eigenvalue distribution of operators induced by continuous kernels. Next we treat Hille-Tamarkin kernels. These are defined by the condition

$$\left(\int_0^1 \left[\int_0^1 |K(\xi, \eta)|^q \, d\eta\right]^{p/q} d\xi\right)^{1/p} < \infty .$$

In this context we also deal with kernels which may have a weak singularity on the diagonal $\xi = \eta$. The main results of this chapter are the eigenvalue theorems for kernels possessing certain smoothness properties which are described in the language of Sobolev and Besov spaces.

Almost all proofs are based on the following factorization technique. We consider appropriate Banach spaces E_0 and E_1 such that the given kernel K defines an absolutely p-summing operator from E_1 into E_0 and such that E_0 embeds into E_1. Then the induced operator K_{op} acts on every intermediate Banach space E:

$$E \xrightarrow{I} E_1 \xrightarrow{K_{op}} E_0 \xrightarrow{I} E .$$

In view of the principle of related operators, the eigenvalues of K_{op} are the same for any choice of E. Furthermore, in all cases under consideration it turns out that the smoothness properties of K are completely reflected by the properties of the embedding operator I from E_0 into E_1. This means that the special shape of the kernel in question is not essential.

In particular, the basic results apply to convolution operators induced by periodic functions. Then, up to factor 2π, the eigenvalues coincide with the Fourier coefficients. In this way we obtain various classical and modern theorems showing how the properties of a given function are inherited by the asymptotic behaviour of their Fourier coefficients. Moreover, convolution operators yield the most important examples for proving that a result obtained in the general setting the is best possible. This means that the fine index w in the conclusion $(\lambda_n(K)) \in l_{r,w}$ cannot be improved.

Finally, we investigate traces and determinants of integral operators. The classical case of continuous kernels is of special importance. Unfortunately, apart from the theory of Hilbert-Schmidt operators presented in F. Smithies's monograph (SMI), almost nothing is known about regularized Fredholm denominators.

6.1. Continuous kernels

6.1.1.* We denote by $[C(X), E]$ the set of all continuous E-valued functions Φ defined on a compact Hausdorff space X. It is well-known that $[C(X), E]$ becomes a Banach space under the norm

$$\|\Phi \mid [C, E]\| := \sup \{\|\Phi(\xi)\| : \xi \in X\} .$$

6.1.2.* We begin by stating an auxiliary result; see (KÖT, 44.7.2).

Lemma. The set of all functions Φ of the form

$$\Phi(\xi) = \sum_{i=1}^{n} x_i f_i(\xi) \quad \text{for} \quad \xi \in X$$

with $x_1, \ldots, x_n \in E$ and $f_1, \ldots, f_n \in C(X)$ is dense in $[C(X), E]$.

6.1.3.* The following fact is now obvious.

Proposition. Let $\Phi \in [C(X), E]$. Then

$$\Phi_{\text{op}} : a \to \langle \Phi(\cdot), a \rangle$$

defines an approximable operator from E' into $C(X)$. Moreover,

$$\|\Phi_{\text{op}}\| = \|\Phi \mid [C, E]\| .$$

6.1.4.* Let X and Y be compact Hausdorff spaces. A kernel K defined on $X \times Y$ is said to be of **Fredholm type** $[C(X), C(Y)]$ if the function-valued function

$$K_X : \xi \to K(\xi, \cdot)$$

belongs to $[C(X), C(Y)]$.

Remark. Obviously, the property just described is nothing other than an awkward transcription of the fact that the kernel K is continuous on $X \times Y$. However, this approach turns out to be quite useful in the following sections when the type of the kernels becomes more involved.

6.1.5.* Let $K \in [C(X), C(X)]$, and consider any finite Borel measure μ defined on the compact Hausdorff space X. Then

$$K_{\text{op}} : g(\eta) \to \int_X K(\xi, \eta)\, g(\eta)\, d\mu(\eta)$$

yields an approximable operator from $L_1(X, \mu)$ into $L_\infty(X, \mu)$. This implies that K_{op} acts on every Banach space E lying between $L_\infty(0, 1)$ and $L_1(0, 1)$:

$$E \xrightarrow{\ I\ } L_1(X, \mu) \xrightarrow{\ K_{\text{op}}\ } L_\infty(X, \mu) \xrightarrow{\ I\ } E.$$

Of course, K_{op} can also be viewed as an operator on $C(X)$. Since the associated **eigenvalue sequence** does not depend on the choice of E, it is simply denoted by $(\lambda_n(K))$.

6.1.6.* **Eigenvalue theorem for Fredholm kernels** (I. Schur 1909: a). Let μ be any finite Borel measure on a compact Hausdorff space X. Then

$$K \in [C(X), C(X)] \quad \text{implies} \quad (\lambda_n(K)) \in l_2 .$$

This result is the best possible.

Proof. We consider the factorization

$$K_{\text{op}} : L_\infty(X, \mu) \xrightarrow{\ I\ } L_1(X, \mu) \xrightarrow{\ K_{\text{op}}\ } L_\infty(X, \mu).$$

By 1.3.9,

$$I \in \mathfrak{P}_1(L_\infty(X, \mu), L_1(X, \mu)).$$

Hence K_{op} is an absolutely 1-summing operator on $L_\infty(X, \mu)$, and we see from Theorem 3.7.1 that $(\lambda_n(K)) \in l_2$.

The convolution operator derived from Example 6.5.11 shows that this result cannot be improved. This observation also follows from 3.8.1 and 6.6.8.

6.2. Hille-Tamarkin kernels

Throughout this section we assume that $1 \leqq p < \infty$ and $1 \leqq q \leqq \infty$. Furthermore, (X, μ) denotes a σ-finite measure space.

6.2.1.* An E-valued function Φ defined on X is said to be **simple** if it can be written in the form

$$\Phi(\xi) = \sum_{i=1}^{n} x_i h_i(\xi) \quad \text{for almost all } \xi \in X,$$

where $x_1, \ldots, x_n \in E$ and $h_1, \ldots, h_n$ are characteristic functions of measurable subsets $X_1, \ldots, X_n$.

Remark. In the following we only consider those simple E-valued functions which are obtained from subsets with finite measures.

6.2.2.* An E-valued function Φ defined on X is called **measurable** if there exists a sequence of simple functions Φ_n with

$$\lim_n \|\Phi(\xi) - \Phi_n(\xi)\| = 0 \quad \text{for almost all } \xi \in X.$$

6.2.3. Next we recall an important criterion; see (DIL, p. 42), (DUN, III.6.11) and (HIP, 3.5.3).

Proposition (B. J. Pettis 1938). An E-valued function Φ is measurable if and only if the following conditions are satisfied:

(1) All complex-valued functions $\langle \Phi(\cdot), a \rangle$ with $a \in E'$ are measurable.
(2) There exists a subset X_0 of measure zero such that $\Phi(X \setminus X_0)$ is separable.

Remark. A function Φ having property (1) is said to be **weakly measurable**, while property (2) means that it is **almost separably-valued.**

6.2.4.* The following result is obvious.

Lemma. For every measurable E-valued function Φ the real-valued function $\|\Phi(\cdot)\|$ is measurable, as well.

6.2.5.* We denote by $[L_p(X, \mu), E]$ the collection of all (equivalence classes of) measurable E-valued functions Φ such that

$$\|\Phi \mid [L_p, E]\| := \left(\int_X \|\Phi(\xi)\|^p \, d\mu(\xi) \right)^{1/p}$$

is finite. Such functions are called **absolutely p-integrable.** We know that $[L_p(X, \mu), E]$ is a Banach space; see (DUN, III.6.6).

Remark. Note that $[L_1(X, \mu), E]$ just consists of all **Bochner integrable** E-valued functions; see (DIL, p. 45).

6.2.6.* For a proof of the next result we refer to (DUN, III.3.8).

Lemma. The set of simple E-valued functions is dense in $[L_p(X, \mu), E]$.

6.2.7.* We now establish an analogue of 5.3.2.

Proposition. Let $\Phi \in [L_p(X, \mu), E]$. Then

$$\Phi_{\text{op}} : a \to \langle \Phi(\cdot), a \rangle$$

defines an absolutely p-summing operator from E' into $L_p(X, \mu)$. Moreover,

$$\|\Phi_{\text{op}} \mid \mathfrak{P}_p\| \leqq \|\Phi \mid [L_p, E]\|.$$

Proof. Given $a_1, \ldots, a_n \in E'$, we have

$$\left(\sum_{i=1}^{n} \|\langle \Phi(\cdot), a_i \rangle \mid L_p\|^p\right)^{1/p} = \left(\sum_{i=1}^{n} \int_X |\langle \Phi(\xi), a_i \rangle|^p \, d\mu(\xi)\right)^{1/p}$$

$$\leqq \left(\int_X \|\Phi(\xi)\|^p \, d\mu(\xi)\right)^{1/p} \sup \left\{\left(\sum_{i=1}^{n} |\langle x, a_i \rangle|^p\right)^{1/p} : x \in U\right\}.$$

6.2.8.* Let (X, μ) and (Y, ν) be measure spaces. A kernel K defined on $X \times Y$ is said to be of **Hille-Tamarkin type** $[L_p(X, \mu), L_q(Y, \nu)]$ if the function-valued function

$$K_X : \xi \to K(\xi, \cdot)$$

belongs to $[L_p(X, \mu), L_q(Y, \nu)]$. For $1 \leqq q < \infty$, this implies that

$$\|K \mid [L_p, L_q]\| := \left(\int_X \left[\int_Y |K(\xi, \eta)|^q \, d\nu(\eta)\right]^{p/q} d\mu(\xi)\right)^{1/p}$$

is finite. In the limiting case $q = \infty$ the same holds for the expression

$$\|K \mid [L_p, L_\infty]\| := \left(\int_X \operatorname*{ess-sup}_{\eta} |K(\xi, \eta)|^p \, d\mu(\xi)\right)^{1/p}.$$

6.2.9. Let $(X \times Y, \mu \times \nu)$ be the product of the measure spaces (X, μ) and (Y, ν). For $1 \leqq q < \infty$, we denote by $[\![L_p(X, \mu), L_q(Y, \nu)]\!]$ the collection of all (equivalence classes of) measurable kernels K defined on $X \times Y$ such that the iterated integral

$$\|K \mid [\![L_p, L_q]\!]\| := \left(\int_X \left[\int_Y |K(\xi, \eta)|^q \, d\nu(\eta)\right]^{p/q} d\mu(\xi)\right)^{1/p}$$

is finite. In the limiting case $q = \infty$ the same is assumed for the expression

$$\|K \mid [\![L_p, L_\infty]\!]\| := \left(\int_X \operatorname*{ess-sup}_{\eta} |K(\xi, \eta)|^p \, d\mu(\xi)\right)^{1/p}.$$

Note that $[\![L_p(X, \mu), L_q(Y, \nu)]\!]$ is a Banach space; see (JÖR, p. 169).
In the case $p = q = 2$ we get the famous Hilbert-Schmidt kernels defined in 1.4.8.

Remark. The formal difference from the concept introduced in the previous paragraph is that the kernel K satisfies a different measurability condition.

6.2.10. By a **simple rectangular kernel** K defined on $X \times Y$ we mean a function which can be written in the form

$$K(\xi, \eta) = \sum_{i=1}^{n} \lambda_i R_i(\xi, \eta),$$

where $\lambda_1, \ldots, \lambda_n \in \mathbb{C}$ and $R_1, \ldots, R_n$ are characteristic functions of measurable rectangles $X_1 \times Y_1, \ldots, X_n \times Y_n$.

Remark. In the following we only consider those simple rectangular kernels which are obtained from rectangles with finite measure.

6.2.11. Lemma. Let $1 \leqq p, q < \infty$. Then the set of simple rectangular kernels is dense in $[\![L_p(X, \mu), L_q(Y, \nu)]\!]$.

Proof. Without loss of generality, we may assume that the underlying measures are finite. Furthermore, it can easily be seen that the linear span of all characteristic functions K corresponding to arbitrary measurable subsets Z of $X \times Y$ is dense.

As is shown in (JÖR, p. 163), given $\varepsilon > 0$, there exist pairwise disjoint rectangles $X_1 \times Y_1, \ldots, X_n \times Y_n$ such that the measure of the symmetric difference

$$D := Z \triangle \bigcup_{i=1}^{n} X_i \times Y_i$$

is less than ε. Let $R_1, \ldots, R_n$ denote the corresponding characteristic functions. Then

$$K_0 := \left| K - \sum_{i=1}^{n} R_i \right|$$

is the characteristic function of D.

We now treat the cases $p \leqq q$ and $p \geqq q$ separately. First, applying Hölder's inequality with respect to the variable ξ, we obtain

$$\begin{aligned} \left\| K - \sum_{i=1}^{n} R_i \mid [\![L_p, L_q]\!] \right\| &= \left(\int_X \left[\int_Y |K_0(\xi, \eta)|^q \, d\nu(\eta) \right]^{p/q} d\mu(\xi) \right)^{1/p} \\ &\leqq \mu(X)^{1/p-1/q} \left(\int_X \int_Y |K_0(\xi, \eta)|^q \, d\mu(\xi) \, d\nu(\eta) \right)^{1/q} \\ &\leqq \mu(X)^{1/p-1/q} \, \varepsilon^{1/q}. \end{aligned}$$

Secondly, an application of Hölder's inequality with respect to the variable η yields

$$\begin{aligned} \left\| K - \sum_{i=1}^{n} R_i \mid [\![L_p, L_q]\!] \right\| &= \left(\int_X \left[\int_Y |K_0(\xi, \eta)|^q \, d\nu(\eta) \right]^{p/q} d\mu(\xi) \right)^{1/p} \\ &\leqq \nu(Y)^{1/q-1/p} \left(\int_X \int_Y |K_0(\xi, \eta)|^p \, d\mu(\xi) \, d\nu(\eta) \right)^{1/p} \\ &\leqq \nu(Y)^{1/q-1/p} \, \varepsilon^{1/p}. \end{aligned}$$

Thus every measurable characteristic function K can be approximated by simple rectangular kernels. This completes the proof.

6.2.12. We are now prepared to give a useful characterization of Hille-Tamarkin kernels.

Proposition (I. Inglis; see J. Kupka 1980). Let $1 \leqq p, q < \infty$. Then the map which assigns to every kernel K the function-valued function

$$K_X : \xi \to K(\xi, \cdot)$$

yields a metric isomorphism between

$$[\![L_p(X, \mu), L_q(Y, \nu)]\!] \quad \text{and} \quad [L_p(X, \mu), L_q(Y, \nu)].$$

Proof. Without loss of generality, we may assume that the underlying measures are finite. Given $K \in [\![L_p(X, \mu), L_q(Y, \nu)]\!]$, Fubini's theorem implies that

$$K(\xi, \cdot) \in L_q(Y, \nu) \quad \text{for almost all } \xi \in X.$$

In view of the preceding lemma, there exists a sequence of simple rectangular kernels K_n with

$$\|K - K_n \mid [\![L_p, L_q]\!]\| \leqq 2^{-n}.$$

Then it follows from

$$\int_X \left[\sum_{n=1}^{\infty} \|K(\xi, \cdot) - K_n(\xi, \cdot) \mid L_q\|\right] d\mu(\xi) < \infty$$

that

$$\lim_n \|K(\xi, \cdot) - K_n(\xi, \cdot) \mid L_q\| = 0 \quad \text{for almost all } \xi \in X.$$

Therefore the $L_q(Y, \nu)$-valued function K_X is measurable. This proves that the map $K \to K_X$ acts from $[\![L_p(X, \mu), L_q(Y, \nu)]\!]$ into $[L_p(X, \mu), L_q(Y, \nu)]$. Obviously, it is a metric injection.

As observed in 6.2.6, the set of simple $L_q(Y, \nu)$-valued functions is dense in $[L_p(X, \mu), L_q(Y, \nu)]$. However, each of these functions can be obtained from a kernel $K \in [\![L_p(X, \mu), L_q(Y, \nu)]\!]$. Thus the map $K \to KX$ is indeed onto.

Remark. In the case $q = \infty$ there exists $K \in [\![L_p(0, 1), L_\infty(0, 1)]\!]$ such that the $L_\infty(0, 1)$-valued function K_X fails to be measurable. For example, take

$$K(\xi, \eta) := \begin{cases} 1 & \text{if } 0 \leqq \eta \leqq \xi \leqq 1, \\ 0 & \text{if } 0 \leqq \xi < \eta \leqq 1. \end{cases}$$

Then K_X is not almost separably-valued.

6.2.13.* Given $K \in [L_p(X, \mu), L_q(X, \mu)]$, it follows from 6.2.7 that

$$K_{\text{op}} : g(\eta) \to \int_X K(\xi, \eta)\, g(\eta)\, d\mu(\eta)$$

defines an operator from $L_q(X, \mu)'$ into $L_p(X, \mu)$. If the measure μ is finite and $1/p + 1/q \leqq 1$, then there exists

$$\langle f, g \rangle := \int_X f(\xi)\, g(\xi)\, d\mu(\xi) \quad \text{for} \quad f \in L_p(X, \mu) \quad \text{and} \quad g \in L_q(X, \mu).$$

Using this duality, we can embed $L_p(X, \mu)$ into $L_q(X, \mu)'$. This implies that K_{op} acts on every intermediate Banach space E:

$$E \xrightarrow{I} L_q(X, \mu)' \xrightarrow{K_{\text{op}}} L_p(X, \mu) \xrightarrow{I} E.$$

For example, we may take $E := L_r(X, \mu)$ with $q' \leqq r \leqq p$. Since the associated **eigenvalue sequence** does not depend on the choice of E, it is simply denoted by $(\lambda_n(K))$.

In the case of an arbitrary measure we must assume that $p = q'$ in order to ensure that K_{op} acts on $L_p(X, \mu)$.

6.2.14.* Using the concept of a Hille-Tamarkin kernel, we now generalize 5.3.6. The proof is analogous.

Theorem (W. B. Johnson / H. König / B. Maurey / J. R. Retherford 1979).

$$K \in [L_p(X, \mu), L_{p'}(X, \mu)] \quad \text{implies} \quad (\lambda_n(K)) \in l_{\max(p, 2)}.$$

This result is the best possible.

6.2.15.* Next we establish a counterpart of 5.3.5. The notation $q^+ := \max(q', 2)$ proves to be quite convenient.

Eigenvalue theorem for Hille-Tamarkin kernels. Assume that the measure μ is finite, and let $1/p + 1/q \leqq 1$. Then

$$K \in [L_p(X, \mu), L_q(X, \mu)] \quad \text{implies} \quad (\lambda_n(K)) \in l_{q^+}.$$

This result is the best possible.

Proof. Since $p \geqq q'$, we have

$$[L_p(X, \mu), L_q(X, \mu)] \subseteq [L_{q'}(X, \mu), L_q(X, \mu)].$$

Hence it follows from the preceding theorem that $(\lambda_n(K)) \in l_{q^+}$.

In view of 6.5.20, Examples 6.5.8 and 6.5.11 show that this conclusion cannot be improved.

Remark. If $p = q = 2$, then the assertion follows immediately from the Schur-Carleman theorem 3.5.6. We further note that the case $1 \leqq p < 2$ can be reduced to this classical result via the principle of related operators. However, the remaining case where $2 < p < \infty$ was open for a long time. It was only solved in 1979 with the help of the theory of absolutely p-summing operators. Another approach, given by H. König / L. Weis (1983), uses interpolation techniques, a change of density argument and the Schatten-von Neumann operators. Preparatory work in this direction is due to M. Z. Solomyak (1970), G. E. Karadzhov (1972), J. A. Cochran / C. Oehring (1977), B. Russo (1977), J. J. F. Fournier / B. Russo (1977), W. B. Johnson / L. Jones (1978) and L. Weis (1982). No old-fashioned proof is known.

6.3. Weakly singular kernels

Throughout this section we assume that $1 < r < \infty$ and $1 \leqq w \leqq \infty$.

6.3.1. Let (X, μ) be a measure space. For any measurable complex-valued function f defined on X, the **non-increasing rearrangement** f^* is given by

$$f^*(\tau) := \inf \{c > 0 : \mu(\xi \in X : |f(\xi)| > c) \leqq \tau\} \quad \text{for} \quad \tau \geqq 0.$$

For $1 \leqq w < \infty$, the **Lorentz function space** $L_{r,w}(X, \mu)$ consists of all (equivalence classes of) measurable complex-valued functions f such that

$$\|f \mid L_{r,w}\| := \left(\int_0^\infty [\tau^{1/r - 1/w} f^*(\tau)]^w \, d\tau \right)^{1/w}$$

is finite. In the limiting case $w = \infty$ the same is assumed for

$$\|f \mid L_{r,\infty}\| := \sup_\tau \tau^{1/r} f^*(\tau).$$

In this way we obtain a linear space which is complete with respect to the quasi-norm $\|\cdot \mid L_{r,w}\|$. Since there exist equivalent norms, $L_{r,w}(X, \mu)$ even becomes a Banach space.

In the following we only consider the case when (X, μ) is the unit interval equipped with the Lebesgue measure. The corresponding Lorentz function space is denoted by $L_{r,w}(0, 1)$.

Remark. For further information we refer to (BER), (BUB), (STI) and (TRI).

6.3.2. We now provide an inequality of Lyapunov type. The special value of the constant c_r indicated below turns out to be quite useful in the proof of 6.3.5.

Lemma. Let $2 < r < \infty$ and $0 < \theta < 1$. If $1/r = (1 - \theta)/2 + \theta/\infty$, then

$$\|f \mid L_{r,1}\| \leqq c_r \|f \mid L_2\|^{1-\theta} \|f \mid L_\infty\|^\theta \quad \text{for all } f \in L_\infty(0, 1),$$

where

$$c_r := r + \left(\frac{r}{r-2}\right)^{1/2}.$$

Proof. If $0 < \sigma < \infty$, then it follows from Hölder's inequality that

$$\begin{aligned}
\|f \mid L_{r,1}\| &= \int_0^\infty \tau^{1/r-1} f^*(\tau) \,\mathrm{d}\tau \\
&= \int_0^\sigma \tau^{1/r-1} f^*(\tau) \,\mathrm{d}\tau + \int_\sigma^\infty \tau^{1/r-1} f^*(\tau) \,\mathrm{d}\tau \\
&\leqq \int_0^\sigma \tau^{1/r-1} \,\mathrm{d}\tau \, \|f \mid L_\infty\| + \left(\int_\sigma^\infty \tau^{2/r-2} \,\mathrm{d}\tau\right)^{1/2} \|f \mid L_2\| \\
&\leqq r\sigma^{1/r} \|f \mid L_\infty\| + \left(\frac{r}{r-2}\right)^{1/2} \sigma^{(2-r)/2r} \|f \mid L_2\|.
\end{aligned}$$

Setting

$$\sigma := \left(\frac{\|f \mid L_2\|}{\|f \mid L_\infty\|}\right)^2$$

yields the result we are looking for.

6.3.3. **Example** (H. König / J. R. Retherford / N. Tomczak-Jaegermann 1980). If $2 < r < \infty$, then the embedding operator from $L_\infty(0, 1)$ into $L_{r,1}(0, 1)$ is absolutely $(r, 2)$-summing, and

$$\|I : L_\infty \to L_{r,1} \mid \mathfrak{P}_{r,2}\| \leqq c_r := r + \left(\frac{r}{r-2}\right)^{1/2}.$$

Proof. Define θ by $1/r = (1 - \theta)/2 + \theta/\infty$. Let $f_1, \ldots, f_n \in L_\infty(0, 1)$. Then it follows from 1.3.9 and 6.3.2 that

$$\begin{aligned}
\left(\sum_{i=1}^n \|f_i \mid L_{r,1}\|^r\right)^{1/r} &\leqq c_r \left(\sum_{i=1}^n [\|f_i \mid L_2\|^{1-\theta} \|f_i \mid L_\infty\|^\theta]^r\right)^{1/r} \\
&\leqq c_r \left(\sum_{i=1}^n \|f_i \mid L_2\|^2\right)^{1/r} (\sup \|f_i \mid L_\infty\|)^\theta \\
&\leqq c_r \|I : L_\infty \to L_2 \mid \mathfrak{P}_2\|^{1-\theta} \|(f_i) \mid w_2\|.
\end{aligned}$$

This proves the assertion.

6.3.4. We are now in a position to determine the asymptotic behaviour of the Weyl numbers of the embedding operator from $L_\infty(0, 1)$ into $L_{r,w}(0, 1)$.

Example. Let $2 < r < \infty$ and $1 \leqq w \leqq \infty$. Then

$$x_n(I : L_\infty \to L_{r,w}) \asymp n^{-1/r}.$$

Proof. In view of 2.7.3 and 6.3.3, the upper estimate follows from

$$x_n(I : L_\infty \to L_{r,w}) \leqq x_n(I : L_\infty \to L_{r,1}) \, \| \, I : L_{r,1} \to L_{r,w} \| .$$

Consider the intervals $X_1, \ldots, X_m$ obtained by dividing $[0, 1]$ by the points $\xi_i := i/m$ with $i = 1, \ldots, m-1$, and let $h_1, \ldots, h_m$ denote the corresponding characteristic functions. Define

$$J_m := \sum_{i=1}^{m} e_i \otimes h_i \quad \text{and} \quad Q_m := \sum_{i=1}^{m} h_i \otimes e_i ,$$

where e_i is the i-th unit vector in $l(m)$. Obviously,

$$\| J_m : l_\infty(m) \to L_\infty \| = 1 .$$

Furthermore,

$$\begin{aligned} |\langle f, h_i \rangle| &\leqq \int_{X_i} |f(\xi)| \, d\xi \leqq \int_0^{1/m} f^*(\tau) \, d\tau \\ &\leqq \int_0^{1/m} \tau^{-1/r} \, d\tau \| f \mid L_{r,\infty} \| = r' m^{-1/r'} \| f \mid L_{r,\infty} \| \end{aligned}$$

implies that

$$\| Q_m : L_{r,\infty} \to l_\infty(m) \| \leqq r' m^{-1/r} .$$

We now conclude from the diagram

$$\begin{array}{ccc} L_\infty(0, 1) & \xrightarrow{\quad mI \quad} & L_{r,\infty}(0, 1) \\ J_m \big\uparrow & & \big\downarrow Q_m \\ l_\infty(m) & \xrightarrow[\quad I_m \quad]{} & l_\infty(m) \end{array}$$

and 2.9.11 that

$$\begin{aligned} \left(\frac{m - n + 1}{m} \right)^{1/2} &= a_n(I : l_1(m) \to l_2(m)) = a_n(I : l_2(m) \to l_\infty(m)) \\ &= x_n(I : l_2(m) \to l_\infty(m)) \leqq x_n(I : l_\infty(m) \to l_\infty(m)) \\ &\leqq m \| Q_m \| \, x_n(I : L_\infty \to L_{r,\infty}) \, \| J_m \| \\ &\leqq r' m^{1/r} x_n(I : L_\infty \to L_{r,w}) \, \| I : L_{r,w} \to L_{r,\infty} \| . \end{aligned}$$

Setting $m := 2n$ yields the lower estimate of $x_n(I : L_\infty \to L_{r,w})$.

Remark. In the case $1 < r < 2$ and $1 \leqq w \leqq \infty$ we have

$$x_n(I : L_\infty \to L_{r,w}) \asymp n^{-1/2} .$$

6.3.5. We now treat the limiting case $r = 2$.

Lemma (H. König 1980: d).

$$x_n(I : L_\infty \to L_{2,1}) \prec \left(\frac{1 + \log n}{n}\right)^{1/2}.$$

Proof. For every exponent r with $2 < r < \infty$ it follows from 2.7.3 and 6.3.3 that

$$x_n(I : L_\infty \to L_{2,1}) \leqq x_n(I : L_\infty \to L_{r,1}) \, \|I : L_{r,1} \to L_{2,1}\|$$
$$\leqq n^{-1/r} \|I : L_\infty \to L_{r,1} \mid \mathfrak{P}_{r,2}\| \leqq n^{-1/r}\left[r + \left(\frac{r}{r-2}\right)^{1/2}\right].$$

Setting

$$r := 2 + \frac{1}{1 + \log n},$$

we have $2 < r \leqq 3$ and

$$1/2 - 1/r = (r - 2)/2r < r - 2 = \frac{1}{1 + \log n}.$$

Hence

$$n^{-1/r} = \exp\left(-\frac{1}{r}\log n\right) < \exp\left(-\frac{1}{2}\log n + \frac{\log n}{1 + \log n}\right) < e n^{-1/2}.$$

We further note that

$$r + \left(\frac{r}{r-2}\right)^{1/2} \leqq 3 + [3(1 + \log n)]^{1/2}.$$

Therefore

$$x_n(I : L_\infty \to L_{2,1}) \leqq c \left(\frac{1 + \log n}{n}\right)^{1/2}.$$

6.3.6. A kernel K defined on the unit square $[0, 1] \times [0, 1]$ is said to be **weakly r-singular** if it can be written in the form

$$K(\xi, \eta) = \frac{B(\xi, \eta)}{|\xi - \eta|^{1/r}} \quad \text{if} \quad \xi \neq \eta,$$

where B is measurable and essentially bounded.

Remark. Note that

$$K_X : \xi \to K(\xi, \cdot)$$

is an essentially bounded $L_{r,\infty}(0, 1)$-valued function which, however, may fail to be measurable.

6.3.7. Let K be a weakly r-singular kernel. Then, in view of the duality between $L_{r,\infty}(0, 1)$ and $L_{r',1}(0, 1)$,

$$K_{\mathrm{op}} : g(\eta) \to \int_0^1 K(\xi, \eta) \, g(\eta) \, d\eta$$

defines an operator from $L_{r',1}(0, 1)$ into $L_\infty(0, 1)$. This implies that K_{op} acts on every Banach space E lying between $L_\infty(0, 1)$ and $L_{r',1}(0, 1)$. Since the associated **eigenvalue sequence** does not depend on the choice of E, it is simply denoted by $(\lambda_n(K))$.

6.3.8. We are now in a position to establish the main result of this section.

Eigenvalue theorem for weakly singular kernels (H. König / J. R. Retherford / N. Tomczak-Jaegermann 1980, H. König 1980: d). Suppose that K is weakly r-singular. Then

$$(\lambda_n(K)) \in l_{r',\infty} \quad \text{if} \quad 1 < r < 2 \quad \text{and} \quad (\lambda_n(K)) \in l_2 \quad \text{if} \quad 2 < r < \infty.$$

In the remaining case $r = 2$ we have

$$|\lambda_n(K)| \prec \left(\frac{1 + \log n}{n}\right)^{1/2}.$$

All results are best possible.

Proof. We consider the factorization

$$K_{\mathrm{op}} : L_\infty(0, 1) \xrightarrow{\ I\ } L_{r',1}(0, 1) \xrightarrow{\ K_{\mathrm{op}}\ } L_\infty(0, 1).$$

If $1 < r < 2$, then it follows from 6.3.4 that

$$I \in \mathfrak{L}^{(x)}_{r',\infty}(L_\infty(0, 1), L_{r',1}(0, 1)).$$

Hence, by the eigenvalue theorem 3.6.2, we have $(\lambda_n(K)) \in l_{r',\infty}$. Let $f^{2\pi}_{r,0}$ be the 2π-periodic function defined in 6.5.8. Then the associated convolution operator shows that this result is sharp.

If $2 < r < \infty$, then

$$I \in \mathfrak{P}_2(L_\infty(0, 1), L_{r',1}(0, 1)).$$

Therefore 3.7.1 implies that $(\lambda_n(K)) \in l_2$. Note that this conclusion cannot be improved, even for continuous kernels.

In the case $r = 2$ the estimate

$$|\lambda_n(K)| \prec \left(\frac{1 + \log n}{n}\right)^{1/2}$$

is an easy consequence of 6.3.5 and 3.6.1. A rather involved example constructed by H. König (1980: d) shows that the logarithmic term is indeed necessary; see also (KÖN, 3.a.11–13).

6.4. Besov kernels

The standard monographs on Sobolev and Besov spaces of scalar-valued functions are (ADA), (NIK) and (TRI). However, the theory of E-valued functions has rarely been treated. The following papers are the only ones (known to me) which deal with this case:

P. Grisvard (1966), J. Wloka (1967), J. Kadlec / V. B. Korotkov (1968), G. Sparr (1974), M. Sh. Birman / M. Z. Solomyak (1977: a) and Z. Ciesielski / T. Figiel (1983).

Moreover, the situation is even worse, since these authors have very often omitted proofs claiming that they can be adapted step by step from the scalar-valued setting. Thus, apart from (KÖN), we are not in a position to recommend any rigorous reference. On the other hand, it would be beyond the scope of this book to provide all necessary details. This section is therefore in striking contrast to the rest of the book. It presents the most beautiful applications of the abstract theory of eigenvalue distributions to integral operators, but requires a lot of blind confidence on the part of the reader. Nevertheless, I bet my mathematical reputation (but not my car!) that all the statements are correct.

Throughout this section all functions under consideration are defined on the unit interval [0, 1]. Furthermore, let $\sigma, \tau > 0$, $1 \leqq p, q \leqq \infty$ and $1 \leqq u, v \leqq \infty$.

6.4.1.* For every E-valued function Φ and $0 \leqq \tau \leqq 1$ we put

$$\Delta_\tau \Phi(\xi) := \Phi(\xi + \tau) - \Phi(\xi) \quad \text{if} \quad 0 \leqq \xi \leqq 1 - \tau$$

and $\Delta_\tau \Phi(\xi) := 0$ otherwise. The m-th **forward difference** of Φ with step τ is inductively defined by $\Delta_\tau^{m+1}\Phi := \Delta_\tau^m \Delta_\tau \Phi$ and $\Delta_\tau^1 \Phi := \Delta_\tau \Phi$. Obviously, we have

$$\Delta_\tau^m \Phi(\xi) = \sum_{h=0}^{m} (-1)^{m-h} \binom{m}{h} \Phi(\xi + h\tau) \quad \text{if} \quad 0 \leqq \xi \leqq 1 - m\tau.$$

6.4.2.* Let $0 < \sigma < m$. For $1 \leqq u < \infty$, the **Besov function space** $[B_{p,u}^{\sigma,m}(0, 1), E]$ is the collection of all (equivalence classes of) functions $\Phi \in [L_p(0, 1), E]$ such that

$$\|\Phi \mid [B_{p,u}^{\sigma,m}, E]\| := \|\Phi \mid [L_p, E]\| + \left(\int_0^1 [\tau^{-\sigma} \|\Delta_\tau^m \Phi \mid [L_p, E]\|]^u \frac{d\tau}{\tau} \right)^{1/u}$$

is finite. In the limiting case $u = \infty$ the same is assumed for

$$\|\Phi \mid [B_{p,\infty}^{\sigma,m}, E]\| := \|\Phi \mid [L_p, E]\| + \sup_\tau \{\tau^{-\sigma} \|\Delta_\tau^m \Phi \mid [L_p, E]\|\}.$$

If $p = \infty$, then $[L_\infty(0, 1), E]$ must be replaced by $[C(0, 1), E]$.

At first sight, this definition seems to depend on the choice of m. However, it can be shown that we always get the same Banach space, the norms $\|\cdot \mid [B_{p,u}^{\sigma,m}, E]\|$ being equivalent. Hence the shortened notation $[B_{p,u}^\sigma(0, 1), E]$ is justified.

If E is the complex field, then we simply write $B_{p,u}^\sigma(0, 1)$.

6.4.3. In the case $0 < \sigma < 1$, $p = \infty$, $u = \infty$ and $m = 1$ the above definition yields the **Lipschitz-Hölder function space** $[C^\sigma(0, 1), E]$ which consists of all $\Phi \in [C(0, 1), E]$ such that

$$\|\Phi \mid [C^\sigma, E]\| := \|\Phi \mid [C, E]\| + \sup_{\xi_1 \neq \xi_2} \frac{\|\Phi(\xi_1) - \Phi(\xi_2)\|}{|\xi_1 - \xi_2|^\sigma}$$

is finite.

6.4.4.* The following inclusion is obvious.

Proposition. Let $1 \leqq p_1 \leqq p_0 \leqq \infty$. Then

$$[B_{p_0,u}^\sigma(0, 1), E] \subseteq [B_{p_1,u}^\sigma(0, 1), E].$$

6.4.5.* We denote by $[C^m(0, 1), E]$ the set of all E-valued functions Φ which have derivatives

$$D^h \Phi \in [C(0, 1), E] \quad \text{for} \quad h = 0, \ldots, m.$$

Note that the norm

$$\|\Phi \mid [C^m, E]\| := \max \{\|D^h \Phi \mid [C, E]\| : h = 0, \ldots, m\}$$

turns $[C^m(0, 1), E]$ into a Banach space.

6.4.6.* Let $1 \leqq p < \infty$. The **Sobolev function space** $[W_p^m(0, 1), E]$ is the collection of all (equivalence classes of) E-valued functions Φ which have weak derivatives

$$D^h\Phi \in [L_p(0, 1), E] \quad \text{for} \quad h = 0, \ldots, m.$$

It is well-known that $[W_p^m(0, 1), E]$ becomes a Banach space under the norm

$$\|\Phi \mid [W_p^m, E]\| := \left(\sum_{h=0}^{m} \|D^h\Phi \mid [L_p, E]\|^p\right)^{1/p}.$$

Remark. Since $[C^m(0, 1), E]$ is a dense linear subset of $[W_p^m(0, 1), E]$, the Sobolev function space $[W_p^m(0, 1), E]$ can be obtained as the completion of $[C^m(0, 1), E]$ with respect to the above norm. Defining $[W_p^m(0, 1), E]$ in this way, we may avoid all reference to distribution theory.

6.4.7.* For a proof of the following result we refer to (KÖN, 3.b.7).

Proposition. Let $0 < \sigma < m$ and $\sigma = \theta m$. Then

$$[B_{p,u}^\sigma(0, 1), E] = ([L_p(0, 1), E], [W_p^m(0, 1), E])_{\theta,u} \quad \text{if } 1 \leqq p < \infty$$

and

$$[B_{\infty,u}^\sigma(0, 1), E] = ([C(0, 1), E], [C^m(0, 1), E])_{\theta,u}.$$

The norms on both sides are equivalent.

6.4.8.* In view of the reiteration property, we have the following corollary of the preceding formulas.

Interpolation proposition. Let $0 < \sigma_0 < \sigma_1 < \infty$, $1 \leqq u_0, u_1, u \leqq \infty$ and $0 < \theta < 1$. If $\sigma = (1 - \theta)\,\sigma_0 + \theta\sigma_1$, then

$$([B_{p,u_0}^{\sigma_0}(0, 1), E], [B_{p,u_1}^{\sigma_1}(0, 1), E])_{\theta,u} = [B_{p,u}^\sigma(0, 1), E].$$

The norms on both sides are equivalent.

6.4.9.* Let $0 < \gamma_1 < \ldots < \gamma_n < 1$ and $m = 1, 2, \ldots$ Recall that $\xi_+ := \max(\xi, 0)$.

By an E-valued **spline** of order m with knots $\gamma_1, \ldots, \gamma_n$ we mean a function Σ which can be written in the form

$$\Sigma(\xi) = \sum_{i=1}^{m} x_i\xi^{i-1} + \sum_{j=1}^{n} x_{m+j}(\xi - \gamma_j)_+^{m-1},$$

where $x_1, \ldots, x_{m+n} \in E$. This representation is unique.

In the case $m \geqq 2$ we have the following characterization: The E-valued function Σ is $(m - 2)$ times continuously differentiable on $[0, 1]$, and the restrictions to the intervals $(0, \gamma_1), \ldots, (\gamma_i, \gamma_{i+1}), \ldots, (\gamma_n, 1)$ are E-valued polynomials of degree at most $(m - 1)$.

An E-valued spline of order 1 is piecewise constant and left-continuous at the knots.

Remark. For further information about real-valued splines we recommend L. Schumaker's monograph (SCU).

6.4.10.* From now on we use the dyadic knots

$$\gamma_{hi} := \frac{2i - 1}{2^{h+1}} \quad \text{with} \quad h = 0, 1, \ldots \quad \text{and} \quad i = 1, \ldots, 2^h.$$

That is

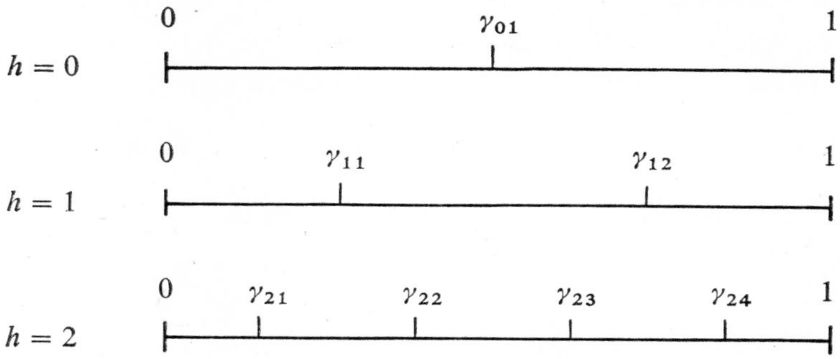

etc.

Let S_n be the set of all complex-valued polynomials of degree at most $(n-1)$. Furthermore, we denote by S^m_{hi} the set of all complex-valued splines of order m with knots $\gamma_{01}; \ldots; \gamma_{h1}, \ldots, \gamma_{hi}$. Note that

$$\dim(S_n) = n \quad \text{and} \quad \dim(S^m_{hi}) = m + 2^h + i - 1\,.$$

Therefore

$$S_1 \subset \ldots \subset S_m \subset S^m_{01} \subset \ldots \subset S^m_{h1} \subset \ldots \subset S^m_{hi} \subset \ldots$$

is a sequence of linear subspaces of the Hilbert space $L_2(0, 1)$ such that the dimension increases by one at every step. Next, by Schmidt's procedure, we construct the orthonormal **Ciesielski system:**

$$s_1, \ldots, s_m, s^m_{01}, \ldots, s^m_{h1}, \ldots, s^m_{hi}, \ldots$$

Each of these functions belongs to the corresponding subspace S_n and S^m_{hi}, respectively. If $m = 1$, then we obtain the classical **Haar system**, while $m = 2$ yields the so-called **Franklin system**; see A. Haar (1910) and Ph. Franklin (1928).

6.4.11.* We are now prepared to describe an isomorphism between Besov function spaces and Besov sequence spaces which is basic in what follows. For a complete proof of this extremely deep result the reader is referred to the original papers of the Gdansk school. Another approach via interpolation theory is due to H. Triebel (1981).

We define the **Ciesielski transform**

$$C_m : f \to (\alpha_n) \oplus (\alpha^m_{hi}),$$

where

$$\alpha_n := \int_0^1 f(\xi)\, s_n(\xi)\, d\xi \quad \text{and} \quad \alpha^m_{hi} := \int_0^1 f(\xi)\, s^m_{hi}(\xi)\, d\xi\,.$$

The desired result now reads as follows:

Theorem (S. Ropela 1976, Z. Ciesielski / T. Figiel 1983). Let $m > \sigma + 1 - 1/p$. Then C_m yields an isomorphism between

$$B^\sigma_{p,u}(0, 1) \quad \text{and} \quad l_p(m) \oplus b^{\sigma-1/p+1/2}_{p,u}\,.$$

Remark. We conjecture that this statement remains true in the case of E-valued functions and sequences.

6.4.12.* The following result is analogous to 5.4.4.

Lemma. Let $\sigma + \tau > 1/p + 1/q - 1$. Then there exists

$$\langle f, g \rangle := \int_0^1 f(\xi)\, g(\xi)\, d\xi \quad \text{for} \quad f \in B^{\sigma}_{p,u}(0, 1) \quad \text{and} \quad g \in B^{\tau}_{q,v}(0, 1).$$

Using this duality, we can embed $B^{\sigma}_{p,u}(0, 1)$ into $B^{\tau}_{q,v}(0, 1)'$.

Proof. In view of $1 > (1/p - \sigma) + (1/q - \tau)$, we may choose p_0 and q_0 with

$$1 > 1/p_0 + 1/q_0, \quad 1/p_0 > 1/p - \sigma, \quad 1/q_0 > 1/q - \tau.$$

Then it follows from the classical embedding theorem that

$$B^{\sigma}_{p,u}(0, 1) \subset L_{p_0}(0, 1) \quad \text{and} \quad B^{\tau}_{q,v}(0, 1) \subset L_{q_0}(0, 1).$$

Finally, Hölder's inequality yields the existence of the bounded bilinear form $\langle f, g \rangle$.

6.4.13.* We now describe how the embedding operator I_B from $B^{\sigma}_{p,u}(0, 1)$ into $B^{\tau}_{q,v}(0, 1)'$ is related to embedding operators I_m and I_b acting between sequence spaces.

Proposition. Let

$$\sigma + \tau > 1/p + 1/q - 1 \quad \text{and} \quad m > \max(\sigma + 1 - 1/p, \tau + 1 - 1/q).$$

Then

$$\begin{array}{ccc} B^{\sigma}_{p,u}(0, 1) & \xrightarrow{\quad I_B \quad} & B^{\tau}_{q,v}(0, 1)' \\ \Big\downarrow C_m & & \Big\uparrow C'_m \\ l_p(m) \oplus b^{\sigma-1/p+1/2}_{p,u} & \xrightarrow[\quad I_m \oplus I_b \quad]{} & l_q(m)' \oplus (b^{\tau-1/q+1/2}_{q,v})'. \end{array}$$

Proof. The assertion is an immediate consequence of Parseval's formula applied to the Ciesielski system.

More precisely, if $C_m f = (\alpha_n) \oplus (\alpha^m_{hi})$ and $C_m g = (\beta_n) \oplus (\beta^m_{hi})$, then

$$\int_0^1 f(\xi)\, g(\xi)\, d\xi = \sum_{n=1}^{m} \alpha_n \beta_n + \sum_{h=0}^{\infty} \sum_{i=1}^{2^h} \alpha^m_{hi} \beta^m_{hi}.$$

Remark. In fact, it is superfluous to prove the existence of the embedding operator I_B in advance, since it follows directly from the above diagram and Lemma 5.4.4.

6.4.14.* We now transfer Proposition 5.4.6 into the context of Besov function spaces. For another, self-contained proof we refer to (KÖN, 3.c.5 and 3.c.7).

Proposition (A. Pietsch 1980: d). Let $\sigma + \tau > 1/p + 1/q - 1 \geqq 0$. Then

$$x_n(I : B^{\sigma}_{p,u}(0, 1) \to B^{\tau}_{q,v}(0, 1)') \prec n^{-\varrho},$$

where

$$\varrho := \sigma + \tau + \begin{cases} 0 & \text{if} \quad 1 \leqq p \leqq q' \leqq 2, \\ 1/2 - 1/q & \text{if} \quad 1 \leqq p \leqq 2 \leqq q' \leqq \infty, \\ 1 - 1/p - 1/q & \text{if} \quad 2 \leqq p \leqq q' \leqq \infty. \end{cases}$$

6.4.15. Adaptation to 6.2.7 yields the proof of the following result.

Proposition. Let $\Phi \in [W_p^m(0, 1), E]$. Then

$$\Phi_{\text{op}} : a \to \langle \Phi(\cdot), a \rangle$$

defines an absolutely p-summing operator from E' into $W_p^m(0, 1)$. Moreover,

$$\|\Phi_{\text{op}} \mid \mathfrak{P}_p\| \leqq \|\Phi \mid [W_p^m, E]\| .$$

6.4.16.* An analogous conclusion holds for Besov function spaces.

Proposition (A. Pietsch 1980: d). Let $\Phi \in [B_{p,u}^\sigma(0, 1), E]$ and $r := \max(p, u)$. Then

$$\Phi_{\text{op}} : a \to \langle \Phi(\cdot), a \rangle$$

defines an absolutely r-summing operator from E' into $B_{p,u}^\sigma(0, 1)$. Moreover,

$$\|\Phi_{\text{op}} \mid \mathfrak{P}_r\| \leqq \|\Phi \mid [B_{p,u}^\sigma, E]\| .$$

Proof. Given $a_1, \ldots, a_n \in E'$, Jessen's inequality C.3.10 yields

$$\left(\sum_{i=1}^n \left[\int_0^1 |\langle \Phi(\xi), a_i \rangle|^p \, d\xi\right]^{r/p}\right)^{1/r} \leqq \left(\int_0^1 \left[\sum_{i=1}^n |\langle \Phi(\xi), a_i \rangle|^r\right]^{p/r} d\xi\right)^{1/p} .$$

Therefore

$$\left(\sum_{i=1}^n \|\langle \Phi(\cdot), a_i \rangle \mid L_p\|^r\right)^{1/r} \leqq \|\Phi \mid [L_p, E]\| \; \|(a_i) \mid [w_r, E']\| .$$

Applying this result to $\Delta_\tau^m \Phi$, we obtain

$$\left(\sum_{i=1}^n \|\langle \Delta_\tau^m \Phi(\cdot), a_i \rangle \mid L_p\|^r\right)^{1/r} \leqq \|\Delta_\tau^m \Phi \mid [L_p, E]\| \; \|(a_i) \mid [w_r, E']\| .$$

Hence

$$\left[\sum_{i=1}^n \left(\int_0^1 [\tau^{-\sigma} \|\langle \Delta_\tau^m \Phi(\cdot), a_i \rangle \mid L_p\|]^u \frac{d\tau}{\tau}\right)^{r/u}\right]^{1/r} \leqq$$

$$\leqq \left[\int_0^1 \left(\sum_{i=1}^n [\tau^{-\sigma} \|\langle \Delta_\tau^m \Phi(\cdot), a_i \rangle \mid L_p\|]^r\right)^{u/r} \frac{d\tau}{\tau}\right]^{1/u}$$

$$\leqq \left[\int_0^1 (\tau^{-\sigma} \|\Delta_\tau^m \Phi \mid [L_p, E]\|)^u \frac{d\tau}{\tau}\right]^{1/u} \|(a_i) \mid [w_r, E']\| .$$

Finally, we conclude from the preceding inequalities that

$$\left(\sum_{i=1}^n \|\langle \Phi(\cdot), a_i \rangle \mid B_{p,u}^\sigma\|^r\right)^{1/r} \leqq$$

$$\leqq \left(\sum_{i=1}^n \|\langle \Phi(\cdot), a_i \rangle \mid L_p\|^r\right)^{1/r} + \left[\sum_{i=1}^n \left(\int_0^1 [\tau^{-\sigma} \|\Delta_\tau^m \langle \Phi(\cdot), a_i \rangle \mid L_p\|]^u \frac{d\tau}{\tau}\right)^{r/u}\right]^{1/r}$$

$$\leqq \left(\|\Phi \mid [L_p, E]\| + \left[\int_0^1 (\tau^{-\sigma} \|\Delta_\tau^m \Phi \mid [L_p, E]\|)^u \frac{d\tau}{\tau}\right]^{1/u}\right) \|(a_i) \mid [w_r, E']\|$$

$$= \|\Phi \mid [B_{p,u}^\sigma, E]\| \; \|(a_i) \mid [w_r, E']\| .$$

This shows that Φ_{op} is absolutely r-summing.

Remark. The result just proved could be deduced immediately from Proposition 5.4.7 provided that the Ciesielski isomorphism would be available in the setting of E-valued functions and sequences.

6.4.17.* A kernel K defined on the unit square $[0, 1] \times [0, 1]$ is said to be of **Besov type** $[B^\sigma_{p,u}(0, 1), B^\tau_{q,v}(0, 1)]$ if the function-valued function

$$K_X : \xi \to K(\xi, \cdot)$$

belongs to $[B^\sigma_{p,u}(0, 1), B^\tau_{q,v}(0, 1)]$.

6.4.18.* Given $K \in [B^\sigma_{p,u}(0, 1), B^\tau_{q,v}(0, 1)]$, it follows from 6.4.16 that

$$K_{\text{op}} : g(\eta) \to \int_0^1 K(\xi, \eta)\, g(\eta)\, d\eta$$

defines an operator from $B^\tau_{q,v}(0, 1)'$ into $B^\sigma_{p,u}(0, 1)$. Here g denotes a distribution. If $\sigma + \tau > 1/p + 1/q - 1$, then $B^\sigma_{p,u}(0, 1)$ is embedded into $B^\tau_{q,v}(0, 1)'$. This implies that K_{op} acts on every intermediate Banach space E. Since the associated **eigenvalue sequence** does not depend on the choice of E, it is simply denoted by $(\lambda_n(K))$.

6.4.19.*We now give the most important application of the abstract theory of eigenvalue distributions to integral operators. Recall that $q^+ := \max(q', 2)$.

Eigenvalue theorem for Besov kernels (A. Pietsch 1980: d). Suppose that $\sigma + \tau > 1/p + 1/q - 1$ and

$$1/r := \sigma + \tau + 1/q^+ .$$

Then

$$K \in [B^\sigma_{p,u}(0, 1), B^\tau_{q,v}(0, 1)] \quad \text{implies} \quad (\lambda_n(K)) \in l_{r,u}.$$

This result is the best possible.

Proof (isomorphic to 5.4.10). First we assume that $p \leqq q'$ and $u = 1$. Define

$$\varrho := \sigma + \tau + \begin{cases} 0 & \text{if } \; 1 \leqq p \leqq q' \leqq 2, \\ 1/2 - 1/q & \text{if } \; 1 \leqq p \leqq 2 \leqq q' \leqq \infty, \\ 1 - 1/p - 1/q & \text{if } \; 2 \leqq p \leqq q' \leqq \infty, \end{cases}$$

and $s := \max(p, 2)$. We know from 6.4.14 that

$$x_n(I : B^\sigma_{p,1}(0, 1) \to B^\tau_{q,v}(0, 1)') \prec n^{-\varrho}.$$

Hence

$$I \in \mathfrak{L}^{(x)}_{1/\varrho, \infty}(B^\sigma_{p,1}(0, 1), B^\tau_{q,v}(0, 1)').$$

Furthermore, by 6.4.16 and 2.7.4,

$$K_{\text{op}} \in \mathfrak{L}^{(x)}_{s,\infty}(B^\tau_{q,v}(0, 1)', B^\sigma_{p,1}(0, 1)).$$

Thus, in view of the factorization

$$K_{\text{op}} : B^\tau_{q,v}(0, 1)' \xrightarrow{K_{\text{op}}} B^\sigma_{p,1}(0, 1) \xrightarrow{I} B^\tau_{q,v}(0, 1)',$$

we conclude from the multiplication theorem 2.4.18 that

$$K_{\text{op}} \in \mathfrak{L}^{(x)}_{r,\infty}(B^\tau_{q,v}(0, 1)'),$$

where $1/r = \varrho + 1/s$ takes precisely the values given in the above theorem.

If $p > q'$ and $u = 1$. Then we see from 6.4.4 that

$$K \in [B^{\sigma}_{p,1}(0,1), B^{\tau}_{q,v}(0,1)] \quad \text{implies} \quad K \in [B^{\sigma}_{q',1}(0,1), B^{\tau}_{q,v}(0,1)].$$

In this way the second case is reduced to the first one.

So far we have shown that the map $K \to K_{\mathrm{op}}$, which assigns to every kernel the corresponding operator, acts as follows:

$$\mathrm{op}: [B^{\sigma}_{p,1}(0,1), B^{\tau}_{q,v}(0,1)] \to \mathfrak{L}^{(x)}_{r,\infty}(B^{\tau}_{q,v}(0,1)').$$

This result can be improved by interpolation. To this end, choose σ_0, σ_1 and θ such that

$$\sigma = (1-\theta)\,\sigma_0 + \theta\sigma_1 \quad \text{and} \quad 0 < \sigma_0 < \sigma_1 < \infty.$$

Since $\sigma + \tau > 1/p + 1/q - 1$, it is further possible to satisfy the condition $\sigma_0 + \tau > 1/p + 1/q - 1$. We now apply the formula

$$([B^{\sigma_0}_{p,1}(0,1), E], [B^{\sigma_1}_{p,1}(0,1), E])_{\theta,u} = [B^{\sigma}_{p,u}(0,1), E]$$

for $E := B^{\tau}_{q,v}(0,1)$. Next replacing σ by σ_0 and σ_1, we define r_0 and r_1 in the same way as r. Then it follows from $1/r = (1-\theta)/r_0 + \theta/r_1$ that

$$(\mathfrak{L}^{(x)}_{r_0,\infty}, \mathfrak{L}^{(x)}_{r_1,\infty})_{\theta,u} \subseteq \mathfrak{L}^{(x)}_{r,u}.$$

Hence the interpolation property yields

$$\mathrm{op}: [B^{\sigma}_{p,u}(0,1), B^{\tau}_{q,v}(0,1)] \to \mathfrak{L}^{(x)}_{r,u}(B^{\tau}_{q,v}(0,1)').$$

By the eigenvalue theorem 3.6.2, we therefore obtain $(\lambda_n(K)) \in l_{r,u}$.

In view of 6.5.24, Examples 6.5.15 and 6.5.16 show that this conclusion cannot be improved. See also Theorem 6.5.17.

Remark. The concept of a Besov function space $[B^{\sigma}_{p,u}(0,1), E]$ can be extended to arbitrary real exponents σ; see G. Sparr (1974). In this setting the above theorem remains true under the condition $\sigma + \tau > (1/p + 1/q - 1)_+$. Furthermore, we may replace the unit interval by any smoothly bounded domain of the d-dimensional euclidean space. If $(\sigma + \tau)/d > (1/p + 1/q - 1)_+$, then the conclusion holds for $1/r := (\sigma + \tau)/d + 1/q^+$; see Z. Ciesielski / T. Figiel (1983) and (KÖN, 3.d.5).

6.4.20. The n-th **spline approximation number** of order m of an E-valued function $\Phi \in [L_p(0,1), E]$ is defined by

$$a_n(\Phi \mid [L_p, E]_m) := \inf \|\Phi - \Sigma \mid [L_p, E]\|,$$

the infimum being taken over all E-valued splines

$$\Sigma(\xi) = \sum_{i=1}^{n-1} x_i \xi^{i-1} \qquad \text{if} \quad n \leqq m+1$$

and

$$\Sigma(\xi) = \sum_{i=1}^{m} x_i \xi^{i-1} + \sum_{2^h + i \leqq n-m} x_{hi}(\xi - \gamma_{hi})_+^{m-1} \quad \text{if} \quad n > m+1.$$

6.4.21. We now investigate the relationship between approximation numbers of functions and of operators.

Lemma. Let $\Phi \in [L_p(0,1), E]$. Then

$$a_n(\Phi_{\mathrm{op}}: E' \to L_p(0,1) \mid \mathfrak{P}_p) \leqq a_n(\Phi \mid [L_p, E]_m).$$

Proof. The assertion is an immediate consequence of the fact that every E-valued spline Σ occuring in the definition of $a_n(\Phi \mid [L_p, E]_m)$ induces an operator from E' into $L_p(0, 1)$ such that

$$\|\Phi_{\text{op}} - \Sigma_{\text{op}} \mid \mathfrak{P}_p\| \leqq \|\Phi - \Sigma \mid [L_p, E]\| \quad \text{and} \quad \operatorname{rank}(\Sigma_{\text{op}}) < n.$$

Remark. The above inequality remains true for the smaller approximation numbers obtained from E-valued splines with *free knots.*

6.4.22. The following criterion shows that, just like the operator ideals $\mathfrak{L}^{(a)}_{p,u}$, the Besov function spaces $[B^\sigma_{p,u}(0, 1), E]$ can be characterized by means of certain approximation rates. For a proof of this result we refer to Z. Ciesielski / T. Figiel (1983). The scalar-valued case is treated in (SCU, 6.27 and 6.55).

Proposition. Let $\Phi \in [L_p(0, 1), E]$ and $m > \sigma + 1 - 1/p$. Then

$$\Phi \in [B^\sigma_{p,u}(0, 1), E] \quad \text{if and only if} \quad (a_n(\Phi \mid [L_p, E]_m)) \in l_{1/\sigma, u}.$$

6.4.23. A kernel K defined on the unit square $[0, 1] \times [0, 1]$ is said to be of **Besov-Hille-Tamarkin type** $[B^\sigma_{p,u}(0, 1), L_q(0, 1)]$ if the function-valued function

$$K_X : \xi \to K(\xi, \cdot)$$

belongs to $[B^\sigma_{p,u}(0, 1), L_q(0, 1)]$.

Kernels of **Hille-Tamarkin-Besov type** $[L_p(0, 1), B^\tau_{q,v}(0, 1)]$ are defined analogously.

6.4.24. Eigenvalue theorem for Besov-Hille-Tamarkin kernels (A. Pietsch 1981: a). Let $\sigma > 1/p + 1/q - 1$ and

$$1/r := \sigma + 1/q^+ .$$

Then

$$K \in [B^\sigma_{p,u}(0, 1), L_q(0, 1)] \quad \text{implies} \quad (\lambda_n(K)) \in l_{r,u}.$$

This result is the best possible.

Proof. Since

$$B^0_{q,1}(0, 1) \subset L_q(0, 1) \subset B^0_{q,\infty}(0, 1),$$

the assertion can be deduced from the extended version of the eigenvalue theorem 6.4.19 (Remark). We prefer, however, to give here a different proof which works under the stronger assumption that $1/p + 1/q \leqq 1$. Then

$$K \in [B^\sigma_{q',u}(0, 1), L_q(0, 1)].$$

It follows from 6.4.21 and 6.4.22 that

$$K_{\text{op}} \in (\mathfrak{P}_{q'})^{(a)}_{1/\sigma, u}(L_{q'}(0, 1)).$$

Thus, by 1.2.4, 2.7.4 and 2.8.15, we have

$$K_{\text{op}} \in \mathfrak{L}^{(x)}_{r,u}(L_{q'}(0, 1)),$$

and the eigenvalue theorem for operators of Weyl type implies that $(\lambda_n(K)) \in l_{r,u}$.

In view of 6.5.22, Examples 6.5.15 and 6.5.16 show that this result cannot be improved.

6.4.25. In order to establish a counterpart of the preceding theorem, an auxiliary result from interpolation theory is required. For a proof we refer to (KÖN, 3.d.1). See also (TRI, 1.18.4).

Proposition (M. Sh. Birman / M. Z. Solomyak 1977: a). Let (E_0, E_1) be any interpolation couple. If $p \leqq w$, then

$$[L_p(0, 1), (E_0, E_1)_{\theta, w}] \subseteqq ([L_p(0, 1), E_0], [L_p(0, 1), E_1])_{\theta, w}.$$

6.4.26. In Theorem 6.4.19 the asymptotic behaviour of eigenvalues does not depend on the fine index v. We now describe a situation where this is not so. A more detailed proof of the following result is to be found in (KÖN, 3.d.7).

Eigenvalue theorem for Hille-Tamarkin-Besov kernels (M. Sh. Birman / M. Z. Solomyak 1977: a, A. Pietsch 1981: a). Let $\tau > 1/p + 1/q - 1$, $p \leqq v$ and

$$1/r := \tau + 1/q^+.$$

Then

$$K \in [L_p(0, 1), B^{\tau}_{q, v}(0, 1)] \quad \text{implies} \quad (\lambda_n(K)) \in l_{r, v}.$$

This result is the best possible.

Proof. Without loss of generality, we may assume that $p \leqq q'$. Then, reasoning similarly as in the proof of 6.4.19, it follows that the map $K \to K_{\text{op}}$, which assigns to every kernel the corresponding operator, acts as follows:

$$\text{op} : [L_p(0, 1), B^{\tau}_{q, v}(0, 1)] \to \mathfrak{L}^{(x)}_{r, \infty}(L_p(0, 1)).$$

This result can be improved by interpolation. To this end, we apply 6.4.25 to the interpolation couple $(B^{\tau_0}_{q, v_0}(0, 1), B^{\tau_1}_{q, v_1}(0, 1))$. The interpolation property now implies that

$$\text{op} : [L_p(0, 1), B^{\tau}_{q, v}(0, 1)] \to \mathfrak{L}^{(x)}_{r, v}(L_p(0, 1)).$$

By the eigenvalue theorem 3.6.2, we therefore obtain $(\lambda_n(K)) \in l_{r, v}$.

In view of 6.5.23, Examples 6.5.15 and 6.5.16 show that this result cannot be improved.

Remark. In the case $p > v$ the precise version of the above eigenvalue theorem is unknown. However, if $\tau > 1/v + 1/q - 1$, then the required condition $p \leqq v$ can always be satisfied by setting $p := v$.

6.4.27. A kernel K defined on the unit square $[0, 1] \times [0, 1]$ is said to be of **Sobolev type** $[W^m_p(0, 1), W^n_q(0, 1)]$ if the function-valued function

$$K_X : \xi \to K(\xi, \cdot)$$

belongs to $[W^m_p(0, 1), W^n_q(0, 1)]$.

6.4.28. Let $1 \leqq p, q < \infty$. We denote by $[\![W^m_p(0, 1), W^n_q(0, 1)]\!]$ the collection of all (equivalence classes of) measurable kernels K defined on the unit square $[0, 1] \times [0, 1]$ which have weak partial derivatives

$$D^h_\xi D^k_\eta K \in [\![L_p(0, 1), L_q(0, 1)]\!] \quad \text{for} \quad h = 0, \ldots, m \quad \text{and} \quad k = 0, \ldots, n.$$

Obviously, $[\![W^m_p(0, 1), W^n_q(0, 1)]\!]$ becomes a Banach space under a suitable norm.

6.4.29. In the context of Sobolev spaces Proposition 6.2.12 reads as follows.

Proposition. Let $1 \leqq p,\ q < \infty$. Then the map which assigns to every kernel K the function-valued function

$$K_X : \xi \to K(\xi, \cdot)$$

yields an isomorphism between

$$\llbracket W_p^m(0, 1), W_q^n(0, 1) \rrbracket \quad \text{and} \quad [W_p^m(0, 1), W_q^n(0, 1)].$$

6.4.30. The interpolation formula stated in 6.4.25 admits the following generalization.

Proposition. Let (E_0, E_1) be any interpolation couple. If $p \leqq w$, then

$$[W_p^m(0, 1), (E_0, E_1)_{\theta, w}] \subseteq ([W_p^m(0, 1), E_0], [W_p^m(0, 1), E_1])_{\theta, w}.$$

6.4.31. We conclude this section with a result which originates in the work of M. G. Krejn (1937).

Eigenvalue theorem for Sobolev kernels. Let $1 \leqq p < \infty$, $1 < q < \infty$,

$$1/r := m + n + 1/q^+ \quad \text{and} \quad w := \min(q, 2).$$

Then

$$K \in [W_p^m(0, 1), W_q^n(0, 1)] \quad \text{implies} \quad (\lambda_n(K)) \in l_{r, w}.$$

This result is the best possible.

Proof. Assume first that $1 < q < 2$. Reasoning as in the proof of Theorem 6.4.19, we see from the diagram

$$\begin{array}{ccccc} K_{\text{op}} : W_p^m(0, 1) & \xrightarrow{\quad I \quad} & W_q^n(0, 1)' & \xrightarrow{\quad K_{\text{op}} \quad} & W_p^m(0, 1) \\ I \downarrow & & \uparrow I & & \\ B_{p, \infty}^m(0, 1) & \xrightarrow[\quad I \quad]{} & B_{q, \infty}^n(0, 1)' & & \end{array}$$

that

$$K_{\text{op}} \in \mathfrak{L}_{r, \infty}^{(x)}(W_p^m(0, 1)).$$

Therefore the map $K \to K_{\text{op}}$, which assigns to every kernel the corresponding operator, acts as follows:

$$\text{op} : [W_p^m(0, 1), W_q^n(0, 1)] \to \mathfrak{L}_{r, \infty}^{(x)}(W_p^m(0, 1)).$$

This result can be improved by interpolation. Choose q_0, q_1 and θ such that

$$1/q = (1 - \theta)/q_0 + \theta/q_1 \quad \text{and} \quad 1 < q_0 < q < q_1 < 2.$$

Then

$$W_q^n(0, 1) = (W_{q_0}^n(0, 1), W_{q_1}^n(0, 1))_{\theta, q}.$$

Without loss of generality, we may assume that $p \leqq q$. In this case, it follows from 6.4.30 that

$$[W_p^m(0, 1), W_q^n(0, 1)] \subseteq ([W_p^m(0, 1), W_{q_0}^n(0, 1)], [W_p^m(0, 1), W_{q_1}^n(0, 1)])_{\theta, q}.$$

Letting

$$1/r_0 := m + n + 1/q_0' \quad \text{and} \quad 1/r_1 := m + n + 1/q_1',$$

we have $1/r = (1-\theta)/r_0 + \theta/r_1$. Thus

$$(\mathfrak{L}^{(x)}_{r_0,\infty}, \mathfrak{L}^{(x)}_{r_1,\infty})_{\theta,q} \subseteq \mathfrak{L}^{(x)}_{r,q}.$$

Hence the interpolation property yields

$$\mathrm{op}: [W_p^m(0,1), W_q^n(0,1)] \to \mathfrak{L}^{(x)}_{r,q}(W_p^m(0,1)).$$

By the eigenvalue theorem 3.6.2, we therefore obtain $(\lambda_n(K)) \in l_{r,q}$.

We now assume that $2 \leqq q < \infty$. Obviously, it is enough to treat the case $1 \leqq p \leqq q = 2$. It follows from 6.4.15 and 2.7.2 that

$$K_{\mathrm{op}} \in \mathfrak{L}_2^{(a)}(W_2^n(0,1)', W_p^m(0,1)).$$

Furthermore, with $\sigma := m + n$ we have

$$I \in \mathfrak{L}^{(x)}_{1/\sigma,\infty}(W_p^m(0,1), W_2^n(0,1)').$$

The multiplication theorem 2.4.18 implies that

$$K_{\mathrm{op}} \in \mathfrak{L}^{(x)}_{r,2}(W_p^m(0,1)).$$

By the eigenvalue theorem 3.6.2, we therefore have $(\lambda_n(K)) \in l_{r,2}$.

In view of 6.5.21, the examples described in 6.5.18 show that the conclusions obtained above cannot be improved.

6.5. Fourier coefficients

Throughout this section we consider complex-valued 2π-periodic functions on the real line. In order to indicate this fact, the underlying function spaces are denoted by $C(2\pi)$, $L_r(2\pi)$, $L_{r,w}(2\pi)$, $W_r^l(2\pi)$ and $B^{\varrho}_{r,w}(2\pi)$. The required modifications in the definitions are left to the reader.

6.5.1. By a **trigonometric polynomial** of degree n we mean a function t which can be represented in the form

$$t(\xi) = \sum_{|k| \leqq n} \gamma_k \exp(ik\xi) \quad \text{for all} \quad \xi \in \mathbb{R},$$

where $\gamma_{-n}, \ldots, \gamma_n \in \mathbb{C}$ and $|\gamma_{-n}| + |\gamma_n| > 0$. If so, then we write $\deg(t) = n$.

6.5.2. Every function $f \in L_1(2\pi)$ induces a **convolution operator**

$$C^f_{\mathrm{op}}: g(\eta) \to \int_0^{2\pi} f(\xi - \eta)\, g(\eta)\, d\eta$$

on $C(2\pi)$ and $L_p(2\pi)$ with $1 \leqq p \leqq \infty$; see (DUN, XI.3.1) and (ZYG, II.1.15).

If t is a trigonometric polynomial, then C^t_{op} possesses finite rank. More precisely, we have

$$\mathrm{rank}(C^t_{\mathrm{op}}) = 2\deg(t) + 1.$$

6.5.3. Recall that the Lorentz function spaces are defined for $1 < r < \infty$ and $1 \leqq w \leqq \infty$; see 6.3.1.

Proposition. Let $f \in L_{r,w}(2\pi)$. Then the associated convolution operator C_{op}^f acts from $L_{r',w'}(2\pi)$ into $L_\infty(2\pi)$.

Proof. Given $g \in L_{r',w'}(2\pi)$, by (HAY, p. 278) and Hölder's inequality, for almost all $\xi \in \mathbb{R}$ we have

$$\left|\int_0^{2\pi} f(\xi - \eta)\, g(\eta)\, d\eta\right| \leq \left|\int_0^{2\pi} f^*(\tau)\, g^*(\tau)\, d\tau\right| = \int_0^{2\pi} \tau^{1/r-1/w} f^*(\tau)\, \tau^{1/r'-1/w'} g^*(\tau)\, d\tau$$
$$\leq \left(\int_0^{2\pi} [\tau^{1/r-1/w} f^*(\tau)]^w\, d\tau\right)^{1/w} \left(\int_0^{2\pi} [\tau^{1/r'-1/w'} g^*(\tau)]^{w'}\, d\tau\right)^{1/w'}.$$

This proves that

$$\|C_{\text{op}}^f g \mid L_\infty\| \leqq \|f \mid L_{r,w}\|\, \|g \mid L_{r',w'}\|.$$

6.5.4. For every function $f \in L_1(2\pi)$ and $k \in \mathbb{Z}$ the k-th **Fourier coefficient** is defined by

$$\gamma_k(f) := \frac{1}{2\pi} \int_0^{2\pi} f(\xi) \exp(-ik\xi)\, d\xi.$$

6.5.5. We now describe the relationship between the Fourier coefficients of a function and the eigenvalues of the associated convolution operator.

Proposition (T. Carleman 1918). Let $f \in L_1(2\pi)$. Then the eigenvalue sequence of C_{op}^f can be obtained as the non-increasing rearrangement of $(2\pi\, \gamma_k(f))$.

Proof. Define

$$e_k(\xi) := \exp(ik\xi) \quad \text{for all} \quad \xi \in \mathbb{R}.$$

Then it follows from

$$\int_0^{2\pi} f(\xi - \eta)\, e_k(\eta)\, d\eta = \int_0^{2\pi} f(\eta) \exp(ik\xi - ik\eta)\, d\eta = 2\pi\, \gamma_k(f)\, e_k(\xi)$$

hat $2\pi\, \gamma_k(f)$ is an eigenvalue of C_{op}^f with the associated eigenfunction e_k.

Next, if λ denotes an arbitrary eigenvalue of C_{op}^f, then for any associated eigenfunction $g \in L_1(2\pi)$ we have

$$\lambda(g, e_k) = (C_{\text{op}}^f g, e_k) = \int_0^{2\pi} \int_0^{2\pi} f(\xi - \eta)\, g(\eta) \exp(-ik\xi)\, d\eta\, d\xi$$
$$= \int_0^{2\pi} \int_0^{2\pi} f(\xi)\, g(\eta) \exp(-ik\xi - ik\eta)\, d\xi\, d\eta$$
$$= (f, e_k)\, (g, e_k) = 2\pi\, \gamma_k(f)\, (g, e_k).$$

Hence $\lambda \neq 2\pi\, \gamma_k(f)$ implies $(g, e_k) = 0$. Since $g \neq o$, the eigenvalue must coincide with at least one of the coefficients $2\pi\, \gamma_k(f)$. Finally, we note that the multiplicity of any eigenvalue $\lambda \neq 0$ is the dimension of the subspace

$$N_\infty(\lambda, C_{\text{op}}^f) = N_1(\lambda, C_{\text{op}}^f) = \text{span}\, \{e_k : 2\pi\, \gamma_k(f) = \lambda\}.$$

6.5.6. We now state one of the oldest results about Fourier coefficients.

Theorem (B. Riemann 1854, H. Lebesgue 1903).

$$f \in L_1(2\pi) \quad \text{implies} \quad (\gamma_k(f)) \in c_0(\mathbb{Z}).$$

Proof. We know from the classical Weierstrass approximation theorem that the set of trigonometric polynomials is dense in $L_1(2\pi)$. This shows that C^f_{op} is an approximable operator on $C(2\pi)$. The assertion now follows from 3.2.19 and 6.5.5.

6.5.7. **Theorem** (M. A. Parseval 1805, F. W. Bessel 1828, E. Fischer 1907, F. Riesz 1907).

$$f \in L_2(2\pi) \quad \text{implies} \quad (\gamma_k(f)) \in l_2(\mathbb{Z}).$$

This result is the best possible.

Proof. Obviously, the above implication is an immediate consequence of Bessel's inequality:

$$\sum_{\mathbb{Z}} |\gamma_k(f)|^2 \leqq \|f \mid L_2\|^2.$$

We prefer, however, to employ the tools presented in this monograph. To this end, we consider the factorization

$$C^f_{\text{op}} : L_\infty(2\pi) \xrightarrow{\ I\ } L_2(2\pi) \xrightarrow{\ C^f_{\text{op}}\ } L_\infty(2\pi).$$

Then it follows from

$$I \in \mathfrak{P}_2(L_\infty(2\pi), L_2(2\pi))$$

that C^f_{op} is an absolutely 2-summing operator on $L_\infty(2\pi)$. Hence, in view of 3.7.1 and 6.5.5, we have $(\gamma_k(f)) \in l_2(\mathbb{Z})$.

Conversely, by the classical Fischer-Riesz theorem, every sequence $(\gamma_k) \in l_2(\mathbb{Z})$ determines a function $f \in L_2(2\pi)$. This proves that the above result is sharp.

6.5.8. **Example.** Define

$$f^{2\pi}_{r,\alpha}(\xi) := \sum_{k=1}^{\infty} \frac{1}{k^{1/r'}(1 + \log k)^\alpha} \exp(ik\xi),$$

where $1 < r < \infty$ and $\alpha \geqq 0$. Then

$$f^{2\pi}_{r,\alpha} \in L_{r,w}(2\pi) \quad \text{if and only if} \quad \alpha w > 1$$

and

$$f^{2\pi}_{r,0} \in L_{r,\infty}(2\pi).$$

Proof. The assertion follows immediately from the fact that $f^{2\pi}_{r,\alpha}$ is continuous at all points except for $\xi = 2\pi k$, and that it behaves like

$$f_{r,\alpha}(\xi) := \frac{1}{|\xi|^{1/r}(1 + |\log \xi|)^\alpha}$$

as $\xi \to 0$; see (ZYG, V.2.6).

6.5.9. **Theorem** (W. H. Young 1912, F. Hausdorff 1923, G. H. Hardy/J. E. Littlewood 1931). Let $1 < r < 2$ and $1 \leqq w \leqq \infty$. Then

$$f \in L_{r,w}(2\pi) \quad \text{implies} \quad (\gamma_k(f)) \in l_{r',w}(\mathbb{Z}).$$

This result is the best possible.

Proof. We first treat the case $w = \infty$. Recall from 6.3.3 that

$$I \in \mathfrak{P}_{r',2}(L_\infty(2\pi), L_{r',1}(2\pi)).$$

Therefore, in view of the factorization

$$C^f_{\text{op}} : L_\infty(2\pi) \xrightarrow{\ I\ } L_{r',1}(2\pi) \xrightarrow{\ C^f_{\text{op}}\ } L_\infty(2\pi)$$

and 2.7.4, we have

$$C^f_{\text{op}} \in \mathfrak{P}_{r',2}(L_\infty(2\pi)) \subseteq \mathfrak{L}^{(x)}_{r',\infty}(L_\infty(2\pi)).$$

This proves that the map $f \to C^f_{\text{op}}$, which assigns to every function the associated convolution operator, acts as follows:

$$\text{op} : L_{r,\infty}(2\pi) \to \mathfrak{L}^{(x)}_{r',\infty}(L_\infty(2\pi)).$$

Next we choose r_0, r_1 and θ such that

$$1/r = (1-\theta)/r_0 + \theta/r_1 \quad \text{and} \quad 1 < r_0 < r < r_1 < 2.$$

Then the interpolation formulas

$$(L_{r_0,\infty}(2\pi), L_{r_1,\infty}(2\pi))_{\theta,w} = L_{r,w}(2\pi)$$

and

$$(\mathfrak{L}^{(x)}_{r_0',\infty}, \mathfrak{L}^{(x)}_{r_1',\infty})_{\theta,w} \subseteq \mathfrak{L}^{(x)}_{r',w}$$

imply that

$$\text{op} : L_{r,w}(2\pi) \to \mathfrak{L}^{(x)}_{r',w}(L_\infty(2\pi)).$$

Consequently, by the eigenvalue theorem for operators of Weyl type and 6.5.5, we have $(\gamma_k(f)) \in l_{r',w}(\mathbb{Z})$.

Finally, the preceding example shows that this result cannot be improved.

6.5.10. The following construction goes back to H. S. Shapiro (1951), W. Rudin (1959) and J. J. F. Fournier (1974).

Lemma. Let $\gamma_0, \ldots, \gamma_n \in \mathbb{C}$. Then there exists a trigonometric polynomial

$$t_n(\xi) = \sum_{k=1}^{2^n} \gamma_k(t_n) \exp(ik\xi)$$

such that

$$\gamma_{2^h}(t_n) = \gamma_h \quad \text{for} \quad h = 0, \ldots, n$$

and

$$\|t_n \mid C\| \leq \left(e \sum_{h=0}^{n} |\gamma_h|^2\right)^{1/2}.$$

Proof. By homogeneity, it is enough to treat the case in which

$$\sum_{h=0}^{n} |\gamma_h|^2 \leq 1.$$

Using the complex variable $\zeta = \exp(i\xi)$, we construct two kinds of polynomials.

First let

$$s_0(\zeta) := \zeta \quad \text{and} \quad t_0(\zeta) := \gamma_0 \zeta .$$

Proceeding inductively, we then define

$$s_h(\zeta) := \zeta^{2^{h-1}} s_{h-1}(\zeta) - \gamma_h^* t_{h-1}(\zeta)$$

and

$$t_h(\zeta) := \gamma_h \zeta^{2^{h-1}} s_{h-1}(\zeta) + t_{h-1}(\zeta).$$

Obviously,

$$s_h(\zeta) = \sum_{k=1}^{2^h} \gamma_k(s_h)\, \zeta^k \quad \text{and} \quad t_h(\zeta) = \sum_{k=1}^{2^h} \gamma_k(t_h)\, \zeta^k .$$

Furthermore, for the leading coefficients we have

$$\gamma_{2^h}(s_h) = 1 \quad \text{and} \quad \gamma_{2^h}(t_h) = \gamma_h .$$

Observe that t_h coincides with the 2^h-th partial sum of the Taylor expansion of t_n provided that $h \leqq n$. Hence

$$\gamma_{2^h}(t_n) = \gamma_{2^h}(t_h) = \gamma_h \quad \text{for} \quad h = 0, \ldots, n .$$

For arbitrary complex numbers γ, σ and τ we have

$$|\sigma - \gamma^* \tau|^2 + |\gamma\sigma + \tau|^2 = (1 + |\gamma|^2)\,(|\sigma|^2 + |\tau|^2).$$

Applying this formula to the defining equations of s_h and t_h yields

$$|s_h(\zeta)|^2 + |t_h(\zeta)|^2 = (1 + |\gamma_h|^2)\,(|s_{h-1}(\zeta)|^2 + |t_{h-1}(\zeta)|^2).$$

Hence

$$|s_n(\zeta)|^2 + |t_n(\zeta)|^2 = \prod_{h=0}^{n} (1 + |\gamma_h|^2) \leqq \exp\left(\sum_{h=0}^{n} |\gamma_h|^2\right) \leqq \mathrm{e}.$$

Thus t_n (considered as a function of ξ) possesses all the desired properties.

Remark. If $\gamma_0 = \ldots = \gamma_n = 1$, then we have

$$\gamma_k(t_n) = \pm 1 \quad \text{for} \quad k = 1, \ldots, 2^n \quad \text{and} \quad \|t_n \mid C\| \leqq 2^{n/2} .$$

6.5.11. In view of 6.5.7,

$$f \in C(2\pi) \quad \text{implies} \quad (\gamma_k(f)) \in l_2(\mathbb{Z}).$$

It was T. Carleman (1918) who made the surprising observation that this result cannot be improved.

Example (S. Banach 1930). Given $(\gamma_h) \in l_2$, there exists a function $f \in C(2\pi)$ such that $\gamma_{2^h}(f) = \gamma_h$ for $h = 0, 1, \ldots$

Proof. The sequence $c = (\gamma_h)$ can be represented by an absolutely convergent series

$$c = \sum_{k=1}^{\infty} c_k ,$$

where $c_k = (\gamma_{kh})$ are finite sequences. Choose $n_1 < n_2 < \ldots$ such that

$$\gamma_{kh} = 0 \quad \text{if} \quad h > n_k .$$

By the preceding lemma, there exist trigonometric polynomials t_{n_k} with

$$\gamma_{2^h}(t_{n_k}) = \gamma_{kh} \quad \text{and} \quad \|t_{n_k} \mid C\| \leqq e^{1/2}\|c_k \mid l_2\|.$$

Therefore, setting

$$f := \sum_{k=1}^{\infty} t_{n_k}$$

yields the desired function.

Remark. Another result along these lines states that for $(\gamma_k) \in l_2(\mathbb{Z})$ there exists $f \in C(2\pi)$ with $|\gamma_k(f)| \geqq |\gamma_k|$; see K. de Leeuw / Y. Katznelson / J.-P. Kahane (1977) and (KAS, 9.2.5).

6.5.12. The n-th **trigonometric approximation number** of a function $f \in L_r(2\pi)$ is defined by

$$a_n(f \mid L_r) := \inf\{\|f - t \mid L_r\| : \deg(t) < n\},$$

where t denotes a trigonometric polynomial.

6.5.13. The next result is analogous to 6.4.21.

Lemma. Let $f \in L_r(2\pi)$. Then

$$a_{2n}(C_{\text{op}}^f : L_{r'} \to L_\infty) \leqq a_n(f \mid L_r).$$

Proof. The assertion is an immediate consequence of the fact that every trigonometric polynomial t induces a convolution operator from $L_{r'}(2\pi)$ into $L_\infty(2\pi)$ such that

$$\|C_{\text{op}}^f - C_{\text{op}}^t\| \leqq \|f - t \mid L_r\| \quad \text{and} \quad \text{rank}(C_{\text{op}}^t) = 2\deg(t) + 1.$$

6.5.14. For a proof of the following criterion we refer to (BUS, 4.2.2 and 4.2.3) and (SCI, 3.7.1). See also (NIK, Chap. 5) and (TIM, Chap. 5 and 6).

Proposition (O. V. Besov 1961). Let $f \in L_r(2\pi)$. Then

$$f \in B_{r,w}^{\varrho}(2\pi) \quad \text{if and only if} \quad (a_n(f \mid L_r)) \in l_{1/\varrho, w}.$$

6.5.15. Next we modify the example given in 6.5.8.

Example. Define

$$f_{r,\varrho,\alpha}^{2\pi}(\xi) := \sum_{k=1}^{\infty} \frac{1}{k^{\varrho+1/r'}(1 + \log k)^\alpha} \exp(ik\xi),$$

where $1 < r < \infty$, $0 < \varrho < 1/r$ and $\alpha \geqq 0$. Then

$$f_{r,\varrho,\alpha}^{2\pi} \in B_{r,w}^{\varrho}(2\pi) \quad \text{if and only if } \alpha w > 1$$

and

$$f_{r,\varrho,0}^{2\pi} \in B_{r,\infty}^{\varrho}(2\pi).$$

Proof. Let $0 < \xi \leqq \pi$. Then it follows from

$$\left|\sin \frac{n}{2}\xi\right| \leqq 1 \quad \text{and} \quad \sin \frac{1}{2}\xi \geqq \frac{\xi}{\pi}$$

that the partial sums

$$r_n(\xi) := \sum_{k=1}^{n} \exp(ik\xi) = \frac{\sin\frac{n}{2}\xi}{\sin\frac{1}{2}\xi} \exp\left(i\frac{n+1}{2}\xi\right)$$

satisfy the inequality

$$|r_n(\xi)| \leqq \frac{\pi}{\xi}.$$

We now write

$$f_n(\xi) := \sum_{k=n}^{\infty} \gamma_k \exp(ik\xi) \quad \text{with } \gamma_k := \frac{1}{k^{\varrho+1/r'}(1+\log k)^{\alpha}}.$$

Summation by parts yields

$$|f_n(\xi)| = \left|\sum_{k=n}^{\infty} \gamma_k(r_k(\xi) - r_{k-1}(\xi))\right| \left| -\gamma_n r_{n-1}(\xi) + \sum_{k=n}^{\infty} (\gamma_k - \gamma_{k+1})\, r_k(\xi)\right|$$

$$\leqq \left(\gamma_n + \sum_{k=n}^{\infty} (\gamma_k - \gamma_{k+1})\right)\frac{\pi}{\xi} = \frac{2\pi}{\xi}\gamma_n.$$

Next, for $m > n$, it follows from G.3.2 that

$$|f_n(\xi)| \leqq \sum_{k=n}^{m-1} \gamma_k + |f_m(\xi)| \leqq \left(c_0 m + \frac{2\pi}{\xi}\right)\gamma_m.$$

Note that $|f_n(2\pi - \xi)| = |f_n(\xi)|$. Using G.3.2 again, we now obtain

$$a_n(f^{2\pi}_{r,\varrho,\alpha} \mid L_r) \leqq \|f_n \mid L_r\| = \left(2\int_0^{\pi} |f_n(\xi)|^r \,d\xi\right)^{1/r}$$

$$= \left(2\sum_{m=n+1}^{\infty} \int_{\pi/m}^{\pi/(m-1)} |f_n(\xi)|^r \,d\xi + 2\int_{\pi/n}^{\pi} |f_n(\xi)|^r \,d\xi\right)^{1/r}$$

$$\leqq c_1\left(\sum_{m=n+1}^{\infty} m^{-2}(m\gamma_m)^r + \gamma_n^r \int_{\pi/n}^{\pi} \xi^{-r}\,d\xi\right)^{1/r} \leqq c_2 \frac{1}{n^{\varrho}(1+\log n)^{\alpha}}.$$

Therefore, by the preceding proposition, $f^{2\pi}_{r,\varrho,\alpha}$ belongs to the desired Besov space.
The converse implication follows from Theorem 6.5.17.

6.5.16. Let t_n denote the trigonometric polynomial constructed in 6.5.10 (Remark) such that

$$\gamma_k(t_n) = \pm 1 \quad \text{for} \quad k = 1, \ldots, 2^n \quad \text{and} \quad \|t_n \mid C\| \leqq 2^{n/2}.$$

Example. Define

$$g^{2\pi}_{\varrho,\alpha}(\xi) := \sum_{n=0}^{\infty} \frac{1}{2^{n(\varrho+1/2)}(n+1)^{\alpha}} \exp(i2^n\xi)\, t_n(\xi),$$

where $\varrho > 0$ and $\alpha \geqq 0$. Then

$$g^{2\pi}_{\varrho,\alpha} \in B^{\varrho}_{\infty,w}(2\pi) \quad \text{if and only if} \quad \alpha w > 1$$

and

$$g^{2\pi}_{\varrho,0} \in B^{\varrho}_{\infty,\infty}(2\pi).$$

Moreover, $g^{2\pi}_{\varrho,\alpha}$ has 2^n times the Fourier coefficient $\pm 2^{-n(\varrho+1/2)}(n+1)^{-\alpha}$.
Proof Approximating $g^{2\pi}_{\varrho,\alpha}$ by its partial sums, we see from G.3.2 that

$$a_{2^k+1}(g^{2\pi}_{\varrho,\alpha} \mid C) \leqq \sum_{n=k}^{\infty} \frac{1}{2^{n(\varrho+1/2)}(n+1)^\alpha} \|t_n \mid C\| \leqq c \frac{1}{2^{k\varrho}(k+1)^\alpha}.$$

Hence, by 2.1.10 and 6.5.14, the function $g^{2\pi}_{\varrho,\alpha}$ belongs to the desired Besov space.
The converse implication follows from the next theorem.
Finally, we observe that the trigonometric polynomial $\exp(\mathrm{i}\, 2^n\xi)\, t_n(\xi)$ has Fourier coefficients ± 1 if and only if $2^n < k \leqq 2^{n+1}$.

6.5.17. Recall that $r^+ := \max(r', 2)$.

Theorem (A. Pietsch 1980: a). Let

$$1/s := \varrho + 1/r^+.$$

Then

$$f \in B^\varrho_{r,w}(2\pi) \quad \text{implies} \quad (\gamma_k(f)) \in l_{s,w}(\mathbb{Z}).$$

This result is the best possible.

Proof. In view of 6.5.24, the theorem is a special case of 6.4.19. We prefer, however, to give here a direct proof.
To this end, we consider the factorization

$$C^f_{\mathrm{op}} : L_\infty(2\pi) \xrightarrow{\ I\ } L_{r'}(2\pi) \xrightarrow{\ C^f_{\mathrm{op}}\ } L_\infty(2\pi).$$

Note that

$$I \in \mathfrak{P}_{r'}(L_\infty(2\pi), L_{r'}(2\pi)) \subseteq \mathfrak{L}^{(x)}_{r^+,\infty}(L_\infty(2\pi), L_{r'}(2\pi)).$$

Moreover, by 6.5.13 and 6.5.14, we have

$$C^f_{\mathrm{op}} \in \mathfrak{L}^{(a)}_{1/\varrho,w}(L_{r'}(2\pi), L_\infty(2\pi)).$$

This proves that

$$C^f_{\mathrm{op}} \in \mathfrak{L}^{(x)}_{s,w}(L_\infty(2\pi)).$$

Hence 3.6.2 and 6.5.5 imply $(\gamma_k(f)) \in l_{s,w}(\mathbb{Z})$.
Finally, Examples 6.5.15 and 6.5.16 show that this result cannot be improved.

6.5.18. Theorem. Let $1 < r < \infty$,

$$1/s := l + 1/r^+ \quad \text{and} \quad w := \min(r, 2).$$

Then

$$f \in W^l_r(2\pi) \quad \text{implies} \quad (\gamma_k(f)) \in l_{s,w}(\mathbb{Z}).$$

This result is the best possible.

Proof. As shown in the course of the previous proof, the map $f \to C^f_{\mathrm{op}}$, which assigns to every function the associated convolution operator, acts as follows:

$$\mathrm{op} : B^l_{r,\infty}(2\pi) \to \mathfrak{L}^{(x)}_{s,\infty}(L_\infty(2\pi)).$$

Since $W_r^l(2\pi) \subseteq B_{r,\infty}^l(2\pi)$, we also have

$$\text{op}: W_r^l(2\pi) \to \mathfrak{L}_{r,\infty}^{(x)}(L_\infty(2\pi)).$$

We now deal with the case $1 < r < 2$. Choose r_0, r_1 and θ such that

$$1/r = (1-\theta)/r_0 + \theta/r_1 \quad \text{and} \quad 1 < r_0 < r < r_1 < 2.$$

Let

$$1/s_0 := l + 1/r_0' \quad \text{and} \quad 1/s_1 := l + 1/r_1'.$$

Then

$$1/s = (1-\theta)/s_0 + \theta/s_1 \quad \text{and} \quad 0 < s_1 < s < s_0 < \infty.$$

In view of the interpolation formulas

$$(W_{r_0}^l(2\pi), W_{r_1}^l(2\pi))_{\theta,r} = W_r^l(2\pi)$$

and

$$(\mathfrak{L}_{s_0,\infty}^{(x)}, \mathfrak{L}_{s_1,\infty}^{(x)})_{\theta,r} \subseteq \mathfrak{L}_{s,r}^{(x)},$$

we obtain

$$\text{op}: W_r^l(2\pi) \to \mathfrak{L}_{s,r}^{(x)}(L_\infty(2\pi)).$$

Therefore, if $1 < r < 2$, then $f \in W_r^l(2\pi)$ implies $(\gamma_k(f)) \in l_{s,r}(\mathbb{Z})$. Taking any l-th primitive of the function $f_{r,\alpha}^{2\pi}$ with $\alpha r > 1$, we see that this result is sharp.

In the case $2 \leqq r < \infty$ it follows from $f \in W_r^l(2\pi)$ that $f \in W_2^l(2\pi) = B_{2,2}^l(2\pi)$. Hence, by the preceding theorem, we have $(\gamma_k(f)) \in l_{s,2}(\mathbb{Z})$. Finally, using a result stated in 6.5.11 (Remark), for $0 < w < 2$, we may construct a function $f \in C^l(2\pi)$ such that $(\gamma_k(f)) \notin l_{s,w}(\mathbb{Z})$.

6.5.19. In what follows we investigate how certain properties of a function $f \in L_1(2\pi)$ are inherited by the **convolution kernel**

$$C^f(\xi, \eta) := f(\xi - \eta) \quad \text{for} \quad \xi, \eta \in \mathbb{R}.$$

In this way we obtain examples which show that the results stated in 6.1.6, 6.2.15, 6.4.19 etc. are sharp.

6.5.20. The first observation along these lines is well-known; see (ZYG, II.1.11).

Proposition. Let $1 \leqq q < \infty$. Then

$$f \in L_q(2\pi) \quad \text{implies} \quad C^f \in [C(2\pi), L_q(2\pi)].$$

Proof. The assertion is trivial for continuous functions. Since $C(2\pi)$ is dense in $L_q(2\pi)$, it can easily be extended to all $f \in L_q(2\pi)$.

Remark. The conclusion fails for $q = \infty$. However, it holds when $L_\infty(2\pi)$ is replaced by $C(2\pi)$.

6.5.21. In the context of Sobolev spaces the preceding result reads as follows; see (NIK, 5.6.5).

Proposition. Let $1 \leqq q < \infty$. Then

$$f \in W_q^{m+n}(2\pi) \quad \text{implies} \quad C^f \in [C^m(2\pi), W_q^n(2\pi)].$$

6.5.22. Proposition. Let $1 \leqq q \leqq \infty$. Then

$$f \in B^{\sigma}_{q,u}(2\pi) \quad \text{implies} \quad C^f \in [B^{\sigma}_{\infty,u}(2\pi), L_q(2\pi)].$$

Proof. We know from the previous propositions that the map $f \to C^f$, which assigns to every function the associated convolution kernel, acts as follows:

$$\text{con} : L_q(2\pi) \to [C(2\pi), L_q(2\pi)]$$

and

$$\text{con} : W^m_q(2\pi) \to [C^m(2\pi), L_q(2\pi)].$$

Let $\sigma = \theta m$. Then, by 6.4.7, the interpolation property yields

$$\text{con} : B^{\sigma}_{q,u}(2\pi) \to [B^{\sigma}_{\infty,u}(2\pi), L_q(2\pi)].$$

6.5.23. The next result is taken from (NIK, 5.6.5).

Proposition. Let $1 \leqq v < \infty$. Then

$$f \in B^{\tau}_{q,v}(0, 1) \quad \text{implies} \quad C^f \in [C(0, 1), B^{\tau}_{q,v}(0, 1)].$$

6.5.24. Proposition (A. Pietsch 1981: a).

$$f \in B^{\sigma+\tau}_{q,u}(2\pi) \quad \text{implies} \quad C^f \in [B^{\sigma}_{\infty,u}(2\pi), B^{\tau}_{q,1}(2\pi)].$$

Proof. As stated above, the map $f \to C^f$, which assigns to every function the associated convolution kernel, acts as follows:

$$\text{con} : B^{\tau}_{q,1}(2\pi) \to [C(2\pi), B^{\tau}_{q,1}(2\pi)].$$

Analogously, we have

$$\text{con} : B^{m+\tau}_{q,1}(2\pi) \to [C^m(2\pi), B^{\tau}_{q,1}(2\pi)].$$

Let $\sigma = \theta m$. Then, by 6.4.7, the interpolation property yields

$$\text{con} : B^{\sigma+\tau}_{q,u}(2\pi) \to [B^{\sigma}_{\infty,u}(2\pi), B^{\tau}_{q,1}(2\pi)].$$

This completes the proof.

6.6. Traces and determinants of kernels

6.6.1. To begin with, we generalize 5.5.1.

Proposition (A. Grothendieck 1956: a). Let $\Phi \in [L_1(X, \mu), E']$. Then

$$\Phi_{\text{op}} : x \to \langle x, \Phi(\cdot) \rangle$$

defines a nuclear operator from E into $L_1(X, \mu)$ with

$$\|\Phi_{op} \mid \mathfrak{N}\| = \|\Phi \mid [L_1, E']\|.$$

Furthermore, all operators $T \in \mathfrak{N}(E, L_1(X, \mu))$ can be obtained in this way.

Proof. We first assume that Φ is a simple function. Then there exists a representation

$$\Phi(\xi) = \sum_{i=1}^{n} a_i h_i(\xi) \quad \text{for almost all } \xi \in X,$$

where $a_1, \ldots, a_n \in E'$ and $h_1, \ldots, h_n$ are characteristic functions of pairwise disjoint measurable subsets $X_1, \ldots, X_n$. In this case Φ_{op} has finite rank and

$$\|\Phi_{\text{op}} \mid \mathfrak{N}\| \leqq \sum_{i=1}^{n} \|a_i\| \, \mu(X_i) = \|\Phi \mid [L_1, E']\|.$$

If $\Phi \in [L_1(X, \mu), E']$ is arbitrary, then it can be approximated by a sequence of simple functions Φ_n. It follows from

$$\|(\Phi_m)_{\text{op}} - (\Phi_n)_{\text{op}} \mid \mathfrak{N}\| \leqq \|\Phi_m - \Phi_n \mid [L_1, E']\|$$

that the associated operators $(\Phi_n)_{\text{op}}$ form a Cauchy sequence in $\mathfrak{N}(E, L_1(X, \mu))$. Since $\mathfrak{N}$ is a Banach operator ideal, there must exist a nuclear limit which is nothing other than Φ_{op}. By continuity, we get

$$\|\Phi_{\text{op}} \mid \mathfrak{N}\| \leqq \|\Phi \mid [L_1, E']\|.$$

Finally, we show that every operator $T \in \mathfrak{N}(E, L_1(X, \mu))$ can be obtained in this way. To this end, given $\varepsilon > 0$, we consider a nuclear representation

$$T = \sum_{i=1}^{\infty} a_i \otimes f_i$$

such that

$$\sum_{i=1}^{\infty} \|a_i\| \, \|f_i \mid L_1\| \leqq (1 + \varepsilon) \, \|T \mid \mathfrak{N}\|.$$

Then it follows from

$$\sum_{i=1}^{\infty} \|a_i\| \int_X |f_i(\xi)| \, \mathrm{d}\mu(\xi) < \infty$$

that

$$\Phi(\xi) := \sum_{i=1}^{\infty} a_i f_i(\xi)$$

is defined almost everywhere. Furthermore, we have $\Phi \in [L_1(X, \mu), E']$ and

$$\|\Phi \mid [L_1, E']\| \leqq \sum_{i=1}^{\infty} \|a_i\| \, \|f_i \mid L_1\| \leqq (1 + \varepsilon) \, \|T \mid \mathfrak{N}\|.$$

Letting $\varepsilon \to 0$ yields

$$\|\Phi \mid [L_1, E']\| \leqq \|T \mid \mathfrak{N}\|.$$

Obviously, $T = \Phi_{\text{op}}$.

6.6.2. Specializing the preceding proposition, we arrive at a characterization of the nuclear operators acting on $L_1(X, \mu)$.

Proposition. Every kernel $K \in [L_1(X, \mu), L_\infty(X, \mu)]$ induces an operator

$$K_{\text{op}} \in \mathfrak{N}(L_1(X, \mu)) \quad \text{with} \quad \|K_{\text{op}} \mid \mathfrak{N}\| = \|K \mid [L_1, L_\infty]\|.$$

Furthermore, all operators $T \in \mathfrak{N}(L_1(X, \mu))$ can be obtained in this way.

6.6.3. Since $L_1(X, \mu)$ has the approximation property, every operator $K_{\text{op}} \in \mathfrak{N}(L_1(X, \mu))$ possesses a well-defined **trace** which is simply denoted by $\tau(K)$.

By analogy with 5.5.3 it could be conjectured that this trace is given by the formula

$$\tau(T) = \int_X K(\xi, \xi)\, d\mu(\xi).$$

However, as in the case $X = [0, 1]$, it may happen that the diagonal $\xi = \eta$ is a zero set of $X \times X$ on which K can be changed arbitrarily. Therefore, unless additional assumptions are placed on the underlying kernel, the right-hand integral makes no sense.

6.6.4. Next we deal with continuous kernels; see 6.1.5.

Proposition (A. Grothendieck 1956: a). Given any finite Borel measure μ on a compact Hausdorff space X, every kernel $K \in [C(X), C(X)]$ induces an operator

$$K_{op} \in \mathfrak{N}(C(X)) \quad \text{with} \quad \|K_{op} \mid \mathfrak{N}\| \leqq \mu(X)\, \|K \mid [C, C]\|.$$

Proof. Let $\Phi \in [L_1(X, \mu), E]$. Then by an appropriate dualization of the proof in 6.6.1, we can show that

$$\Phi^{op} : f(\xi) \to \int_X \Phi(\xi) f(\xi)\, d\mu(\xi)$$

defines a nuclear operator from $C(X)$ into E. Hence, taking $E := C(X)$, the assertion follows from the fact that every continuous kernel belongs to $[L_1(X, \mu), C(X)]$.

6.6.5. Since $C(X)$ has the approximation property, every operator $K_{op} \in \mathfrak{N}(C(X))$ induced by a continuous kernel possesses a well-defined **trace** which is simply denoted by $\tau(K)$.

Theorem. Suppose that μ is a finite Borel measure on a compact Hausdorff space X. Let $K \in [C(X), C(X)]$. Then

$$\tau(K) = \int_X K(\xi, \xi)\, d\mu(\xi) \quad \text{and} \quad |\tau(K)| \leqq \mu(X)\, \|K \mid [C, C]\|.$$

Proof. We first consider a degenerate kernel K. Then there exist continuous functions $f_1, \ldots, f_n$ and $g_1, \ldots, g_n$ such that

$$K(\xi, \eta) = \sum_{i=1}^{n} f_i(\xi)\, g_i(\eta) \quad \text{for all } \xi, \eta \in X.$$

Hence

$$\tau(K) = \sum_{i=1}^{n} \int_X f_i(\xi)\, g_i(\xi)\, d\mu(\xi) = \int_X K(\xi, \xi)\, d\mu(\xi).$$

According to the Weierstrass approximation theorem, the set of degenerate kernels is dense in $[C(X), C(X)]$. Thus, by continuity, the desired formula holds for all continuous kernels.

6.6.6. Given any kernel K, we write

$$K\begin{pmatrix} \xi_1, \ldots, \xi_n \\ \eta_1, \ldots, \eta_n \end{pmatrix} := \det \begin{pmatrix} K(\xi_1, \eta_1) & \ldots & K(\xi_1, \eta_n) \\ \vdots & & \vdots \\ K(\xi_n, \eta_1) & \ldots & K(\xi_n, \eta_n) \end{pmatrix}.$$

In particular, observe that $K\binom{\xi}{\eta} = K(\xi, \eta)$.

6.6.7. At the end of this monograph, we establish the very result which was the historical starting point of the whole theory. We are going to describe the canonical **Fredholm denominator**

$$\delta(\zeta, K) = 1 + \sum_{n=1}^{\infty} \alpha_n(K)\, \zeta^n$$

of an operator K_{op} induced by a continuous kernel K.

Replacing infinite sums by integrals, the proof given in 5.5.5 can be adapted to this classical setting.

Theorem (I. Fredholm 1903). Suppose that μ is a finite Borel measure on a compact Hausdorff space X. Let $K \in [C(X), C(X)]$. Then

$$\alpha_n(K) = \frac{1}{n!} \int\limits_X \dots \int\limits_X K\begin{pmatrix}\xi_1, \dots, \xi_n \\ \xi_1, \dots, \xi_n\end{pmatrix} \mathrm{d}\mu(\xi_1) \dots \mathrm{d}\mu(\xi_n)$$

and

$$|\alpha_n(K)| \leqq \frac{n^{n/2}}{n!}\, \mu(X)^n\, \|K \mid [C, C]\|^n.$$

6.6.8. The next result is a counterpart of 5.5.6. It shows that, by the principle of related operators, the classical construction of Fredholm denominators can be transferred to nuclear operators on arbitrary Banach spaces.

Proposition. Every operator $T \in \mathfrak{N}(E)$ is related to an operator $K_{\mathrm{op}} \in \mathfrak{N}(C(0,1))$ induced by a continuous kernel K.

Proof. In view of 1.7.4, there exists a representation

$$T = \sum_{i=1}^{\infty} \tau_i a_i \otimes x_i$$

such that

$$\tau_i > 0, \quad \sum_{i=1}^{\infty} \tau_i = 1, \quad \lim_i \|a_i\| = 0, \quad \lim_i \|x_i\| = 0.$$

Divide $[0, 1]$ into disjoint intervals X_i of length τ_i, and denote the corresponding characteristic functions by h_i. Then the functions $g_i := \tau_i^{-1/2} h_i$ form an orthonormal system. Next, by a small perturbation, we pass to an orthonormal system of continuous trapezoidal functions f_i such that

$$\|f_i \mid C\| \leqq 2\tau_i^{-1/2} \quad \text{and} \quad f_i(\xi) = 0 \quad \text{for} \quad \xi \in [0, 1] \setminus X_i.$$

Define $A \in \mathfrak{L}(E, C(0, 1))$ and $X \in \mathfrak{L}(C(0, 1), E)$ by

$$A := \sum_{i=1}^{\infty} \tau_i^{1/2} a_i \otimes f_i \quad \text{and} \quad X := \sum_{j=1}^{\infty} \tau_j^{1/2} f_j \otimes x_j.$$

Then $T = XA$. Furthermore, the related operator $S := AX$ is induced by the kernel

$$K(\xi, \eta) := \sum_{i=1}^{\infty} \sum_{j=1}^{\infty} \tau_i^{1/2} \tau_j^{1/2} \langle x_j, a_i \rangle\, f_i(\xi)\, f_j(\eta).$$

Note that, for every point (ξ, η), the right-hand expression contains at most one non-vanishing summand. This observation implies that

$$\Big|K(\xi, \eta) - \sum_{i+j<n} \tau_i^{1/2} \tau_j^{1/2} \langle x_j, a_i \rangle\, f_i(\xi)\, f_j(\eta)\Big| \leqq 4 \sup\{\|x_j\|\, \|a_i\| : i + j \geqq n\}.$$

Thus K must be continuous, since it is the uniform limit of a sequence of continuous functions.

6.6.9. The following result is a by-product of the proof given in 6.4.24.

Proposition (W. F. Stinespring 1958, G. E. Karadzhov 1977). Every kernel $K \in [B^{1/2}_{2,1}(0, 1), L_2(0, 1)]$ induces an operator

$$K_{\text{op}} \in \mathfrak{N}(L_2(0, 1)) \quad \text{with} \quad \|K_{\text{op}} \mid \mathfrak{N}\| \leqq c\|K \mid [B^{1/2}_{2,1}, L_2]\| .$$

Remark. Note that

$$[C^\sigma(0, 1), C(0, 1)] \subset [B^{1/2}_{2,1}(0, 1), L_2(0, 1)] \quad \text{if} \quad 1/2 < \sigma < 1 .$$

6.6.10. We now show how the estimates for the coefficients of the Fredholm denominator given in 6.6.7 improve when the kernel satisfies stronger smoothness conditions.

Proposition (I. Fredholm 1903). Let $K \in [C^\sigma(0, 1), C(0, 1)]$ with $0 < \sigma < 1$. Then

$$|\alpha_n(K)| \leqq \frac{n^{n(1/2-\sigma)}}{n!} 2^{\sigma n} \|K \mid [C^\sigma, C]\|^n .$$

Proof. Using standard properties of determinants, we transform

$$K\begin{pmatrix} \xi_1, \dots, \xi_n \\ \xi_1, \dots, \xi_n \end{pmatrix} := \det \begin{pmatrix} K(\xi_1, \xi_1) & \dots & K(\xi_1, \xi_n) \\ \vdots & & \vdots \\ K(\xi_n, \xi_1) & \dots & K(\xi_n, \xi_n) \end{pmatrix}$$

in the following way. If $j = 1, \dots, n - 1$, then the coefficients of the new j-th column are defined by

$$\mu_{ij} := \frac{K(\xi_i, \xi_j) - K(\xi_i, \xi_{j+1})}{|\xi_j - \xi_{j+1}|^\sigma} \quad \text{for} \quad i = 1, \dots, n,$$

while the n-th column remains unchanged:

$$\mu_{in} := K(\xi_i, \xi_n) \qquad\qquad \text{for} \quad i = 1, \dots, n.$$

This yields

$$K\begin{pmatrix} \xi_1, \dots, \xi_n \\ \xi_1, \dots, \xi_n \end{pmatrix} = \det(\mu_{ij}) \prod_{j=1}^{n-1} |\xi_j - \xi_{j+1}|^\sigma .$$

Applying Hadamard's inequality A.4.5, we obtain

$$\left| K\begin{pmatrix} \xi_1, \dots, \xi_n \\ \xi_1, \dots, \xi_n \end{pmatrix} \right| \leqq n^{n/2} \prod_{j=1}^{n-1} |\xi_j - \xi_{j+1}|^\sigma \|K \mid [C^\sigma, C]\|^n .$$

Since the left-hand expression is a symmetric function, we may arrange the variables such that $0 \leqq \xi_1 \leqq \dots \leqq \xi_n \leqq 1$. Then the inequality of means G.1.1 implies that

$$\left(\prod_{j=1}^{n-1} |\xi_j - \xi_{j+1}| \right)^{1/(n-1)} \leqq \frac{1}{n-1} \sum_{j=1}^{n-1} |\xi_j - \xi_{j+1}| \leqq \frac{1}{n-1} \leqq \frac{2}{n} .$$

Hence, in view of $2^n(n-1)^{n-1} \geqq n^n$, we have

$$\left| K \begin{pmatrix} \xi_1, \dots, \xi_n \\ \xi_1, \dots, \xi_n \end{pmatrix} \right| \leqq n^{n(1/2-\sigma)} \, 2^{\sigma n} \| K \mid [C^\sigma, C] \|^n .$$

The desired estimate is now obvious.

Remark. It follows from 4.8.7 and Stirling's formula that $\delta(\zeta, K)$ is an entire function of order r, where $1/r := \sigma + 1/2$.

6.6.11. An operator $T \in \mathfrak{L}(C(2\pi))$ is said to be a **multiplier** if

$$S_\xi T S_\xi^{-1} = T \quad \text{for all} \quad \xi \in \mathbb{R},$$

where

$$S_\xi : f(\alpha) \to f(\alpha + \xi)$$

is the translation operator with step ξ.

6.6.12. Finally, we give a nice application of the previous results.

Proposition (W. Bauhardt 1979). Every function $f \in C(2\pi)$ induces a multiplier

$$C_{\mathrm{op}}^f \in \mathfrak{N}(C(2\pi)) \quad \text{with} \quad \| C_{\mathrm{op}}^f \mid \mathfrak{N} \| = 2\pi \, \| f \mid C \| .$$

Furthermore, all multipliers $T \in \mathfrak{N}(C(2\pi))$ can be obtained in this way. More precisely, we have

$$T = C_{\mathrm{op}}^f \quad \text{with} \quad f(\xi) := \frac{1}{2\pi} \tau(S_\xi T) \quad \text{for all} \quad \xi \in \mathbb{R} .$$

Proof. The first part of the assertion follows from 6.6.4.

Next we show that

$$\int_0^{2\pi} \langle S_\xi f, \mu \rangle \langle S_\xi^{-1} g, \nu \rangle \, d\xi = \int_0^{2\pi} \langle S_\eta f, \nu \rangle \langle S_\eta^{-1} g, \mu \rangle \, d\eta$$

for $f, g \in C(2\pi)$ and $\mu, \nu \in C(2\pi)'$. In the special case of Dirac measures, $\mu = \delta_\alpha$ and $\nu = \delta_\beta$, this formula reduces to the obvious fact that

$$\int_0^{2\pi} f(\alpha + \xi) \, g(\beta - \xi) \, d\xi = \int_0^{2\pi} f(\beta + \eta) \, g(\alpha - \eta) \, d\eta .$$

Recall that the linear span of all Dirac measures is dense in $C(2\pi)'$ with respect to the topology of uniform convergence on the precompact subsets of $C(2\pi)$. Thus the above equation holds in general.

Assume that $T \in \mathfrak{N}(C(2\pi))$ is a multiplier. Take any nuclear representation

$$T = \sum_{i=1}^{\infty} \mu_i \otimes f_i ,$$

where $\mu_1, \mu_2, \ldots \in C(2\pi)'$ and $f_1, f_2, \ldots \in C(2\pi)$. Let $g \in C(2\pi)$ and $\nu \in C(2\pi)'$. Then

$$\begin{aligned}\langle Tg, \nu\rangle &= \frac{1}{2\pi}\int_0^{2\pi} \langle S_\eta T S_\eta^{-1} g, \nu\rangle \, d\eta \\ &= \frac{1}{2\pi}\sum_{i=1}^{\infty}\int_0^{2\pi} \langle S_\eta f_i, \nu\rangle \langle S_\eta^{-1} g, \mu_i\rangle \, d\eta \\ &= \frac{1}{2\pi}\sum_{i=1}^{\infty}\int_0^{2\pi} \langle S_\xi f_i, \mu_i\rangle \langle S_\xi^{-1} g, \nu\rangle \, d\xi \\ &= \frac{1}{2\pi}\int_0^{2\pi} \tau(S_\xi T) \langle S_\xi^{-1} g, \nu\rangle \, d\xi .\end{aligned}$$

Hence

$$Tg = \frac{1}{2\pi}\int_0^{2\pi} \tau(S_\xi T)\, S_\xi^{-1} g \, d\xi .$$

This proves that T coincides with the convolution operator C_{op}^f, where the continuous function f is defined by

$$f(\xi) := \frac{1}{2\pi}\tau(S_\xi T) \quad \text{for all} \quad \xi \in \mathbb{R} .$$

Furthermore,

$$\|f \mid C\| = \sup_\xi |f(\xi)| \leqq \frac{1}{2\pi}\sup_\xi |\tau(S_\xi T)| \leqq \frac{1}{2\pi}\|T \mid \mathfrak{N}\| .$$

6.6.13. It follows from 6.5.17 that all functions belonging to $B_{2,1}^{1/2}(2\pi)$ have absolutely convergent Fourier series. In the context of nuclear operators this result, due to B. S. Stechkin (1951), reads as follows.

Proposition. Every function $f \in B_{2,1}^{1/2}(2\pi)$ induces a multiplier

$$C_{\mathrm{op}}^f \in \mathfrak{N}(L_2(2\pi)) \quad \text{with} \quad \|C_{\mathrm{op}}^f \mid \mathfrak{N}\| \leqq c\|f \mid B_{2,1}^{1/2}\| .$$

Remark. The nuclear multipliers on $L_2(2\pi)$ are characterized as those convolution operators C_{op}^f such that f is the convolution product of two square integrable functions. According to G. H. Hardy/ J. E. Littlewood (1928: b) this criterion goes back to M. Riesz. Another necessary and sufficient condition in terms of modified approximation numbers was given by B. S. Stechkin (1955); see also (BAR, IX.8) and (KAH, II.3).

CHAPTER 7

Historical survey

> Besonders aber wünschte ich meiner Arbeit dadurch einigen Werth zu verleihen, dass ich soviel als möglich bis zu den Originalquellen vorzudringen suchte, um die ersten Erfinder von Methoden und die ersten Entdecker von Lehrsätzen citiren zu können. Solche Citate sind nicht nur ein Opfer, welches die spätere Zeit den frühern Offenbarungen des Genius schuldet, sie bilden ein Stück Geschichte der Wissenschaft und laden zum Studium der hohen Werke ein, aus denen die Wissenschaft aufgebaut ist, und in denen noch immer reiche Schätze ungehoben ruhen.
>
> Richard Baltzer (1857)

> In preparing this lecture, the speaker has assumed that he is expected to talk on a subject in which he had some first-hand experience through his own work. And glancing back over the years he found that the one topic to which he has returned again and again is the problem of eigenvalues and eigenfunctions in its various ramifications.
>
> Hermann Weyl (1950)

The last chapter is devoted to the history of the subject under consideration and to those mathematicians who essentially contributed to the development of this theory. The following memorial list is ordered according to the dates of birth. We have included some famous names who were involved in the creation of matrix theory as well as those who laid the foundation of the functional analytical approach to eigenvalue problems.

Leibniz, Gottfried Wilhelm	1646–1716
Cramer, Gabriel	1704–1752
Euler, Leonhard	1707–1783
D'Alembert, Jean Baptiste le Rond	1717–1783
Lagrange, Joseph Louis	1736–1813
Laplace, Pierre Simon	1749–1827
Gauss, Carl Friedrich	1777–1855
Cauchy, Augustin Louis	1789–1857
Jacobi, Carl Gustav Jacob	1804–1851
Hamilton, William Rowan	1805–1865
Grassmann, Hermann Günther	1809–1877
Sylvester, James Joseph	1814–1897
Weierstrass, Karl	1815–1897
Borchardt, Carl Wilhelm	1817–1880
Cayley, Arthur	1821–1895
Dedekind, Richard	1831–1916
Jordan, Camille	1838–1922
Hill, George William	1838–1914
Frobenius, Georg	1849–1917
Poincaré, Henri	1854–1912
Goursat, Édouard Jean Baptiste	1858–1936
Peano, Giuseppe	1858–1932
Hilbert, David	1862–1943
Dixon, Alfred Cardew	1865–1936
Hadamard, Jacques	1865–1963
Fredholm, Ivar	1866–1927
Koch, Helge von	1870–1924
Plemelj, Josip	1873–1967
Schur, Issai	1875–1941
Fischer, Ernst	1875–1954
Schmidt, Erhard	1876–1959
Riesz, Friedrich	1880–1956
Toeplitz, Otto	1881–1940
Lalesco, Trajan	1882–1929
Mercer, James	1883–1932
Hellinger, Ernst	1883–1950
Weyl, Hermann	1885–1955
Courant, Richard	1888–1972
Tamarkin, Jacob David	1888–1945
Banach, Stefan	1892–1945
Carleman, Torsten	1892–1949
Hille, Einar	1894–1980
Schauder, Pawel Juliusz	1899–1943
Neumann, Johann von	1903–1957

From my personal viewpoint mathematics should be considered as a whole, and priorities of single mathematicians are of secondary importance. Solely, the main streams of mathematical thought are worth considering, and these have indeed been controlled by the great minds. Let us describe one of those lines of development.

At the end of the nineteenth century G. Mittag-Leffler, a student and friend of K. Weierstrass, reprinted G. W. Hill's paper on infinite determinants in Acta Mathematica, and he inspired H. von Koch and I. Fredholm to work on this subject. The latter created his famous determinant theory of integral operators in spring 1899 during a visit in Paris where he attended lectures of H. Poincaré, E. Picard and J. Hadamard. The decisive moment came when E. Holmgren, a visitor from Stockholm, reported I. Fredholm's results in D. Hilbert's seminar (1901) at Göttingen. In the words of H. Weyl, "*Hilbert caught fire at once*", and he focused his attention on this new subject. He also filled several of his famous pupils with enthusiasm: E. Schmidt, O. Toeplitz, E. Hellinger, H. Weyl, R. Courant, J. Plemelj from Yugoslavia and F. Riesz from Hungary. Due to Nazi barbarism, the Hilbert school was scattered to the four winds. As a consequence, R. Courant came to Cambridge in 1934 where he directed F. Smithies's attention to A. Hammerstein's work (1923) on integral equations. For his part F. Smithies had several students who essentially contributed to the development of spectral theory: Shih-Hsun Chang, A. F. Ruston, J. R. Ringrose and T. T. West. After his graduation in Cambridge, on his way back to China, Shih-Hsun Chang visited Princeton and showed his results to H. Weyl who had been interested in eigenvalue problems throughout his life. This triggered off H. Weyl's note in 1949. A few years earlier, J. von Neumann, whose way of mathematical thinking was formed by E. Schmidt and H. Weyl, had created the theory of operator ideals on Hilbert spaces. Among his pupils were J. W. Calkin and R. Schatten. The Swedish tradition was continued by T. Carleman. Finally, it should be mentioned that E. Hille, a former student of Stockholm university, visited Göttingen in 1927. The preliminary version of his important paper on characteristic values of integral equations, jointly written with J. D. Tamarkin, appeared in 1928.

7.1. Classical background

> In dieser Idee der Analogie mit der analytischen Geometrie und allgemeiner überhaupt in der Idee des Übergangs von algebraischen Tatsachen zu solchen der Analysis liegt der Sinn der Lehre von den Integralgleichungen.
>
> Ernst Hellinger, Otto Toeplitz (1927)

> That Hilbert space theory and elementary matrix theory are intimately associated came as a surprise to me and to many colleagues of my generation only after studying the two subjects separately.
>
> Paul Richard Halmos (1942)

For further reading we recommend the monographs of N. Bourbaki (BOU), M. J. Crowe (CRO), M. Kline (KLI), C. C. MacDuffee (MAC) and T. Muir (MUI) which contain extensive contributions to the history of linear algebra. Important aspects of the historical development of matrix theory are presented in a series of papers by T. Hawkins (1975, 1977: a, b).

7.1.1. Determinants

Numerum $bb-ac$, a cuius indole proprietates formae (a, b, c) imprimis pendere, in sequentibus docebimus, determinantem huius formae vocabimus.

Carl Friedrich Gauss (1801)

The concept of a determinant appeared for the first time in a letter from G. W. Leibniz to G. F. A. de l'Hospital (April 23, 1693). However, this idea got lost and was rediscovered by

G. Cramer: *Introduction à l'analyse des lignes courbes algébriques.* Genève 1750.

In the appendix of his famous treatise he established a formula for solving systems of linear equations which became well-known as "*Cramer's rule*". Further important contributions are due to E. Bézout, A. T. Vandermonde, J. L. Lagrange and P. S. Laplace. The latter used the word "*résultant*" and proved the expansion formula. After this pioneering period, it was

A. L. Cauchy: *Mémoire sur les fonctions qui ne peuvent obtenir que deux valeurs égales et de signes contraires par suite des transpositions opérées entre des variables qu'elles renferment.* Journal École Polytechnique, Paris 1815,

who developed the theory as a whole. In particular, he adopted from

C. F. Gauss: *Disquisitiones arithmeticae.* Leipzig 1801,

the name "*determinant*" and introduced the symbol $S(\pm\mu_{11} \dots \mu_{nn})$. The notation $\det(\mu_{ij})$, used throughout this monograph, is due to E. Catalan (1846) and F. Joachimsthal (1849). Special credit goes to C. G. J. Jacobi who popularized determinants as a powerful mathematical tool and made several important contributions to this subject. Finally, it should be mentioned that K. Weierstrass (1886) gave an axiomatic approach to determinant theory based on the concept of an alternating multilinear functional. However, for the purpose of this book the multiplication formula, independently discovered by J. P. M. Binet (1813) and A. L. Cauchy (1815), is more relevant.

7.1.2. Matrices

Die Zusammensetzung der Transformationen hat bekanntlich die größte Analogie mit der Multiplication der Zahlen und soll auch durch das Zeichen dieser Operation angedeutet werden, wenn die Substitutions-Systeme selbst gleichsam als selbständige Größen in der Rechnung auftreten, nur daß man hier die Ordnung der Faktoren nicht vertauschen darf.

Gotthold Max Eisenstein (1852)

The idea of a matrix precedes that of a determinant but historically the order was the reverse and this is why the basic properties of matrices were already clear by the time that matrices were introduced.

Morris Kline (1972)

The word "*matrix*" was first used by J. J. Sylvester (1850) to denote a rectangular array of numbers. Such arrays or systems were considered much earlier. However, before the middle of the nineteenth century the main interest was directed towards the determinant as a value assigned to every square matrix. It is only due to

A. Cayley: *A memoir on the theory of matrices.* Phil. Trans. Roy. Soc. London 1858 that the arrays themselves came into consideration. More important from the viewpoint of operator theory is the fact, already known to C. F. Gauss (1801), that matrices can be employed to describe linear substitutions as well as quadratic and bilinear forms. In this way the formal algebraic manipulations acquired some concrete meaning.

7.1.3. Vector spaces

> Such step-sets could be added or subtracted, by adding or subtracting their component steps, each to or from its own corresponding step, as indicated by the double formula,
>
> $$(b_1, \ldots, b_n) \pm (a_1, \ldots, a_n) = (b_1 \pm a_1, \ldots, b_n \pm a_n);$$
>
> and a step-set could be multiplied by a number α, by the formula
>
> $$\alpha(a_1, \ldots, a_n) = (\alpha a_1, \ldots, \alpha a_n).$$
>
> William Rowan Hamilton (1853)

> Zwei extensive Grössen, die aus demselben System von Einheiten abgeleitet sind, addiren, heisst, ihre zu denselben Einheiten gehörigen Ableitungszahlen addiren, das heisst,
>
> $$\sum \alpha e + \sum \beta e = \sum (\alpha + \beta) e.$$
>
> Eine extensive Größe mit einer Zahl multipliciren heisst, ihre sämmtlichen Ableitungszahlen mit dieser Zahl multipliciren, das heisst
>
> $$\sum \alpha e . \beta = \beta . \sum \alpha e = \sum (\alpha\beta) e.$$
>
> Hermann Günther Grassmann (1862)

The first rigorous approach to the concept of a complex number, by considering ordered couples of reals, was given by W. R. Hamilton (1837). Later on, he generalized this idea in his famous theory of quaternions (1853), observing that the same general view of algebra admitted easily, at least in thought, of an extension of this whole theory, not only from couples to triplets, but also from triplets to sets of moments, steps, and numbers. Similar and even more profound considerations are to be found in

H. Grassmann: *Die (lineale) Ausdehnungslehre*. Leipzig 1844 und Berlin 1862,

an important work which was neglected for a very long time. As indicated by the term "*hyper-complex number*", at this period the main goal was the construction of linear algebras in which there is also defined a multiplication such as in the real or complex field.

7.1.4. Eigenvalues

> It will be convenient to introduce here a notion, namely that of the latent roots of a matrix—latent in a somewhat similar sense as vapour may be said to be latent in water or smoke in a tobacco-leaf.
>
> James Joseph Sylvester (1883)

Quod attinet ipsam ipsius Γ formationem, observo, si signo summatorio S amplectamur expressiones inter se diversas, quae permutatis indicibus 1, 2, 3, ..., n proveniunt, fieri:

$$\begin{aligned} \Gamma = \sum \pm & a_{1,1}a_{2,2} \dots a_{n,n} \\ & -xS\sum \pm a_{1,1}a_{2,2} \dots a_{n-1,n-1} \\ & +x^2S\sum \pm a_{1,1}a_{2,2} \dots a_{n-2,n-2} \\ & \dots\dots\dots\dots\dots\dots \\ & \pm x^{n-2}S\sum \pm a_{1,1}a_{2,2} \\ & \mp x^{n-1}Sa_{1,1} \pm x^n. \end{aligned}$$

Qua in formula expressio

$$S\sum \pm a_{1,1}a_{2,2} \dots a_{m,m}$$

designat summam

$$\frac{n(n-1)\dots(n+1-m)}{1.2\dots m}$$

expressionum, quae e

$$\sum \pm a_{1,1}a_{2,2} \dots a_{m,m}$$

proveniunt, si in

$$a_{1,1}a_{2,2} \dots a_{m,m}$$

loco indicum priorum simul ac posteriorum 1, 2, ..., m scribimus omnibus modis, quibus fieri potest, m alios e numeris 1, 2, 3, ..., n.

Carl Gustav Jacob Jacobi (1834)

In the course of applying the methods of analysis to the vibrating string problem and to the theory of secular perturbations,

J. B. D'Alembert: *Traité de dynamique*. Paris 1743,
J. L. Lagrange: *Méchanique analytique*. Paris 1788,
J. P. Laplace: *Traité de méchanique céleste*. Paris 1799,

were led to the consideration of systems of linear differential equations with constant coefficients. In this context the concept of an eigenvalue arose for the first time. Another equally important source of eigenvalue problems was the principal axis theorem of mechanics of rigid bodies and of the theory of quadratic surfaces. First results in this direction were already mentioned by

L. Euler: *Introductio in analysin infinitorum*. Lausanne 1748.

At the very beginning, the so-called "*secular equation*" $\pi(\lambda, M) = 0$ of a matrix $M = (\mu_{ij})$ was obtained by eliminating the unknowns $\xi_1, \dots, \xi_n$ from

$$\sum_{i=1}^{n} \mu_{ij}\xi_j = \lambda\xi_i \quad \text{for} \quad i = 1, \dots, n$$

and $n = 2, 3, 4$, sometimes 5. It was

A. L. Cauchy: *Sur l'équation a l'aide de laquelle on détermine les inégalités séculaires des mouvements des planètes*. Exer. de Math. 1829,

who identified and treated $\pi(\lambda, M)$ as the determinant of the matrix $\lambda I - M$, the number of coordinates being arbitrary. A few years later,

C. G. J. Jacobi: *De binis quibuslibet functionibus homogeneis secundi ordinis per substitutiones lineares in alias binas transformandis*, ... J. Reine Angew. Math. 1834,

expressed the coefficients of

$$\pi(\lambda, M) = \lambda^n - \alpha_1\lambda^{n-1} + \ldots + (-1)^n \alpha_n$$

by means of principal minors. For example,

$$\alpha_1 = \mu_{11} + \ldots + \mu_{nn}.$$

Nowadays, following A. L. Cauchy (1839), we call $\pi(\lambda, M)$ the "*characteristic polynomial*". The terminology for its zeros $\lambda_1, \ldots, \lambda_n$ was and is multifarious and colourful:

"*characteristic numbers, latent roots, secular values*",

and most of the possible combinations of these words. Curiously enough, the hybrid term

"*eigenvalue*" (German: *Eigenwert*)

prevailed.

The eigenvalues of a matrix M are precisely those complex numbers λ_0 for which $\lambda_0 I - M$ fails to be invertible. G. Frobenius (1878) was the first who considered the resolvent $(\lambda I - M)^{-1}$ as a meromorphic function of the parameter λ and obtained the Laurent expansions about the poles.

7.1.5. Traces

Unter der Spur der Zahl θ verstehen wir die Summe aller mit ihr konjugierten Zahlen; wir bezeichnen diese offenbar rationale Zahl mit $S(\theta)$; dann ist

$$S(\theta) = e_{1,1} + e_{2,2} + \ldots + e_{n,n}.$$

Richard Dedekind (1882)

The word "*trace*" (German: *Spur*) was introduced by R. Dedekind (1882) in the framework of algebraic number theory; see also R. Dedekind/ H. Weber (1882). As mentioned above, C. G. J. Jacobi (1834) observed that the sum of the diagonal elements of a matrix appears as a coefficient of its characteristic polynomial. The first explicit reference to the *trace formula* originates from C. W. Borchardt (1846) who even proved that

$$\operatorname{trace}(M^k) = \lambda_1^k + \ldots + \lambda_n^k$$

for all natural exponents k.

7.1.6. Canonical forms of matrices

Le premier de ces problèmes n'a pas encore été abordé *à* notre connaissance; le deuxième a déjà été traité (dans le cas où n est pair), par M. Kronecker, et le troisième par M. Weierstrass; mais les solutions données par les éminents géomètres de Berlin sont incomplètes, en ce qu'ils ont laissé de côté certains cas exceptionnels, qui ne manquent pas d'intérêt. Leur analyse est en outre assez difficile à suivre, surtout celle de M. Weierstrass.

Camille Jordan (1874)

> Die vorstehenden Entwickelungen zeigen, dass in der Jordanschen Abhandlung die Lösung des ersten Problems nicht eigentlich neu ist, während die des zweiten sich als gänzlich verfehlt und die des dritten als durchaus unzulänglich begründet erwiesen hat. Nimmt man hinzu, dass eben dieses dritte Problem in Wahrheit die beiden ersten als besondere Fälle umfasst, dass ferner dessen vollständige Lösung einestheils unmittelbar aus der Weierstrass'schen Arbeit vom Jahre 1868 folgt und anderntheils mit leichter Mühe aus den Bemerkungen entnommen werden kann, welche ich damals daran angeschlossen habe, so ist wahrlich hinreichender Grund vorhanden, Hrn. Jordan „seine Resultate", soweit sie eben richtig sind, streitig zu machen. Aber nicht um dieses untergeordneten Zweckes willen bin ich hier und in meiner früheren Mittheilung auf die Jordanschen Arbeiten näher eingegangen; es galt vielmehr die wirkliche Bedeutung der darin enthaltenen Methoden und Resultate zu ermitteln und ihre Beziehungen zu den vorher bekannten aufzuklären. Es war also nicht die Feststellung der Priorität, sondern die Feststellung der Wahrheit der eigentliche Zweck meiner Ausführungen, aber sie erfüllen nebenher auch die Bestimmung, es im Voraus zu rechtfertigen, wenn ich mich künftig der Rücksichtnahme auf die bezüglichen Jordan'schen Publicationen enthalte.
>
> Leopold Kronecker (1874)

Beginning with the principal axis theorem, canonical forms played an important role in the theory of matrices and operators. Since the symmetric or hermitian case is not relevant for the purpose of this monograph, we focus our attention on the general situation. The main result was obtained almost simultaneously by

K. Weierstrass: *Zur Theorie der quadratischen und bilinearen Formen.* Monatsber. Preuss. Akad. Wiss. Berlin 1868,

and

C. Jordan: *Traité des substitutions.* Paris 1870.

While the eminent geometer from Berlin looked at matrices from the viewpoint of quadratic and bilinear forms, his French colleague thought of matrices in terms of linear substitutions. At least in functional analysis the latter interpretation is more important. Maybe this was the deciding factor in the naming of the classical canonical form of a matrix the "*Jordan form*". As a first step towards this result, triangular matrices should have been considered. However, there is only a paper of L. Fuchs (1866) in which this concept was used for solving systems of linear differential equations. Thus it was only I. Schur (1909: a) who proved that every matrix is unitarily equivalent to a triangular one.

7.1.7. Axiomatic approach

> Un'operazione R, a eseguirsi su ogni ente a d'un sistema lineare A, dicesi distributiva, se il risultato dell'operazione R sull'ente a, che indicheremo con Ra, è pure un ente d'un sistema lineare, e sono verificate le identità
>
> $$R(a + a') = Ra + Ra', \ R(ma) = m(Ra),$$

> ove a e a' sono enti qualunque del sistema A, ed m un numero reale qualunque.
>
> Giuseppe Peano (1888)

The *axiomatic definitions of linear spaces and linear mappings* are due to

G. Peano: *Calcolo geometrico secondo l'Ausdehnungslehre di H. Grassmann.* Torino 1888.

However, for some decades these coordinate-free notions remained unused. It was only in 1918 that H. Weyl reestablished this concept in his excellent book "*Raum, Zeit, Materie*", where he gave an axiomatic definition of vector spaces in which the "Dimensionsaxiom" was clearly separated from the other ones. Surprisingly enough, up to the thirties, in all famous textbooks of algebra the so-called "*vector spaces*" were viewed as sets of n-tuples. Therefore, when writing his ingenious thesis, S. Banach (1922) was compelled to establish the definitions of all algebraic concepts himself. Thus, without exaggeration, it may be stated that linear algebra, in its present form, was developed as a by-product of functional analysis.

7.2. Spaces

> Die geometrische Deutung der in diesem Kapitel entwickelten Begriffe und Theoreme verdanke ich Kowalewski. Sie tritt noch klarer hervor, wenn $A(x)$ statt als Funktion als Vector in einem Raum von unendlich vielen Dimensionen definiert wird.
>
> Erhard Schmidt (1908)

The historical development of the theory of sequence and function spaces as well as of abstract topological linear spaces is extensively described in the books of N. Bourbaki (BOU), J. Dieudonné (DIU) and A. F. Monna (MON). We also refer to a paper of M. Bernkopf (1966). Concerning spaces of differentiable functions we quote J. Lützen's "*The prehistory of the theory of distributions*" and S. M. Nikol'skij's monograph (NIK). Valuable comments on the history of function spaces and interpolation theory are to be found in a book of P. L. Butzer/ H. Berens (BUB).

7.2.1. Sequence spaces

> Fortan ziehen wir durchweg nur solche Wertsysteme der unendlich vielen Variablen $x_1, x_2, \ldots$ in Betracht, die der Bedingung
>
> $$x_1^2 + x_2^2 + \ldots \leqq 1$$
>
> genügen.
>
> David Hilbert (1906)

> Considérons l'espace hilbertien; nous y entendons l'ensemble des systèmes (x_k) tels que $\sum |x_k|^2$ converge.
>
> Frédéric Riesz (1913)

In view of the fundamental contributions of

D. Hilbert: *Grundzüge einer allgemeinen Theorie der linearen Integralgleichungen (Vierte Mitteilung).* Göttinger Nachr. 1906,

to the theory of linear equations in infinitely many unknowns, the sequence space l_2 and its axiomatically defined descendants are with good reason named after this ingenious mathematician. Nevertheless, it should be emphasized that D. Hilbert himself mainly directed his attention towards the closed unit ball of l_2. It was due to the elegant work of

E. Schmidt: *Über die Auflösung linearer Gleichungen mit unendlich vielen Unbekannten.* Rend. Circ. Mat. Palermo 1908,

that the algebraical and topological structure of l_2 came into consideration. A few years later

F. Riesz: *Les systèmes d'équations linéaires à une infinité d'inconnues.* Paris 1913,

introduced l_p for $1 < p < \infty$. Finally, the one-parameter scale of *classical sequence spaces* was refined by employing a second parameter such that the resulting scale of so-called *Lorentz sequence spaces* $l_{p,u}$, defined even for $0 < p, u \leqq \infty$, is lexicographically ordered. This idea originates from a paper of G. H. Hardy/ J. E. Littlewood (1931) in which the non-increasing rearrangement of a sequence was introduced. Next G. G. Lorentz (1950) defined certain forefathers of the function spaces nowadays denoted by $L_{p,u}(0, 1)$. Afterwards those spaces were built on arbitrary measure spaces; see R. O'Neil (1963) and R. A. Hunt (1966). The fact that Lorentz spaces can be obtained from the classical ones by real interpolation was observed by J. Peetre (1963).

7.2.2. Function spaces

Линейное многообразие всех суммируемых функций $\varphi(x_1, \ldots, x_n)$ имеющих в ограниченной области Ω все обобщенные производные порядка l, суммируемые со степенью $p > 1$, назовем $W_p^{(l)}$.

Сергей Львович Соболев (1950)

Due to the work of C. Arzelà, G. Ascoli and K. Weierstrass, around 1885 mathematicians became interested in statements not only referring to single continuous functions, but concerning certain collections of them. Later on, $C(0, 1)$ was identified by F. Riesz (1909) as a complete normed linear space (modern terminology). Fortunately, at the beginning of this century H. Lebesgue established his famous integral. This cleared the way for the introduction of *spaces of p-integrable functions*. By the use of concepts such as absolute continuity and generalized differentiability it was possible to define *spaces of differentiable functions*. Finally, by means of Lipschitz-Hölder conditions, spaces with fractional order of differentation were introduced. At present we know several equivalent approaches to the theory of so-called *Sobolev-Besov spaces*:

- the classical definition via derivatives and Lipschitz-Hölder conditions or by the completion of spaces of infinitely differentiable functions with respect to certain norms,
- the characterization of the desired functions in terms of their degree of approximation,
- the construction of function spaces by interpolation,
- the decomposition method based on the Fourier transform,
- approaches via singular integrals.

Since we are not in a position here to give a detailed account of the history of function spaces, the following list should be enough to create a rough impression.

spaces	authors
C	R. Riesz (1909)
C^m	S. Banach (1922)
C^σ	J. Schauder (1934)
L_2	F. Riesz (1907), E. Fischer (1907)
L_p	F. Riesz (1910)
W_2^1	O. Nikodym (1933)
W_2^m	S. L. Sobolev (1936)
W_p^m	S. Banach (1922)
W_2^σ	N. Aronszajn (1955)
W_p^σ	L. N. Slobodeckij (1958)
$B_{p,\infty}^\sigma$	G. H. Hardy/ J. E. Littlewood (1928: a), S. M. Nikol'skij (1951)
$B_{p,u}^\sigma$	O. V. Besov (1961), M. H. Taibleson (1964/66)

The chronological disorder between W_p^m and W_2^m is due to the fact that S. L. Sobolev treated functions in several variables, while S. Banach considered the 1-dimensional case only.

7.2.3. Banach spaces

> Thus the two pieces of work, Banach's and my own, came for a time to be known as the theory of Banach-Wiener spaces. For thirty-four years it has remained a popular direction of work. Although many papers have been written on it, only now is it beginning to develop its full effectiveness as a scientific method.
>
> For a short while I kept publishing a paper or two on this topic, but I gradually left the field. At present these spaces are quite justly named after Banach alone.
>
> Norbert Wiener (1956)

Based upon the work of F. Riesz (1918), the *axiomatic theory of Banach spaces* was independently created by

S. Banach: *Sur les opérations dans les ensembles abstraits et leur application aux équations intégrales.* Fundamenta Math. 1922,

and N. Wiener (1923). A particular highlight in the history of mathematics as a whole was the publication of

S. Banach: *Théorie des opérations linéaires.* Warszawa 1932.

Since that time his celebrated monograph has been an indispensable tool for all those working in functional analysis. The concept of an *abstract Hilbert space* was designed by J. von Neumann (1929), relatively late. *Quasi-Banach spaces* and *p-Banach spaces*, as well as the connection between both notions, were defined and investigated by among others T. Aoki (1942), D. G. Bourgin (1943) and S. Rolewicz (1957).

7.2.4. Interpolation theory

A_0 et A_1 étant donnés, construire des espaces intermédiaires, de façon à ce que la propriété d'interpolation par rapport aux applications linéaires ait lieu.

Jacques-Louis Lions, Jaak Peetre (1964)
Les A. ont été soutenus par l'Interpol

Based on the classical ideas of M. Riesz (1926) and G. O. Thorin (1938), *interpolation theory* was founded at the end of the fifties. The most important papers following this pioneer time are

J. L. Lions / J. Peetre: *Sur une class d'espaces d'interpolation.* Inst. Hautes Études Sci. Publ. Math. 1964,

and

A. P. Calderon: *Intermediate spaces and interpolation, the complex method.* Studia Math. 1964.

Subsequently, interpolation theory has become a special branch of functional analysis with important and far-reaching applications. The final version of the *real method,* used throughout this book, is due to

J. Peetre: *Inledning till interpolation.* Föreläsningar, Lund 1964–1966.

This powerful technique was first applied within the theory of operator ideals by H. Triebel (1967), A. Pietsch / H. Triebel (1968), R. Oloff (1970), J. Peetre / G. Spaar (1972) and H. König (1978).

7.3. Operators

Cette théorie mérite donc avec raison, aussi bien par sa valeur esthétique que par la portée de ses raisonnements (même abstraction faite de ses nombreuses applications) l'intérêt de plus en plus croissant que lui prètent les mathématiciens. Aussi on ne s'étonnera pas à l'opinion de M. J. Hadamard, qui considère la théorie des opérations comme une des plus puissantes méthodes de recherche de la mathématique contemporaine.

Stefan Banach (1932)

For further reading we recommend J. Dieudonné's "*History of functional analysis*". The monumental treatise of N. Dunford / J. Schwartz (DUN) contains a lot of historical information, as well. We also quote an excellent survey about spectral theory written by L. A. Steen (1973).

7.3.1. Infinite matrices

Diese beiden Sätze können leicht durch eine Betrachtung, die der Eingangs dieses Paragraphen geführten analogisch ist, streng nachgewiesen werden, indem man zeigt, dass, weil dieselben gelten für ein System linearer Gleichungen von endlicher Anzahl, und unverändert Geltung behalten, wenn die Anzahl der Gleichungen um eine endliche Zahl wächst, sie auch gelten müssen, wenn die Anzahl der Gleichungen und Unbekannten ins Unbegrenzte wächst.

Theodor Kötteritzsch (1870)

Es handelt sich dabei um eine Reihe von Konvergenzhilfssätzen, die im wesentlichen in der einen Tatsache gipfeln, daß man mit passender Vorsicht für unendliche Matrizen (Bilinearformen von unendlichvielen Veränderlichen) einen ebensolchen Kalkül aufstellen kann, wie er sich seit Cayley und Frobenius in der Algebra der Bilinearformen von endlichvielen Veränderlichen eingebürgert hat.

Ernst Hellinger, Otto Toeplitz (1910)

General *infinite matrices* were considered for the first time by T. Kötteritzsch (1870) when he tried to solve *systems of linear equations in infinitely many unknowns*, written as

$$\overset{0}{\underset{\infty}{m}}\left|\sum_0^\infty {}_p f_0(m,p)\, x_p = \varphi_m\right|$$

in his notation. Later on, this concept became basic for the theory of infinite determinants. D. Hilbert (1906: a) used infinite matrices to define *bilinear forms in infinitely many variables* and, finally, it was F. Riesz (1913) who stressed the fact that every infinite matrix induces an *operator on a suitable sequence space.*

There does not exist any intrinsic description of those infinite matrices determining the operators on l_p with $1 < p < \infty$. Thus it was quite natural to take into account some classes of infinite matrices $M = (\mu_{ij})$ which can be defined by simple conditions such as

$$\sum_{i=1}^{\infty} \sum_{j=1}^{\infty} |\mu_{ij}|^2 < \infty \quad \text{(D. Hilbert 1906: a)}$$

and

$$\sum_{i=1}^{\infty} \sup_j |\mu_{ij}| < \infty \quad \text{(H. von Koch 1901, A. C. Dixon 1902).}$$

Both of these inequalities are included as special cases in a 2-parameter scale of conditions first introduced by E. Hille / J. D. Tamarkin (1934) in the setting of kernels; see also S. Bóbr (1921) and L. W. Cohen (1930). The definition of *infinite matrices of Besov type* $[b^\sigma_{p,u}, b^\tau_{q,v}]$ was given by A. Pietsch (1980: d).

E-valued sequences came under consideration when I. M. Gel'fand (1938) and R. S. Phillips (1940) observed that these are especially suited to represent operators, for example, from l_1 into the Banach space E.

7.3.2. Integral operators

Why should one study integral operators? The traditional answers are that integral equations have important applications outside of mathematics, and that they are the proper extension to analysis of the concepts and methods of the classical algebraic theory of linear equations. A third possible answer is that the theory of integral operators is the source of all modern functional analysis and remains to this day a rich source of non-trivial examples. Since the main obstacle to progress in many parts of operator theory is the dearth of concrete examples whose properties can be explicitly determined, a systematic theory of integral operators offers new hope for new insights.

Paul R. Halmos, Viakalathur Sh. Sunder (1978)

A glaring example of lack of perspective is given by the Hellinger-Toeplitz article of 1927 in the Enzyklopädie der Math. Wiss., which gives undue emphasis to integral equations.

Jean Alexandre Dieudonné (1981)

Following D. Hilbert (1904) a function $K(\xi, \eta)$ defined for $0 \leqq \xi, \eta \leqq 1$ is said to be a "*kernel*" if it is used to determine an *integral operator*

$$K_{\mathrm{op}} : g(\eta) \to \int_0^1 \mathrm{K}(\xi, \eta)\, g(\eta)\, \mathrm{d}\eta .$$

As in the case of infinite matrices, certain classes of kernels proved to be of special importance. This statement is in particular true for the *Hilbert-Schmidt kernels*:

$$\int_0^1 \int_0^1 |K(\xi, \eta)|^2 \, \mathrm{d}\xi \, \mathrm{d}\eta < \infty .$$

Generalizing this notion, E. Hille / J. D. Tamarkin (1934), F. Smithies (1937: b) and I. A. Itskovich (1948) considered kernels for which

$$\left(\int_0^1 \left(\int_0^1 |K(\xi, \eta)|^q \, \mathrm{d}\eta \right)^{p/q} \mathrm{d}\xi \right)^{1/p} < \infty .$$

Such conditions have not only been put on the kernels themselves but also on their derivatives. This yields a large variety of different types of kernels which, from the viewpoint of the theory of function spaces, are defined as functions with mixed degrees of differentiability. The most important examples are the *kernels of Besov type* $[B^{\sigma}_{p,u}, B^{\tau}_{q,v}]$ introduced by A. Pietsch (1980: d). We would also like to mention the so-called *weakly singular integral operators* which arise in potential theory and have played an essential role as basic examples for applications.

The fact that certain operators on function spaces can be represented by means of E-valued functions was discovered by N. Dunford, B. J. Pettis, R. S. Phillips and I. M. Gel'fand between 1935 and 1940; see (DUN, Vol. I, pp. 542–551).

7.3.3. Abstract operators

Die in der Arbeit gemachte Einschränkung auf stetige Funktionen ist nicht von Belang. Der in den neueren Untersuchungen über diverse Funktionenräume bewanderte Leser wird die allgemeinere Verwendbarkeit der Methode sofort erkennen.

Friedrich Riesz (1918)

The concept of an *abstract operator* between *Banach spaces* is due to S. Banach (1922). In another paper (1929) he introduced dual Banach spaces and dual operators. However, even in his monograph there is no reference to the set $\mathfrak{L}(E, F)$ constituted by all operators acting from E into F. The definition of the product of two operators is also missing. Clearly, S. Banach must have been well acquainted with these notions. However, it was not until the work of I. M. Gel'fand (1941) that "*Normierte Ringe*" became attractive.

7.3.4. Operator ideals on Hilbert spaces

Since that time, I have had a great respect and use for a subject that may be called "the hard analysis of compact operators in Hilbert space".

Barry Simon (1979)

In the late twenties, J. von Neumann (1929) stated the fact that the set of all operators acting on the separable infinite dimensional Hilbert space H can be viewed algebraically as a ring. Next, in a small but important paper (1937), he introduced the norms $\|\cdot \mid \mathfrak{S}_r\|$ with $1 \leqq r < \infty$ for finite matrix operators. Ten years later the generalization to the infinite setting was carried out by him jointly with one of his pupils,

R. Schatten / J. von Neumann: *The cross-space of linear transformations.* Ann. of Math. 1946 and 1948.

These papers contain the basic facts from what is now called the *theory of Schatten-von Neumann ideals*. In the meantime, another one of his students,

J. W. Calkin: *Two-sided ideals and congruences in the ring of bounded operators in Hilbert space.* Ann. of Math. 1941,

had investigated the ideal structure of $\mathfrak{L}(H)$ from an abstract viewpoint. The outcome was a famous theorem which asserts that $\mathfrak{K}(H)$ is the largest proper ideal in $\mathfrak{L}(H)$. Thus, at the end of the forties, the *theory of operator ideals on the Hilbert space*, mainly developed in Princeton, was ready for application to eigenvalue problems. Who should do it? Of course, H. Weyl (1949), who was also in Princeton since 1933.

A masterly presentation of this subject was given by

I. C. Gohberg / M. G. Krejn: *Introduction to the theory of non-self-adjoint operators in Hilbert space* (*Russian*). Moscow 1965.

7.3.5. Operator ideals on Banach spaces

Unangenehm ist mir nur immer Dedekinds Terminologie gewesen, welche aller Anschaulichkeit entbehrt. Er nennt diese Gesamtheiten Ideale. Er hätte von „Realen" sprechen sollen.

Felix Klein (1915)

Ideal theory in $L(X)$, where X denotes a generic Banach space, is likely to be a zoo.

Barry Simon (1979)

The *theory of operator ideals on Banach spaces* emerged as a by-product of the *theory of tensor products* which was created by

R. Schatten: *A theory of cross-spaces.* Princeton 1950,
A. F. Ruston: *Direct products of Banach spaces and linear functional equations.* Proc. London Math. Soc. 1951,
A. Grothendieck: *Résumé de la théorie métrique des produits tensoriels topologiques.* Bol. Soc. Mat. São Paulo 1956.

Clearly, special ideals such as those of *finite, approximable and compact operators* had been considered much earlier. In the period from 1965 to 1975, the theory of operator ideals developed into a separate discipline of functional analysis, which provides powerful methods for other branches of mathematics. Important applications have been made to the geometry of Banach spaces, to Brownian motion and, in particular, to eigenvalue problems. An extensive treatment was given by

A. Pietsch: *Operator ideals*. Berlin 1978.

7.3.6. *s*-Numbers

Es sei $K(s, t)$ eine für $a \leqq s \leqq b$, $a \leqq t \leqq b$ definierte reelle stetige Funktion, die nicht als symmetrisch vorausgesetzt werden soll. Wenn dann die beiden reellen oder komplexen stetigen nicht identisch verschwindenden Funktionen $\varphi(s)$ und $\psi(s)$ den Gleichungen

$$\varphi(s) = \lambda \int_b^a K(s, t)\, \psi(t)\, \mathrm{d}t$$

$$\psi(s) = \lambda \int_b^a K(t, s)\, \varphi(t)\, \mathrm{d}t$$

genügen, so sollen sie als ein Paar zum betreffenden Eigenwert λ gehöriger adjungierter Eigenfunktionen des Kernes $K(s, t)$ bezeichnet werden.

Erhard Schmidt (1907)

The concept of *s-numbers* was invented by

E. Schmidt: *Entwicklung willkürlicher Funktionen nach Systemen vorgeschriebener*. Math. Ann. 1907,

when he established his famous representation for integral operators induced by arbitrary continuous kernels. However, contrary to modern terminology, he referred to these quantities as eigenvalues. Later on, F. Smithies (1937: a) used the term "*singular value*" which was finally changed into the present notation.

For a long time the sequence of *s*-numbers of an operator acting on a Hilbert space had been defined as the sequence of the eigenvalues of the positive operator $|T| := (T^*T)^{1/2}$. From the viewpoint of our present knowledge this approach is quite troublesome. However, it was not until 1957 that D. Eh. Allakhverdiev, in an exotic journal, proved that

$$s_n(T) = \inf \{\|T - L\| : L \in \mathfrak{F}(H), \operatorname{rank}(L) < n\};$$

see also M. Fiedler / V. Pták (1962). Only then did it become possible to extend this useful notion to operators acting on Banach spaces. This was carried out by A. Pietsch (1963: b) when he invented the so-called *approximation numbers*. Subsequently, it turned out that there exist various other possibilities for generalizing the *s*-numbers to the Banach space setting. Thus, based on A. N. Kolmogorov's notion of diameters, the concept of *Kolmogorov numbers* was defined by I. A. Novosel'skij (1964). The dual concept of *Gel'fand numbers*, first considered by H. Triebel (1970: a), originates from the work of V. M. Tikhomirov (1965) who was stimulated by a proposal of his famous teacher. In view of

$$c_n(T) = \min \{\max [\|Tx\| : x \in M, \|x\| \leqq 1] : \operatorname{codim}(M) < n\},$$

the definition of these numbers is closely related to a well-known *minimax principle* attributed to E. Fischer (1905) and R. Courant (1920).

In order to systematize the theory of all these s-numbers, A. Pietsch (1972: a, 1974: a) developed an *axiomatic approach*. Later on, W. Bauhardt (1977) studied the so-called *Hilbert numbers*. A temporary keystone was laid by A. Pietsch (1980: c) when he introduced the *Weyl numbers* and their dual counterparts.

According to a suggestion of J. R. Retherford, generalized s-numbers with respect to an arbitrary quasi-Banach operator ideal were considered for the first time by H. König (1978). Nevertheless, it is worth while to stress the fact that already E. Schmidt (1907: a) had determined the best approximation of an integral operator by those of rank less than or equal to n with respect to the Hilbert-Schmidt norm (Approximationstheorem).

Beginning with the *Schatten-von Neumann classes*

$$\mathfrak{S}_r(H) := \{T \in \mathfrak{L}(H) : (s_n(T)) \in l_r\},$$

many useful operator ideals have been defined by means of s-numbers. The fine index w, well-known from the theory of Lorentz sequence spaces, was introduced by H. Triebel (1967). Afterwards it was no problem to extend these constructions to operators acting on Banach spaces. The quasi-Banach operator ideals

$$\mathfrak{L}^{(s)}_{r,w} := \{T \in \mathfrak{L} : (s_n(T)) \in l_{r,w}\}$$

have now become the indispensable background of the modern theory of eigenvalue distributions.

7.3.7. Absolutely summing operators

> Soit u une application semi-intégrale droite d'un espace de Banach E dans un autre F, de norme semi-intégrale droite $\leqq 1$. Alors on peut trouver un espace compact muni d'une mesure positive μ de norme $\leqq 1$, une application linéaire α de norme $\leqq 1$ de E dans $L^\infty(\mu)$ et une application linéaire γ de norme $\leqq 1$ de $L^2(\mu)$ dans F, tels que l'on ait $u = \gamma\varphi\alpha$, où φ est l'application identique de $L^\infty(\mu)$ dans $L^2(\mu)$.
>
> Alexandre Grothendieck (1955)

The *theory of absolutely summing operators* originates from a paper of W. Orlicz (1933) who proved that in L_p with $1 \leqq p \leqq 2$ all summable sequences are absolutely 2-summable. Later on,

A. Grothendieck: *Produits tensoriels topologiques et espaces nucléaires.* Memoirs Amer. Math. Soc. 1955,

characterized the so-called "*applications semi-intégrales à droite*" by the property that the image of every summable sequence is even absolutely summable. The main part of the theory of absolutely p-summing operators is due to

A. Pietsch: *Absolut p-summierende Abbildungen in normierten Räumen.* Studia Math. 1967.

J. R. Holub (1970/74) established the ε-tensor stability of the operator ideals $\mathfrak{P}_r$.

Absolutely (r, s)-summing operators have been defined by B. S. Mityagin / A. Pełczyński (1966). Important contributions are due to S. Kwapień (1968).

7.3.8. Nuclear operators

> $(H \otimes {}_\lambda H)^*$ represents the "trace class" of linear transformations on H, that is, all linear transformations on H which can be represented as a product of two linear transformations on H belonging to the Schmidt class.
>
> Robert Schatten, Johann von Neumann (1948)

At the beginning of this century, E. Schmidt (1907: a) proved his famous *representation theorem* which states, in modern terms, that every approximable operator on a Hilbert space can be written in the form

$$T = \sum_{i=1}^{\infty} \tau_i x_i^* \otimes y_i,$$

where (x_i) and (y_i) are extended orthonormal sequences and $(\tau_i) \in c_0$. The *trace class operators*, defined by R. Schatten / J. von Neumann (1946), can be characterized by the property $(\tau_i) \in l_1$. This means that the Schmidt series, formed by the 1-dimensional operators $\tau_i x_i{}^* \otimes y_i$, is absolutely summable. Based on this fact, A. F. Ruston (1951: a) and A. Grothendieck (1951) independently introduced the concept of a trace class or *nuclear operator* which also makes sense in the Banach space setting:

$$T = \sum_{i=1}^{\infty} a_i \otimes y_i.$$

Modifying the assumption $(\|a_i\| \, \|y_i\|) \in l_1$, which guarantees the absolute convergence of the right-hand series, several types of nuclear operators have been considered. These are, for example, the *p-nuclear operators* introduced by A. Grothendieck (GRO, Chap. II, p. 3) as "*opérateurs de puissance p. ème sommable*" and the $(r, 2)$-*nuclear operators* first defined by P. Saphar (1966).

7.4. Eigenvalues

> Die Werte $\lambda_1, \lambda_2, \ldots$ und die zugehörigen Funktionen $\varphi_1(s), \varphi_2(s), \ldots$ sind wesentlich durch den Kern $K(s, t)$ bestimmt; ich habe sie Eigenwerte bzw. Eigenfunktionen des Kerns $K(s, t)$ genannt.
>
> David Hilbert (1906)

There does not exist any book or paper devoted to the history of eigenvalue distributions of abstract operators. We may only refer to a report of H. König (1981) which surveys the development around 1977, the year of the great revolution.

7.4.1. Riesz theory

> F. Riesz's 1918 paper is one of the most beautiful ever written: it is entirely geometric in language and spirit, and so perfectly adapted to its goal that it has never been superseded and that Riesz's proofs can still be transcribed almost verbatim.
>
> Jean Alexandre Dieudonné (1981)

The first *determinant-free* treatments of linear equations in infinitely many unknowns were carried out by A. C. Dixon (1902), D. Hilbert (1906: a) and E. Schmidt

(1908). An analogous approach to the theory of integral equations is due to E. Schmidt (1907: b). However, the real birth of modern *spectral theory of compact (non-self-adjoint) operators* took place with the publication of

F. Riesz: *Über lineare Funktionalgleichungen.* Acta Math. 1918 (Hungarian version: *Linearis függvényegyenletekröl.* 1916).

This basic branch of functional analysis was completed by J. Schauder (1930) who proved the missing statements about dual operators. In the following period it was observed that the typical results, established by F. Riesz, remain true even for non-compact operators which have some compact power. Finally,

A. F. Ruston: *Operators with a Fredholm theory.* Journal London Math. Soc. 1954,

found necessary and sufficient conditions characterizing those operators which share all the nice properties with the compact ones. Among others he gave a useful criterion by means of the meromorphic nature of the Fredholm resolvent. The geometric definition of an *iteratively compact operator*, used as a starting point in this monograph, is due to A. Pietsch (1961). Later on, M. R. F. Smyth (BAS, p. 11) gave a characterization of *Riesz operators* in the same spirit. The fact that the Riesz properties of an operator are inherited by its dual and conversely was proved by T. T. West (1966).

7.4.2. Related operators

> The latent roots of the product of two matrices, it may be added, are the same in whichever order the factors be taken.
>
> James Joseph Sylvester (1883)

One hundred years ago J. J. Sylvester (1883) observed that, for two matrices A and B, the products AB and BA have the same eigenvalues. This result extends to arbitrary linear mappings and eigenvalues $\lambda \neq 0$. That fact was tacitly used by A. Grothendieck (GRO, Chap. II, p. 15) in order to determine the eigenvalue type of the ideal of nuclear operators. The explicit formulation of the *principle of related operators* as well as several important applications are due to A. Pietsch (1963: a, 1972: b).

7.4.3. Eigenvalues and s-numbers

> I have so far been unable to establish any direct connection between the orders of magnitude of the eigenvalues and the singular values when the kernel is not symmetric.
>
> Frank Smithies (1937)

We now describe the history of those results concerning the *asymptotic behaviour of eigenvalues* which can be obtained by determinant-free methods. More precisely, we deal with the relationship between eigenvalues and s-numbers which gave rise to the title of this monograph.

The first observation along this line was made by

I. Schur: *Über die charakteristischen Wurzeln einer linearen Substitution mit einer Anwendung auf die Theorie der Integralgleichungen.* Math. Ann. 1909.

Using his theorem on the triangular form of matrices, he established the inequality

$$\sum_{n=1}^{\infty} |\lambda_n(K)|^2 \leqq \int_0^1 \int_0^1 |K(\xi, \eta)|^2 \, d\xi \, d\eta$$

for arbitrary continuous kernels. Later on, T. Carleman (1921) proved the natural fact that this formula also holds for Hilbert-Schmidt kernels. This means that the operator ideal $\mathfrak{S}_2$ is of eigenvalue type l_2. There was then a long stagnation period until 1949, when

H. Weyl: *Inequalities between the two kinds of eigenvalues of a linear transformation.* Proc. Nat. Acad. Sci. U.S.A. 1949,

found his famous inequalities. This eminent mathematician, who was fascinated by eigenvalue problems again and again, became stimulated anew by the thesis of Shih-Hsun Chang (1949) in which the optimum eigenvalue type of $\mathfrak{S}_r$ with $0 < r \leqq 2$ was obtained via determinants. A preliminary version of this result, formulated, however, in terms of the order of Fredholm denominators had already been given by S. A. Gheorghiu (1927/28). Immediately after H. Weyl's paper there followed several publications devoted to this subject by Ky Fan (1949/50, 1951), G. Pólya (1950), A. Horn (1950) and J. P. O. Silberstein (1953); see also A. S. Markus (1964). Subsequently, there was again a long break until 1978 when, using approximation numbers, Gel'fand and Kolmogorov numbers,

H. König: *Interpolation of operator ideals with an application to eigenvalue distribution problems.* Math. Ann. 1978,

proved a modified *Weyl inequality* for operators on Banach spaces. A first step in this direction had been taken by A. S. Markus / V. I. Matsaev (1971). Finally, a more or less complete form of this theory is due to

A. Pietsch: *Weyl numbers and eigenvalues of operators in Banach spaces.* Math. Ann. 1980.

7.4.4. Eigenvalues of nuclear operators

> Le déterminant de Fredholm d'un noyau de Fredholm dans $E' \mathbin{\overline{\otimes}} E$ est d'ordre 2, et la suite de ses valeurs propres est carré sommable.
>
> Alexandre Grothendieck (1955)

Having in mind that every nuclear operator is related to an integral operator induced by a continuous kernel, we can claim that the *eigenvalue distribution of nuclear operators* was already determined by I. Fredholm (1903) and I. Schur (1909 : a). The explicit statement of the fact that the operator ideal $\mathfrak{N}$ has eigenvalue type l_2 is to be found in the famous thesis of A. Grothendieck (GRO, Chap. II, pp. 13–17). According to an example given by T. Carleman (1918) this is the best possible result in general Banach spaces. For operators acting on the Hilbert function space, the same problem had already been treated by T. Lalesco (1915). He stated, but without a rigorous proof, that the eigenvalue sequence of any operator which is the product of two Hilbert-Schmidt operators must be absolutely summable. Nowadays it is known from the fundamental papers of W. B. Johnson / H. König / B. Maurey / J. R. Retherford (1979) and H. König / J. R. Retherford / N. Tomczak-Jaegermann (1980) that the worst possible asymptotic behaviour of the eigenvalues of a nuclear operator $T \in \mathfrak{N}(E)$ heavily depends on the geometric structure of the underlying Banach space E. A complete clarification of this interplay is, however, still pending.

7.4.5. Eigenvalues of absolutely summing operators

In the case of a matrix of complex numbers, the preceding proposition reads ($p \geqq 2$)

$$\Big(\sum_i |\lambda_i(T)|^p\Big)^{1/p} \leqq \Big(\sum_j \Big[\sum_k |t_{jk}|^{p'}\Big]^{p/p'}\Big)^{1/p}.$$

We do not know a simpler proof of this inequality exept for $p = 2$.

..., Bernard Maurey, ... (1979)

The first result about *eigenvalue distributions of absolutely summing operators* is the famous *Schur-Carleman inequality* for Hilbert-Schmidt kernels. Using the principle of related operators, A. Pietsch (1963: a, 1967) proved that the operator ideal $\mathfrak{P}_r$ with $1 \leqq r \leqq 2$ has eigenvalue type l_2. The much more complicated case when $2 \leqq r < \infty$ was solved by

W. B. Johnson / H. König / B. Maurey / J. R. Retherford: *Eigenvalues of p-summing and l_p-type operators in Banach spaces.* J. Funct. Anal. 1979.

Very recently A. Pietsch (1986) developed the *tensor stability technique* which yields a new and simpler approach. Further contributions to this subject, which is closely related to the theory of Hille-Tamarkin operators and to the Hausdorff-Young theorem, are due to M. Z. Solomyak (1970), G. E. Karadzhov (1972), J. A. Cochran / C. Oehring (1977), B. Russo (1977), J. J. F. Fournier / B. Russo (1977) and H. König / L. Weis (1983).

The precise asymptotic behaviour of the eigenvalues of absolutely $(r, 2)$-summing operators was determined in a fundamental paper of

H. König / J. R. Retherford / N. Tomczak-Jaegermann: *On the eigenvalues of $(p, 2)$-summing operators and constants associated with normed spaces.* J. Funct. Anal. 1980.

7.4.6. Summary

The most important results concerning the optimum eigenvalue type of operator ideals as well as their discoverer are recorded in the following table.

operator ideals	eigenvalue types	authors
$\mathfrak{S}_2$	l_2	I. Schur (1909:a), T. Carlemann (1921)
$\mathfrak{S}_1 = (\mathfrak{S}_2)^2$	l_1	T. Lalesco (1915), S. A. Gheorghiu (1927/28)
$\mathfrak{S}_r$	l_r	S. H. Chang (1949), H. Weyl (1949)
$\mathfrak{N}$	l_2	A. Grothendieck (1955)
$\mathfrak{N}_p$, $0 < p < 1$	$l_{r,p}$ $1/r = 1/p + 1/2$	A. Grothendieck (1955), H. König (1977)
$\mathfrak{N}_{r,2}$	l_r	P. Saphar (1966)
$\mathfrak{P}_2$	l_2	A. Pietsch (1963:a, 1967)
$(\mathfrak{P}_2)^m$	$l_{2/m}$	W. B. Johnson et al. (1979)
$\mathfrak{P}_r$, $2 < r < \infty$	l_r	W. B. Johnson et al. (1979)
$\mathfrak{P}_{r,2}$	$l_{r,\infty}$	H. König et al. (1980)
$\mathfrak{L}^{(a)}_{r,w}$	$l_{r,w}$	H. König (1978)
$\mathfrak{L}^{(x)}_{r,w}$	$l_{r,w}$	A. Pietsch (1980:c)
$(\mathfrak{P}_2)^{(a)}_{s,w}$	$l_{r,w}$ $1/r = 1/s + 1/2$	B. Carl and H. König (unpublished), A. Pietsch (1981:a)

7.5. Determinants

> Most scientific discoveries are made when "their time is fulfilled"; sometimes, but seldom, a genius lifts the veil decades earlier than could have been expected. Fredholm's discovery has always seemed to me one that was long overdue when it came. What could be more natural than the idea that a set of linear equations connected with a discrete set of mass points gives way to an integral equation when one passes to the limit of a continuum?
>
> Hermann Weyl (1944)

The history of determinants of infinite matrices and integral operators is extensively presented in the Enzyklopädie-Artikel by E. Hellinger / O. Toeplitz (HEL) and in a paper by M. Bernkopf (1968). A survey given by R. Sikorski (1961) contains several historical remarks about determinants of abstract operators on Banach spaces. We further quote a nicely written examination-paper of P. Mäurer (1981) which is unfortunately unpublished.

7.5.1. Determinants of infinite matrices

> Pour que le déterminant des A_{ik} et tous ses mineurs convergent absolument, il suffit que le produit des éléments diagonaux converge absolument et que les éléments non-diagonaux de chaque ligne soient moindres en valeur absolue que les termes d'une série donnée absolument convergente.
>
> Helge von Koch (1901)

Apart from an insufficient attempt of T. Kötteritzsch (1870), *determinants of infinite matrices* were investigated, for the first time, by the American astronomer and mathematician G. W. Hill (1877). Subsequently, H. Poincaré (1886) put these ideas into a rigorous form. Very extensive contributions to the theory of infinite determinants are due to H. von Koch who published a whole series of papers beginning in 1890. From the viewpoint assumed in this monograph his most important result concerns those matrices $M = (\mu_{ij})$ for which

$$\sum_{i=1}^{\infty} \sup_{j} |\mu_{ij}| < \infty .$$

As we now know, this definition anticipated the concept of a nuclear operator acting on the sequence space l_1. Surprisingly enough, H. von Koch was not interested in eigenvalues, and, in contrast to his colleague I. Fredholm (1903), he failed to look at $\det(I + \zeta M)$ as an entire function of the complex variable ζ. Another peculiarity of H. von Koch's approach is that he never used Hadamard's inequality for verifying the convergence of infinite determinants. Clearly, at the very beginning of his work, this tool, established only in 1893, was not yet available. But later on it would have yielded significant improvements; see O. Szász (1912).

7.5.2. Determinants of integral operators

> Si le déterminant D_f d'une équation fonctionnelle de la forme
>
> $$\varphi(x) + \int_0^1 f(x, s)\, \varphi(s)\, \mathrm{d}s = \psi(x),$$

où $f(x, s)$ et $\psi(x)$ sont des fonctions finies et intégrables, est différent de zéro, il existe une et une seule fonction $\varphi(x)$ satisfaisant à cette équation.

Cette fonction est donnée par l'équation:

$$\varphi(x) = \psi(x) - \int_0^1 \frac{D_f\binom{x}{y}}{D_f}\, \psi(y)\, dy.$$

Ivar Fredholm (1903)

Certainly, *Fredholm's determinant theory of integral operators* induced by continuous kernels is the first milestone in the history of functional analysis. The fundamental paper

I. Fredholm: *Sur une classe d'équations fonctionelles.* Acta Math. 1903,

contains so may new ideas that even nowadays it is worth while to have a look at this original treatise. The contributions of

J. Plemelj: *Zur Theorie der Fredholmschen Funktionalgleichungen.* Monatsh. Math. Phys. 1904,

are also extremely important. In particular, he expressed the coefficients of Fredholm's determinants by very elegant formulas which now bear his name.

For applications to potential theory it turned out to be necessary to cover those kernels which have singularities on the diagonal. A first successful step in this direction was made by D. Hilbert (1904) who modified Fredholm's coefficients by writing zeros instead of the undefined values $K(\xi, \xi)$. Later on,

T. Carleman: *Zur Theorie der linearen Integralgleichungen.* Math. Z. 1921,

extended this approach to his famous *determinant theory of Hilbert-Schmidt operators.* By employing *Plemelj's formulas,* further improvements were made by F. Smithies (1941) who also wrote a fine textbook (SMI) on this subject. Regularized Fredholm denominators of order $m > 2$ were first considered by H. Poincaré (1910).

7.5.3. Determinants of abstract operators

It happens rather often in Mathematics that some problems are open for a long time and suddenly they are solved independently by several mathematicians at the same time. This has also happened in the case of the theory of determinants in Banach spaces.

Roman Sikorski (1961)

As previously stated, the development of functional analysis has heavily been stimulated by Fredholm's determinant theory of integral operators (1903). Subsequently, it took only a few years until E. Schmidt (1907: b) and F. Riesz (1918) established the determinant-free approach. The fundamental work of the latter may also be considered as the starting point of abstract operator theory. Taking into account J. Schauder's contributions (1930), we are entitled to claim that the spectral theory of compact operators was more or less complete by the end of the twenties. A first representation is to be found in S. Banach's monograph (BAN). Afterwards for a long period, no progress was made towards an abstract version of determinant theory. It was only in the early fifties that

A. F. Ruston: *On the Fredholm theory of integral equations for operators belonging to the trace class of a general Banach space*. Proc. London Math. Soc. (2) 1951,

T. Leżański: *The Fredholm theory of linear equations in Banach spaces*. Studia Math. 1953,

A. Grothendieck: *La théorie de Fredholm*. Bull. Soc. Math. France 1956,

almost simultaneously defined *determinants for nuclear or integral operators* on Banach spaces. Because of the vexed approximation problem, however, their theories were quite sophisticated. As a consequence, this subject has been treated in books exclusively in the Hilbert space setting. When P. Enflo (1973) discovered his famous counter-example to the approximation property, it became clear that, on arbitrary Banach spaces, a satisfactory determinant theory could be expected only for classes of operators strictly smaller than that of the nuclear ones. Contributions along this line are due to A. Grothendieck (GRO, Chap. II, pp. 18–19), A. Pietsch (PIE, 27.4.9) and H. König (1980: a). Taking the work of A. D. Michal / R. S. Martin (1934), S. G. Mikhlin (1944), R. Sikorski (1953, 1961), S. T. Kuroda (1961), A. Pietsch (1963: a), J. J. Grobler / H. Raubenheimer / P. van Eldik (1982) and J. C. Engelbrecht / J. J. Grobler (1983) as a basis, A. Pietsch designed an *axiomatic approach* to determinant theory which is published for the first time in this monograph.

Regularized Fredholm denominators were investigated by P. Saphar (1966), H. J. Brascamp (1969), B. Simon (1977) and H. König (1980: b); see also (DUN, XI.9) and (GOH, IV.2).

Concerning the history of holomorphic functions defined on Banach spaces we refer to the reports of A. E. Taylor (1970, 1971).

7.5.4. Eigenvalues and zeros of entire functions

Herr Fredholm hat eine gewisse, durch $K(s, t)$ eindeutig definierte ganze transzendente Funktion

$$D(x) = 1 + d_1 x + d_2 x^2 + \dots$$

angegeben, die nur für die Eigenwerte λ des Kerns $K(s, t)$ verschwindet. Ist λ eine n-fache Nullstelle von $D(x)$, so heißt n die Ordnung des Eigenwerts λ. Diese Zahl n hat noch folgende Bedeutung: es lassen sich n, aber nicht mehr als n, linear unabhängige (stetige) Funktionen $\varphi_1(s), \varphi_2(s), \dots, \varphi_n(s)$ bestimmen, für die

$$\int_a^b K(s, t)\,\varphi_\alpha(t)\,\mathrm{d}t = c_{\alpha 1}\varphi_1(s) + \dots + c_{\alpha,\alpha-1}\varphi_{\alpha-1}(s) + \frac{1}{\lambda}\varphi_\alpha(s),$$

wird, wo die $c_{\alpha\beta}$ Konstanten sind.

Issai Schur (1909)

Das Laguerresche Geschlecht dieser ganzen Funktion kann den Wert 2 nicht übersteigen, vermutlich ist es niemals 2, sondern im allgemeinen gerade zu 1, bei durchwegs stetigem $f(st)$ vielleicht sogar stets 0.

Josip Plemelj (1904)

$D^*(\lambda)$ ist höchstens vom Geschlechte Eins und besitzt die Produktdarstellung

$$D^*(\lambda) = \prod_{\nu=1}^{\infty} \left(1 - \frac{\lambda}{\lambda_\nu}\right) e^{\frac{\lambda}{\lambda_\nu}}.$$

Torsten Carleman (1921)

I. Fredholm (1903) had already observed that the eigenvalues $\lambda \neq 0$ of an integral operator induced by a continuous kernel and the zeros ζ of its canonical Fredholm denominator are related via the formula $\lambda\zeta = -1$. The geometric interpretation of the order of those zeros as the dimension of the linear space spanned by the principal functions associated with λ is due to J. Plemelj (1904), E. Goursat (1908) and I. Schur (1909: b). Thanks to the fundamental contributions of K. Weierstrass (1876), J. Hadamard (1893: b), E. Borel (1897), J. L. W. V. Jensen (1899), A. Pringsheim (1904) and E. Lindelöf (1903, 1905) the theory of entire functions was well-developed at the turn of the century. In particular, the interplay between the growth of the maximum modulus and the asymptotic behaviour of the zeros had become transparent. These results have been applied in order to get information about the eigenvalue distributions of integral operators determined by kernels with certain properties such as integrability, differentiability etc. This was done by T. Lalesco (1907, LAL), S. Mazurkiewicz (1915), T. Carleman (1917, 1921), A. O. Gel'fond (1957), D. W. Swann (1971) and, in particular, by

E. Hille / J.D. Tamarkin: *On the characteristic values of linear integral equations*. Act. Math. 1931.

The powerful tool of determinants had also been used by S. A. Gheorghiu (1927/28) and Shih-Hsun Chang (1949) when they investigated the connection between the eigenvalues and the s-numbers of Hilbert-Schmidt integral operators.

7.6. Traces

Bei aller Kürze und Einfachheit unserer auf die Spur bezüglichen Betrachtungen sind dieselben mathematisch nicht einwandfrei. Wir haben nämlich Reihen ohne Rücksicht auf ihre Konvergenz betrachtet, ineinander umgeformt (umsummiert) – kurzum alles getan, was man korrekterweise nicht tun soll. Zwar kommen derartige Nachlässigkeiten in der theoretischen Physik auch sonst vor, und die vorliegende wird in unseren quantenmechanischen Anwendungen kein Unheil anrichten – es muß aber doch festgestellt werden, daß es sich um eine Nachlässigkeit handelt.

Johann von Neumann (1932)

Concerning the history of trace theory, we refer to an excellent report given by J. R. Retherford (1980).

7.6.1. Operator ideals with a trace

It should be observed that, although to each $A \in B \otimes_\gamma B^*$ there corresponds a unique operator of the trace class, we have not proved in general that to each operator of the trace class there corresponds a unique element of $B \otimes_\gamma B^*$.

Anthony Francis Ruston (1951)

On notera que faute d'avoir résolu le problème de biunivocité, il n'est pas possible en général de parler de la trace de l'operateur dans E défini par un élément u de $E' \overline{\overline{\otimes}} E$.

Alexandre Grothendieck (1955)

Clearly, expressions such as

$$\sum_{i=1}^{\infty} \mu_{ii} \quad \text{and} \quad \int_0^1 K(\xi, \xi)\, d\xi$$

appeared early in the history of infinite matrices and integral operators. Nevertheless, J. von Neumann (1932) was the first to deal with the abstract concept of a *trace*. At the very beginning he restricted himself to positive operators on a Hilbert space. In the subsequent period, jointly with his student R. Schatten (1946/48), he established a theory of so-called *trace class operators* which were originally defined as the product of two Hilbert-Schmidt operators. Later on, the approach via s-numbers was used. Based on this work, A. F. Ruston (1951: a) and A. Grothendieck (1951) independently introduced the notion of a trace class or *nuclear operator* between Banach spaces. Their main goal was in fact the creation of a determinant theory. However, to this end a trace was required as well. Unfortunately, because of the negative answer to the approximation problem, such a well-defined trace cannot exist for arbitrary nuclear operators. Thus the extremely powerful concept of the trace class failed for the purpose it was originally designed for. In view of those difficulties A. Grothendieck (GRO, Chap. II, pp. 18–19) reduced the ideal of nuclear operators to that of 2/3-*nuclear operators,* and for this smaller class he got all the desired results. Later on, A. Pietsch (1963: b) showed that every operator of approximation type l_1, acting from an arbitrary Banach space into itself, possesses a well-defined trace. The *axiomatic approach* to traces which live on certain operator ideals is due to the same author (1981: b).

7.6.2. The trace formula

Пусть оператор C имеет след. Тогда, каков бы ни был ортонормированный базис $\varphi_i (i = 1, 2, \ldots)$ в H, справедливо равенство

$$\sum_{i=1}^{\infty} (C\varphi_i, \varphi_i) = \sum_{s=1}^{\infty} \lambda_s,$$

λ_s – собственные значения C.

Виктор Борисович Лидскии (1959)

A classical theorem of J. Mercer (1909) states that

$$\int_0^1 K(\xi, \xi)\, d\xi = \sum_{n=1}^{\infty} \lambda_n(K)$$

for every positive definite continuous kernel. Later on, using *Carleman's determinant formula* (1921), the Czech mathematicians B. Hostinsky (1921) and J. Kaucky (1921) showed that, for every Hilbert-Schmidt kernel $K(\xi, \eta)$,

$$\int_0^1 \int_0^1 K(\xi, \eta)\, K(\eta, \xi)\, d\xi\, d\eta = \sum_{n=1}^{\infty} \lambda_n(K)^2.$$

It was not until the end of the fifties that

V. B. Lidskij: *Non-self-adjoint operators with a trace (Russian)*. Doklady Akad. Nauk SSSR 1959,

succeeded in proving the *trace formula* for trace class operators on an abstract Hilbert space. A few years earlier, A. Grothendieck (GRO, Chap. II, p. 19) had obtained this result for 2/3-nuclear operators on Banach spaces, and, referring to H. Weyl (1949), he also claimed that the canonical Fredholm denominator of every nuclear operator on a Hilbert space has genus zero (GRO, Chap. II, p. 13). This fact immediately implies the trace formula which he used without comment in another paper (1961).

Both authors, A. Grothendieck and V. B. Lidskij, approached this problem indirectly, via determinant theory. All subsequent proofs, given by J. Weidmann (1965), (DUN, XI.9.19), (RIN, p. 139) and (REE, Vol. IV, p. 328), proceeded along the same line. Likewise employing this method,

H. König: *s-Numbers, eigenvalues and the trace theorem in Banach spaces*. Studia Math. 1980,

verified the trace formula for those operators on a Banach space which have summable approximation numbers. In another paper (1980: b) he successfully treated the operator ideal $(\mathfrak{P}_2)^2$ which, in the hidden form $\mathfrak{H} \circ \mathfrak{N}$, was already considered by A. Grothendieck (1961) and Chung-Wei Ha (1975).

Determinant-free proofs of the trace formula are due to J. A. Erdos (1968, 1974) and S. C. Power (1983: a, b). Their methods apply, however, to the Hilbert space setting exclusively. The comparatively elementary approach presented in this monograph is taken from H. Leiterer / A. Pietsch (1982).

7.7. Applications

> Ich schrieb diese Gleichungen nicht hin, als ob sie etwa das Problem lösten oder doch der Lösung näher führten, sie sollten nur ein Beispiel unter zahllosen sein, dafür, dass man bei Randproblemen der linearen partiellen Differentialgleichungen beständig vor dieselbe Gattung von Aufgaben gestellt wird, welche jedoch, wie es scheint, für die heutige Analysis im Allgemeinen unüberwindliche Schwierigkeiten darbieten. Ich meine die zweckmäßig Integralgleichungen zu nennenden Aufgaben, welche darin bestehen, dass die zu bestimmende Function, ausser ihrem sonstigen Vorkommen, in ihnen unter bestimmten Integralen enthalten ist.
>
> Paul du Bois-Reymond (1888)

The standard reference for the history of integral equations and equations in infinitely many unknowns is the famous Enzyklopädie-Artikel of E. Hellinger / O. Toeplitz (HEL). Many historical remarks concerning eigenvalue distributions of integral operators are to be found in J. A. Cochran's monograph (COC). Excellent reports about s-numbers of integral operators on Hilbert spaces were written by M. Sh. Birman / M. Z. Solomyak (1974, 1977: a). Concerning the asymptotic behaviour of eigenvalues of differential operators we refer to C. Clark (1967) and M. Sh. Birman / M. Z. Solomyak (1977: b). A complete survey of results from the theory of diameters, obtained by Soviet mathematicians, is given in V. M. Tikhomirov's book (TIK). L. Schumaker (SCU) provides many valuable comments to the history of spline functions. The best reference about trigonometric series, also from the historical viewpoint, is still A. Zygmund's treatise (ZYG).

7.7.1. Diagonal and embedding operators

Лично я пришел к вопросам теории вложения в связи с давно интересовавшими меня идеями классической теории приближения функций полиномами, прежде всево тригонометрическими полиномами и их непериодическими аналогами – целыми функциями экспоненциального типа.

Сергей Михайлович Николский (1969)

If $1 \leqq p \leqq q \leqq \infty$, then every bounded sequence (τ_k) determines a *diagonal operator* acting from l_p into l_q. In the case $1 \leqq q < p \leqq \infty$, it follows from classical results of O. Hölder (1889) and E. Landau (1907) that the same statement holds if and only if $(\tau_k) \in l_r$ with $1/r = 1/q - 1/p$.

On the other hand, apart from the inclusion $L_p(0, 1) \subseteqq L_q(0, 1)$ for $1 \leqq q \leqq p \leqq \infty$, according to

S. L. Sobolev: *Applications of functional analysis in mathematical physics (Russian).* Moscow 1950,

we have the *embedding* $W_p^\sigma(0, 1) \subset L_q(0, 1)$ provided that $\sigma > 1/p - 1/q$ and $1 \leqq p < q \leqq \infty$.

For a long time nobody could see any connection between the results just mentioned. However, with the help of spline bases Z. Ciesielsky and his pupils were able to construct an isomorphism between $B_{p,p}^\sigma(0, 1)$ and l_p. For a fairly complete treatment of this extremely involved subject we refer to

Z. Ciesielski / T. Figiel: *Spline bases in classical function spaces on compact C^∞ manifolds.* Studia Math. 1983.

Employing this result, A. Pietsch (1980: d) obtained the diagram

$$\begin{array}{ccc} B_{p,p}^\sigma(0,1) & \xrightarrow{\;I\;} & B_{q,q}^\tau(0,1)' \\ \downarrow\uparrow & & \downarrow\uparrow \\ l_p & \xrightarrow{\;D_\alpha\;} & l_q' \end{array},$$

where D_α is the diagonal operator induced by $(k^{-\alpha})$ with

$$\alpha := \sigma + \tau + 1 - 1/p - 1/q > 0.$$

It now became transparent why certain properties of diagonal and embedding operators are analogous. In a less explicit form this relationship had already been described by H. König (1974) and V. E. Majorov (1975).

7.7.2. s-Numbers of diagonal operators

Приведем для примера следующий изящный результат В. И. Мацаева: если M – компакт в l_1, состоящий из всех последовательностей $x = \{x_j\}$ таких, что $|x_j| \leqq \varepsilon_j$, где $\varepsilon_j \geqq \varepsilon_{j+1} \geqq 0$ и $\sum \varepsilon_j < \infty$, то n-поперечник компакта M определяется равенством

$$d_n(M) = \sum_n^\infty \varepsilon_j.$$

Александр Семенович Маркус (1966)

The first asymptotic estimate for the *approximation numbers of diagonal operators* from l_∞ into l_1 was obtained by A. Pietsch (1963: b). In 1974, employing a method due to V. D. Mil'man (1970), the same author computed the precise values of these numbers not only for diagonal operators from l_∞ into l_1 but also from l_p into l_q with $1 \leqq q \leqq p \leqq \infty$. The opposite case $1 \leqq p < q \leqq \infty$ is much more complicated. Here, due to the work of B. S. Stechkin (1954) and L. B. Sofman (1969, 1973), the approximation numbers are known for diagonal operator from l_1 into l_2. For all remaining combinations of the parameters p and q we must be content with some information about their asymptotic behaviour. The situation is similar for the Gel'fand and Kolmogorov numbers. The most striking result is due to B. S. Kashin (1977) who estimated the Kolmogorov numbers of the identity operator from $l_2(m)$ into $l_\infty(m)$ by means of probabilistic methods.

Generalizing classical theorems of W. Orlicz (1933) and A. Grothendieck (1956: b), G. Bennett (1973) and B. Carl (1974) proved independently that the identity operator from l_p into l_q is absolutely $(r, 2)$-summing for $1/r = 1/p - 1/q$ and $1 \leqq p < q \leqq 2$. This result was basic for the later investigations of Weyl numbers begun by A. Pietsch (1980: d) and completed by C. Lubitz (1982).

We have mentioned only those results which are at present relevant for applications in the theory of eigenvalue distributions. There are, however, further important contributions along this line, in particular, that of E. D. Gluskin (1981, 1983) and K. Höllig (1979). For additional information the reader is referred to survey papers of R. S. Ismagilov (1977) and R. Linde (1986); see also (PIN).

7.7.3. *s*-Numbers of embedding operators

Die Menge aller Funktionen $\varphi = c_1\varphi_1 + c_2\varphi_2 + \ldots + c_n\varphi_n$ mit festen Funktionen $\varphi_1, \varphi_2, \ldots, \varphi_n$, bildet einen n-dimensionalen linearen Unterraum Φ_n des Raumes R aller in Betracht kommenden Funktionen. $E_n(f)$ ist die Entfernung des Punktes f von der Menge Φ_n. Für eine Klasse F von Funktionen f bezeichnen wir weiter mit $E_n(F)$ die obere Grenze von $E_n(f)$ für alle f aus F. Die Größe $E_n(F)$ ist also ein natürliches Maß der Abweichung der Menge F von dem linearem Raum Φ_n. Wir stellen jetzt eine neue Aufgabe: bei gegebenen F und n durch die Wahl von Funktionen $\varphi_1, \varphi_2, \ldots, \varphi_n$, das Minimum von $E_n(F)$ zu erreichen. Die untere Grenze $D_n(F)$ der Größen $E_n(F)$ kann man, folglich, als die n-te Breite der Menge F zu bezeichnen.

Nikolay Andreevich Kolmogorov (1936)

Обычно под теоремами вложения понимают "качественные" утверждения о непрерывности или компактности вложения одного класса W_p^α в другой такой класс. Нашей главной целью является "количественное" исследование соответствующих операторов вложения, которые, вообще говоря, представляют собой компактные операторы.

Михаил Шлемович Бирман, Михаил Захарович Соломяк (1974)

Until the discovery of the *Ciesielski isomorphism*, *s*-numbers of embedding operators had been estimated by direct methods, for example, by trigonometric or piecewise-polynomial approximation. There is a vast literature devoted to this problem. The most important contributions were made in the USSR. We quote, for example,

V. M. Tikhomirov: *Diameters of sets in functional spaces and the theory of best approximation (Russian)*. Uspehi Mat. Nauk 1960,
M. Sh. Birman / M. Z. Solomyak: *Piecewise-polynomial approximations of functions of the classes W_p^α (Russian)*. Mat. Sb. 1967,
M. Z. Solomyak / V. M. Tikhomirov: *Geometric characteristics of the embedding from W_p^α into C (Russian)*. Izv. Vyssh. Uchebn. Zaved. Mat. 1967.

In contrast to the approach used in this monograph, the Soviet mathematicians take another viewpoint. They consider the closed unit ball of the first space as a precompact subset of the second one and look at its diameters. In this setting, however, the powerful concept of s-numbers gets lost.

7.7.4. Eigenvalues of infinite matrices

Man suche die linearen Gleichungen

$$Wx_k - \sum_l H(kl)\, x_l = 0$$

zu lösen; das ist nur möglich für gewisse Werte des Parameters W, nämlich $W = W_n$, wo W_n wieder die Eigenwerte (Energiewerte) bedeuten.

Max Born, Werner Heisenberg, Pascual Jordan (1926)

The literature concerned with eigenvalue distributions of operators induced by infinite matrices is very poor. Clearly, the Schur-Carleman inequality, proved for Hilbert-Schmidt kernels, could be rephrased in these terms. The main reason for the underdevelopment of this branch of functional analysis seems to be the fact that there was no need to deal with very special operators which are covered by the general spectral theory of operators acting on abstract Hilbert or Banach spaces. Nowadays, within the framework of a refined theory, matrix operators play an important role not only as simple patterns for the more complicated case of integral operators but also in their own right.

7.7.5. Eigenvalues of integral operators

What can be said about the distribution of the characteristic values of the Fredholm integral equation

$$y(x) = \lambda \int_a^b K(x, \xi)\, y(\xi)\, d\xi$$

on the basis of the general analytic properties of the kernel $K(x, \xi)$ such as integrability, continuity, differentiability, analyticity and the like?

Einar Hille, Jacob David Tamarkin (1931)

The first observation concerning the *asymptotic behaviour of eigenvalues of integral operators* is implicitly contained in I. Fredholm's paper from 1903 in which he determined the order of $\delta(\zeta, K)$ for continuous and Lipschitz-Hölder continuous kernels. This result was turned into an explicit statement about eigenvalue distributions by T. Lalesco (LAL, p. 88). The most basic theorem, obtained by determinant-free methods, is due to I. Schur (1909: a) and T. Carleman (1921).

In order to get the desired information about the eigenvalues of integral operators the following techniques have been employed:

- estimates of the growth of the Fredholm denominator,
- the Fischer-Courant minimax principle,
- estimates of the s-numbers.

As shown in this monograph, the last method is by far the most powerful.

In the course of proving those results for certain types of kernels the following stages of development have usually been passed through:

- $(\lambda_n) \in l_{r+\varepsilon}$ for $\varepsilon > 0$,
- $(\lambda_n) \in l_{r,\infty}$,
- $(\lambda_n) \in l_{r,w}$ with the best possible fine index w.

The following list, where the exponent of convergence is given by

$$1/r := m + n + 1/q^+ \quad \text{or} \quad 1/r := \sigma + \tau + 1/q^+,$$

respectively, yields a relatively complete historical survey.

kernels	eigenvalues	authors
$[C, C]$	l_2	I. Schur (1909:a)
$[C^m, C]$	$l_{r,\infty}$	H. Weyl (1912), S. Mazurkiewicz (1915), A. O. Gel'fond (1957)
$[C^\sigma, C]$	$l_{r,\infty}$	I. Fredholm (1903), T. Lalesco (1912)
$[L_2, L_2]$	l_2	T. Carleman (1921)
$[L_p, L_q]$	l_{q^+}	W. B. Johnson et al. (1979)
$[L_2, W_2^n]$	$l_{r,\infty}$	M. G. Krejn (1937), S. H. Chang (1952), I. C. Gohberg / M. G. Krejn (1965), V. I. Paraska (1965)
$[L_p, W_q^n]$	$l_{r,\infty}$	E. Hille / J. D. Tamarkin (1931)
$[L_p, B_{q,\infty}^\tau]$	$l_{r,\infty}$	E. Hille / J. D. Tamarkin (1931), F. Smithies (1937)
$[W_p^m, W_q^n]$	$l_{r,\infty}$	M. Sh. Birman / M. Z. Solomyak (1967/69)
$[B_{p,u}^\sigma, B_{q,v}^\tau]$	$l_{r,u}$	A. Pietsch (1980:d)

The above references do not mean that the authors have solved the problem for all possible choices of the parameters.

Up to 1977 almost all investigations about s-numbers of integral operators have been carried out in the Hilbert space setting. The initiators were M. G. Krejn (1937) and F. Smithies (1937: a). Some results along this line may be found in the famous monograph of I. C. Gohberg / M. G. Krejn (GOH). The most important contributions are due to the Leningrad school: M. Sh. Birman / M. Z. Solomyak (1967/69, 1970, 1974, 1977: a), M. Z. Solomyak (1970), G. P. Kostometov / M. Z. Solomyak (1971), G. P. Kostometov (1974, 1977) and G. E. Karadzhov (1972, 1977). We also refer to the work of V. I. Paraska (1965), P. E. Sobolevskij (1967), H. Triebel (1967, 1970: b), J. Kadlec / V. B. Korotkov (1968), S. L. Blyumin / B. D. Kotlyar (1970), J. A. Cochran (1975, 1976), J. Weidmann (1975) and B. D. Kotlyar / T. N. Semirenko (1981). Various criteria of nuclearity were estabilished by Shih-Hsun Chang (1947), W. F. Stinespring (1958), J. Weidmann (1966), M. Sh. Birman / M. Z. Solomyak (1969), J. A. Cochran (1974, 1977) and G. E. Karadzhov (1977). The special case of Hankel operators was treated by J. S. Howland (1971) and V. V. Peller (1980).

All authors just quoted are concerned with integral operators acting on a Hilbert function space. This technical assumption requires some additional conditions on the underlying kernel which sometimes look quite unnatural. The trouble disappears in the Banach space setting. The main results in this new branch of the theory of eigenvalue distributions are due to W. B. Johnson / H. König / B. Maurey / J. R. Retherford (1979), H. König / J. R. Retherford / N. Tomczak-Jaegermann (1980), H. König (1980: d) and A. Pietsch (1980: d).

7.7.6. Eigenvalues of differential operators

> In einer vollkommen spiegelnden Hülle können sich stehende elektromagnetische Schwingungen ausbilden, ähnlich den Tönen einer Orgelpfeife. Hierbei entsteht das mathematische Problem, zu beweisen, daß die Anzahl der genügend hohen Obertöne zwischen n und $n + dn$ unabhängig von der Gestalt der Hülle und nur ihrem Volumen proportional ist.
>
> Hendrik Antoon Lorentz (1910)

> In der vorliegenden Arbeit habe ich mir die Aufgabe gestellt, mit den Methoden der Integralgleichungstheorie folgenden Satz zu beweisen: Schwingungsvorgänge, deren Gesetzmäßigkeit sich in einer linearen Differentialgleichung vom Typus der gewöhnlichen Schwingungsgleichung ausspricht, besitzen, unabhängig von der geometrischen Gestalt und physikalischen Beschaffenheit der Räume, in denen sie sich abspielen, im Gebiet der hohen Schwingungszahlen alle wesentlich ein und dasselbe "Spektrum".
>
> Hermann Weyl (1912)

The interest in the asymptotic behaviour of eigenvalues goes back to a conjecture stated by H. A. Lorentz in Göttingen in 1910 when he gave a series of lectures "*Über alte und neue Fragen der Physik*". Only after a few months

H. Weyl: *Das asymptotische Verteilungsgesetz der Eigenwerte linearer partieller Differentialgleichungen (mit einer Anwendung auf die Theorie der Hohlraumstrahlung)*. Math. Ann. 1912,

solved this problem by the method of integral equations. Later on,

R. Courant: *Über die Eigenwerte bei den Differentialgleichungen der mathematischen Physik*. Math. Z. 1920,

found another approach based on the minimax principle. Finally, the Tauberian techniques were introduced by

T. Carleman: *Über die asymptotische Verteilung der Eigenwerte partieller Differentialgleichungen*, Ber. Sächs. Akad. Wiss. Leipzig 1936.

Since all problems concerning *eigenvalue distributions of differential operators* can be transformed, via the Green function, into those for integral operators, one could guess that both subjects are closely mixed up with each other; see S. Agmon (1965). Surprisingly enough, this is not so! Each of these branches has developed its own types of results and methods. Thus it remains a problem for future work to bring together these estranged relatives.

7.7.7. Fourier coefficients

> Ob nun die Coefficienten der Reihe zuletzt unendlich klein werden, wird in vielen Fällen nicht aus ihrem Ausdrucke durch bestimmte Integrale, sondern auf anderem Wege entschieden

werden müssen. Es verdient indess ein Fall hervorgehoben zu werden, wo sich dies unmittelbar aus der Natur der Function entscheiden läßt, wenn nämlich die Function $f(x)$ durchgehends endlich bleibt und eine Integration zuläßt.

Bernhard Riemann (1854)

Vor einiger Zeit haben Sie die Frage aufgeworfen, ob die Ordnung des zu einem stetigen Kern gehörigen FREDHOLM-schen Nenners die obere Schranke 2 erreichen kann. Ich werde im folgenden ein Beispiel angeben, für das diese Grenze tatsächlich erreicht wird. Das Beispiel hängt übrigens mit einer Konvergenzfrage betreffend die Fourierkoeffizienten einer stetigen Funktion zusammen, welche vielleicht auch an und für sich nicht ohne Interesse ist.

Torsten Carleman (1918)
in a letter to A. Wiman

Many basic notions and results of the theory of functions have been obtained by mathematicians while working on trigonometric series.

Antoni Zygmund (1958)

The fact that the *Fourier coefficients* of a periodic function f coincide up to the factor 2π with the eigenvalues of the associated *convolution operator*,

$$C_{\text{op}}^f : g(\eta) \to \int_0^{2\pi} f(\xi - \eta)\, g(\eta)\, d\eta,$$

was realized rather late. The first remark concerning this connection is due to

T. Carleman: *Über die Fourierkoeffizienten einer stetigen Funktion.* Acta Math. 1918.

Thus it happened that corresponding results for eigenvalues and Fourier coefficients were proved independently and at quite different dates. For example, the precise asymptotic behaviour of the eigenvalues of a Lipschitz-Hölder continuous kernel was already known to I. Fredholm (1903), while S. N. Bernstein proved not until 1914 that all functions $f \in C^\sigma(2\pi)$ with $\sigma > 1/2$ have absolutely and uniformly convergent Fourier series. Later on, the theory of trigonometric series became an extremely rich source for counterexamples in the theory of eigenvalue distributions.

The following list sketches the historical development of the most important results related to the question of how the sequence of Fourier coefficients reflects the properties of the underlying function.

functions	Fourier coefficients	authors
L_1	c_0	B. Riemann (1854), H. Lebesgue (1903)
L_2	l_2	F. Riesz (1907), E. Fischer (1907)
$C^\sigma, \sigma > 1/2$	l_1	S. N. Bernstein (1914)
$B_{\infty,1}^{1/2}$	l_1	S. N. Bernstein (1934)
$B_{2,1}^{1/2}$	l_1	O. Szász (1946), S. B. Stechkin (1951)
$L_r, 1 < r < 2$	$l_{r',r}$	W. H. Young (1912), F. Hausdorff (1923), G. H. Hardy / J. E. Littlewood (1931)
$B_{r,w}^\varrho$	$l_{s,w}$ $1/s = \varrho + 1/r^+$	H. Weyl (1917), O. Szász (1922, 1928), A. A. Konyuskov (1958), A. Pietsch (1980: a)

7.7.8. Practical applications

Teoria cum praxi.

Gottfried Wilhelm Leibniz (1700)

Le but unique de la science, c'est l'honeur de l'esprit humain.

Carl Gustav Jacob Jacobi (1830)

Mathematicians are people who devote their lives to what seems to me a wonderful kind of play.

Constance Reid (1980)

Though this be madness, yet there is method in't.

William Shakespeare (1602)

As already indicated, results about the asymptotic behaviour of eigenvalues of differential and integral operators have some significance in physics. Nevertheless, at present we do not know any application in the real world. Hopefully, this will change in the future. On the other hand, mathematics in its own right is an important and indispensable part of human culture. Bearing in mind these facts, the following motto was used when writing the monograph in hand:

If a mathematical theory has not yet proved useful,
then it should at least be beautiful. •

Appendix

At the time when finishing the work on this monograph, I did only know examples of such quasi-Banach operator ideals which admit either none or at most one continuous trace. Luckily enough, this gap was filled by N. J. Kalton during the "International Conference on Banach Spaces and Classical Analysis" (Kent, Ohio, summer 1985). He constructed quasi-Banach operator ideals supporting a large amount of different continuous traces. This fact shed a new light on many aspects of the trace and determinant theory. In particular, some open problems have become much more urgent than before.

In what follows we present a new approach to Kalton's example. His original paper "*Unusual traces on operator ideals*", essentially based on Hilbert space techniques, is going to appear in the journal "Mathematische Nachrichten".

A.X.1. Throughout this appendix $\mathfrak{r} = (\varrho_k)$ denotes a **weight sequence** which, by definition, satisfies the following condition:

(W₁) $\quad \varrho_k > 0 \quad \text{and} \quad \sum_{k=0}^{\infty} \varrho_k = 1.$

(W₂) There exists a constant $c_\mathfrak{r} > 1$ such that $\varrho_k \leqq c_\mathfrak{r}\varrho_{k+1}$ for $k = 0, 1, \dots$

A.X.2. Operators $T_0, T_1, \dots \in \mathfrak{F}(E, F)$ are said to constitute an $\mathfrak{r}$-**decomposition** if

$$\operatorname{rank}(T_k) \leqq 2^k \quad \text{and} \quad \|T_k\| \leqq c\varrho_k \quad \text{for} \quad k = 0, 1, \dots$$

where $c > 0$ is a constant. Then we let

$$\|(T_k) \mid \mathfrak{M}_\mathfrak{r}\| := \sup_k \varrho_k^{-1} \|T_k\|.$$

Every $\mathfrak{r}$-decomposition (T_k) determines an operator

$$T = \sum_{k=0}^{\infty} T_k$$

such that

$$\|T\| \leqq \sum_{k=0}^{\infty} \|T_k\| \leqq \|(T_k) \mid \mathfrak{M}_\mathfrak{r}\| \sum_{k=0}^{\infty} \varrho_k = \|(T_k) \mid \mathfrak{M}_\mathfrak{r}\|.$$

Hence (T_k) is also called an $\mathfrak{r}$-decomposition of $T \in \mathfrak{L}(E, F)$.

A.X.3. We denote by $\mathfrak{M}_\mathfrak{r}(E, F)$ the collection of all operators $T \in \mathfrak{L}(E, F)$ admitting an $\mathfrak{r}$-decomposition

$$T = \sum_{k=0}^{\infty} T_k.$$

Let

$$\|T \mid \mathfrak{M}_\mathfrak{r}\| := \inf \|(T_k) \mid \mathfrak{M}_\mathfrak{r}\|.$$

the infimum being taken over all possible choices of (T_k).

A.X.4. The following result is obtained by standard techniques.

Theorem. $\mathfrak{M}_\mathfrak{r}$ is a quasi-Banach operator ideal.

Proof. Obviously,

$$\|T\| \leqq \|T \mid \mathfrak{M}_\mathfrak{r}\| \quad \text{for} \quad T \in \mathfrak{M}_\mathfrak{r}(E, F).$$

Moreover, if $\operatorname{rank}(T) = 1$, then we obtain an $\mathfrak{r}$-decomposition (T_k) by putting $T_k := \varrho_k T$. Hence

$$\|T \mid \mathfrak{M}_\mathfrak{r}\| \leqq \|(T_k) \mid \mathfrak{M}_\mathfrak{r}\| = \|T\|.$$

This proves that

$$a \otimes y \in \mathfrak{M}_\mathfrak{r}(E, F) \quad \text{and} \quad \|a \otimes y \mid \mathfrak{M}_\mathfrak{r}\| = \|a\| \, \|y\| \quad \text{for } a \in E' \text{ and } y \in F.$$

Given $S, T \in \mathfrak{M}_\mathfrak{r}(E, F)$ and $\varepsilon > 0$, we choose $\mathfrak{r}$-decompositions

$$S = \sum_{k=0}^{\infty} S_k \quad \text{and} \quad T = \sum_{k=0}^{\infty} T_k$$

such that

$$\|(S_k) \mid \mathfrak{M}_\mathfrak{r}\| \leqq (1 + \varepsilon) \|S \mid \mathfrak{M}_\mathfrak{r}\| \quad \text{and} \quad \|(T_k) \mid \mathfrak{M}_\mathfrak{r}\| \leqq (1 + \varepsilon) \|T \mid \mathfrak{M}_\mathfrak{r}\|.$$

Define

$$R_0 := O \quad \text{and} \quad R_{k+1} := S_k + T_k \quad \text{for } k = 0, 1, \ldots$$

Then

$$\operatorname{rank}(R_{k+1}) \leqq \operatorname{rank}(S_k) + \operatorname{rank}(T_k) \leqq 2^{k+1}$$

and

$$\|R_{k+1}\| \leqq \|S_k\| + \|T_k\| \leqq [\|(S_k) \mid \mathfrak{M}_\mathfrak{r}\| + \|(T_k) \mid \mathfrak{M}_\mathfrak{r}\|] \, c_\mathfrak{r} \varrho_{k+1}.$$

Hence (R_k) is an $\mathfrak{r}$-decomposition of $S + T$ with

$$\|(R_k) \mid \mathfrak{M}_\mathfrak{r}\| \leqq (1 + \varepsilon) \, c_\mathfrak{r}[\|S \mid \mathfrak{M}_\mathfrak{r}\| + \|T \mid \mathfrak{M}_\mathfrak{r}\|].$$

Letting $\varepsilon \to 0$, we obtain

$$S + T \in \mathfrak{M}_\mathfrak{r}(E, F) \quad \text{and} \quad \|S + T \mid \mathfrak{M}_\mathfrak{r}\| \leqq c_\mathfrak{r}[\|S \mid \mathfrak{M}_\mathfrak{r}\| + \|T \mid \mathfrak{M}_\mathfrak{r}\|].$$

If $X \in \mathfrak{L}(E_0, E)$, $T \in \mathfrak{M}_\mathfrak{r}(E, F)$ and $Y \in \mathfrak{L}(F, F_0)$, then it follows immediately that

$$YTX \in \mathfrak{M}_\mathfrak{r}(E_0, F_0) \quad \text{and} \quad \|YTX \mid \mathfrak{M}_\mathfrak{r}\| \leqq \|Y\| \, \|T \mid \mathfrak{M}_\mathfrak{r}\| \, \|X\|.$$

So far we have shown that $\mathfrak{M}_\mathfrak{r}$ is a quasi-normed operator ideal. In order to check its completeness, we consider a sequence of operators $T^{(h)} \in \mathfrak{M}_\mathfrak{r}(E, F)$ such that

$$\sum_{h=0}^{\infty} c_\mathfrak{r}^{h+1} \|T^{(h)} \mid \mathfrak{M}_\mathfrak{r}\| < \infty.$$

Given $\varepsilon > 0$, there exist $\mathfrak{r}$-decompositions

$$T^{(h)} = \sum_{k=0}^{\infty} T_{hk} \text{ such that } \|(T_{hk}) \mid \mathfrak{M}_\mathfrak{r}\| \leqq (1 + \varepsilon) \|T^{(h)} \mid \mathfrak{M}_\mathfrak{r}\|.$$

Define

$$T_0 := O \quad \text{and} \quad T_{i+1} := \sum_{h+k=i} T_{hk} \quad \text{for} \quad i = 0, 1, \ldots$$

Then

$$\operatorname{rank}(T_{i+1}) \leqq \sum_{h+k=i} \operatorname{rank}(T_{hk}) \leqq \sum_{k=0}^{i} 2^k < 2^{i+1}$$

and

$$\|T_{i+1}\| \leqq \sum_{h+k=i} \|T_{hk}\| \leqq \sum_{h+k=i} \|(T_{hk}) \mid \mathfrak{M}_{\mathfrak{r}}\| \, \varrho_k$$
$$\leqq \sum_{h+k=i} \|(T_{hk}) \mid \mathfrak{M}_{\mathfrak{r}}\| \, c_{\mathfrak{r}}^{i+1-k} \, \varrho_{i+1}$$
$$\leqq (1+\varepsilon) \sum_{h=0}^{i} c_{\mathfrak{r}}^{h+1} \, \|T^{(h)} | \, \mathfrak{M}_{\mathfrak{r}}\| \, \varrho_{i+1}.$$

Hence (T_i) is an $\mathfrak{r}$-decomposition of

$$T := \sum_{h=0}^{\infty} T^{(h)} = \sum_{h=0}^{\infty} \sum_{k=0}^{\infty} T_{hk} = \sum_{i=0}^{\infty} T_i$$

with

$$\|(T_i) \mid \mathfrak{M}_{\mathfrak{r}}\| \leqq (1+\varepsilon) \sum_{h=0}^{\infty} c_{\mathfrak{r}}^{h+1} \, \|T^{(h)} \mid \mathfrak{M}_{\mathfrak{r}}\|.$$

Letting $\varepsilon \to 0$, we obtain

$$T \in \mathfrak{M}_{\mathfrak{r}}(E, F) \quad \text{and} \quad \|T \mid \mathfrak{M}_{\mathfrak{r}}\| \leqq \sum_{h=0}^{\infty} c_{\mathfrak{r}}^{h+1} \, \|T^{(h)} \mid \mathfrak{M}_{\mathfrak{r}}\|.$$

Finally, given any Cauchy sequence (T_n) in $\mathfrak{M}_{\mathfrak{r}}(E, F)$, we choose an increasing sequence of natural numbers n_h such that

$$\|T_m - T_n \mid \mathfrak{M}_{\mathfrak{r}}\| \leqq (2c_{\mathfrak{r}})^{-h-1} \quad \text{for} \quad m, n \geqq n_h.$$

Let

$$T^{(0)} := T_{n_0} \quad \text{and} \quad T^{(h+1)} := T_{n_{h+1}} - T_{n_h} \quad \text{for} \quad h = 0, 1, \ldots$$

Then

$$\sum_{h=0}^{\infty} c_{\mathfrak{r}}^{h+1} \, \|T^{(h)} \mid \mathfrak{M}_{\mathfrak{r}}\| \leqq c_{\mathfrak{r}}(\|T_{n_0} \mid \mathfrak{M}_{\mathfrak{r}}\| + 1) < \infty.$$

Hence

$$T := \sum_{h=0}^{\infty} T^{(h)} \in \mathfrak{M}_{\mathfrak{r}}(E, F)$$

and

$$\|T - T_{n_k} \mid \mathfrak{M}_{\mathfrak{r}}\| = \|T - \sum_{h=0}^{k} T^{(h)} \mid \mathfrak{M}_{\mathfrak{r}}\| \leqq \sum_{h=k+1}^{\infty} c_{\mathfrak{r}}^{h+1} \, \|T^{(h)} \mid \mathfrak{M}_{\mathfrak{r}}\| \leqq 2^{-k} c_{\mathfrak{r}}.$$

Therefore the subsequence (T_{n_k}) converges to T with respect to the metrizable topology of $\mathfrak{M}_{\mathfrak{r}}(E, F)$, and so does (T_n).

Remark. Suppose that the weight sequence $\mathfrak{r} = (\varrho_k)$ satisfies the condition

$$\sum_{h=k}^{\infty} \varrho_h \leqq c_0 \varrho_k \quad \text{for} \quad k = 0, 1, \ldots$$

where $c_0 > 1$ is a constant. Then it can easily be seen that an operator $T \in \mathfrak{L}(E, F)$ belongs to $\mathfrak{M}_{\mathfrak{r}}(E, F)$ if and only if there exists a constant $c > 0$ such that

$$a_{2^k}(T) \leqq c\varrho_k \quad \text{for} \quad k = 0, 1, \ldots$$

In the particular case

$$\mathfrak{r} = (\varrho 2^{-k/r}),$$

where

$$\varrho := \left[\sum_{k=0}^{\infty} 2^{-k/r}\right]^{-1} \quad \text{and} \quad 0 < r < \infty,$$

it follows from 2.3.8 that $\mathfrak{M}_\mathfrak{r}$ consists of all operators having approximation type $l_{r,\infty}$. In other words, $\mathfrak{M}_\mathfrak{r} = \mathfrak{L}^{(a)}_{r,\infty}$.

A.X.5. Next we prove an auxiliary result.

Lemma. Let $T \in \mathfrak{L}(E)$. Then

$|\text{trace}\,(T)| \leqq n\,\|T\|$ whenever $\text{rank}\,(T) \leqq n$.

Proof. Without loss of generality, we may assume that $\text{rank}\,(T) = n$. In this case, by Auerbach's lemma 1.7.6, there exist $x_1, \ldots, x_n \in M(T)$ and $a_1, \ldots, a_n \in E'$ such that

$$\|x_i\| = 1, \quad \|a_i\| = 1 \quad \text{and} \quad \langle x_i, a_j\rangle = \delta_{ij}.$$

Then it follows from

$$Tx = \sum_{i=1}^{n} \langle Tx, a_i\rangle\, x_i \quad \text{for} \quad x \in E$$

that

$$T = \sum_{i=1}^{n} T'a_i \otimes x_i.$$

Hence

$$|\text{trace}\,(T)| = \left|\sum_{i=1}^{n} \langle x_i, T'a_i\rangle\right| \leqq \sum_{i=1}^{n} \|x_i\|\,\|T'a_i\| \leqq n\,\|T\|.$$

A.X.6. A weight sequence $\mathfrak{r} = (\varrho_k)$ is said to be **strongly convergent** if

$$\sum_{k=0}^{\infty} 2^k \varrho_k < \infty.$$

A.X.7. Theorem. The quasi-Banach operator ideal $\mathfrak{M}_\mathfrak{r}$ admits a continuous trace if and only if $\mathfrak{r} = (\varrho_k)$ is strongly convergent.

Proof. We consider first a strongly convergent weight sequence $\mathfrak{r} = (\varrho_k)$.
If $T \in \mathfrak{M}_\mathfrak{r}(E)$, then for every $\mathfrak{r}$-decomposition

$$T = \sum_{k=0}^{\infty} T_k$$

we have

$$\sum_{k=0}^{\infty} |\text{trace}\,(T_k)| \leqq \sum_{k=0}^{\infty} 2^k \|T_k\| \leqq \sum_{k=0}^{\infty} 2^k \varrho_k \|(T_k) \mid \mathfrak{M}_\mathfrak{r}\|.$$

Hence the right-hand side in the following definition makes sense:

$$\tau(T) := \sum_{k=0}^{\infty} \text{trace}\,(T_k).$$

Of course, it must be shown that this quantity is uniquely determined. In order to verify this fact, we consider different $\mathfrak{r}$-decompositions

$$T = \sum_{k=0}^{\infty} T_k^{(1)} \quad \text{and} \quad T = \sum_{k=0}^{\infty} T_k^{(2)}.$$

Next we define an $\mathfrak{r}$-decomposition (T_k) by putting

$$T_0 := O \quad \text{and} \quad T_{k+1} := T_k^{(1)} - T_k^{(2)} \quad \text{for} \quad k = 0, 1, \ldots$$

Note that

$$\sum_{k=0}^{\infty} T_k = O \quad \text{and} \quad \operatorname{rank}\left(\sum_{k=0}^{n} T_k\right) \leqq 2^{n+1}.$$

Thus the preceding lemma implies that

$$\begin{aligned} \left|\sum_{k=0}^{n} \operatorname{trace}(T_k)\right| &= \left|\operatorname{trace}\left(\sum_{k=0}^{n} T_k\right)\right| \leqq 2^{n+1} \left\|\sum_{k=0}^{n} T_k\right\| \\ &= 2^{n+1} \left\|\sum_{k=n+1}^{\infty} T_k\right\| \leqq 2^{n+1} \sum_{k=n+1}^{\infty} \|T_k\| \\ &\leqq \sum_{k=n+1}^{\infty} 2^k \varrho_k \, \|(T_k) \mid \mathfrak{M}_{\mathfrak{r}}\|. \end{aligned}$$

Hence, letting $n \to \infty$, we obtain

$$\sum_{k=0}^{\infty} \operatorname{trace}(T_k^{(1)}) - \sum_{k=0}^{\infty} \operatorname{trace}(T_k^{(2)}) = \sum_{k=0}^{\infty} \operatorname{trace}(T_k) = 0.$$

This proves that the above definition is correct. Furthermore, it can easily be seen that τ is indeed a continuous trace on $\mathfrak{M}_{\mathfrak{r}}$. In particular, we have

$$|\tau(T)| \leqq \sum_{k=0}^{\infty} 2^k \varrho_k \, \|T \mid \mathfrak{M}_{\mathfrak{r}}\| \quad \text{for} \quad T \in \mathfrak{M}_{\mathfrak{r}}(E).$$

Conversely, we assume that $\mathfrak{M}_{\mathfrak{r}}$ supports a continuous trace τ. Then there exists a constant $c \geqq 1$ such that

$$|\tau(T)| \leqq c \|T \mid \mathfrak{M}_{\mathfrak{r}}\| \quad \text{for} \quad T \in \mathfrak{M}_{\mathfrak{r}}(E).$$

Let

$$D_{\mathfrak{r}}^{(m)} := \sum_{k=0}^{m} \varrho_k P_k,$$

where P_k denotes the orthogonal projection from l_2 onto the subspace spanned by the unit sequences e_i with $2^k \leqq i < 2^{k+1}$. We now conclude from

$$\operatorname{trace}(D_{\mathfrak{r}}^{(m)}) = \sum_{k=0}^{m} 2^k \varrho_k \quad \text{and} \quad \|D_{\mathfrak{r}}^{(m)} \mid \mathfrak{M}_{\mathfrak{r}}\| \leqq 1$$

that

$$\sum_{k=0}^{m} 2^k \varrho_k \leqq c \quad \text{for} \quad m = 0, 1, \ldots$$

Hence the weight sequence $\mathfrak{r} = (\varrho_k)$ must be strongly convergent.

Remark. We know from 4.2.26 that the quasi-Banach operator ideal $\mathfrak{L}_1^{(a}$ admits a unique continuous trace. Moreover, it follows from 2.3.8 that $\mathfrak{M}_\mathfrak{r} \subset \mathfrak{L}_1^{(a)}$ for every strongly convergent weight sequence $\mathfrak{r} = (\varrho_k)$. Hence the desired continuous trace on $\mathfrak{M}_\mathfrak{r}$ can be obtained by restriction.

A.X.8. Suppose that the quasi-Banach operator ideal $\mathfrak{A}$ supports different (continuous) traces τ_1 and τ_2. Then $\tau := \tau_1 - \tau_2$ still possesses the properties (T_2), (T_3) and (T_4), stated in 4.2.1, while (T_1) changes into

(T_1^0) $\quad \tau(a \otimes x) = 0 \quad$ for $\quad a \in E'$ and $x \in E$.

A function τ satisfying these modified conditions is called a (continuous) **zero trace.** Of course, we are only interested in the non-trivial case when τ does not vanish identically.

A.X.9. A strongly convergent weight sequence $\mathfrak{r} = (\varrho_k)$ is said to be **singular** if there exists an increasing sequence of natural numbers n_i such that

$$\lim_i \frac{2^{n_i} \sum_{k=n_i}^{\infty} \varrho_k}{\sum_{k=n_i}^{\infty} 2^k \varrho_k} = 0.$$

A.X.10. We are now in a position to establish the main result of this appendix.

Theorem. Let $\mathfrak{r} = (\varrho_k)$ be a singular strongly convergent weight sequence. Then $\mathfrak{M}_\mathfrak{r}$ admits a continuous non-trivial zero trace.

Proof. First of all, we fix an ultrafilter $\mathscr{U}$ on $\mathbb{N}_0 := \{0, 1, \ldots\}$ such that

$$\lim_{\mathscr{U}} \frac{2^{n} \sum_{k=n}^{\infty} \varrho_k}{\sum_{k=n}^{\infty} 2^k \varrho_k} = 0.$$

This condition is satisfied if $\mathscr{U}$ contains all tails $\{n_i, n_{i+1}, \ldots\}$ of the sequence described in A.X.9.

If $T \in \mathfrak{M}_\mathfrak{r}(E)$, then for every $\mathfrak{r}$-decomposition

$$T = \sum_{k=0}^{\infty} T_k$$

we have

$$\sum_{k=n}^{\infty} |\text{trace}\,(T_k)| \leqq \sum_{k=n}^{\infty} 2^k \|T_k\| \leqq \sum_{k=n}^{\infty} 2^k \varrho_k \, \|(T_k) \mid \mathfrak{M}_\mathfrak{r}\|.$$

Hence the right-hand side in the following definition makes sense:

$$\tau_{\mathscr{U}}(T) := \lim_{\mathscr{U}} \frac{\sum_{k=n}^{\infty} \text{trace}\,(T_k)}{\sum_{k=n}^{\infty} 2^k \varrho_k}.$$

Of course, it must be shown that this quantity is uniquely determined. In order to verify this fact, we consider different $\mathfrak{r}$-decompositions

$$T = \sum_{k=0}^{\infty} T_k^{(1)} \quad \text{and} \quad T = \sum_{k=0}^{\infty} T_k^{(2)}.$$

Next we define an $\mathfrak{r}$-decomposition (T_k) by putting

$$T_0 := O \quad \text{and} \quad T_{k+1} := T_k^{(1)} - T_k^{(2)} \quad \text{for} \quad k = 0, 1, \ldots$$

Note that

$$\sum_{k=0}^{\infty} T_k = O \quad \text{and} \quad \operatorname{rank}\left(\sum_{k=0}^{n} T_k\right) \leq 2^{n+1}.$$

As shown in A.X.7, we have

$$\sum_{k=0}^{\infty} \operatorname{trace}(T_k) = 0$$

Thus Lemma A.X.5 implies that

$$\begin{aligned}\left|\sum_{k=n+1}^{\infty} \operatorname{trace}(T_k)\right| &= \left|\sum_{k=0}^{n} \operatorname{trace}(T_k)\right| = \left|\operatorname{trace}\left(\sum_{k=0}^{n} T_k\right)\right| \\ &\leq 2^{n+1}\left\|\sum_{k=0}^{n} T_k\right\| = 2^{n+1}\left\|\sum_{k=n+1}^{\infty} T_k\right\| \leq 2^{n+1}\sum_{k=n+1}^{\infty}\|T_k\| \\ &\leq 2^{n+1}\sum_{k=n+1}^{\infty} \varrho_k \|(T_k) \mid \mathfrak{M}_{\mathfrak{r}}\|.\end{aligned}$$

Hence

$$\begin{aligned}\left|\frac{\sum_{k=n}^{\infty}\operatorname{trace}(T_k^{(1)})}{\sum_{k=n}^{\infty} 2^k\varrho_k} - \frac{\sum_{k=n}^{\infty}\operatorname{trace}(T_k^{(2)})}{\sum_{k=n}^{\infty} 2^k\varrho_k}\right| &= \frac{\left|\sum_{k=n+1}^{\infty}\operatorname{trace}(T_k)\right|}{\sum_{k=n}^{\infty} 2^k\varrho_k} \\ &\leq \frac{2^{n+1}\sum_{k=n+1}^{\infty}\varrho_k}{\sum_{k=n}^{\infty} 2^k\varrho_k}\|(T_k) \mid \mathfrak{M}_{\mathfrak{r}}\|.\end{aligned}$$

Passing to the limit with respect to the ultrafilter $\mathscr{U}$, we conclude that the above definition is correct. Furthermore, it can easily be seen that $\tau_{\mathscr{U}}$ is indeed a continuous zero trace on $\mathfrak{M}_{\mathfrak{r}}$. In particular, we have

$$|\tau_{\mathscr{U}}(T)| \leq \|T \mid \mathfrak{M}_{\mathfrak{r}}\| \quad \text{for} \quad T \in \mathfrak{M}_{\mathfrak{r}}(E).$$

Finally, we define

$$D_{\mathfrak{r}} := \sum_{k=0}^{\infty} \varrho_k P_k,$$

where P_k denotes the orthogonal projection from l_2 onto the subspace spanned by the unit sequences e_i with $2^k \leq i < 2^{k+1}$. Then

$$D_{\mathfrak{r}} \in \mathfrak{M}_{\mathfrak{r}}(l_2) \quad \text{and} \quad \tau_{\mathscr{U}}(D_{\mathfrak{r}}) = 1.$$

Therefore $\tau_{\mathscr{U}}$ does not vanish identically.

A.X.11. Example. Let

$$\varrho_k := \varrho 2^{-k}(k+1)^{-\mu},$$

where

$$\varrho := \left[\sum_{k=0}^{\infty} 2^{-k}(k+1)^{-\mu}\right]^{-1} \quad \text{and} \quad \mu > 1.$$

Then $\mathfrak{r} = (\varrho_k)$ is a singular strongly convergent weight sequence.

Proof. Obviously,

$$\sum_{k=0}^{\infty} 2^k \varrho_k = \varrho \sum_{k=0}^{\infty} (k+1)^{-\mu} < \infty.$$

We further observe that

$$2^n \sum_{k=n}^{\infty} \varrho_k = 2^n \varrho \sum_{k=n}^{\infty} 2^{-k}(k+1)^{-\mu} \leqq 2\varrho(n+1)^{-\mu}$$

and

$$\sum_{k=n}^{\infty} 2^k \varrho_k = \varrho \sum_{k=n}^{\infty} (k+1)^{-\mu} \geqq c(n+1)^{1-\mu}.$$

Hence $\mathfrak{r} = (\varrho_k)$ is singular.

A.X.12. Example. Let $0 < r < 1$. Then there exists a singular strongly convergent weight sequence $\mathfrak{r} = (\varrho_k)$ such that $\varrho_k \leqq c 2^{-k/r}$.

Proof. Fix $\mu > 1$, $\varrho > 0$ and

$$0 = m_0 < n_0 < m_1 < n_1 < \dots$$

Let

$$a_i := n_i - m_i \quad \text{and} \quad s_i := \sum_{j=0}^{i} a_j,$$

$$\alpha_i := \varrho 2^{(s_i - n_i)(\mu-1)} \quad \text{and} \quad \beta_i := \varrho 2^{s_i(\mu-1)},$$

$$A_i := \{k : m_i \leqq k < n_i\} \quad \text{and} \quad B_i := \{k : n_i \leqq k < m_{i+1}\}.$$

We now define

$$\varrho_k := \begin{cases} \alpha_i 2^{-k} & \text{if} \quad k \in A_i, \\ \beta_i 2^{-k\mu} & \text{if} \quad k \in B_i. \end{cases}$$

Note that

$$\varrho_{n_i} = \alpha_i 2^{-n_i} = \beta_i 2^{-n_i \mu} \quad \text{and} \quad \varrho_{m_{i+1}} = \alpha_{i+1} 2^{-m_{i+1}} = \beta_i 2^{-m_{i+1}\mu}.$$

This means that the different pieces used in the definition of $\mathfrak{r} = (\varrho_k)$ agree with each other.

Since

$$2\varrho_{k+1} = \varrho_k \quad \text{for} \quad k \in A_i \quad \text{and} \quad 2^\mu \varrho_{k+1} = \varrho_k \quad \text{for} \quad k \in B_i,$$

we have

$$\varrho_k \leqq 2^\mu \varrho_{k+1} \quad \text{and} \quad 2\varrho_{k+1} \leqq \varrho_k \quad \text{for} \quad k = 0, 1, \dots$$

Hence $\mathfrak{r} = (\varrho_k)$ is indeed a weight sequence provided that

$$\sum_{k=0}^{\infty} \varrho_k = 1.$$

This condition can be satisfied by a proper choice of the parameter ϱ.

We now specialize the preceding construction by setting

$$\mu := 2/r - 1, \quad m_i := 2(i+1)\,i \quad \text{and} \quad n_i := 2(i+1)^2 - 1.$$

Then

$$a_i = 2i + 1 \quad \text{and} \quad s_i = (i+1)^2.$$

Note that

$$\mu - 1 = 2(1/r - 1) = 2(\mu - 1/r) > 0 \quad \text{and} \quad 2s_i - n_i = 1.$$

If $m_i \leqq k < n_i$, then

$$\begin{aligned}\log_2 \varrho_k/\varrho &= (s_i - n_i)(\mu - 1) - k = (s_i - n_i)(\mu - 1) + k(1/r - 1) - k/r\\ &\leqq (s_i - n_i)(\mu - 1) + n_i(1/r - 1) - k/r\\ &= (2s_i - n_i)(1/r - 1) - k/r = (1/r - 1) - k/r.\end{aligned}$$

The same estimate holds for $n_i \leqq k < m_{i+1}$:

$$\begin{aligned}\log_2 \varrho_k/\varrho &= s_i(\mu - 1) - k\mu = s_i(\mu - 1) + k(1/r - \mu) - k/r\\ &\leqq s_i(\mu - 1) + n_i(1/r - \mu) - k/r\\ &= (2s_i - n_i)(\mu - 1/r) - k/r = (\mu - 1/r) - k/r.\end{aligned}$$

Hence

$$\varrho_k \leqq c\, 2^{-k/r} \quad \text{for} \quad k = 0, 1, \ldots$$

Moreover,

$$\sum_{k=m_i}^{\infty} 2^k \varrho_k > \sum_{A_i} 2^k \varrho_k = \alpha_i a_i = (2i+1)\, 2^{m_i} \varrho_{m_i}$$

and

$$2^{m_i} \sum_{k=m_i}^{\infty} \varrho_k \leqq 2^{m_i} \sum_{k=m_i}^{\infty} 2^{m_i - k} \varrho_{m_i} = 2^{m_i+1} \varrho_{m_i}.$$

Thus $\mathfrak{r} = (\varrho_k)$ is indeed a singular strongly convergent weight sequence with the required property.

A.X.13. Summarizing the preceding considerations, we obtain the following result.

Theorem. Let $0 < r < 1$. Then there exists a quasi-Banach operator ideal $\mathfrak{M}_{\mathfrak{r}}$ with different continuous traces which is contained in $\mathfrak{L}^{(a)}_{r,\infty}$.

A.X.14. The counterpart of the concept of a zero trace is that of a one determinant.

Suppose that the quasi-Banach operator ideal $\mathfrak{A}$ supports different (continuous) determinants δ_1 and δ_2. Then $\delta := \delta_1/\delta_2$ still possesses the properties (D_2), (D_3) and (D_4), stated in 4.3.1, while (D_1) changes into

(D_1^0) $\quad \delta(I_E + a \otimes x) = 1$ for $a \in E'$ and $x \in E$.

A function δ satisfying these modified conditions is called a (continuous) **one determinant**. Of course, we are only interested in the non-trivial case when δ takes values different from one.

Remark. Let δ^* be the function which assigns to every operator $I_E + T$ with $T \in \mathfrak{A}(E)$ the complex-conjugate of $\delta(I_E + T)$. Then, with the above notation, the function $\delta^*(I_E + \zeta T)\,\delta_2(I_E + \zeta T)$ may fail to be holomorphic. This proves that the property (D_4) is independent of (D_1), (D_2) and (D_3).

Open problems

In what follows we present a collection of open problems, most of which have already been mentioned in the text. Hopefully, the reader will be stimulated to do his own research work in this interesting branch of functional analysis.

The problems marked with (*) are of special importance.

O.P.1 (see 2.3.7). Find a non-sophisticated expression for a p-norm on $\mathfrak{L}_r^{(a)}$, where $1/p = 1 + 1/r$ and $0 < r < \infty$.

O.P.2 (see 4.2.25 and 4.2.26). Prove that the quasi-Banach operator ideals $\mathfrak{L}_1^{(c)}$ and $\mathfrak{L}_1^{(x)}$ fail to be approximative.

O.P.3 (see 4.5.15). Show that $(\mathfrak{P}_3)^3$ is approximative.

O.P.4 (see 2.10.7). Let $0 < p < 2$ and $1/q = 1/p - 1/2$. Is it true that

$$\mathfrak{L}_{p,w}^{(x)} \subset \mathfrak{L}_{q,w}^{(a)}?$$

O.P.5* (see 2.10.7). Let $0 < p < 1$ and $1/q = 1/p - 1$. Do we have

$$\mathfrak{L}_{p,w}^{(h)} \subset \mathfrak{L}_{q,w}^{(a)}?$$

O.P.6 (see 2.8.18). Let $0 < s < \infty$ and $1/r = 1/s + 1/2$. Prove that the inclusion $(\mathfrak{P}_2)_{s,w}^{(a)} \subseteq \mathfrak{L}_{r,w}^{(x)}$ is strict.

O.P.7* (see 3.6.1 and 3.6.2). Does there exist a constant $c > 1$ such that

$$|\lambda_n(T)| \leqq c \left[\prod_{k=1}^{n} x_k(T)\right]^{1/n} \quad \text{for} \quad n = 1, 2, \ldots$$

and every operator $T \in \mathfrak{L}(E)$, where E is any finite dimensional Banach space?

O.P.8 (see 3.6.7). Determine the optimum eigenvalue type of $\mathfrak{S}_{r,w}^{\text{weak}}$ for $0 < r < 2$.

O.P.9 (see 3.9.4). Find the optimum eigenvalue type of $\mathfrak{L}_{r,w}^{(x)} + \mathfrak{L}_{r,w}^{(y)}$ for $2 \leqq r < \infty$.

O.P.10 (see 3.9.7). Suppose that $\mathfrak{A}_1, \ldots, \mathfrak{A}_n \in \mathbb{E}_{r,w}$. What can be said about the eigenvalue type of $\mathfrak{A}_1 + \ldots + \mathfrak{A}_n$?

O.P.11 (see 3.9.1). Find examples of operator ideals which are maximal elements in the partially ordered class $\mathbb{E}_{r,w}$.

O.P.12 (see 3.8.4). Characterize those Banach spaces E for which $T \in \mathfrak{N}(E)$ implies $(\lambda_n(T)) \in l_{r,w}$, where $0 < r < 2$.

O.P.13* (see 4.2.13). Construct a quasi-Banach operator ideal admitting a continuous trace and having the optimum eigenvalue type l_2.

O.P.14 (see 4.2.25 and 4.5.15). Does there exist a continuous (spectral) trace on the quasi-Banach operator ideals $\mathfrak{L}_1^{(x)}$ and $(\mathfrak{P}_3)^3$?

O.P.15* (see 4.2.14). Must every operator ideal $\mathfrak{A} \in \mathbb{E}_1$ admit a spectral trace?

O.P.16 (see 4.2.5). Suppose that $\mathfrak{A}$ and $\mathfrak{B}$ are approximative quasi-Banach operator ideals supporting continuous traces α and β, respectively. Is it true that $\alpha(T) = \beta(T)$ for $T \in \mathfrak{A}(E) \cap \mathfrak{B}(E)$?

O.P.17* (see 4.2.37). Let $\mathfrak{A}$ be a quasi-Banach operator ideal with a continuous trace τ. Show that the restriction of τ on $\mathfrak{A}^2$ is always spectral.

O.P.18 (see 4.2.38 and A.X.13). Modify the definition of a continuous trace by adding such conditions that the new object is uniquely determined by its axiomatic properties. In the special case when the underlying quasi-Banach operator ideal is of eigenvalue type l_1 one should obtain the spectral trace.

O.P.19 (see 4.2.14). Let $\mathfrak{A}$ be a quasi-Banach operator ideal with a continuous trace τ. Suppose that the operators $S \in \mathfrak{A}(E)$ and $T \in \mathfrak{A}(F)$ have the same non-zero eigenvalues. Does it follow that $\tau(S) = \tau(T)$?

O.P.20 (see A.X.5). Let $T \in \mathfrak{L}(E)$. Prove that

$$|\text{trace}\,(T)| \leqq cn^{1/2}\, \|T\,|\mathfrak{P}_2 + \mathfrak{P}_2'\| \quad \text{whenever} \quad \text{rank}\,(T) \leqq n,$$

where $c \geqq 1$ is a constant.

O.P.21 (see 5.3.6 and 7.4.5). Give an elementary proof of the inequality

$$\Big(\sum_{k=1}^{n} |\lambda_k(M)|^p\Big)^{1/p} \leqq \Big(\sum_{i=1}^{n} \Big[\sum_{j=1}^{n} |\mu_{ij}|^{p'}\Big]^{p/p'}\Big)^{1/p}$$

which holds for every (n, n)-matrix $M = (\mu_{ij})$ and $2 < p < \infty$. The limiting case $p = 2$ was already treated by I. Schur (1909: a).

O.P.22 (see 5.5.10). Let $M \in [l_p, l_{p'}]$ and $2 < p < m < \infty$. Find explicite formulas for the Taylor coefficients of the regularized Fredholm denominator $\delta_m(\zeta, M)$.

O.P.23 (see 5.5.10). Let $M \in [l_p, l_{p'}]$ and $2 < p < m < \infty$. Prove that $(\lambda_n(M)) \in l_{p,\infty}$ by estimating the growth of $\delta_m(\zeta, M)$ as $\zeta \to \infty$.

O.P.24 (see 6.4.11). Show that the Ciesielski isomorphism extends to the case of E-valued functions and sequences.

O.P.25 (see 6.4.19). Parallel with the theory of Besov function spaces $B^{\sigma}_{p,u}(0, 1)$ there exists a theory of so-called Triebel-Lizorkin spaces $F^{\sigma}_{p,u}(0, 1)$ which, among others, include the Bessel potential spaces $H^{\sigma}_p(0, 1)$; see (TRI, 2.3.1, 2.3.3 and 4.2.1). Let $\sigma + \tau > (1/p + 1/q - 1)_+$ and $1/r = \sigma + \tau + 1/q^+$. Find the best possible value of the fine index w in the following assertion:

$$K \in [F^{\sigma}_{p,u}(0, 1), F^{\tau}_{q,v}(0,1)] \quad \text{implies} \quad (\lambda_n(K)) \in l_{r,w}.$$

Epilogue

Finally, we list some subjects, related to eigenvalues and s-numbers, which are not treated in this monograph.

First of all, we have to mention the theory of *entropy numbers* and their applications to eigenvalue distributions. The most striking result in this direction is *Carl's inequality*

$$|\lambda_n(T)| \leqq \sqrt{2}\, e_n(T) \quad \text{for all} \quad T \in \mathfrak{L}(E).$$

Further important topics, originating from classical theorems on normal forms of matrices, are the *invariant subspace problem* and the *problem to determine the closed linear span of all principal elements* of a given operator.

Concerning applications we have restricted most of the considerations to integral operators defined on function spaces over the unit interval. The extension of the results to bounded domains in the n-dimensional euclidean space does not cause any new phenomena provided that the boundary is sufficiently smooth. However, in the case of unbounded domains the situation may become rather involved. Here one deals with special types of *kernels characterized by means of weight functions.* Another interesting subject is the theory of *integral operators induced by holomorphic kernels.*

We now give a small selection of references.

Entropy numbers:
(PIE, Chap. 12), B. Carl (1981: a, b, 1982: a, b, 1984), B. Carl / Th. Kühn (1984), B. Carl / A. Pietsch (1977), B. Carl / H. Triebel (1980), Th. Kühn (1984), B. S. Mityagin / A. Pełczyński (1966).

Invariant subspaces and triangular form:
(DOW, pp. 54–66), (DUN, XI.10.1), (RIN, pp. 165–179), (TAY, V.7.16), V. I. Lomonosov (1973), H. Radjavi / P. Rosenthal (1982), J. R. Ringrose (1962).

Completeness of principal elements:
(DUN, XI.6.29, XI.10.15), (GOH, V.6.1, V.6.2), H. E. Benzinger (1973), H. König (1980: b), A. S. Markus (1966), F. Reuter (1981).

Weighted kernels:
M. Sh. Birman / M. Z. Solomyak (1967/69, 1977: a, b), A. Pietsch (1983), H. Triebel (1970: b).

Holomorphic kernels:
A. O. Gel'fond (1931), E. Hille / J. D. Tamarkin (1931), H. König / S. Richter (1984).

Bibliography

The bibliography is divided into two parts:

A) Textbooks and monographs,
B) Research papers.

Books are quoted by a symbol consisting of three typical letters from the name(s) of their author(s). The other items are cited by the name(s) of their author(s) and the year of publication.

The numbers at the end of each entry indicate the paragraphs in the text where this book or paper is quoted. For example, 7.0.0 and 7.1.0 refer to the introductions of Chapter 7 and Section 7.1, respectively.

A) Textbooks and monographs

Adams, R. A.
(ADA) *Sobolev spaces*. New York: Academic Press 1975.
6.4.0.

Baltzer, R.
(BAL) *Theorie und Anwendungen der Determinanten*. Leipzig 1857.
7.0.0.

Banach, S.
(BAN) *Théorie des opérations linéaires*. Monografie Matematyczne, Warszawa 1932.
1.7.6, 7.2.3, 7.3.0, 7.5.3.

Bari, N. K.
(BAR) *A treatise on trigonometric series*. Moscow: Fizmatgiz 1961, and New York: Pergamon Press 1964.
6.6.13.

Barnes, B. A.; Murphy, G. J.; Smyth, M. R. F.; West, T. T.
(BAS) *Riesz and Fredholm theory in Banach algebras*. Research Notes in Math. **67**, Boston, London, Melbourne: Pitman 1982.
3.2.0, 3.2.14, 7.4.1.

Bergh, J.; Löfström, J.
(BER) *Interpolation spaces*. Berlin, Heidelberg, New York: Springer-Verlag 1976.
F.0.0, F.3.2, F.3.3, F.3.4, 1.1.6, 2.1.14, 5.4.3, 6.3.1.

Boas, R. P., Jr.
(BOA) *Entire functions*. New York: Academic Press 1954.
4.8.0, 4.8.9.

Bôcher, M.
(BOC) *An introduction to the study of integral equations*. Cambridge Univ. Press 1909.
Preface.

Bourbaki, N.
(BOU) *Eléments d'histoire des mathématiques*. Paris: Hermann 1960.
7.1.0, 7.2.0.

Butzer, P. L.; Berens, H.
(BUB) *Semi-groups of operators and approximation*. Berlin, Heidelberg, New York: Springer-Verlag 1967.
F.0.0, 6.3.1, 7.2.0.

Butzer, P. L.; Scherer, K.
(BUS) *Approximationsprozesse und Interpolationsmethoden*. Mannheim, Zürich: Bibliographisches Inst. 1968.
6.5.14.

Caradus, S. R.; Pfaffenberger, W. E.; Yood, B.
(CAR) *Calkin algebras and algebras of operators on Banach spaces.* Lecture Notes in Pure and Appl. Math. **9**, New York: Dekker 1974.
3.2.0, 4.1.6.

Cochran, J. A.
(COC) *The analysis of linear integral equations.* New York: Mc Graw-Hill 1972.
7.7.0.

Cramer, G.
(CRA) *Introduction à l'analyse des lignes courbes algébriques.* Genève 1750.
7.1.1.

Crowe, M. J.
(CRO) *A history of vector analysis.* London: Univ. Notre Dame Press 1967.
7.1.0.

D'Alembert, J. B.
(DAL) *Traité de dynamique.* Paris 1743.
7.1.4.

Diestel, J.
(DIE) *Sequences and series in Banach spaces.* New York, Berlin, Heidelberg, Tokyo: Springer-Verlag 1984.
1.6.4.

Diestel, J.; Uhl, J. J. (Junior)
(DIL) *Vector measures.* Providence: Amer. Math. Soc. 1977.
E.0.0, 1.5.4, 4.7.1, 6.2.3, 6.2.5.

Dieudonné, J.
(DIU) *History of functional analysis.* Math. Studies 49. Amsterdam, New York, Oxford: North-Holland 1981.
7.3.0, 7.3.2, 7.4.1.

Dowson, H. R.
(DOW) *Spectral theory of linear operators.* London, New York, San Francisco: Academic Press 1978.
3.2.0, 4.1.6, Epilogue.

Dunford, N.; Schwartz, J. T.
(DUN) *Linear operators, Vol. I and II.* New York, London: Interscience Publ. 1958/63.
B.0.0, B.2.5, B.2.11, D.0.0, D.2.3, 1.1.7, 1.3.5, 1.3.6, 2.4.2, 2.4.12, 2.9.6, 3.2.24, 4.2.18, 6.2.3, 6.2.5, 6.2.6, 7.3.0, 7.3.2, 7.5.3, 7.6.2, Epilogue.

Euler, L.
(EUL) *Introductio in analysin infinitorum.* Lausanne 1748.
7.1.4.

Gauss, C. F.
(GAU) *Disquisitiones arithmeticae.* Leipzig 1801.
7.1.1, 7.1.2.

Gohberg, I. C.; Krejn, M. G.
(GOH) *Introduction to the theory of linear non-self-adjoint operators in Hilbert space.* Moscow: Nauka 1965, and Providence: Amer. Math. Soc. 1969.
D.0.0, 7.3.4, 7.5.3, 7.7.5, Epilogue.

Grassmann, H.
(GRA) *Die (lineale) Ausdehnungslehre.* Leipzig: Wigand 1844, and Berlin: Enslin 1862.
7.1.3.

Grothendieck, A.
(GRO) *Produits tensoriels topologiques et espaces nucléaires.* Mem. Amer. Math. Soc. **16**, Providence 1955.
1.7.3, 3.8.1, 4.2.9, 4.7.1, 7.3.7, 7.3.8, 7.4.2, 7.4.4, 7.4.6, 7.5.3, 7.6.1, 7.6.2.

Halmos, P. R.
(HAL) *Finite dimensional vector spaces.* Princeton Univ. Press 1942, and Princeton, New York, Toronto, London: van Nostrand 1958.
A.0.0, 7.1.0.

(HAM) *Measure theory.* Princeton, New York, Toronto, London: van Nostrand 1950.
C.0.0, 1.3.6.

Halmos, P. R.; Sunder, V. S.
(HAS) *Bounded integral operators on L^2 spaces.* Berlin, Heidelberg, New York: Springer-Verlag 1978.
7.3.2.

Hamilton, W. R.
(HAT) *Lectures on quaternions.* Dublin 1853.
7.1.3.

Hardy, G. H.; Littlewood, J. E.; Pólya, G.
(HAY) *Inequalities.* Cambridge Univ. Press 1934.
C.3.10, F.3.5, G.0.0, G.1.1, 2.1.7, 6.5.3.

Hellinger, E.; Toeplitz, O.
(HEL) *Integralgleichungen und Gleichungen mit unendlich vielen Unbekannten.* Encyklopädie Math. Wiss. II.C.13. Leipzig: Teubner 1927, and New York: Chelsea 1953.
7.1.0, 7.5.0, 7.7.0, Preface.

Hilbert, D.
(HIL) *Grundzüge einer allgemeinen Theorie der linearen Integralgleichungen.* Leipzig: Teubner 1912, and New York: Chelsea 1952.
7.2.1, 7.3.1, 7.3.2, 7.4.0, 7.4.1, 7.5.2, Preface.

Hille, E.; Phillips, R. S.
(HIP) *Functional analysis and semi-groups.* Providence: Amer. Math. Soc. 1957.
6.2.3.

Jarchow, H.
(JAR) *Locally convex spaces.* Stuttgart: Teubner 1981.
D.0.0, E.0.0.

Jordan, C.
(JOR) *Traité des substitutions.* Paris: Gauthier-Villars 1870.
7.1.6.

Jörgens, K.
(JÖR) *Lineare Integralgleichungen.* Stuttgart: Teubner 1970.
6.2.9, 6.2.11.

Kahane, J.-P.
(KAH) *Séries de Fourier absolument convergentes.* Berlin, Heidelberg, New York: Springer-Verlag 1970.
6.6.13.

Kashin, B. S.; Saakyan, A. A.
(KAS) *Orthogonal series* (Russian). Moscow: Nauka 1984.
6.5.11.

Klein, F.
(KLE) *Vorlesungen über die Entwicklung der Mathematik im 19. Jahrhundert.* Berlin: Springer-Verlag 1926/27.
7.3.5.

Kline, M.
(KLI) *Mathematical thought from ancient to modern times.* New York: Oxford Univ. Press. 1972.
7.1.0, 7.1.2.

König, H.
(KÖN) *Eigenvalue distribution of compact operators*. Basel: Birkhäuser 1986.
1.5.4, 6.3.8, 6.4.0, 6.4.7, 6.4.14, 6.4.19, 6.4.25, 6.4.26.

Köthe, G.
(KÖT) *Topological linear spaces. Vol. I and II*. New York, Heidelberg, Berlin: Springer-Verlag 1969/79.
E.0.0, 6.1.2.

Kowalewski, G.
(KOW) *Einführung in die Determinantentheorie*. Leipzig: Veit 1909.
Preface.

Lagrange, J. L.
(LAG) *Méchanique analytique*. Paris 1788.
7.1.4.

Lalesco, T.
(LAL) *Introduction à la théorie des équations intégrales*. Paris: Hermann 1912.
7.5.4, 7.7.5.

Landau, E.
(LAN) *Darstellung und Begründung einiger neuer Ergebnisse der Funktionentheorie*. Berlin: Springer-Verlag 1916.
4.8.8.

Laplace, J. P.
(LAP) *Traité de méchanique céleste*. Paris 1799.
7.1.4.

Levin, B. Ya.
(LEV) *Distributions of zeros of entire functions*. Moscow: Gostekhizdat 1956, Berlin: Akademie-Verlag 1962, and Providence: Amer. Math. Soc. 1964.
4.8.0, 4.8.9.

Lindenstrauss, J.; Tzafriri, L.
(LIN) *Classical Banach spaces, Vol. I and II*. Berlin, Heidelberg, New York: Springer-Verlag 1977/79.
1.6.4, 4.7.5.

Lützen, J.
(LÜT) *The prehistory of the theory of distributions*. New York, Heidelberg, Berlin: Springer-Verlag 1982.
7.2.0.

MacDuffee, C. C.
(MAC) *The theory of matrices*. Berlin: Springer-Verlag 1933.
7.1.0.

Monna, A. F.
(MON) *Functional analysis in historical perspective*. Utrecht: Oesthoek 1973.
7.2.0.

Muir, Th.
(MUI) *The theory of determinants in the historical order of development, Vol. I–IV*. London: Macmillan 1906/23.
7.1.0.

Neumann, J. von
(NEU) *Mathematische Grundlagen der Quantenmechanik*. Berlin: Springer-Verlag 1932.
7.6.0, 7.6.1.

Nikol'skij, S. M.
(NIK) *Approximation of functions of several variables and imbedding theorems.* Moscow: Nauka 1969, and New York, Heidelberg: Springer-Verlag 1975.
6.4.0, 6.5.14, 6.5.21, 6.5.23, 7.2.0, 7.7.1.

Peano, G.
(PEA) *Calcolo geometrico secondo l'Ausdehnungslehre di H. Grassmann.* Torino: Fratelli Bocca 1888.
7.1.7.

Pietsch, A.
(PIE) *Operator ideals.* Berlin: VEB Deutscher Verlag der Wissenschaften 1978, and Amsterdam, New York, Oxford: North-Holland 1980.
D.0.0, D.1.3, D.1.4, D.1.8, D.1.9, D.1.10, D.2.4, D.3.3, 1.6.4, 1.7.3, 1.7.6, 1.7.9, 2.3.16, 2.5.6, 2.9.8, 2.9.11, 2.11.36, 4.2.9, 4.7.5, 7.3.5, 7.5.3, Epilogue.

Pinkus, A.
(PIN) *n-Widths in approximation theory.* Berlin, Heidelberg, New York, Tokyo: Springer-Verlag 1985.
7.7.2.

Reed, M.; Simon, B.
(REE) *Methods of modern mathematical physics, Vol. I–IV.* New York, San Francisco, London: Academic Press 1972/79.
D.0.0, 7.6.2.

Riesz, F.
(RIE) *Les systèmes d'équations linéaires à une infinité d'inconnues.* Paris: Gauthier-Villars 1913.
4.1.3, 7.2.1, 7.3.1.

Ringrose, J. R.
(RIN) *Compact non-self-adjoint operators.* London: van Nostrand 1971.
D.0.0, 2.11.14, 7.6.2, Epilogue.

Schatten, R.
(SCA) *A theory of cross-spaces.* Annals of Math. Studies 26, Princeton Univ. Press 1950.
7.3.5.
(SCE) *Norm ideals of completely continuous operators.* Berlin, Heidelberg, New York: Springer-Verlag 1960.
D.0.0.

Schmeisser, H. J.; Triebel, H.
(SCI) *Topics in Fourier analysis and function spaces.* Leipzig: Akad. Verlagsgesellschaft Geest & Portig 1987, and Chichester: Wiley 1987.
6.5.14.

Schumaker, L. L.
(SCU) *Spline functions, basic theory.* New York: Wiley 1981.
6.4.9, 6.4.22, 7.7.0.

Simon, B.
(SIM) *Trace ideals and their applications.* London Math. Soc. Lecture Notes 35, Cambridge Univ. Press 1979.
D.0.0, 2.11.37, 7.3.4, 7.3.5.

Smithies, F.
(SMI) *Integral equations.* Cambridge Univ. Press. 1958.
5.5.11, 6.0.0, 7.5.2, Preface.

Sobolev, S. L.
(SOB) *Applications of functional analysis in mathematical physics.* Izdat. Leningrad. Univ. 1950, Providence: Amer. Math. Soc. 1963, and Berlin: Akademie-Verlag 1964.
7.2.2, 7.7.1.

Stein, E. M.
(STE) *Singular integrals and differentiability properties of functions.* Princeton Univ. Press 1970.
6.4.0.

Stein, E. M.; Weiss, G.
(STI) *Fourier analysis on euclidean spaces.* Princeton Univ. Press 1971.
6.3.1.

Taylor, A. E.; Lay, D. C.
(TAY) *Introduction to functional analysis.* New York, Chichester, Brisbane, Toronto: Wiley 1980.
B.0.0, B.2.5, B.2.11, D.2.3, 1.1.7, 1.3.5, 1.3.6, 3.2.24, Epilogue.

Tikhomirov, V. M.
(TIK) *Some topics in approximation theory* (Russian). Moscow: Izdat. Moskov. Univ. 1976.
7.7.0.

Timan, A. F.
(TIM) *Theory of approximation of functions of a real variable.* Moscow: Fizmatgiz. 1960, and Oxford: Pergamon Press 1963.
6.5.14.

Titchmarsh, E. C.
(TIT) *The theory of functions.* Oxford: Clarendon Press 1932.
G.3.3, 4.8.0.

Triebel, H.
(TRI) *Interpolation theory, function spaces, differential operators.* Berlin: VEB Deutscher Verlag der Wissenschaften 1978, and Amsterdam, New York, Oxford: North-Holland 1978.
F.0.0, F.3.2, F.3.3, F.3.4, 1.1.6, 2.1.14, 5.4.3, 6.3.1, 6.4.0, 6.4.25, O.P.25.

Valiron, G.
(VAL) *General theory of integral functions.* Toulouse: Librairie Univ. 1923.
4.8.0, 4.8.9.

Weyl, H.
(WEY) *Raum, Zeit, Materie.* Berlin: Springer-Verlag 1918.
7.1.7.

Wiener, N.
(WIE) *I am a mathematician.* New York: Doubleday 1956.
7.2.3.

Zaanen, A. C.
(ZAN) *Linear Analysis.* Groningen: North-Holland 1953.
5.5.11.

Zygmund, A.
(ZYG) *Trigonometric series, Vol. I and II.* Cambridge Univ. Press 1959.
6.5.2, 6.5.8, 7.7.0, 7.7.7.

B) Research papers

Agmon, S.
(1965) *On kernels, eigenvalues, and eigenfunctions of operators related to elliptic problems.* Comm. Pure Appl. Math. **18**, 627–663.
7.7.6.

Allakhverdiev, D. Eh.
(1957) *On the rate of approximation of completely continuous operators by finite dimensional operators* (Russian). Azerbajzhan. Gos. Univ. Uchen. Zap. (Baku) **2**, 27–37.
2.11.6, 7.3.6.

Aoki, T.
(1942) *Locally bounded linear topological spaces*. Proc. Imp. Acad. Tokyo **18**, 588–594.
7.2.3.

Aronszajn, N.
(1955) *Boundary values of functions with finite Dirichlet integral*. Conf. Part. Diff. Equations, Univ. of Kansas, Techn. Report **14**, 77–94.
7.2.2.

Banach, S.
(1922) *Sur les opérations dans les ensembles abstraits et leur application aux équations intégrales*. Fund. Math. **3**, 133–181.
7.1.7, 7.2.2, 7.2.3, 7.3.3.
(1929) *Sur les fonctionnelles linéaires*. Studia Math. **1**, 211–216 and 223–229.
7.3.3.
(1930) *Über einige Eigenschaften der lakunären trigonometrischen Reihen*. Studia Math. **2**, 207–220.
6.5.11.

Bauhardt, W.
(1977) *Hilbert Zahlen von Operatoren in Banachräumen*. Math. Nachr. **79**, 181–187.
2.6.2, 2.6.3, 2.6.9, 7.3.6.
(1979) *Nuclear multipliers on compact groups*. Math. Nachr. **93**, 293–303.
6.6.12.

Bennett, G.
(1973) *Inclusion mappings between l^p-spaces*. J. Funct. Anal. **13**, 20–27.
1.6.6, 1.6.7, 7.7.2.

Benzinger, H. E.
(1973) *Completeness of eigenvectors in Banach spaces*. Proc. Amer. Math. Soc. **38**, 319–324.
Epilogue.

Bernkopf, M.
(1966) *The development of function spaces with particular reference to their origins in integral equation theory*. Arch. Hist. Exact Sci. **3**, 1–96.
7.2.0.
(1968) *A history of infinite matrices*. Arch. Hist. Exact Sci. **4**, 308–358.
7.5.0.

Bernstein, S.
(1914) *Sur la convergence absolue des séries trigonométriques*. C. R. Acad. Sci. Paris **158**, 1661–1664.
7.7.7.
(1934) *Sur la convergence absolue des séries trigonométriques*. C. R. Acad. Sci. Paris **199**, 397–400.
7.7.7.

Besov, O. V.
(1961) *Investigation of a family of function spaces in connection with theorems of imbedding and extension* (Russian). Trudy Mat. Inst. Steklov **60**, 42–81.
6.5.14, 7.2.2.

Bessel, F. W.
(1828) *Über die Bestimmung des Gesetzes einer periodischen Erscheinung*. Astron. Nachr. **6**, 333–348.
6.5.7.

Binet, J. P. M.
(1813) *Mémoire sur un système de formules analytiques, et leur application à des considérations géométriques*. J. École Polytechn. **9**: 16, 280–302.
7.1.1.

Birman, M. Sh.; Solomyak, M. Z.
(1967) *Piecewise-polynomial approximations of functions of the classes* W_p^α (Russian). Mat. Sb. **73**, 331–355.
7.7.3.
(1967/69) *Estimates of singular numbers of integral operators* (Russian). Vestnik Leningrad. Univ., Mat. Mekh. Astronom. **22**: 7, 43–53, **22**: 13, 21–28 and **24**: 1, 35–48.
7.7.5, Epilogue.
(1969) *Remarks on the nuclearity of integral operators and the boundedness of pseudodifferential operators* (Russian). Isv. Vyssh. Uchebn. Zaved. Mat. **1969**: 9, 11–17.
7.7.5.
(1970) *Asymptotics of the spectrum of weakly polar integral operators* (Russian). Izv. Akad. Nauk SSSR, Ser. Mat. **34**, 1142–1158.
7.7.5.
(1974) *Quantitative analysis in Sobolev imbedding theorems and applications to spectral theory.* 10th Math. School, Kiev 1974, and Amer. Math. Soc. Translations (2) **114**, Providence 1980.
7.7.0, 7.7.3, 7.7.5.
(1977: a) *Estimates for the singular numbers of integral operators* (Russian). Uspehi Mat. Nauk **32**: 1, 17–84.
6.4.0, 6.4.25, 6.4.26, 7.7.0, 7.7.5, Epilogue.
(1977: b) *Asymptotics of the spectrum of differential equations* (Russian). Itogi Nauki, Seriya "Matematika" **14**, 5–58.
7.7.0, Epilogue.

Bljumin, S. L.; Kotljar, B. D.
(1970) *Hilbert-Schmidt operators and the absolute convergence of Fourier series* (Russian). Izv. Akad. Nauk SSSR, Ser. Mat. **34**, 209–217.
7.7.5.

Bóbr, St.
(1921) *Eine Verallgemeinerung des v. Kochschen Satzes über die absolute Konvergenz der unendlichen Determinanten.* Math. Z. **10**, 1–11.
7.3.1.

Bois-Reymond, P. du
(1888) *Bemerkungen über* $\Delta z = o$. J. Reine Angew. Math. **103**, 204–229.
7.7.0.

Borchardt, C. W.
(1846) *Neue Eigenschaften der Gleichung mit deren Hülfe man die seculären Störungen von Planeten bestimmt.* J. Reine Angew. Math. **30**, 38–45.
7.1.5.

Borel, E.
(1897) *Sur les zéros des fonctions entières.* Acta Math. **20**, 357–396.
4.8.8, 7.5.4.

Born, M.; Heisenberg, W.; Jordan, P.
(1926) *Zur Quantenmechanik.* Z. Phys. **35**, 557–615.
7.7.4.

Bourgin, D. G.
(1943) *Linear topological spaces.* Amer. J. Math. **65**, 637–659.
7.2.3.

Brascamp, H. J.
(1969) *The Fredholm theory of integral equations for special types of compact operators on a separable Hilbert space.* Compositio Math. **21**, 59–80.
7.5.3.

Calderon, A. P.
(1964) *Intermediate spaces and interpolation, the complex method.* Studia Math. **24,** 113–190.
7.2.4.

Calkin, J. W.
(1941) *Two-sided ideals and congruences in the ring of bounded operators in Hilbert space.* Ann. of Math. **42,** 839–873.
2.11.11, 7.3.4.

Carl, B.
(1974) *Absolut $(p, 1)$-summierende identische Operatoren von l_u nach l_v.* Math. Nachr. **63,** 353–360.
1.6.6, 1.6.7, 7.7.2.
(1976) *Distribution of the eigenvalues of nuclear operators in Banach spaces* (Russian). Vestnik Moskov. Univ., Ser. Mat. Mekh. **31**, 3–10.
3.8.6.
(1981: a) *Entropy numbers, s-numbers, and eigenvalue problems.* J. Funct. Anal. **41,** 290–306.
Epilogue.
(1981: b) *Entropy numbers of diagonal operators with an application to eigenvalue problems.* J. Approx. Theory **32**, 135–150.
Epilogue.

(1982: a) *Inequalities between geometric quantities of operators in Banach spaces.* Integral Equations Operator Theory **5**, 759–773.
Epilogue.
(1982: b) *On a characterization of operators from l_q into a Banach space of type p with some applications to eigenvalue problems.* J. Funct. Anal. **48**, 394–407.
5.3.5, Epilogue.
(1984) *On the entropy of absolutely summing operators.* Arch. Math. (Basel) **43,** 183–186.
Epilogue.

Carl, B.; Kühn, Th.
(1984) *Entropy and eigenvalues of certain integral operators.* Math. Ann. **268,** 127–136.
Epilogue.

Carl, B.; Pietsch, A.
(1977) *Entropy numbers of operators in Banach spaces.* Lecture Notes in Math. **609**, 21–33.
Epilogue.
(1978) *Some contributions to the theory of s-numbers.* Comment. Math., Prace Mat. **21**, 65–76.
2.9.19.

Carl, B.; Triebel, H.
(1980) *Inequalities between eigenvalues, entropy numbers, and related quantities of compact operators in Banach spaces.* Math. Ann. **251**, 129–133.
Epilogue.

Carleman, T.
(1917) *Sur le genre du dénominateur $D(\lambda)$ de Fredholm.* Arkiv för Mat., Astr. och Fysik, **12**: 15, 1–5.
7.5.4.
(1918) *Über die Fourierkoeffizienten einer stetigen Funktion.* Acta Math. **41**, 377–384.
6.5.5, 6.5.11, 7.4.4, 7.7.7.
(1921) *Zur Theorie der linearen Integralgleichungen.* Math. Z. **9**, 196–217.
3.5.6, 7.4.3, 7.4.6, 7.4.19, 7.5.2, 7.5.4, 7.6.2, 7.7.5.
(1936) *Über die asymptotische Verteilung der Eigenwerte partieller Differentialgleichungen.* Ber. Sächs. Akad. Wiss. Leipzig **88**, 119–132.
7.7.6.

Catalan, E.
(1846) *Recherches sur les déterminants.* Bull. Acad. Royal Belgique **13**, 534–555.
7.1.1.

Cauchy, A. L.

(1815) *Mémoire sur les fonctions qui ne peuvent obtenir que deux valeurs égales et de signes contraires par suite de transpositions opérées entre les variables qu'elles renferment.* J. École Polytechn. **10**: 17, 29–112.
7.1.1.

(1829) *Sur l'équation a l'aide de laquelle on détermine les inégalités séculaires des mouvements des planètes.* Exer. de Math. **4**.
7.1.4.

(1839) *Mémoire sur l'intégration des équations linéaires.* C. R. Acad. Sci. Paris **8**, 827–830.
7.1.4.

Cayley, A.

(1858) *A memoir on the theory of matrices.* Phil. Trans. Roy. Soc. London **148**, 17–37.
7.1.2.

Chang, Shih-Hsun

(1947) *A generalization of a theorem of Lalesco.* J. London Math. Soc. **22**, 185–189.
7.7.5.

(1949) *On the distribution of the characteristic values and singular values of linear integral equations.* Trans. Amer. Math. Soc. **67**, 351–367.
7.4.3, 7.4.6, 7.5.4.

(1952) *A generalization of a theorem of Hille and Tamarkin with applications.* Proc. London Math. Soc. (3) **2**, 22–29.
7.7.5.

Ciesielski, Z.; Figiel, T.

(1983) *Spline bases in classical function spaces on compact C^∞ manifolds.* Studia Math. **76**, 1–58 and 95–136.
6.4.0, 6.4.11, 6.4.22, 7.7.1.

Clark, C.

(1967) *The asymptotic distribution of eigenvalues and eigenfunctions for elliptic boundary value problems.* SIAM Rev. **9**, 627–646.
7.7.0.

Cochran, J. A.

(1974) *The nuclearity of operators generated by Hölder continuous kernels.* Proc. Cambridge Philos. Soc. **75**, 351–356.
7.7.5.

(1975) *Growth estimates for the singular values of square integrable kernels.* Pacific J. Math. **56**, 51–58.
7.7.5.

(1976) *Summability of singular values of L^2 kernels. Analogies with Fourier series.* Enseign. Math. **22**, 141–157.
7.7.5.

(1977) *Composite integral operators and nuclearity.* Ark. Mat. **15**, 215–222.
7.7.5.

Cochran, J. A.; Oehring, C.

(1977) *Integral operators and an analogue of the Hausdorff-Young theorem.* J. London Math. Soc. (2) **15**, 511–520.
6.2.15, 7.4.5.

Cohen, J. S.

(1973) *Absolutely p-summing, p-nuclear operators and their conjugates.* Math. Ann. **201**, 177–200.
1.3.13.

Cohen, L. W.

(1930) *A note on a system of equations with infinitely many unknowns.* Bull. Amer. Math. Soc. **36**, 563–572.
7.3.1.

Courant, R.
(1920) *Über die Eigenwerte bei den Differentialgleichungen der mathematischen Physik*. Math. Z. **7**, 1–57.
7.3.6, 7.7.6.

Dedekind, R.
(1882) *Über die Diskriminante endlicher Körper*. Abhandl. Wiss. Gesell. Göttingen **29**, 1–56.
7.1.5.

Dedekind, R.; Weber, H.
(1882) *Theorie der algebraischen Funktionen einer Veränderlichen*. J. Reine Angew. Math. **92**, 181–290.
7.1.5.

Dixon, A. C.
(1902) *On a class of matrices of infinite order, and on the existence of "matricial" functions on a Riemann surface*. Cambridge Trans. **19**, 190–233.
7.3.1, 7.4.1.

Dunford, N.; Pettis, B. J.
(1940) *Linear operations on summable functions*. Trans. Amer. Math. Soc. **47**, 323–392.
7.3.2.

Eisenstein, G. M.
(1852) *Über die Vergleichung von solchen ternären quadratischen Formen, welche verschiedene Determinanten haben*. Monatsber. Preuss. Akad. Wiss. Berlin, 350–389.
7.1.2.

Enflo, P.
(1973) *A counterexample to the approximation problem in Banach spaces*. Acta Math. **130**, 309–317.
4.0.0, 4.7.5, 7.5.3.

Engelbrecht, J. C.; Grobler, J. J.
(1983) *Fredholm theory for operators in an operator ideal with a trace*. Integral Equations Operator Theory **6**, 21–30.
4.2.36, 4.4.11, 4.5.12, 7.5.3.

Erdos, J. A.
(1968) *Operators of finite rank in nest algebras*. J. London Math. Soc. **43**, 391–397.
7.6.2.
(1974) *On the trace of a trace class operator*. Bull. London Math. Soc. **6**, 47–50.
7.6.2.

Fiedler, M.; Pták, V.
(1962) *Sur la meilleure approximation des transformations linéaires par des transformation de rang prescrit*. C. R. Acad. Sci. Paris **254**, 3805–3807.
7.3.6.

Fischer, E.
(1905) *Über quadratische Formen mit reellen Koeffizienten*. Monatsh. Math. Phys. **16**, 234–249.
7.3.6.
(1907) *Sur la convergence en moyenne*. C. R. Acad. Sci. Paris **144**, 1022–1024.
6.5.7, 7.2.2, 7.7.7.

Fournier, J. J. F.
(1974) *An interpolation problem for coefficients of H^∞ functions*. Proc. Amer. Math. Soc. **42**, 402–408.
6.5.10.

Fournier, J. J. F.; Russo, B.
(1977) *Abstract interpolation and operator-valued kernels*. J. London Math. Soc. (2) **16**, 283–289.
6.2.15, 7.4.5.

Franklin, Ph.
(1928) *A set of continuous orthogonal functions*. Math. Ann. **100**, 522–529.
6.4.10.

Fredholm, I.
(1903) *Sur une classe d'équations fonctionelles*. Acta Math. **27**, 365–390.
6.6.7, 6.6.10, 7.4.4, 7.5.1, 7.5.2, 7.5.3, 7.5.4, 7.7.5, 7.7.7.

Frobenius, G.
(1878) *Über lineare Substitutionen und bilineare Formen*. J. Reine Angew. Math. **84**, 1–63.
7.1.4.

Fuchs, L.
(1866) *Zur Theorie der linearen Differentialgleichungen mit veränderlichen Koeffizienten*. J. Reine Angew. Math. **66**, 121–160.
7.1.6.

Garling, D. J. H.
(1970) *Absolutely p-summing operators in Hilbert space*. Studia Math. **38**, 319–331.
1.3.16.

Garling, D. J. H.; Gordon, Y.
(1971) *Relations between some constants associated with finite dimensional Banach spaces*. Israel J. Math. **9**, 346–361.
1.7.17.

Gel'fand, I. M.
(1938) *Abstrakte Funktionen und lineare Operatoren*. Mat. Sb. **4**, 235–286.
7.3.1, 7.3.2.
(1941) *Normierte Ringe*. Mat. Sb. **9**, 3–24.
7.3.3.

Gel'fond, A. O.
(1931) *Sur l'ordre de $D(\lambda)$*. C. R. Acad. Sci. Paris **192**, 828–831.
Epilogue.
(1957) *On the growth of eigenvalues of homogeneous integral equations* (Russian). Appendix: W. V. Lovitt, Linear integral equations, 329–352. Gostekh. Izdat. Moscow 1957.
7.5.4, 7.7.5.

Gheorghiu, S. A.
(1927/28) *Sur la croissance du dénominateur $D(\lambda)$ de Fredholm*. C. R. Acad. Sci. Paris **184**, 864–865, 1309–1311, and **186**, 838–840.
7.4.3, 7.4.6, 7.5.4.

Gluskin, E. D.
(1981) *Some finite dimensional problems in the theory of widths* (Russian). Vestnik Leningrad. Univ., Mat. Mekh. Astronom. **36**: 13, 5–10.
7.7.2.
(1983) *Norms of random matrices and diameters of finite dimensional sets* (Russian). Mat. Sb. **120**, 180–189.
7.7.2.

Goursat, E.
(1908) *Recherches sur les équations intégrales linéaires*. Ann. Fac. Sci. Univ. Toulouse (2) **10**, 5–98.
7.5.4.

Grisvard, P.
(1966) *Commutativité de deux foncteurs d'interpolation et applications*. J. Math. Pures Appl. **45**, 143–206.
6.4.0.

Grobler, J. J.; Raubenheimer, H.; Eldik, P. van
(1982) *Fredholm theory for operators in an operator ideal with a trace.* Integral Equations Operator Theory **5**, 774–790.
4.4.7, 4.4.10, 4.5.9, 4.5.11, 7.5.3.

Grothendieck, A.
(1951) *Sur une notion de produit tensoriel topologique d'espaces vectoriels topologiques, et une class remarquable d'espaces vectoriels liée à cette notion.* C. R. Acad. Sci. Paris **233**, 1556–1558.
7.3.8, 7.6.1.
(1956: a) *La théorie de Fredholm.* Bull. Soc. Math. France **84**, 319–384.
4.7.3, 4.7.9, 5.5.1, 6.6.1, 6.6.4, 7.5.3.
(1956: b) *Résumé de la théorie métrique des produits tensoriels topologiques.* Bol. Soc. Mat. São Paulo **8**, 1–79.
1.6.4, 1.7.15, 7.3.5, 7.7.2.
(1956: c) *Sur certaines classes de suites dans les espaces de Banach, et le théorème de Dvoretzky-Rogers.* Bol. Soc. Mat. São Paulo **8**, 81–110.
1.7.18.
(1961) *The trace of certain operators.* Studia Math. **20**, 141–143.
7.6.2.

Ha, Chung-Wei
(1975) *On the trace of a class of nuclear operators.* Bull. Inst. Math. Acad. Sinica **3**, 131–137.
7.6.2.

Haagerup, U.
(1982) *The best constants in the Khintchine inequality.* Studia Math. **70**, 231–283.
G.2.3.

Haar, A.
(1910) *Zur Theorie der orthogonalen Funktionensysteme.* Math. Ann. **69**, 331–371.
6.4.10.

Hadamard, J.
(1893: a) *Résolution d'une question relative aux déterminants.* Bull. Sci. Math. (2) **17**, 240–246.
A.4.5, 5.1.7, 7.5.1.
(1893: b) *Étude sur les propriétés des fonctions entières et en particulier d'une fonction considérée par Riemann.* J. de Math. (4) **9**, 171–215.
4.8.6, 7.5.4.

Hamilton, W. R.
(1837) *Theory of conjugate functions.* Trans. Roy. Irish. Acad. **17**, 293–422.
7.1.3.

Hammerstein, A.
(1923) *Über die Entwicklung des Kerns linearer Integralgleichungen nach Eigenfunktionen.* Sitzungsber. Deut. Akad. Wiss. Berlin, Phys. Math. Kl., 181–184.
7.0.0.

Hardy, G. H.; Littlewood, J. E.
(1928: a) *Some properties of fractional integrals.* Math. Z. **27**, 565–606.
7.2.2.
(1928: b) *On the absolute convergence of Fourier series.* J. London Math. Soc. **3**, 250–253.
6.6.13.
(1931) *Some new properties of Fourier constants.* J. London Math. Soc. **6**, 3–9.
6.5.9, 7.2.1, 7.7.7.

Hasumi, M.
(1958) *The extension property of complex Banach spaces.* Tôhoku Math. J. (2) **10**, 135–142.
C.4.2.

Hausdorff, F.
(1923) *Eine Ausdehnung des Parsevalschen Satzes über Fourierreihen.* Math. Z. **16**, 163–169. 6.5.9, 7.7.7.

Hawkins, T.
(1975) *Cauchy and the spectral theory of matrices.* Historia Math. **2**, 1–29. 7.1.0.
(1977: a) *Another look at Cayley and the theory of matrices.* Arch. Internat. Hist. Sci. **27**, 82–112. 7.1.0.
(1977: b) *Weierstrass and the theory of matrices.* Arch. Hist. Exact Sci. **17**, 119–163. 7.1.0.

Hellinger, E.; Toeplitz, O.
(1910) *Grundlagen für eine Theorie der unendlichen Matrizen.* Math. Ann. **69**, 289–330. 7.3.1.

Hilbert, D.
(1904) *Grundzüge einer allgemeinen Theorie der linearen Integralgleichungen (Erste Mitteilung).* Nachr. Wiss. Gesell. Göttingen, Math.-Phys. Kl., 49–91. 7.3.2, 7.5.2.
(1906: a) *Grundzüge einer allgemeinen Theorie der linearen Integralgleichungen (Vierte Mitteilung).* Nachr. Wiss. Gesell. Göttingen, Math.-Phys. Kl., 157–227. 7.2.1, 7.3.1, 7.4.1.
(1906: b) *Grundzüge einer allgemeinen Theorie der linearen Integralgleichungen (Fünfte Mitteilung).* Nachr. Wiss. Gesell. Göttingen, Math.-Phys. Kl., 439–480. 7.4.0.

Hill, G. W.
(1877) *On the part of the motion of the lunar perigee which is a function of the mean motions of the sun and the moon.* Cambridge (Mass.) 1877 and Acta Math. **8** (1886), 1–36. 7.5.1.

Hille, E.; Tamarkin, J. D.
(1931) *On the characteristic values of linear integral equations.* Acta Math. **57**, 1–76. 7.5.4, 7.7.5, Epilogue.
(1934) *On the theory of linear integral equations.* Ann. of Math. **35**, 445–455. 7.3.1, 7.3.2.

Hölder, O.
(1889) *Über einen Mittelwertsatz.* Nachr. Wiss. Gesell. Göttingen, Math.-Phys. Kl. 38–47. 7.7.1.

Höllig, K.
(1979) *Approximationszahlen von Sobolev-Einbettungen.* Math. Ann. **242**, 273–281. 7.7.2.

Holub, J. R.
(1970/74) *Tensor product mappings.* Math. Ann. **188**, 1–12, and Proc. Amer. Math. Soc. **42**, 437–441. 1.3.11, 1.7.12, 7.3.7.

Horn, A.
(1950) *On the singular values of a product of completely continuous operators.* Proc. Nat. Acad. Sci. USA **36**, 374–375. 2.11.23, 7.4.3.

Hostinsky, B.
(1921) *Notes sur l'equation de Fredholm.* Publ. Fac. Sci. Univ. Masaryk (Brno), **1921**: 1, 3–14. 7.6.2.

Howland, J. S.
(1971) *Trace class Hankel operators.* Quart. J. Math. Oxford (2) **22**, 147–159.
7.7.5.

Hunt, R. A.
(1966) *On L(p, q) spaces.* Enseign. Math. (2), **12**, 249–276.
7.2.1.

Hutton, C. V.
(1974) *On the approximation numbers of an operator and its adjoint.* Math. Ann. **210**, 277–280.
2.3.16.

Hutton, C. V.; Morrell, J. S.; Retherford, J. R.
(1976) *Diagonal operators, approximation numbers, and Kolmogoroff diameters.* J. Approx. Theory **16**, 48–80.
2.10.2.

Ismagilov, R. S.
(1977) *Diameters of compact sets in linear normed spaces* (Russian). Collection of papers on "Geometry of linear spaces and operator theory", 75–113, Jaroslavl.
7.7.2.

Itskovich, I. A.
(1948) *On Fredholm series* (Russian). Doklady Akad. Nauk **59**, 423–425.
7.3.2.

Jacobi, C. G. J.
(1834) *De binis quibuslibet functionibus homogeneis secundi ordinis per substitutiones linearis in alias binas transformandis, ...* J. Reine Angew. Math. **12**, 1–69.
7.1.4, 7.1.5.

Jarchow, H.; Ott, R.
(1982) *On trace ideals.* Math. Nachr. **108**, 23–37.
1.7.15.

Jensen, J. L. W. V.
(1899) *Sur un nouvel et important théorème de la théorie des fonctions.* Acta Math. **22**, 359–364.
4.8.5, 4.8.6, 7.5.4.

Jessen, B.
(1931) *Om Uligheder imellem Potensmiddelvaerdier.* Mat. Tidsskrift B, No. 1.
C.3.10.

Joachimsthal, F.
(1849) *Sur quelques applications des déterminants à la géométrie.* J. Reine Angew. Math. **11**, 21–47.
7.1.1.

John, F.
(1948) *Extremum problems with inequalities as subsidiary conditions.* Courant Anniversary Volume, 187–204, New York: Interscience.
1.5.5.

Johnson, W. B.; Jones, L.
(1978) *Every L_p-operator is an L_2-operator.* Proc. Amer. Math. Soc. **72**, 309–312.
6.2.15.

Johnson, W. B.; König, H.; Maurey, B.; Retherford, J. R.
(1979) *Eigenvalues of p-summing and l_p-type operators in Banach spaces.* J. Funct. Anal. **32**, 353–380.
3.7.2, 3.7.3, 3.8.3, 3.8.4, 5.3.6, 6.2.14, 7.4.4, 7.4.5, 7.4.6, 7.7.5.

Jordan, C.
(1874) *Mémoire sur les formes bilinéaires.* J. de Math. (2) **19**, 35–54.
7.1.6.

Kadets, M. I.; Snobar, M. G.
(1971) *Certain functionals on the Minkowski compactum* (Russian). Mat. Zametki **10**, 453–458.
1.5.5.

Kadlec, J.; Korotkov, V. B.
(1968) *Estimates of the s-numbers of imbedding operators and operators which increase smoothness.* Czechoslovak. Math. J. **18**, 678–699.
6.4.0, 7.7.5.

Kaiser, R. J.; Retherford, J. R.
(1983) *Eigenvalue distribution of nuclear operators.* Proc. Conf. Functional Analysis/Banach Space Geometry, 245–287, Univ. Essen 1982.
3.8.2.
(1984) *Nuclear cyclic diagonal mappings.* Math. Nachr. **119**, 129–135.
3.8.2.

Karadzhov, G. E.
(1972) *The inclusion of integral operators in the classes S_p for $p \geqq 2$* (Russian). Problemy Matematicheskogo Analiza (Leningrad) **3**, 28–33.
6.2.15, 7.4.5, 7.7.5.
(1977) *The inclusion of integral operators in the classes S_q for $0 < q < 2$* (Russian). Serdica **3**, 52–70.
6.6.9, 7.7.5.

Kashin, B. S.
(1977) *Diameters of some finite dimensional sets and of some classes of smooth functions* (Russian). Izv. Akad. Nauk SSSR, Ser. Mat. **41**, 334–351.
2.9.13, 7.7.2.

Kaucký, J.
(1922) *Contribution à la théorie de l'equation de Fredholm.* Publ. Fac. Sci. Univ. Masaryk (Brno) **1922**: 13, 3–8.
7.6.2.

Khintchine, A. Ya.
(1923) *Über dyadische Brüche.* Math. Z. **18**, 109–116.
G.2.3.

Koch, H. von
(1896) *Sur la convergence des déterminants d'ordre infini.* Bihang till Kongl. Svenska Vetenskaps-Academiens Handlingar **22**, I:4, 1–31.
7.5.1.
(1901) *Sur quelques points de la théorie des déterminants infinis.* Acta Math. **24**, 89–122.
5.5.5, 7.3.1, 7.5.1.
(1909) *Sur la convergence des déterminants infinis.* Rend. Circ. Mat. Palermo **28**, 255–266.
7.5.1.
(1910) *Sur les systèmes d'une infinité d'equations linéares à une infinité d'inconnues.* Proc. Intern. Congress Math. 43–61, Stockholm.
7.5.1.

Kolmogorov, A. N.
(1936) *Über die beste Annäherung von Funktionen einer gegebenen Funktionenklasse.* Ann. of Math. (2) **37**, 107–110.
7.3.6, 7.7.3.

König, H.
(1974) *Grenzordnungen von Operatorenidealen.* Math. Ann. **212**, 51–64 and 65–77.
7.7.1.
(1977) *Eigenvalues of p-nuclear operators.* Proc. Intern. Conf. "Operator Algebras, Ideals, ...", Leipzig 1977, 106–113, Teubner-Texte Math. 18, Leipzig.
3.8.5, 7.4.6.

(1978) *Interpolation of operator ideals with an application to eigenvalue distribution problems.* Math. Ann. **233**, 35–48.
1.2.6, 2.2.10, 2.3.14, 3.6.3, 7.2.4, 7.3.6, 7.4.3, 7.4.6.

(1979) *On the spectrum of products of operator ideals.* Math. Nachr. **93**, 223–232.
2.11.29, 2.11.35, 3.6.7.

(1980: a) *s-numbers, eigenvalues and the trace theorem in Banach spaces.* Studia Math. **67**, 157–172.
4.2.26, 7.5.3, 7.6.2.

(1980: b) *A Fredholm determinant theory for p-summing maps in Banach spaces.* Math. Ann. **247**, 255–274.
4.2.6, 4.2.30, 7.5.3, 7.6.2, Epilogue.

(1980: c) *Weyl-type inequalities for operators in Banach spaces.* Math. Stud. **38**, 297–317, Amsterdam, New York, Oxford: North-Holland.
2.7.1, 2.7.2, 2.7.6, 2.11.29, 2.11.30.

(1980: d) *Some remarks on weakly singular integral operators.* Integral Equations Operator Theory **3**, 397–407.
6.3.5, 6.3.8, 7.7.5.

(1981) *On the eigenvalue spectrum of certain operator ideals.* Colloq. Math. **44**, 1–28.
7.4.0.

(1984: a) *On the tensor stability of s-number ideals.* Math. Ann. **269**, 77–93.
2.3.15.

(1984: b) *Some inequalities for the eigenvalues of compact operators.* Internat. Ser. Numer. Math. **71**, 213–219, Basel: Birkhäuser.
3.6.2.

König, H.; Retherford, J. R.; Tomczak-Jaegermann, N.
(1980) *On the eigenvalues of (p, 2)-summing operators and constants associated to normed spaces.* J. Funct. Anal. **37**, 88–126.
3.7.6, 3.8.4, 6.3.3, 6.3.8, 7.4.4, 7.4.5, 7.4.6, 7.7.5.

König, H.; Richter, S.
(1984) *Eigenvalues of integral operators defined by analytic kernels.* Math. Nachr. **119**, 141–155.
Epilogue.

König, H.; Weis, L.
(1983) *On the eigenvalues of order bounded integral operators.* Integral Equations Operator Theory **6**, 706–729.
6.2.15, 7.4.5.

Konyushkov, A. A.
(1958) *Best approximation by trigonometric polynomials and Fourier coefficients* (Russian). Mat. Sb. **44**, 53–84.
7.7.7.

Kostometov, G. P.
(1974) *Asymptotic behaviour of the spectrum of integral operators with a singularity on the diagonal* (Russian). Mat. Sb. **94**, 444–451.
7.7.5.

(1977) *On the asymptotics of the spectrum of integral operators with polar kernels* (Russian). Vestnik Leningrad. Univ., Mat. Mekh. Astronom. **32**: 13, 166–167.
7.7.5.

Kostometov, G. P.; Solomyak, M. Z.
(1971) *Estimates of singular values of integral operators with a weak singularity* (Russian). Vestnik Leningrad. Univ., Mat. Mekh. Astronom. **26**: 1, 28–39.
7.7.5.

Kotlyar, B. D.; Semirenko, T. N.
(1981) *On the spectrum of operators that increase smoothness* (Russian). Ukrain. Mat. Zh. **33**, 765–770.
7.7.5.

Kötteritzsch, Th.
(1870) *Ueber die Auflösung eines Systems von unendlich vielen linearen Gleichungen.* Z. f. Mathematik u. Physik **15**, 1–15 and 229–268.
7.3.1, 7.5.1.

Krejn, M. G.
(1937) *On characteristic numbers of differentiable symmetric kernels* (Russian). Mat. Sb. **2**, 725–730.
6.4.31, 7.7.5.

Kronecker, L.
(1874) *Über Scharen von quadratischen und bilinearen Formen.* Monatsber. Preuss. Akad. Wiss. Berlin 59–76, 149–156 and 206–232.
7.1.6.

Kühn, Th.
(1984) *Entropy numbers of matrix operators in Besov sequence spaces.* Math. Nachr. **119**, 165–174.
Epilogue.

Kupka, J.
(1980) *$L_{p,q}$ spaces.* Dissertationes Math. **164**, Warszawa.
6.2.12.

Kuroda, S. T.
(1961) *On a generalization of the Weinstein-Aronszajn formula and the infinite determinant.* Sci. Papers College Gen. Ed. Univ. Tokyo (Mathematics) **11**, 1–12.
7.5.3.

Kwapień, S.
(1968) *Some remarks on (p, q)-absolutely summing operators in l_p-spaces.* Studia Math. **29**, 327–337.
1.2.3, 1.2.4, 1.2.5, 2.11.28, 7.3.7.
(1970) *On a theorem of L. Schwartz and its applications to absolutely summing operators.* Studia Math. **38**, 193–201.
1.3.15.
(1972) *On operators factorable through L_p space.* Bull. Soc. Math. France, Mém. **31/32**, 215–225.
2.11.36.

Ky Fan
(1949/50) *On a theorem of Weyl concerning eigenvalues of linear transformations.* Proc. Nat. Acad. Sci. USA **35**, 652–655, and **36**, 31–35.
2.11.13, 7.4.3.
(1951) *Maximum properties and inequalities for the eigenvalues of completely continuous operators.* Proc. Nat. Acad. Sci. USA **37**, 760–766.
2.11.12, 7.4.3.

Lacey, H. E.
(1963) *Generalizations of compact operators in locally convex topological linear spaces.* Thesis, New Mexico State Univ.
2.4.10.

Lalesco, T.
(1907) *Sur l'ordre de la fonction entière $D(\lambda)$ de Fredholm.* C. R. Acad. Sci. Paris **145**, 906–907.
7.5.4.
(1915) *Un théorème sur les noyaux composés.* Bull. Sect. Sci. Acad. Roumaine **3**, 271–272.
7.4.4, 7.4.6.

Landau, E.
(1907) *Über einen Konvergenzsatz.* Nachr. Wiss. Gesell. Göttingen, Math.-Phys. Kl., 25–27.
7.7.1.

Lebesgue, H.
(1903) *Sur les séries trigonométriques.* Ann. Sci. École Norm. Sup. (3) **20**, 453–485.
6.5.6, 7.7.7.

Leeuw, K. de; Katznelson, Y.; Kahane, J. P.
(1977) *Sur les coefficients de Fourier des functions continues*. C. R. Acad. Sci. Paris **285**, 1001–1003.
6.5.11.

Leiterer, H.; Pietsch, A.
(1982) *An elementary proof of Lidskij's trace formula*. Wiss. Z. Friedrich-Schiller-Univ. Jena, Math.-Natur. Reihe **31**, 587–594.
3.6.5, 4.2.24, 4.2.25, 7.6.2.

Leżański, T.
(1953) *The Fredholm theory of linear equations in Banach spaces*. Studia Math. **13**, 244–276.
7.5.3.

Lidskij, V. B.
(1959) *Non-self-adjoint operators with a trace* (Russian). Doklady Akad. Nauk SSSR **125**, 485–487.
4.7.14, 4.7.15, 7.6.2.

Linde, R.
(1986) *s-Numbers of diagonal operators and Besov embeddings*. Proc. 13-th Winter School, Suppl. Rend. Circ. Mat. Palermo (2).
7.7.2.

Lindelöf, E.
(1903) *Sur la détermination de la croissance des fonctions entières défines par un développement de Taylor*. Bull. Sci. Math. (2) **27**, 213–232.
4.8.7, 7.5.4.
(1905) *Sur les fonctions entières d'ordre entier*. Ann. Sci. École Norm. Sup. (3) **22**, 369–395.
4.8.9, 7.5.4.

Lions, J. L.; Peetre, J.
(1964) *Sur une classe d'espaces d'interpolation*. Inst. Hautes Études Sci. Publ. Math. **19**, 5–68.
1.1.6, 7.2.4.

Littlewood, J. E.
(1930) *On bounded bilinear forms in an infinite number of variables*. Quart. J. Math. Oxford **1**, 164–174.
5.1.7.

Lomonosov, V. I.
(1973) *Invariant subspaces for the family of operators which commute with a completely continuous operator* (Russian). Funktsional. Anal. i Prilozhen. **7**: 3, 55–56.
Epilogue.

Lorentz, G. G.
(1950) *Some new functional spaces*. Ann. of Math. (2) **51**, 37–55.
7.2.1.

Lorentz, H. A.
(1910) *Alte und neue Fragen der Physik*. Physikal. Z. **11**, 1234–1257.
7.7.6.

Lubitz, C.
(1982) *Weylzahlen von Diagonaloperatoren und Sobolev-Einbettungen*. Thesis, Univ. Bonn, Bonner Math. Schriften **144**.
2.9.17, 2.9.18, 7.7.2.

Majorov, V. E.
(1975) *Discretization of the problem of diameters* (Russian). Uspehi Mat. Nauk **30**: 6, 179–180.
7.7.1.

Markus, A. S.
(1964) *Eigenvalues and singular values of the sum and the product of linear operators* (Russian). Uspehi Mat. Nauk **19**: 4, 93–123.
7.4.3.

(1966) *Some criteria for the completeness of a system of root vectors of a linear operator in a Banach space* (Russian). Mat. Sb. **70**, 526–561.
7.7.2, Epilogue.

Markus, A. S.; Matsaev, V. I.
(1971) *Analogues of the Weyl inequalities and the trace theorem in a Banach space* (Russian). Mat. Sb. **86**, 299–313.
7.4.3.

Mäurer, P.
(1981) *Die Geschichte der Fredholmschen Determinantentheorie*. Staatsexamensarbeit, Univ. Bonn.
7.5.0.

Mazurkiewicz, S.
(1915) *Sur le déterminant de Fredholm*. Sprawozdania, Towarzystwo Naukowe Warszawskie, Wyd. III, **8**, 656–662 and 805–810.
7.5.4, 7.7.5.

Mercer, J.
(1909) *Functions of positive and negative type, and their connection with the theory of integral equations*. Phil. Trans. Roy. Soc. London (A) **209**, 415–446.
7.6.2.

Michal, A. D.; Martin, R. S.
(1934) *Some expansions in vector space*. J. de Math. **13**, 69–91.
7.5.3.

Mikhlin, S. G.
(1944) *On the convergence of Fredholm series* (Russian). Doklady Akad. Nauk SSSR **42**, 373–376.
7.5.3.

Milman, V. D.
(1970) *Operators of class C_0 and C_0^** (Russian). Teor. Funktsij. Funktsional. Anal. i Prilozhen. **10**, 15–26.
2.9.6, 7.7.2.

Mityagin, B. S.; Pełczyński, A.
(1966) *Nuclear operators and approximative dimension*. Proc. ICM, 366–372, Moscow.
7.3.7, Epilogue.

Neumann, J. von
(1929) *Allgemeine Eigenwerttheorie Hermitescher Funktionaloperatoren*. Math. Ann. **102**, 49–131.
7.2.3, 7.3.4.
(1937) *Some matrix-inequalities and metrization of matrix-space*. Tomsk Univ. Rev. **1**, 286–300.
7.3.4.

Nikodym, O.
(1933) *Sur une classe de fonctions considérée dans l'étude du problème de Dirichlet*. Fund. Math. **21**, 129–150.
7.2.2.

Nikol'skij, S. M.
(1951) *Inequalities for entire functions of finite order and their applications to the theory of differentiable functions in several variables* (Russian). Trudy Mat. Inst. Steklov **38**, 244–278.
7.2.2.

Novoselskij, I. A.
(1964) *About certain asymptotical characteristics of linear operators* (Russian). Izv. Akad. Nauk Moldav. SSR Ser. Fiz. Tehn. Mat. Nauk **6**, 85–90.
2.5.5, 7.3.6.

Oloff, R.
(1969) *p-Normideale von Operatoren in Banachräumen.* Wiss. Z. Friedrich-Schiller-Univ. Jena, Math.-Nat. Reihe **18**, 259–262.
1.7.2, 2.11.26.
(1970) *Interpolation zwischen den Klassen S_p von Operatoren in Hilberträumen.* Math. Nachr. **46**, 209–218.
7.2.4.
(1972) *p-normierte Operatorenideale.* Beiträge Anal. **4**, 105–108.
2.11.26.

O'Neil, R.
(1963) *Convolution operators and $L(p, q)$ spaces.* Duke Math. J. **30**, 129–142.
7.2.1.

Orlicz, W.
(1933) *Über unbedingte Konvergenz in Funktionenräumen.* Studia Math. **4**, 33–37 and 41–47.
1.6.5, 7.3.7, 7.7.2.

Paraska, V. I.
(1965) *On asymptotics of eigenvalues and singular numbers of linear operators which increase smoothness* (Russian). Mat. Sb. **68**, 623–631.
7.7.5.

Parseval, M. A.
(1805) *Intégration générale et complète de deux équations importantes dans la mécaniques des fluids.* Mémoires présentés à l'Institut par divers Savans **1**, 524–545.
6.5.7.

Peetre, J.
(1963) *Nouvelles propriétés d'espaces d'interpolation.* C. R. Acad. Sci. Paris **256**, 1424–1426.
2.1.14, 7.2.1.
(1964/66) *Inledning till interpolation.* Föreläsningar, Lund.
7.2.4.

Peetre, J.; Sparr, G.
(1972) *Interpolation of normed Abelian groups.* Ann. Mat. Pura Appl. (4) **92**, 217–262.
2.3.14, 7.2.4.

Pełczyński, A.
(1967) *A characterization of Hilbert-Schmidt operators.* Studia Math. **28**, 355–360.
1.3.16.

Peller, V. V.
(1980) *Hankel operators of the class S_p and their applications* (Russian). Mat. Sb. **113**, 538–581.
7.7.5.

Persson, A.
(1969) *On some properties of p-nuclear and p-integral operators.* Studia Math. **33**, 213–222.
1.3.13.

Persson, A.; Pietsch, A.
(1969) *p-Nukleare und p-integrale Abbildungen in Banachräumen.* Studia Math. **33**, 19–62.
1.3.4, 1.3.9, 1.6.3.

Pettis, B. J.
(1938) *On integration in vector spaces.* Trans. Amer. Math. Soc. **44**, 277–304.
6.2.3.

Phillips, R. S.
(1940) *On linear transformations.* Trans. Amer. Math. Soc. **48**, 516–541.
1.1.12, 7.3.1, 7.3.2.

Pietsch, A.
(1961) *Quasi-präkompakte Endomorphismen und ein Ergodensatz in lokalkonvexen Vektorräumen.* J. Reine Angew. Math. **207**, 16–30.
3.2.1, 3.2.9, 7.4.1.
(1963: a) *Zur Fredholmschen Theorie in lokalkonvexen Räumen.* Studia Math. **22**, 161–179.
3.3.1, 3.3.4, 3.7.1, 5.5.6, 7.4.2, 7.4.5, 7.4.6, 7.5.3.
(1963: b) *Einige neue Klassen von kompakten linearen Abbildungen.* Rev. Roumaine Math. Pures Appl. **8**, 427–447.
2.3.6, 2.3.9, 2.3.11, 2.3.12, 7.3.6, 7.6.1, 7.7.2.
(1967) *Absolut p-summierende Abbildungen in normierten Räumen.* Studia Math. **28**, 333–353.
1.2.2, 1.3.2, 1.3.3, 1.3.5, 1.3.8, 1.3.10, 1.3.16, 1.4.5, 1.5.1, 3.7.1, 7.3.7, 7.4.5, 7.4.6.
(1968) *Hilbert-Schmidt-Abbildungen in Banach-Räumen.* Math. Nachr. **37**, 237–245.
1.7.14.
(1970) *Ideale von S_p-Operatoren in Banachräumen.* Studia Math. **38**, 59–69.
2.11.32, 2.11.33.
(1971) *Interpolationsfunktoren, Folgenideale und Operatorenideale.* Czechoslovak. Math. J. **21**, 644–652.
2.11.18.
(1972: a) *Theorie der Operatorenideale (Zusammenfassung).* Wiss. Beiträge Friedrich-Schiller-Univ. Jena.
2.2.5, 2.2.9, 2.3.3, 2.3.4, 2.4.3, 2.4.8, 2.5.3, 7.3.6.
(1972: b) *Eigenwertverteilungen von Operatoren in Banachräumen*, Hausdorff-Festband: Theory of sets and topology. 391–402, Berlin: Akademie-Verlag.
3.4.6, 3.7.1, 7.4.2.
(1972: c) *Absolutely p-summing operators in L_r-spaces.* Bull. Soc. Math. France, Mém. **31/32**, 285–315.
1.3.14.
(1974: a) *s-Numbers of operators in Banach spaces.* Studia Math. **51**, 201–223.
2.2.12, 2.4.7, 2.5.6, 2.9.4, 2.9.8, 2.10.2, 2.11.9, 7.3.6, 7.7.2.
(1974: b) *Ultraprodukte von Operatoren in Banachräumen.* Math. Nachr. **61**, 123–132.
2.4.12.
(1976) *Extensions of operator ideals.* Math. Nachr. **75**, 31–39.
2.11.32, 2.11.33.
(1980: a) *Über die Verteilung von Fourierkoeffizienten und Eigenwerten.* Wiss. Z. Friedrich-Schiller-Univ. Jena, Math.-Nat. Reihe **29**, 203–211.
6.5.17, 7.7.7.
(1980: b) *Factorization theorems for some scales of operator ideals.* Math. Nachr. **97**, 15–19.
2.3.8, 2.3.13.
(1980: c) *Weyl numbers and eigenvalues of operators in Banach spaces.* Math. Ann. **247**, 149–168.
2.4.14, 2.4.17, 2.4.19, 2.5.9, 2.5.12, 2.7.3, 2.7.4, 2.7.5, 2.7.7, 2.7.8, 2.8.15, 2.10.4, 2.10.8, 3.6.1, 3.6.2, 7.3.6, 7.4.3, 7.4.6.
(1980: d) *Eigenvalues of integral operators.* Math. Ann. **247**, 169–178.
5.4.6, 5.4.7, 5.4.10, 6.4.14, 6.4.16, 6.4.19, 7.3.1, 7.3.2, 7.7.1, 7.7.2, 7.7.5.
(1981: a) *Approximation spaces.* J. Approx. Theory **32**, 115–134.
2.3.8, 2.3.9, 2.3.10, 2.8.8, 2.8.10, 2.8.13, 6.4.24, 6.4.26, 6.5.24, 7.4.6.
(1981: b) *Operator ideals with a trace.* Math. Nachr. **100**, 61–91.
2.8.17, 2.8.18, 2.8.21, 3.7.4, 4.2.3, 4.2.7, 4.2.12, 4.2.29, 4.2.31, 4.7.6, 7.6.1.
(1982: a) *Distribution of eigenvalues and nuclearity.* Banach Center Publ. **8**, 361–365. Warsaw.
3.4.3, 3.4.5.
(1982: b) *Tensor products of sequences, functions, and operators.* Arch. Math. (Basel) **38**, 335–344.
2.3.15.
(1983) *Eigenvalues of integral operators.* Math. Ann. **262**, 343–376.
Epilogue.

(1986) *Eigenvalues of absolutely r-summing operators.* Aspects of mathematics and its applications (Ed. J. A. Barroso), 607–617. Amsterdam: Elsevier Sci. Publ.
3.4.8, 3.4.9, 7.4.5.

Pietsch, A.; Triebel, H.
(1968) *Interpolationstheorie für Banachideale von beschränkten linearen Operatoren.* Studia Math. **31**, 95–109.
7.2.4.

Pisier, G.
(1979) *Some applications of the complex interpolation method to Banach lattices.* J. Analyse Math. **35**, 264–281.
3.8.3.
(1983) *Counterexamples to a conjecture of Grothendieck.* Acta Math. **151**, 181–208.
3.8.2.

Plemelj, J.
(1904) *Zur Theorie der Fredholmschen Funktionalgleichung.* Monatsh. Math. Phys. **15**, 93–128.
4.4.10, 7.5.2, 7.5.4.

Poincaré, H.
(1886) *Sur les déterminants d'ordre infini.* Bull. Soc. Math. France **14**, 77–90.
7.5.1.
(1910) *Remarques diverses sur l'équation de Fredholm.* Acta Math. **33**, 57–86.
5.5.10, 7.5.2.

Pólya, G.
(1950) *Remark on Weyl's note: Inequalities between the two kinds of eigenvalues of a linear transformation.* Proc. Nat. Acad. Sci. USA **36**, 49–51.
7.4.3.

Power, S. C.
(1983: a) *Nuclear operators in nest algebras.* J. Operator Theory **10**, 337–352.
7.6.2.
(1983: b) *Another proof of Lidskii's theorem on the trace.* Bull. London Math. Soc. **15**, 146–148.
7.6.2.

Pringsheim, A.
(1904) *Elementare Theorie der ganzen transcendenten Funktionen endlicher Ordnung.* Math. Ann. **58**, 257–342.
4.8.7, 7.5.4.

Radjavi, H.; Rosenthal, P.
(1982) *The invariant subspace problem.* Math. Intelligencer **4**: 1, 33–37.
Epilogue.

Retherford, J. R.
(1980) *Trace.* Proc. Sem. Ges. Math. Datenverarb., Special topics of applied mathematics, 47–56, Bonn 1979, Amsterdam: North-Holland.
7.6.0.

Reuter, F.
(1981) *Resolventenwachstum und Vollständigkeit meromorpher Operatoren in normierten Räumen.* Math. Z. **178**, 387–397.
Epilogue.

Riemann, B.
(1854) *Ueber die Darstellbarkeit einer Funktion durch eine trigonometrische Reihe.* Habilitationsschrift, Univ. Göttingen.
6.5.6, 7.7.7.

Riesz, F.
(1907) *Sur les systèmes orthogonaux de fonctions*. C. R. Acad. Sci. Paris **144**, 615–619.
6.5.7, 7.2.2, 7.7.7.
(1909) *Sur les opérations fonctionelles linéaires*. C. R. Acad. Sci. Paris **149**, 974–977.
7.2.2.
(1910) *Untersuchungen über Systeme integrierbarer Funktionen*. Math. Ann. **69**, 449–497.
7.2.2.
(1918) *Über lineare Funktionalgleichungen*. Acta Math. **41**, 71–98.
3.1.14, 3.2.3, 3.2.9, 3.2.19, 4.1.5, 7.2.3, 7.3.3, 7.4.1, 7.5.3.

Riesz, M.
(1926) *Sur les maxima des formes bilinéaires et sur les fonctionelles linéaires*. Acta Math. **49**, 465–497.
7.2.4.

Ringrose, J. R.
(1962) *Super-diagonal forms of compact linear operators*. Proc. London Math. Soc. (3) **12**, 367–384.
Epilogue.

Rolewicz, S.
(1957) *On a certain class of linear metric spaces*. Bull. Acad. Polon. Sci., Cl. III, **5**, 471–473.
7.2.3.

Ropela, S.
(1976) *Spline bases in Besov spaces*. Bull. Acad. Polon. Sci., Ser. Math. Astronom. Phys. **24**, 319–325.
6.4.11.

Rotfeld, S. J.
(1967) *Remarks on the singular values of a sum of completely continuous operators* (Russian). Funktsional. Anal. i Prilozhen. **1**: 3, 95–96.
2.11.20.

Rudin, W.
(1959) *Some theorems on Fourier coefficients*. Proc. Amer. Math. Soc. **10**, 855–859.
6.5.10.

Russo, B.
(1977) *On the Hausdorff-Young theorem for integral operators*. Pacific J. Math. **68**, 241–253.
6.2.15, 7.4.5.

Ruston, A. F.
(1951: a) *On the Fredholm theory of integral equations for operators belonging to the trace class of a general Banach space*. Proc. London Math. Soc. (2) **53**, 109–124.
2.11.25, 4.7.9, 7.3.8, 7.5.3, 7.6.1.
(1951: b) *Direct products of Banach spaces and linear functional equations*. Proc. London Math. Soc. (3) **1**, 327–384.
7.3.5.
(1953) *Formulae of Fredholm type for compact linear operators on a general Banach space*. Proc. London Math. Soc. (3) **3**, 368–377.
4.4.1.
(1954) *Operators with a Fredholm theory*. J. London Math. Soc. **29**, 318–326.
4.1.6, 7.4.1.
(1962) *Auerbach's theorem and tensor products of Banach spaces*. Proc. Cambridge Phil. Soc. **58**, 476–480.
1.7.6, 1.7.7.
(1967) *Fredholm formulae and the Riesz theory*. Compositio Math. **18**, 25–48.
4.4.1.

Saphar, P.
(1966) *Applications à puissance nucléaire et applications de Hilbert-Schmidt dans les espaces de Banach*. Ann. Sci. École Norm. Sup. (3) **83**, 113–151.
3.8.6, 7.3.8, 7.4.6, 7.5.3.
(1970) *Applications p-sommantes et p-décomposantes*, C. R. Acad. Sci. Paris (A) **270**, 1093–1096.
1.3.13.

Schatten, R.
(1946) *The cross-space of linear transformations*. Ann. of Math. (2) **47**, 73–84.
7.3.4, 7.6.1.

Schatten, R.; Neumann, J. von
(1946) *The cross-space of linear transformations*. Ann. of Math. (2) **47**, 608–630.
1.4.1, 1.4.3, 1.4.10, 7.3.4, 7.3.8, 7.6.1.
(1948) *The cross-space of linear transformations*. Ann. of Math. (2) **49**, 557–582.
2.11.20, 7.3.4, 7.3.8, 7.6.1.

Schauder, J.
(1930) *Über lineare vollstetige Funktionaloperationen*. Studia Math. **2**, 183–196.
3.2.25, 7.4.1, 7.5.3.
(1934) *Über lineare elliptische Differentialgleichungen zweiter Ordnung*. Math. Z. **38**, 257–282.
7.2.2.

Schmidt, E.
(1907: a) *Entwicklung willkürlicher Funktionen nach Systemen vorgeschriebener*. Math. Ann. **63**, 433–476.
1.4.4, 2.11.8, 7.3.6, 7.3.8.
(1907: b) *Auflösung der allgemeinen linearen Integralgleichung*. Math. Ann. **64**, 161–174.
7.4.1, 7.5.3.
(1908) *Über die Auflösung linearer Gleichungen mit unendlich vielen Unbekannten*. Rend. Circ. Mat. Palermo **25**, 53–77.
7.2.0, 7.2.1, 7.4.1.

Schottky, F.
(1904) *Über den Picard'schen Satz und die Borel'schen Ungleichungen*. Sitzungsber. Preuss. Akad. Wiss. Berlin **42**, 1244–1262.
4.8.8.

Schur, I.
(1909: a) *Über die charakteristischen Wurzeln einer linearen Substitution mit einer Anwendung auf die Theorie der Integralgleichungen*. Math. Ann. **66**, 488–510.
3.5.6, 6.1.6, 7.1.6, 7.4.3, 7.4.4, 7.4.6, 7.7.5.
(1909: b) *Zur Theorie der linearen homogenen Integralgleichungen*. Math. Ann. **67**, 306–339.
7.5.4.

Shapiro, H. S.
(1951) *Extremal problems for polynomials and power series*. S. M. Thesis, MIT, Cambridge, Mass.
6.5.10.

Sikorski, R.
(1953) *On Leżański's determinants of linear equations in Banach spaces*. Studia Math. **14**, 24–48.
7.5.3.
(1961) *The determinant theory in Banach spaces*. Colloq. Math. **8**, 141–198.
7.5.0, 7.5.3.

Silberstein, J. P. O.
(1953) *On eigenvalues and inverse singular values of compact linear operators in Hilbert spaces*. Proc. Cambridge Phil. Soc. **49**, 201–212.
7.4.3.

Simon, B.
(1977) *Notes on infinite determinants of Hilbert space operators*. Adv. in Math. **24**, 244–273.
7.5.3.

Slobodeckij, L. N.
(1958) *Sobolev spaces of fractional order and their applications to boundary value problems for partial differential equations* (Russian). Doklady Akad. Nauk SSSR **118**, 243–246.
7.2.2.

Smithies, F.
(1937: a) *The eigen-values and singular values of integral equations*. Proc. London Math. Soc. (2) **43**, 255–279.
7.3.6, 7.4.3, 7.7.5.
(1937: b) *A note on completely continuous transformations*. Ann. of Math. (2) **38**, 626–630.
7.3.2.
(1941) *The Fredholm theory of integral equations*. Duke Math. J. **8**, 107–130.
7.5.2.

Sobolev, S. L.
(1936) *Méthode nouvelle à resoudre le problème de Cauchy pour les équations linéaires hyperboliques normales*. Math. Sb. **1**, 39–72.
7.2.2.

Sobolevskij, P. E.
(1967) *On the s-numbers of integral operators* (Russian). Uspehi Mat. Nauk **22**: 2, 114–116.
7.7.5.

Sofman, L. B.
(1969) *Diameters of octahedra* (Russian). Mat. Zametki **5**, 429–436.
7.7.2.
(1973) *Diameters of an infinite dimensional octahedron* (Russian). Vestnik Moskov. Univ., Mat. Mekh. **28**: 5, 54–56.
7.7.2.

Solomyak, M. Z.
(1970) *Estimates of singular numbers of integral operators* (Russian). Vestnik Leningrad Univ., Mat. Mekh. Astronom. **25**: 1, 76–87.
6.2.15, 7.4.5, 7.7.5.

Solomyak, M. Z.; Tikhomirov, V. M.
(1967) *Geometric characteristics of the embedding from W_p^α into C* (Russian). Izv. Vyssh. Uchebn. Zaved. Mat. **1967**: 10, 76–82.
7.7.3.

Sparr, G.
(1974) *Interpolation of several Banach spaces*. Ann. Mat. Pura Appl. (4) **99**, 247–316.
6.4.0.

Stechkin, S. B.
(1951) *On the absolute convergence of orthogonal series* (Russian). Mat. Sb. **29**, 225–232.
6.6.13, 7.7.7.
(1954) *On the best approximation of given classes of functions by arbitrary polynomials* (Russian). Uspehi Mat. Nauk **9**: 1, 133–134.
2.9.11, 7.7.2.
(1955) *On the absolute convergence of orthogonal series* (Russian). Doklady Akad. Nauk SSSR **102**, 37–40.
6.6.13.

Steen, L. A.
(1973) *Highlights in the history of spectral theory*. Amer. Math. Monthly **80**, 359–381.
7.3.0.

Stinespring, W. F.
(1958) *A sufficient condition for an integral operator to have a trace*. J. Reine Angew. Math. **200**, 200–207.
6.6.9, 7.7.5.

Swann, D. W.
(1971) *Some new classes of kernels whose Fredholm determinants have order less than one.* Trans. Amer. Math. Soc. **160**, 427–435.
7.5.4.

Sylvester, J. J.
(1850) *Additions to the articles "On a new class of theorems" and "On Pascal's theorem".* Phil. Mag. (3) **37**, 363–370.
7.1.2.
(1883) *On the equation to the secular inequalities in the planetary theory.* Phil. Mag. (5) **16**, 267–269.
7.1.4, 7.4.2.

Szankowski, A.
(1981) *B(H) does not have the approximation property.* Acta Math. **147**, 89–108.
B.4.7.

Szarek, S. J.
(1976) *On the best constants in the Khintchine inequality.* Studia Math. **58**, 197–208.
G.2.3.

Szász, O.
(1912) *A végtelen determinánsok elméletéhez.* Math. és Phys. Lapok **21**, 224–295.
7.5.1.
(1922) *Über den Konvergenzexponenten der Fourierschen Reihen gewisser Funktionenklassen.* Sitzungsber. Bayrische Akad. Wiss. München, Math. Phys. Kl., 135–150.
7.7.7.
(1928) *Über die Fourierschen Reihen gewisser Funktionenklassen.* Math. Ann. **100**, 530–536.
7.7.7.
(1946) *On the absolute convergence of trigonometric series.* Ann. of Math. (2) **47**, 213–220.
7.7.7.

Taibleson, M. H.
(1964/66) *On the theory of Lipschitz spaces of distributions on Euclidean n-space.* J. Math. Mech. **13**, 407–479, **14**, 821–839, and **15**, 973–981.
7.2.2.

Taylor, A. E.
(1947) *A geometric theorem and its application to biorthogonal systems.* Bull. Amer. Math. Soc. **53**, 614–616.
1.7.6.
(1966) *Theorems on ascent, descent, nullity and defect of linear operators.* Math. Ann. **163**, 18–49.
3.1.14.
(1970) *Historical notes on analyticity as a concept in functional analysis.* Problems in analysis (Ed. R. C. Gunning), 325–343, Princeton Univ. Press.
7.5.3.
(1971) *Notes on the history of the uses of analyticity in operator theory.* Amer. Math. Monthly **78**, 331–342.
7.5.3.

Thorin, G. O.
(1938) *An extension of a convexity theorem due to M. Riesz.* Kungl. Fysiogr. Sällsk. i Lund Förh. **8**, 166–170.
7.2.4.

Tichomirov, V. M.
(1960) *Diameters of sets in functional spaces and the theory of best approximation* (Russian). Uspehi Mat. Nauk **15**: 3, 81–120.
7.7.3.
(1965) *A remark on n-dimensional diameters of sets in Banach spaces* (Russian). Uspehi Mat. Nauk **20**: 1, 227–230.
7.3.6.

Triebel, H.
(1967) *Über die Verteilung der Approximationszahlen kompakter Operatoren in Sobolev-Besov-Räumen*. Invent. Math. **4**, 275–293.
7.2.4, 7.3.6, 7.7.5.
(1970: a) *Interpolationseigenschaften von Entropie- und Durchmesseridealen kompakter Operatoren*. Studia Math. **34**, 89–107.
7.3.6.
(1970: b) *Über die Verteilung der Approximationszahlen von Integraloperatoren in Sobolev-Besov-Räumen*. J. Math. Mech. **19**, 783–796.
7.7.5, Epilogue.
(1981) *Spline bases and spline representation in function spaces*. Arch. Math. (Basel) **36**, 348–359.
6.4.11.

Walsh, J. L.
(1923) *A closed set of normal orthogonal functions*. Amer. J. Math. **55**, 5–24.
5.1.7.

Weidmann, J.
(1965) *Ein Satz über nukleare Operatoren im Hilbertraum*. Math. Ann. **158**, 69–78.
7.6.2.
(1966) *Integraloperatoren der Spurklasse*. Math. Ann. **163**, 340–345.
7.7.5.
(1975) *Verteilung der Eigenwerte für eine Klasse von Integraloperatoren in $L_2(a, b)$*. J. Reine Angew. Math. **276**, 213–220.
7.7.5.

Weierstrass, K.
(1868) *Zur Theorie der quadratischen und bilinearen Formen*. Monatsber. Preuss. Akad. Wiss. Berlin, 310–338.
7.1.6.
(1876) *Zur Theorie der eindeutigen analytischen Funktionen*. Abhandl. Preuss. Akad. Wiss. Berlin, 11–60.
4.8.4, 7.5.4.
(1886) *Zur Determinantentheorie* (Vorlesungsausarbeitung von P. Günther). Coll. Works III, 271–286.
7.1.1.

Weis, L.
(1982) *Integral operators and changes of density*. Indiana Univ. Math. J. **31**, 83–96.
6.2.15.

West, T. T.
(1966) *Riesz operators in Banach spaces*. Proc. London Math. Soc. (3) **16**, 131–140.
3.2.26, 7.4.1.

Weyl, H.
(1912) *Das asymptotische Verteilungsgesetz der Eigenwerte linearer partieller Differentialgleichungen (mit einer Anwendung auf die Theorie der Hohlraumstrahlung)*. Math. Ann. **71**, 441–479.
7.7.5, 7.7.6.
(1917) *Bemerkungen zum Begriff des Differentialquotienten gebrochener Ordnung*. Vierteljahrschrift Naturforsch. Gesell. Zürich **62**, 296–302.
7.7.7.
(1944) *David Hilbert and his mathematical work*. Bull. Amer. Math. Soc. **50**, 612–654.
7.5.0.
(1949) *Inequalities between the two kinds of eigenvalues of a linear transformation*. Proc. Nat. Acad. Sci. USA **35**, 408–411.
3.5.1, 3.5.4, 3.5.5, 7.3.4, 7.4.3, 7.4.6, 7.6.2.
(1950) *Ramifications, old and new, of the eigenvalue problem*. Bull. Amer. Math. Soc. **56**, 115–139.
7.0.0.

Wiener, N.
(1923) *Note on a paper of Banach.* Fund. Math. **4**, 136–143.
7.2.3.

Wloka, J.
(1967) *Vektorwertige Sobolev-Slobodeckijsche Distributionen.* Math. Z. **98**, 303–318.
6.4.0.

Yosida, K.
(1939) *Quasi-completely-continuous linear functional equations,* Japan J. Math. **15**, 297–301.
3.2.1, 3.2.9.

Young, W. H.
(1912) *Sur la généralisation du théorème de Parseval.* C. R. Acad. Sci. Paris **155**, 30–33.
6.5.9, 7.7.7.

Index

List of special symbols

Throughout this monograph we use the standard notations

$$\mathbb{N} := \{1, 2, \ldots\}, \quad \mathbb{Z} := \{0, \pm 1, \ldots\} \quad \mathbb{Z}_n := \{1, \ldots, n\}.$$

Furthermore, $\mathbb{R}$ and $\mathbb{C}$ denote the real and complex fields, respectively. The positive part of a real number ξ is defined by $\xi_+ := \max(\xi, 0)$, and ζ^* denotes the conjugate of a complex number ζ.

If $1 \leqq r \leqq \infty$, then the dual exponent r' is determined by the equation $1/r + 1/r' = 1$, and we write $r^+ := \max(r', 2)$.

Set-theoretic inclusions are denoted by $\subseteq$, while $\subset$ means that equality is excluded. The symbols $\prec$ and $\asymp$ are used to denote asymptotic estimates; see G.3.1.

The letter c (without or with index) is reserved to denote numerical constants which may or may not depend on some fixed parameters.

All operator diagrams are assumed to be commutative.

1. Banach and Hilbert spaces

Unless the contrary is explicitly stated, E and F (without or with index) denote arbitrary complex Banach spaces, while H and K are complex Hilbert spaces. Elements are usually denoted by x and y, functionals by a and b.

$\|x\|$, $\|x \mid E\|$ B.1.1
f_x B.3.3, C.4.3
$\langle x, a\rangle$ B.3.1
$\|a\|$ B.3.2
$x \otimes y$ E.1.2
$\|x \otimes y \mid \alpha\|$ E.2.1
y^* B.3.7
E' B.3.2
E'' B.3.3
U B.1.2
U^0 B.3.2
$E_1 \oplus \ldots \oplus E_n$ C.2.1
$[l_r(I), E_i]$ C.2.1
$\mathfrak{L}(E, F)$ B.2.1
$\varepsilon(E, F)$ E.1.1
$E \otimes F$ E.1.2
$E \tilde{\otimes}_\alpha F$ E.2.3
$\dim(M)$ B.4.1
$\operatorname{codim}(M)$ B.4.2
$M^\perp$ B.2.12
$K(\tau, x, E_0, E_1)$, $K(\tau, x)$ F.3.1
$(E_0, E_1)_{\theta, w}$ F.3.2
$(E_0, E_1)_\Phi$ F.2.1

2. Operators

Operators (which may or may not belong to special ideals) are mostly denoted by S and T.

$\|T\|$, $\|T: E \to F\|$ B.1.9
T^* B.3.8
T' B.3.4
T'' B.3.5
T^{-1} B.2.5
T^m B.2.4
$\operatorname{rank}(T)$ B.4.4
$\varrho(T)$ 4.1.1
$F(\zeta, T)$, $F_m(\zeta, T)$ 4.1.1, 4.5.6
$\lambda_n(T)$ 3.2.20
$\lambda(T)$ 4.2.14
$\pi(I + T)$ 4.3.21
$\operatorname{trace}(T)$, $\tau(T)$ 4.2.2, 4.2.1
$\tau(\zeta, T)$, $\tau_m(\zeta, T)$ 4.4.7, 4.5.9
$\det(I + T)$, $\delta(I + T)$ 4.3.3, 4.3.1
$\delta(\zeta, T)$, $\delta^\bullet(\zeta, T)$ $\delta_m(\zeta, T)$ 4.4.1, 4.4.5, 4.5.4
$D(\zeta, T)$, $D_m(\zeta, T)$ 4.4.1, 4.5.11
$\delta^\bullet(T)$ 4.3.5
$\varepsilon_m(T)$ 4.5.2
$\alpha_n(T)$, $\alpha_n^{(m)}(T)$ 4.4.3, 4.5.11
$A_n(T)$, $A_n^{(m)}(T)$ 4.4.3, 4.5.11
$M(T)$, $M_h(T)$, $M_\infty(T)$ B.2.2, 3.1.1
$N(T)$, $N_k(T)$, $N_\infty(T)$ B.2.2, 3.1.1
$m(T)$, $n(T)$ 3.1.1.
$d(T)$, $d_M(T)$, $d_N(T)$ 3.1.11, 3.1.6, 3.1.2
$\exp(T)$ 4.3.12
$\log(I + T)$ 4.3.16, 4.6.2
$a \otimes y$, $x^* \otimes y$ B.4.3, B.3.7
I, I_E, I_n A.2.4, B.2.3
K_E B.3.3
J_M^E, J_k B.2.7, C.2.2
Q_N^E, Q_k B.2.9, C.2.2
ST B.2.4
$\varepsilon(S, T)$ E.1.3
$S \otimes T$ E.1.4
$S \tilde{\otimes}_\alpha T$ E.2.4
D_t 1.6.1, 2.9.1

3. s-Numbers

$s_n(T)$, $s_n(T: E \to F)$ 2.2.1
$a_n(T)$, $a_n(T \mid \mathfrak{A})$ 2.3.1, 2.8.1
$c_n(T)$ 2.4.2
$d_n(T)$ 2.5.2
$h_n(T)$ 2.6.1
$x_n(T)$ 2.4.13
$y_n(T)$ 2.5.8

4. Operator ideals

Generic operator ideals are denoted by $\mathfrak{A}$ and $\mathfrak{B}$.

$\|T \mid \mathfrak{A}\|$, $\|T: E \to F \mid \mathfrak{A}\|$ D.1.2
$\mathfrak{A}(E, F)$ D.1.1
$\mathfrak{A}'$ D.1.12
$\mathfrak{A} \circ \mathfrak{B}$ D.1.10
$\mathfrak{A} + \mathfrak{B}$ D.1.11
$\mathfrak{A}^m$ D.1.10
$(\mathfrak{A}_0, \mathfrak{A}_1)_\Phi$ F.4.1
$\mathfrak{A}^{(a)}$, $\mathfrak{A}_r^{(a)}$, $\mathfrak{A}_{r,w}^{(a)}$ D.1.13, 2.8.4
$\mathfrak{F}$ D.2.1
$\mathfrak{G}$ D.2.2
$\mathfrak{H}$ D.2.4
$\mathfrak{I}$ 4.2.8
$\mathfrak{K}$ D.2.3
$\mathfrak{L}$ B.2.1
$\mathfrak{L}^{(s)}$, $\mathfrak{L}_r^{(s)}$, $\mathfrak{L}_{r,w}^{(s)}$ 2.2.11, 2.2.4
$\mathfrak{M}_\mathfrak{r}$ A.X.3
$\mathfrak{N}$, $\mathfrak{N}_p$, $\mathfrak{N}_{r,2}$ 1.7.1, 1.7.8
$\mathfrak{P}$, $\mathfrak{P}_r$, $\mathfrak{P}_{r,s}$ 1.3.1, 1.2.1
$\mathfrak{S}$, $\mathfrak{S}_r$, $\mathfrak{S}_{r,w}$ 1.4.2, 2.11.15
$\mathfrak{S}_r^{\text{weak}}$, $\mathfrak{S}_{r,w}^{\text{weak}}$ 2.11.31
$\mathbb{E}_r$, $\mathbb{E}_{r,w}$ 3.4.1

5. Sequences and matrices

Complex-valued families $x = (\xi_i)$ and E-valued families (x_i) are defined on an arbitrary index set I. In the case when this set is finite or countably infinite we speak of vectors or sequences, respectively. As far as possible, the letters i, j, h, k and m, n are reserved to denote indices (in particular, natural numbers).

Finite or infinite matrices are denoted by $M = (\mu_{ij})$.

$\mathfrak{F}(I)$ C.1.2
e_k 1.6.0
card (x) 2.1.1
$a_n(x)$ 2.1.2
$\langle x, y\rangle$ C.1.10, 5.4.4
$(x_i)_{\text{op}}$ 1.1.12
M_{op} A.2.2, 5.3.4
$\lambda_n(M)$ 5.3.4
trace (M) A.3.1
det (M) A.4.1
$\tau(M)$ 5.5.3
$\delta(\zeta, M)$, $\alpha_n(M)$ 5.5.5
$M\begin{pmatrix} i_1, \dots, i_n \\ j_1, \dots, j_n \end{pmatrix}$ 5.5.4
$D_t(n)$ 5.1.3
$F(n)$ 5.1.6
$I(n)$ A.2.4, 5.1.2
$I_{r,\alpha}$ 5.2.2
$S(n)$ 5.1.4
$S_t(n)$ 5.1.5
$W(2^h)$ 5.1.7
$W_{r,\alpha}$ 5.2.3

6. Spaces of sequences and matrices

$l(n)$ A.1.1
$c_0(I)$, c_0 C.1.4, C.1.7
$l_\infty(I)$, l_∞, $l_\infty(n)$ C.1.3, C.1.7
$l_r(I)$, l_r, $l_r(n)$ C.1.6, C.1.7
U_r, $U_r(n)$ C.1.7
$[l_r(I), E]$, $[l_r, E]$ 1.1.3
$[w_r(I), E]$, $[w_r, E]$ 1.1.7
$l_{r,w}(I)$, $l_{r,w}$ 2.1.4
$b_{r,w}^\varrho$, $[b_{r,w}^\varrho, E]$ 5.4.1
$[b_{p,u}^\sigma, b_{q,v}^\tau]$ 5.4.8
$[l_p, l_q]$ 5.3.3

7. Functions and kernels

Complex-valued functions f and E-valued functions Φ are defined either on a compact Hausdorff space X or on a σ-finite measure space (X, μ). On the unit interval we always use the Lebesgue measure. Variables are usually denoted by ξ and η.

By a kernel K we mean a function of two variables which is used to determine an integral operator.

f^* 6.3.1
$\gamma_k(f)$ 6.5.4
$a_n(f \mid L_r)$ 6.5.12
$\langle f, g \rangle$ C.3.9
C^f 6.5.19
C^f_{op} 6.5.2
ϱ_k G.2.1
$f^{2\pi}_{r,\alpha}, f^{2\pi}_{r,\varrho,\alpha}$ 6.5.8, 6.5.15
$g^{2\pi}_{\varrho,\alpha}$ 6.5.16
$s_k, s^m_{h,i}$ 6.4.10
$a_n(\Phi \mid [L_r, E]_m)$ 6.4.20
$\Delta_\tau \Phi, \Delta^m_\tau \Phi$ 6.4.1
Φ_{op} 6.1.3
K_X 6.1.4
K_{op} 6.1.5
$\lambda_n(K)$ 6.1.5
$\tau(K)$ 6.6.3, 6.6.5
$\delta(\zeta, K), \alpha_n(K)$ 6.6.7
$K\begin{pmatrix}\xi_1, \ldots, \xi_n \\ \eta_1, \ldots, \eta_n\end{pmatrix}$ 6.6.6

8. Spaces of functions and kernels

$C(X), C(0, 1)$ C.3.2
$L_\infty(X, \mu), L_\infty(0, 1)$ C.3.5, C.3.7
$L_r(X, \mu), L_r(0, 1)$ C.3.6, C.3.7
$L_{r,w}(X, \mu), L_{r,w}(0, 1)$ 6.3.1
$[C(X), E]$ 6.1.1
$[L_r(X, \mu), E]$ 6.2.5
$[C^m(0, 1), E]$ 6.4.5
$[C^\sigma(0, 1), E]$ 6.4.3
$[W^l_r(0, 1), E]$ 6.4.6
$[B^\varrho_{r,w}(0, 1), E]$ 6.4.2
$[C(X), C(Y)]$ 6.1.4
$[L_p(X, \mu), L_q(Y, \nu)]$ 6.2.8
$[\![L_p(X, \mu), L_q(Y, \mu)]\!]$ 6.2.9
$[W^m_p(0, 1), W^n_q(0, 1)]$ 6.4.27
$[\![W^m_p(0, 1), W^n_q(0, 1)]\!]$ 6.4.28
$[B^\sigma_{p,u}(0, 1), B^\tau_{q,v}(0, 1)]$ 6.4.17
$[B^\sigma_{p,u}(0, 1), L_q(0, 1)]$ 6.4.23
$[L_p(0, 1), B^\tau_{q,v}(0, 1)]$ 6.4.23

The fact that a space consists of periodic functions is indicated by the symbol 2π. For example, we write $C(2\pi)$ instead of $C(0, 1)$.